Jürgen Stückrad Wolfgang Vogel

Buchsbaum Rings and Applications

An Interaction Between Algebra, Geometry and Topology

Springer Basel AG

Dr. Jürgen Stückrad
Karl-Marx-Universität Leipzig
Department of Mathematics
DDR - 7010 Leipzig

Prof. Dr. Wolfgang Vogel
Martin-Luther-Universität Halle—Wittenberg
Department of Mathematics
DDR - 4020 Halle

With 3 Figures

Mathematics Subject Classifikation (1980): 14M05, 13H10, 13H15, 05A20, 55U99

ISBN 978-3-662-02502-4

Library of Congress Cataloging-in-Publication Data
Stückrad, Jürgen, 1948—
Buchsbaum Rings and Applications.
Bibliography: p.
Includes index.
1. Buchsbaum Rings. 2. Geometry, Algebraic. 3. Algebraic topology.
4. Commutative algebra. I. Vogel, Wolfgang, 1940—
II. Title.
QA 251.3.S77 1986 512′.24 86-17880
ISBN 978-3-662-02502-4 ISBN 978-3-662-02500-0 (eBook)
DOI 10.1007/978-3-662-02500-0

Originally published by VEB Deutscher Verlag der Wissenschaften, Berlin in 1986
Softcover reprint of the hardcover 1st edition 1986

2141/3140-543210

Preface

Da die algebraische Geometrie wesentlich
vom Fundamentalsatz der Algebra ausgeht,
den man nur deshalb in der gewohnten
allgemeinen Form aussprechen kann,
weil man dabei die Vielfachheit der Lösungen
in Betracht zieht, so muß man auch bei
jedem Resultat der algebraischen Geometrie
beim Zurückschreiten die gemeinsame Quelle
wiederfinden. Das wäre aber nicht mehr möglich,
wenn man auf dem Wege das Werkzeug verlöre,
welches den Fundamentalsatz fruchtbar
uud bedeutungsreich macht.

Francesco Severi

Abh. Math. Sem. Hansischen Univ.
15 (1943), p. 100

This book describes interactions between algebraic geometry, commutative and homological algebra, algebraic topology and combinatorics. The main object of study are Buchsbaum rings. The basic underlying idea of a Buchsbaum ring is a continuation of the well-known concept of a Cohen-Macaulay ring, its necessity being created by open questions of algebraic geometry and algebraic topology. The theory of Buchsbaum rings started from a negative answer to a problem of David A. Buchsbaum. The concept of this theory was introduced in our joint paper published in 1973.

In presenting our treatment of algebraic geometry, it is a pleasure to acknowledge the help and encouragement which we have had from all sides. Some decisive results came from the applications of homological algebra to derived categories, an approach we learned in joint discussions with Reinhardt Kiehl. A further development of these ideas, with a view towards their topological applications, came in our long collaboration with Peter Schenzel; to both colleagues go our special thanks. In addition, without using the theory of derived categories, we will describe a different approach to these applications with respect to liaison and combinatorics.

This book has profited from the research investigations of a number of doctoral theses. In particular some ideas used here first appeared in the theses of Markus Brodmann, Juan C. Migliore, Peter Schenzel, Philip W. Schwartau and Ngo Viet Trung. They are presently widely available in standard mathematical journals.

Our treatment has also profited from many publications of our Japanese colleagues. For example, we have gained greatly from the work of Shiro Goto, Yoichi Aoyama, Shin Ikeda, Yasuhiro Shimoda and Naoyoshi Suzuki. Among others whose suggestions have served us well, we note David A. Buchsbaum, David Eisenbud, Heisuke Hironaka and Balwant Singh. To all these and others who have helped us, we express our sincerest thanks.

In the late stages of polishing the manuscript we received valuable suggestions from Henrik Bresinsky and Günther Eisenreich.

We would like to thank the staff of Deutscher Verlag der Wissenschaften, especially Erika Arndt for her enormous patience and skill in converting a very rough manuscript into book form.

Leipzig and Halle,
Spring, 1986

Jürgen Stückrad
Wolfgang Vogel

Table of Contents

Introduction and some examples

The aim of this introduction is to present the ideas of our theory of Buchsbaum modules with a wealth of background material. The different viewpoints which are treated in the works of Lasker-Macaulay-Gröbner and Severi-van der Waerden-Weil concerning the multiplicity theory for Bezout's theorem are basic to the understanding of the theory of Buchsbaum modules. The simplest case of Bezout's theorem is the following very simple but fundamental principle in the field of complex numbers.

Fundamental principle. The number of roots of a polynomial $f(x)$ in one variable, counted with their multiplicities, equals the degree of $f(x)$.

The definition of this multiplicity is well-known and clear.

The next simple case to consider is that of plane curves. The problem of the intersection of two algebraic plane curves was already tackled by Newton; he and Leibniz had a clear idea of "elimination" processes which describe the fact that two algebraic equations in one variable have a common root. Using such a process, Newton observed in his "Geometria analytica", published in 1680, that the abcissas (for instance) of the intersection points of two curves of respective degrees m, n, are given by an equation of degree $m \cdot n$. This result was gradually improved upon during the 18th century, until Bezout, using a refined elimination process, was able to prove in general that the equation giving the intersection had exactly the degree $m \cdot n$; however, no general attempt was yet made during that period to attach to each intersection point an integer measuring the "multiplicity" of the intersection, in such a way that the sum of multiplicities would always be $m \cdot n$. Therefore the classical theorem of Bezout states, that two plane curves of degree m and n, intersect in at most $m \cdot n$ different points, unless they have infinitely many points in common. In this form the theorem was also stated by Maclaurin in his "Geometrica organica", published in 1720 (see p. 67/68) but the first correct proof was given by Bezout. An interesting fact, usually not mentioned in the literature, is that Bezout proved in 1764 not only the above-mentioned theorem, but already the following n-dimensional version: Let X be an algebraic projective variety of a projective n-space. If X is a complete intersection of dimension zero then the degree of X is equal to the product of the degrees of the polynomials defining X. The proof can be found in the papers of Bezout [1, 3] and [2]. In his book "Théorie générale des équations algébriques", published in 1779, a statement of this theorem can be found already in the foreword. We quote from page XII:

"Le degré de l'équation finale résultante d'un nombre quelconque d'équations complettes, renfermant un pareil nombre d'inconnues, & de degrés quelconques, est égal au produit des exposans des degrés de ces équations. Theorème dont la verité n'étoit connue et démontrée que pour deux équations seulement."

The theorem appears again on page 32 as Theorem 47. The special cases $n = 2, 3$ are interpreted geometrically on page 33 in Section 3^0 and it is being mentioned there, that these results are already known from geometry. (For these historical remarks see also Renschuch [1], Dieudonné [1], Vogel [6].) To-day we have the following modern statement of Bezout's Theorem, where the degree of a variety $X \subset \mathbf{P}^n$ of dimension d, denoted by $\deg(X)$, is the number of points in which almost all linear subspaces $L \subset \mathbf{P}^n$ of dimension $n - d$ meet X.

Bezout's Theorem. *Let X, Y be unmixed (i.e. each component of X and Y has the dimension of X or Y, resp.) varieties of the projective n-space $\mathbf{P}_K^n$ over an algebraically closed field K such that $\dim(X \cap Y) = \dim(X) + \dim(Y) - n$. Letting C run over all proper components of the intersection $X \cap Y$ (i.e. (irreducible) components with dimension equal $\dim(X \cap Y)$) we get that there exist 'intersection multiplicities', say $i(X, Y; C)$, of X and Y along C such that the following 'Anzahl-Formel' is true:*

$$\deg(X) \cdot \deg(Y) = \sum_C i(X, Y; C) \cdot \deg(C).$$

Here the number $i(X, Y; C)$ itself measures the degree of contact of X and Y along C. It represents fairly sophisticated concepts in full generality and it has taken a century or two and a lot of work to be developed, see also S. Kleiman [1, 2].

To get equality in the above equation one may follow different approaches to arrive at several different multiplicity theories. It is well-known that there is no loss of generality in assuming for projective varieties that one variety is a complete intersection in order to define local intersection multiplicities. This statement does indeed follow from Samuel's book [1], p. 81.

However, it is not clear to us that it is possible to prove a global statement like Bezout's Theorem by reducing in a simple fashion to the case in which one of the two intersecting varieties is a complete intersection. Therefore we would like to present the high points of the proof of the following theorem:

Theorem 1. *Let X, Y be arbitrary varieties of $\mathbf{P}_K^n$ with $\dim(X \cap Y) \geq 0$. Then there exist varieties X', Y' of $\mathbf{P}_K^{2n+1}$ such that one variety, say X', is a complete intersection with*

$$\deg(X) \cdot \deg(Y) = \deg(X') \cdot \deg(Y')$$

and there is a 1-1 correspondence between the components C of $X \cap Y$ (in $\mathbf{P}_K^n$) with $\dim(C) = \dim(X \cap Y)$ and the components C' of $X' \cap Y'$ (in $\mathbf{P}_K^{2n+1}$) with $\dim(C') = \dim(X' \cap Y')$ and that this correspondence preserves dimensions and degrees.

Proof (see also Vogel [6]): We will apply idealtheoretic methods. Let X, $Y \subset \mathbf{P}_K^n$ be our projective varieties with defining ideals $\mathfrak{a}$ and $\mathfrak{b}$ in $K[x_0, \ldots, x_n] =: R_x$ and dimensions d and δ, resp. We introduce a second copy $K[y_0, \ldots, y_n] =: R_y$ and denote by $\mathfrak{b}'$ the ideal in R_y corresponding to $\mathfrak{b}$. We consider the polynomial ring $R := K[x_0, \ldots, x_n, y_0, \ldots, y_n]$ and the ideal $\mathfrak{c} = (x_0 - y_0, \ldots, x_n - y_n)$. Let $h_0(\ldots)$ be the (rectified) leading coefficient of the Hilbert polynomial of the homogeneous ideal $(\ldots)$.

Claim. $h_0(\mathfrak{a} \cdot R_x) \cdot h_0(\mathfrak{b} \cdot R_x) = h_0(\mathfrak{a} \cdot R + \mathfrak{b}' \cdot R)$,
where $\mathfrak{a} \cdot R$ is the extension ideal of $\mathfrak{a}$ in R.

This follows from

$$R/(\mathfrak{a} + \mathfrak{b}') \cdot R \cong R_x/\mathfrak{a} \cdot R_x \otimes_K R_y/\mathfrak{b}' \cdot R_y,$$

i.e. the Hilbert function $H(n, (\mathfrak{a}+\mathfrak{b}')\cdot R)$ of $(\mathfrak{a}+\mathfrak{b}')\cdot R$ can be expressed in terms of the Hilbert functions of $\mathfrak{a}\cdot R_x$ and $\mathfrak{b}'\cdot R_y$, that is

$$H(n, (\mathfrak{a}+\mathfrak{b}')\cdot R) = \sum_{i+j=n} H(i, \mathfrak{a}\cdot R_x)\cdot H(j, \mathfrak{b}'\cdot R_y).$$

The degree and the leading coefficient of the Hilbert polynomial of $(\mathfrak{a}+\mathfrak{b}')\cdot R$ are given by $d+\delta+1$ and $h_0((\mathfrak{a}+\mathfrak{b}')\cdot R)$, resp. We choose an integer r such that the Hilbert functions $H(i, \mathfrak{a}\cdot R_x) =: H_i$ and $H(i, \mathfrak{b}'\cdot R_y) =: H'_i$ are given by their Hilbert polynomials h_i and h'_i, resp. for $i > r$. Then we can decompose for $n \gg 0$ $(n > 2r)$:

$$\sum_{i=0}^{n} H_i\cdot H'_{n-i} = \sum_{i=0}^{n} h_i\cdot h'_{n-i} + \sum_{i=0}^{r} (H_i - h_i)\cdot h'_{n-i} + \sum_{i=n-r}^{n} h_i(H'_{n-i} - h'_{n-i}).$$

Some calculations involving the coefficient of $n^{d+\delta+1}$ therefore yield:

$$\sum_{i=0}^{n} H_i\cdot H'_{n-i} = h_0(\mathfrak{a})\cdot h_0(\mathfrak{b}')\left[\sum_{i=0}^{n}\binom{i}{d}\cdot\binom{n-i}{\delta}\right] + \text{(other terms)}$$

$$= h_0(\mathfrak{a})\cdot h_0(\mathfrak{b}')\cdot\binom{n}{d+\delta+1} + \text{(lower order terms)},$$

and our claim follows.

It is well-known that the degree of a projective variety is equal to the (rectified) leading coefficient of its Hilbert polynomial (see, e.g., D. Mumford [5], p. 112, Theorem 6.25). Therefore we get our first statement of Theorem 1 where X' and Y' are defined by $\mathfrak{c}$ and $(\mathfrak{a}+\mathfrak{b}')$ in R, resp. The second statement follows from the fact that there is a 1-1 correspondence between the isolated prime ideals $\mathfrak{a}+\mathfrak{b}$ in R_x and the isolated prime ideals of $(\mathfrak{a}+\mathfrak{b}')\cdot R + \mathfrak{c}\cdot R$ in R. This correspondence is given by

$$R_x \supset \mathfrak{p} \mapsto (\mathfrak{p}+\mathfrak{c}) \subset R$$

and it therefore preserves dimensions and degrees, q.e.d.

S. Kleiman [3] pointed out to us that the reduction of Theorem 1 may be accomplished by replacing the original varieties X, Y of $\mathbf{P}^n$ by the varieties X', Y' of $\mathbf{P}^{2n+1}$ which may be described as follows. Take three n-planes in $\mathbf{P}^{2n+1}$ in general position. Embbed X in the first, Y in the second and take Y' to be their join. Take X' to be the third plane. The claim in Theorem 1 is apparently proven by showing that the degree of the join Y' is equal to the product of the degrees of X and Y. We note that the reduction of Theorem 1 is not only to the case in which one of the varieties is a complete intersection but even a linear space.

At the beginning of this century one investigated the notion of the length of a primary ideal in order to define another intersection multiplicity. This multiplicity is defined as follows:

Let $X, Y \subset \mathbf{P}^n$ be arbitrary varieties. Let C be a component of $X\cap Y$ such that $\dim(C) = \dim(X\cap Y)$. Denote by $A(X, C)$ the local ring of X at C.

We set

$$\mu(X, Y; C) := \text{length of } A(X, C)/I(Y)\cdot A(X, C),$$

where $I(Y)$ is the defining ideal of Y. This length $\mu(X, Y; C)$ is well-defined and is called the *idealtheoretic intersection multiplicity* of X and Y at C. For instance, this multiplicity is the intersection multiplicity as set forth in the beginning in the case of projective plane curves. Prior to 1928 most mathematicians hoped that this multiplicity would provide for Bezout's Theorem always the correct intersection multiplicity. In 1928 B. L. van der Waerden studied Macaulay's famous space curve (see Macaulay [1], p. 98) to show, that this idealtheoretic intersection multiplicity does not yield the correct multiplicity for Bezout's Theorem to be valid in projective spaces $\mathbf{P}^n$ with $n \geq 4$. We quote van der Waerden [1], p. 770: "In these cases we must reject the notion length and try to find another definition of multiplicity".

Nowadays it is of course well-known that

$$\mu(X, Y; C) = i(X, Y; C)$$

if and only if the local rings $A(X, C)$ of X at C and $A(Y, C)$ of Y at C are Cohen-Macaulay rings for all proper components C of $X \cap Y$ where $\dim(X \cap Y) = \dim(X) + \dim(Y) - n$, see J.-P. Serre [2], p. V-20.

Without loss of generality we may now suppose by applying our Theorem 1 that one of the two intersecting varieties X and Y is a complete intersection, say Y. With this assumption we get that

$$\mu(X, Y; C) \geq i(X, Y; C)$$

for each proper component C.

Let Y be a complete intersection. Then there arises another problem posed by D. A. Buchsbaum [1] in 1965, as follows:

Problem (from the viewpoint of the theory of intersection multiplicities). Is it true that $\mu(X, Y; C) - i(X, Y; C)$ is independent of Y; that is, does there exist an invariant, say $I(A)$, of the local ring $A := A(X, C)$ of X at C such that

$$\mu(X, Y; C) - i(X, Y; C) = I(A)?$$

From the viewpoint of local algebra we get the problem in its original form:

Let A be a local ring of dimension $d \geq 1$ with maximal ideal $\mathfrak{m}$. Let $\mathfrak{q}$ be an $\mathfrak{m}$-primary ideal which is generated by a system of parameters. Denote by $e_0(\mathfrak{q}; A)$ the multiplicity of the ideal $\mathfrak{q}$ (see, e.g. Zariski-Samuel [1], Vol. II, Chap. VIII, § 10) and by $l(A/\mathfrak{q})$ the length of $A/\mathfrak{q}$ over A. Is it then true that $l(A/\mathfrak{q}) - e_0(\mathfrak{q}; A)$ is independent of $\mathfrak{q}$?

For instance, is

$$l(A/\mathfrak{q}) - e_0(\mathfrak{q}; A) = \dim(A) - \operatorname{depth}(A)?$$

We will show that this is not always the case. From the viewpoint of the theory of intersection multiplicities we want to study our first counter-example.

Example 1. Let $X' \subset \mathbf{P}_K^3$ be the non-singular curve given parametrically by

$$\{u^5, u^4v, uv^4, v^5\}.$$

Let $X \subset \mathbf{P}_K^4$ be the projective cone over X'. We consider the surfaces Y and Y' in projective 4-space defined by the two hypersurfaces

$$x_1 = x_4 = 0 \quad \text{and} \quad x_2 = x_1^2 + x_4^2 = 0, \quad \text{resp.}$$

Let $\mathfrak{p}$ be the defining ideal of X in $K[x_0, x_1, x_2, x_3, x_4]$. Then

$$\mathfrak{p} = (x_1x_4 - x_2x_3,\ x_1^3x_3 - x_2^4,\ x_1^2x_3^2 - x_2^3x_4,\ x_1x_3^3 - x_2^2x_4^2,\ x_3^4 - x_2x_4^3)$$

and Hilbert's characteristic polynomial of $\mathfrak{p}$ is given by

$$H(n, \mathfrak{p}) = 5 \cdot \binom{n}{2} + 6 \cdot \binom{n}{1} - 3$$

(see, e.g. Renschuch [2], 8.2); that is, the degree of X is 5.

Now, we see that $X \cap Y$ and $X \cap Y'$ intersect at the vertex C: $(0, 0, 0, 0)$ of X, and Bezout's Theorem therefore gives us

$$i(X, Y; C) = 5 \quad \text{and} \quad i(X, Y'; C) = 10.$$

Since $\big(\mathfrak{p} + (x_1, x_4)\big) = (x_1, x_4, x_2x_3, x_2^4, x_3^4)$ it is easy to see that $\mu(X, Y; C) = 7$.

It remains to calculate the length $\mu(X, Y'; C)$. We have

$$\big(\mathfrak{p} + (x_2, x_1^2 + x_4^2)\big) = (x_2, x_1^2 + x_4^2, x_1x_4, x_1^2x_3^2, x_1x_3^3, x_3^4) =: \mathfrak{q}_1$$

and we can construct the following chain of primary ideals belonging to (x_1, x_2, x_3, x_4):

$$\begin{aligned}\mathfrak{q}_1 &\subset (\mathfrak{q}_1, x_3^3x_4) =: \mathfrak{q}_2 \subset (\mathfrak{q}_2, x_3^2x_4) =: \mathfrak{q}_3 \subset (\mathfrak{q}_3, x_1x_3^2) =: \mathfrak{q}_4 \\ &\subset (\mathfrak{q}_4, x_1^2x_3) =: \mathfrak{q}_5 \subset (\mathfrak{q}_5, x_3x_4) =: \mathfrak{q}_6 \subset (\mathfrak{q}_6, x_1x_3) =: \mathfrak{q}_7 \subset (\mathfrak{q}_7, x_3^3) =: \mathfrak{q}_8 \\ &\subset (\mathfrak{q}_8, x_3^2) =: \mathfrak{q}_9 \subset (\mathfrak{q}_9, x_3) =: \mathfrak{q}_{10} \subset (\mathfrak{q}_{10}, x_1^2) =: \mathfrak{q}_{11} \subset (\mathfrak{q}_{11}, x_1) =: \mathfrak{q}_{12} \\ &\subset (x_1, x_2, x_3, x_4) =: \mathfrak{q}_{13}.\end{aligned}$$

It follows that $\mu(X, Y'; C) = 13$; that is,

$$2 = \mu(X, Y; C) - i(X, Y; C) \neq \mu(X, Y'; C) - i(X, Y'; C) = 3.$$

The theory of Buchsbaum modules started from such a negative answer to the above problem of D. A. Buchsbaum (see Vogel [4]). The concept of Buchsbaum modules was introduced in Stückrad-Vogel [2] and [3], and the theory is now developing rapidly; see, for example, the following Symposium: Study of Buchsbaum rings and generalized Cohen-Macaulay rings. Proceedings of a Symposium held at the Research Institute for Mathematical Sciences, Kyoto University, Kyoto 1982.

The basic underlying idea of a Buchsbaum module generalizes the well-known concept of a Cohen-Macaulay module, its necessity being created by open questions in Commutative Algebra and Algebraic Geometry. For instance, such a necessity to investigate generalized Cohen-Macaulay structure occurs when one classifies algebraic curves in $\mathbf{P}^3$ or when one studies singularities of algebraic varieties. Furthermore S. Goto and Y. Shimoda [1] discovered that the Cohen-Macaulay property of Rees algebras of parameter systems can be described by certain Buchsbaum rings. Also it was shown that interesting and extensive classes of Buchsbaum rings do exist.

We now introduce the definition of a Buchsbaum module and we describe a first geometrical interpretation of Buchsbaum modules.

Let A be a Noetherian local ring with maximal ideal $\mathfrak{m}$. Let M be a finitely generated A-module.

Definition 1. A sequence $a_1, \ldots, a_r$ of elements of $\mathfrak{m}$ is a *weak M-sequence* if for each $i = 1, \ldots, r$

$$(a_1, \ldots, a_{i-1}) \cdot M : a_i = (a_1, \ldots, a_{i-1}) \cdot M : \mathfrak{m}$$

(for $i = 1$ we set $(a_1, \ldots, a_{i-1}) = (0)$ in A).

We denote by $e_0(\mathfrak{q}; M)$ the multiplicity of M relative to a *parameter ideal* $\mathfrak{q}$ of M, i.e., an ideal of A generated by a system of parameters for M. Then as our first important result which explains the notion of Buchsbaum modules in connection with the above problem of D. A. Buchsbaum (see Chap. I, § 1) we get

Theorem 2 and definition. *The following conditions are equivalent:*

(i) *M is a Buchsbaum module.*

(ii) *The difference of length $l_A(M/\mathfrak{q} \cdot M)$ and multiplicity $e_0(\mathfrak{q}; M)$ of M is an integer, say $I(M)$, of M not depending on the choice of the parameter ideal $\mathfrak{q}$ of M.*

(iii) *Every system of parameters for M is a weak M-sequence.*

Note that a Noetherian local ring is said to be a *Buchsbaum ring* if it is a Buchsbaum module over itself. We want to give some simple examples.

Example 2. A finitely generated module M is Cohen-Macaulay if and only if M is Buchsbaum and $I(M) = 0$, that is the class of Buchsbaum modules contains the Cohen-Macaulay modules.

Example 3. Let A be a local ring with maximal ideal $\mathfrak{m}$. Let M be a Cohen-Macaulay module over A of dimension $d \geq 1$. Let N be a submodule of M such that the factor module M/N is a finite-dimensional vector space over $A/\mathfrak{m}$, say $\bigoplus_t A/\mathfrak{m}$. Then N is a Buchsbaum module over A with the invariant $I(N) = (d - 1) \cdot t$.

Proof: Let $\mathfrak{q} \subset A$ be an ideal generated by a system of parameters with respect to N. Take the exact sequence $0 \to N \to M \to M/N \to 0$. Since M is a Cohen-Macaulay module and $\mathfrak{m} \cdot M \subseteq N$ we get the following exact sequence by applying the functor $A/\mathfrak{q} \otimes_A$:

$$0 \to \operatorname{Tor}_1^A(A/\mathfrak{q}, M/N) \to N/\mathfrak{q} \cdot N \to M/\mathfrak{q} \cdot M \to M/N \to 0.$$

This is an exact sequence of A-modules with finite length. We obtain therefore that $\mathfrak{q}$ is also a parameter ideal of M and that

$$l_A(N/\mathfrak{q} \cdot N) = l(M/\mathfrak{q} \cdot M) - (d - 1) \cdot t$$

since $l_A(\operatorname{Tor}_1(A/\mathfrak{q}, M/N)) = t \cdot l_A(\operatorname{Tor}_1(A/\mathfrak{q}, A/\mathfrak{m})) = t \cdot l_A(\mathfrak{q}/\mathfrak{q} \cdot \mathfrak{m}) = t \cdot d$. On the other hand the additivity property of the multiplicity symbol (see Chap. 0, § 1) implies $e_0(\mathfrak{q}; N) = e_0(\mathfrak{q}; M)$. Thus the difference $l_A(N/\mathfrak{q} \cdot N) - e_0(\mathfrak{q}; N) = (d - 1) \cdot t$ does not depend on the choice of $\mathfrak{q}$ since $e_0(\mathfrak{q}; M) = l(M/\mathfrak{q} \cdot M)$. Therefore N is a Buchsbaum module with $I(N) = (d - 1) \cdot t$, q.e.d.

This example has some useful applications.

(3a) Let A be a regular or Cohen-Macaulay local ring of dimension $d \geq 2$. Then the maximal ideal $\mathfrak{m}$ of A is a Buchsbaum module which is not Cohen-Macaulay.

(3b) Take Macaulay famous curve X in $\mathbf{P}^3$ given parametrically by $\{s^4, s^3t, st^3, t^4\}$ (see Macaulay [1], p. 98). This curve was studied by F. S. Macaulay as early as 1916. His purpose was to show that not every prime ideal in a polynomial ring is perfect. We will show that X is arithmetically Buchsbaum; that is, the local ring of the (affine) cone over X at the vertex is a Buchsbaum ring.

Let K be a field and $R = K[[s, t]]$ a formal power series ring in s and t. We put $S = K[[s^4, s^3t, st^3, t^4]]$ in R. Take the normalization T of S. Then T is a Cohen-Macaulay ring of dimension 2 and we have $T = K[[s^4, s^3t, s^2t^2, st^3, t^4]]$. It is easily seen that the conductor of S in T is the maximal ideal of S; that is, T/S is a vector space generated by the element s^2t^2. Hence we get that S is a Buchsbaum ring with invariant $I(S) = 1$.

(3c) (See also Herrmann-Schmidt [1].) Let K be a field with char $K \neq 2$ and $S = K[x, y]$ a polynomial ring. We put $R = \{f \in S \mid f(1, 0) = f(-1, 0)\}$. Then R is the finitely generated subring $K[1 - x^2, xy, y, x - x^3]$ of S. So $X = \operatorname{Spec}(R)$ is realized as a surface in the affine space $\mathbf{A}_K^4$ which is non-singular in codimension 1, but with an isolated singularity at the origin. Applying the above statement we see immediately that the local ring of X at the origin is a Buchsbaum ring.

Note that this example (3c) is a very simple example of Buchsbaum rings obtained by glueing (see Goto [4]).

Example 4. M is a Buchsbaum module over A if and only if $\hat{M}$ is a Buchsbaum module over $\hat{A}$ in which case $I(M) = I(\hat{M})$. Here denotes ^ the $\mathfrak{m}$-adic completion. (See also Lemma I.1.13.)

In order to describe examples from the viewpoint of the theory of intersection multiplicities we need some special results on Buchsbaum modules. We summarize these assertions in the following theorem. The proofs are given in Chapter I under even more general conditions.

Let A be a local ring and let $\mathfrak{a} \subset A$ be an ideal of A. Denote by $U(\mathfrak{a})$ the intersection of the primary ideals $\mathfrak{q}$ belonging to $\mathfrak{a}$ with $\dim(\mathfrak{q}) = \dim(\mathfrak{a})$.

Theorem 3. *Let A be a local ring of dimension $d \geq 1$ with maximal ideal $\mathfrak{m}$. The following two conditions are equivalent:*

(i) *A is a Buchsbaum ring.*

(ii) *For each part $a_1, \ldots, a_k$ of a system of parameters of A we have*

$$\mathfrak{m} \cdot U\big((a_1, \ldots, a_k)\big) \subseteq (a_1, \ldots, a_k) \qquad \text{for every } k = 0, \ldots, d-1.$$

(notice the case $d = 1$!)

Furthermore, we have the following statements.

(iii) *Let $d \geq 2$. Let $\mathfrak{a}, \mathfrak{b}$ be ideals of A such that the intersection $\mathfrak{a} \cap \mathfrak{b}$ is the zero ideal of A. Assume that $A/\mathfrak{a}$ and $A/\mathfrak{b}$ are Cohen-Macaulay rings of dimension d and* $\dim\big(A/(\mathfrak{a} + \mathfrak{b})\big) = 0$, *then $\mathfrak{m} = \mathfrak{a} + \mathfrak{b}$ if and only if A is a Buchsbaum ring.*

(iv) *Let $d >$* $\operatorname{depth}(A) \geq 1$. *$A$ is a Buchsbaum ring if and only if there exists in A a non-zero divisor $x \in \mathfrak{m}^2$ such that the ring $A/(x)$ is a Buchsbaum ring.*
(Lifting Buchsbaum mod a non-zero divisor.)

As an application of the statements (i), (ii) of Theorem 3 we get the following examples:

Example 5. Let K be any field and set $A := K[[x, y]]/(x) \cap (x^2, y)$. Then A is a Buchsbaum non-Cohen-Macaulay ring. If $A := K[[x, y]]/(x) \cap (x^3, y)$ then A is not a Buchsbaum ring.

From the vantage point of the theory of intersection multiplicities we can construct the following examples by use of the statements (iii) or (iv) of Theorem 3.

Example 6. Let X be the union of two planes in four-dimensional affine space $\mathbf{A}^4$ meeting at the point $P: x_1 = x_2 = x_3 = x_4 = 0$; that is the ideal of X is $(x_1, x_2) \cap (x_3, x_4)$. Let A be the local ring of X at P; i.e.

$$A = K[x_1, x_2, x_3, x_4]_{x_1, \ldots, x_4} / (x_1, x_2) \cap (x_3, x_4).$$

Our statement (iii) implies that A is a Buchsbaum ring. It is well-known that A is not a Cohen-Macaulay ring.

Example 7. Let X be the rational twisted cubic curve in $\mathbf{P}^3$ with defining ideal $\mathfrak{p} = (x_0x_2 - x_1^2, x_0x_3 - x_1x_2, x_1x_3 - x_2^2)$ in $K[x_0, x_1, x_2, x_3]$ for any field K. It is well-known that $\mathfrak{p}$ is perfect. We will show that $\mathfrak{p}^2$ is Buchsbaum; that is, we claim that the local ring of the affine cone over X, counted with multiplicity 3 at the vertex, is a Buchsbaum ring with invariant 1.

Proof: Note that $\mathfrak{p}^2$ is the defining ideal of the curve X counted with multiplicity 3. Furthermore, we have that $\mathfrak{p}^2$ is a primary ideal (see, e.g., Achilles-Schenzel-Vogel [1]). Therefore we can apply the statement (iv) of Theorem 3. By using localization it is not hard to calculate the intersection $U(\mathfrak{p}^2, x_2^2)$ of the primary ideals belonging to $\mathfrak{p}^2 + (x_2^2)$ of dimension 1. It then follows that $(x_0, x_1, x_2, x_3) \cdot U(\mathfrak{p}^2, x_2^2) \subseteq \mathfrak{p}^2 + (x_2^2)$. The statements (ii) and (iv) yield our assertion. It is also not too difficult to calculate the invariant by using a system of parameters.

Having this example we can construct new irreducible and reduced arithmetical Buchsbaum curves by using the theory of residual intersection. Therefore we will apply liaison among curves in $\mathbf{P}^3$ coupled with our deep result from Theorem III.1.2, on liaison, obtained by applying the theory of dualizing complexes. This result, or the theory of residual intersection for the special case of curves, results in the following.

Theorem 4. *Assume that the scheme theoretic union of two curves X_1, X_2 of $\mathbf{P}_K^3$ over an algebraically closed field K is the complete intersection of two hypersurfaces, and that the curves have no components in common* (X_1 and X_2 are then said to be *linked geometrically*, see C. Peskine, L. Szpiro [3] and A. P. Rao [1]). *X_1 is arithmetically Buchsbaum (i.e. the local ring of the affine cone over X_1 at the vertex is a Buchsbaum ring) if and only if X_2 is arithmetically Buchsbaum.*

Take, for example, Macaulay's curve $X = X_1$ and the curve X_2 with defining ideal $\mathfrak{a} = (x_0, x_1) \cap (x_2, x_3)$. Then it is not hard to show that

$$\mathfrak{p} \cap \mathfrak{a} = (x_0x_3 - x_1x_2, x_0x_2^2 - x_1^2x_3).$$

Applying Theorem 4 and Example 6 it therefore follows again that Macaulay's curve is arithmetically Buchsbaum. Notice, that we have used only a simple calculation on $\mathfrak{p} \cap \mathfrak{a}$. This liaison was discovered by G. Salmon [1], p. 40, already in 1848 and a little later again by J. Steiner [1], p. 138, in 1857.

Now, consider our Example 7. Let Y be the curve X counted with multiplicity 3. It follows from K. Rohn [2] that Y is linked for instance, to an irreducible and reduced curve C_{43}^{213} of degree 43 and genus 213 by two hypersurfaces of degree 4 and 13. We thus get that the liaison equivalence class corresponding to a vector space of dimension 1 contains also the curve C_{43}^{213}.

Using the theory of liaison and Buchsbaum rings we also obtain new statements concerning the classification of algebraic curves in projective 3-space $\mathbf{P}_K^3$. For instance, let C_6^3 be an irreducible and non-singular curve in $\mathbf{P}_K^3$ of degree 6 and genus 3 with defining ideal $I(C_6^3)$ in $S := K[x_0, x_1, x_2, x_3]$. It follows from M. Noether [1], p. 87, (a_3) and (a_3') and Theorem 4 that either
C_6^3 is arithmetically Cohen-Macaulay, or
C_6^3 is arithmetically Buchsbaum.
We note that in 1881 F. Schur [1] discovered a first difference between the curves C_6^3 to distinguish between them. Nowadays the resolution of the curve C_6^3 is well-known if C_6^3 is arithmetically Cohen-Macaulay. It is (see also G. Ellingsrud [1] or L. Gruson and C. Peskine [1]):

$$0 \to S^3(-4) \to S^4(-3) \to S \to S/I(C_6^3) \to 0.$$

If C_6^3 is arithmetically Buchsbaum then we have obtained in addition the resolution of C_6^3:

$$0 \to S(-6) \to S^4(-5) \to S(-2) \oplus S^3(-4) \to S \to S/I(C_6^3) \to 0.$$

(See also Chap. III, § 1.) For arithmetically Buchsbaum curves C in $\mathbf{P}^3$ we get that the invariant $I(A)$ of the local ring A of the affine cone over C at the vertex is given by

$$I(A) = \dim \Big(\bigoplus_v H^1\big(\mathbf{P}^3, \mathcal{J}_C(v)\big)\Big)$$

where $\mathcal{J}_C$ is the ideal sheaf of the curve C.

Having this arithmetically Buchsbaum property of a variety in $\mathbf{P}^n$ we want to describe another geometric interpretation of the invariant $I(A)$ of the corresponding (local) Buchsbaum ring A of surfaces. If X is any projective variety, the finite-dimensional vector spaces $H^i(X, \mathcal{O}_X)$ are important invariants of X. One of the most interesting is the alternating sum of their dimensions, obtaining the so-called arithmetic genus:

$$p_a(X) = \dim H^n(X, \mathcal{O}_X) - \dim H^{n-1}(X, \mathcal{O}_X) + \dots + (-1)^{n-1} \dim H^1(X, \mathcal{O}_X),$$

where n is the dimension of X. One drops the term $\dim H^0(X, \mathcal{O}_X) = 1$, which has the advantage that for X a curve,

$$p_a(X) = \dim H^1(X, \mathcal{O}_X) = \text{usual genus of } X.$$

But, when X is a surface, we get

$$p_a(X) = \dim H^2(X, \mathcal{O}_X) - \dim H^1(X, \mathcal{O}_X).$$

The point here is that the Italian geometers regarded $\dim H^2(X, \mathcal{O}_X)$ as the dominant term, and called it, for non-singular surfaces, the geometric genus; while $\dim H^1(X, \mathcal{O}_X)$ was considered a "correction" term, and was called the irregularity. Now, consider any arithmetical normal irregular (i.e. the irregularity $\neq 0$) surface F, such that we have $H^1\big(F, \mathcal{O}_F(p)\big) \neq 0$ only for $p = 0$. Let A be the local ring of the vertex of the (affine) cone over F. Applying our cohomological investigations from Chapter I, § 2, we get: A is a normal non-Cohen-Macaulay-Buchsbaum ring where the invariant $I(A)$ of A is given by the irregularity of F, that is

$$I(A) = \dim H^1(F, \mathcal{O}_F).$$

It is not hard to construct an arithmetically normal irregular surface F free of singularities. For instance, this will be the case if F is the Segre embedding of pairs of points $E \times G$ of any two plane curves E, G free of singularities, where at least one of them has a positive genus. For example, let $E = \mathbf{P}^1$, and let G be the cubic defined by $x_0^3 + x_1^3 + x_2^3$ in $\mathbf{P}^2$. Let F be $E \times G$ in its Segre embedding in $\mathbf{P}^5$.

For such surfaces F we get that the vertex of the affine cone over F is a normal Buchsbaum singularity which is not a Cohen-Macaulay singularity. This gives an answer to a question posed by H. Hironaka in a discussion (at the University of Halle in 1974), who asked whether we can construct normal non-Cohen-Macaulay-Buchsbaum singularities, and which aroused our interest in the subject of singularities (see also Chap. V, § 2). To motivate the study of other Buchsbaum singularities we quote the following remark made by D. Mumford in [2] on p. 42:

"Incidentally, one should regard the depth of O (a local ring) itself, for example, as a measure of the topological complexity of the singularity at the closed point of Spec(O): if the depth is maximal, i.e., equals the dimension of O, then O is in a weak sense, nonsingular, while if the depth is much less then the dimension, the singularity is very bad."

As mentioned before, D. A. Buchsbaum (see the above problem) and also A. Seidenberg [2], on p. 620 considered the difference between the dimension and depth of any local ring taking it as a measure of the deviation from the Cohen-Macaulay property. With our theory of Buchsbaum rings we will study examples of projective varieties in Chapter I (see Examples 4.14) and Chapter V, § 2, which show that this measure does not describe the non-Cohen-Macaulay property satisfactorily.

Shiro Goto and Yasuhiro Shimoda [1] have introduced another aspect to the study of Buchsbaum singularities. Here certain Buchsbaum rings are characterized by the behaviour of Rees algebras relative to parameter ideals. We conclude our considerations by briefly discussing a main result from Goto-Shimoda [1].

Theorem 5. *Let A be a local ring with maximal ideal $\mathfrak{m}$. Let $H^i_{\mathfrak{m}}(A)$ denote the ith local cohomology module of A with respect to $\mathfrak{m}$. The following two conditions are equivalent:*

(i) *A is a Buchsbaum ring and $H^i_{\mathfrak{m}}(A) = (0)$ for $i \neq 1, \dim(A)$.*

(ii) *The Rees algebra $R(\mathfrak{q}) = \bigoplus_{n \geq 0} \mathfrak{q}^n$ is a Cohen-Macaulay ring for every parameter ideal $\mathfrak{q}$ of A.*

This striking theorem gives a complete answer to a problem of Rees algebras of powers of parameter ideals. It is stated as follows (see Goto-Shimoda [1]):

Corollary 6. *Let A be a local ring and assume that* $\operatorname{depth}(A) \neq 1$. *Then A is a Cohen-Macaulay ring if and only if the Rees algebra $R(\mathfrak{q}^n)$ is Cohen-Macaulay for every parameter ideal $\mathfrak{q}$ of A, for all integer $n > 0$.*

Also, Shiro Goto [5] obtains the following result. Let A be a local ring of dimension $d > 0$. Then $A/H^0_{\mathfrak{m}}(A)$ is a Buchsbaum local ring if and only if Proj $R(\mathfrak{q})$ is a Cohen-Macaulay scheme for every parameter ideal $\mathfrak{q}$ auf A. In a letter to one of the authors, dated February 8, 1980, he underlines its significance: "I believe that the above blowing-up characterization of Buchsbaum rings really clarifies the importance of the concept of Buchsbaum singularities." Therefore the main object of our study of Chapter IV is the stability of Rees rings and form rings with respect to the Buchsbaum property of parameter ideals. This chapter has its origin in the effort to extend Hironaka's desingularization to a more general situation due to G. Faltings [1] and M. Brodmann [1] (see also the introduction of Chapter IV).

We conclude our introduction by briefly discussing an application of the theory of Buchsbaum modules to the so-called Upper Bound Conjecture which establishes some interesting connections among algebraic topology, commutative algebra and combinatorics (see Chap. II). Let Δ denote an abstract finite simplicial complex with vertices $x_1, \ldots, x_n$. That is a family of subsets of $\{x_1, \ldots, x_n\}$ such that if $\sigma \in \Delta$ and $\tau \subseteq \sigma$ then $\tau \in \Delta$, and such that the vertices are in Δ. We call the elements of Δ *faces*. If the largest face of Δ has d elements, then we say $\dim \Delta = d - 1$. The f-vector of Δ is $f = (f_{-1}, f_0, \ldots, f_{d-1})$ where exactly f_i faces of Δ have $i + 1$ elements and $f_{-1} = 1$. The Upper Bound Conjecture now states that the number of i-dimensional faces f_i of Δ is less than or equal to a certain number $c_i(n, d)$. In 1975 R. P. Stanley [1] solved the Upper Bound Conjecture for spheres and, more generally, for all simplicial complexes Δ such that the associated graded k-algebra $k[\Delta]$ for an arbitrary field k is a Cohen-Macaulay ring. In 1976 G. A. Reisner described those simplicial complexes for which $k[\Delta]$ is a Cohen-Macaulay ring. We now define Δ to be a Buchsbaum complex if $k[\Delta]$ is a Buchsbaum ring. For instance, if the geometric realization $X = |\Delta|$ of Δ is a connected manifold, then Δ is a Buchsbaum complex.

Let $|\Delta|$, for example, be the torus. We know that a major objection to the use of simplicial complexes in computing topological invariant of compact polyhedra is that the dissection of the polyhedron may require an uncomfortably large number of simplexes. Thus the most obvious dissection of the torus, which is pictured as

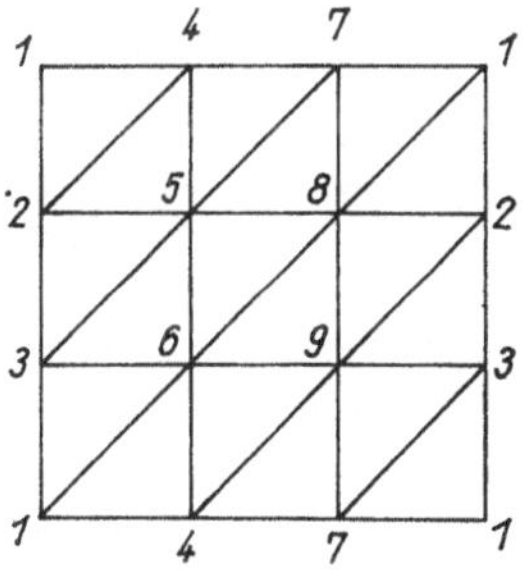

(see Hilton-Wylie [1], p. 49)

requires 9 vertices, 27 edges and 18 triangles. But we do get from such triangulations our graded k-algebras $k[\Delta]$. We obtain for the torus:

$$k[\Delta] = k[x_1, \ldots, x_9]/\mathfrak{a},$$

where

$$\mathfrak{a} = (x_1x_5,\ x_1x_9,\ x_2x_6,\ x_2x_7,\ x_3x_4,\ x_3x_8,\ x_4x_8,\ x_5x_9,\ x_6x_7,\ x_1x_2x_3,\ x_4x_5x_6,\ x_7x_8x_9,\ x_1x_6x_8,\ x_2x_4x_9,\ x_3x_5x_7,\ x_1x_4x_7,\ x_2x_5x_8,\ x_3x_6x_9).$$

We can see that the torus is a (non-Cohen-Macaulay) Buchsbaum complex. For the invariant $I(k[\Delta])$ of the Buchsbaum ring $k[\Delta]$ we have $I(k[\Delta]) = 2$ since there are 1-cycles z_1^1, z_1^2 in any simplicial decomposition of the torus, and as it turns out that $\{z_1^1\} \neq 0$ and $\{z_1^2\} \neq 0$ generate the homology group $H_1(\Delta)$ freely.

2*

Applying some fundamental principles of Buchsbaum rings our investigations now result in a particular solution of the Upper Bound Conjecture, which was extended to arbitrary manifolds by V. Klee [1] in 1964. For instance, we will prove the following statement in Chapter II, § 4:

If the geometric realization $X = |\Delta|$ of Δ is a connected manifold then we have:

$$f_v \leq \binom{n}{v+1} - \binom{d}{v+1} \sum_{i=0}^{v-1} \binom{v}{i+1} \dim_k \tilde{H}^i(\Delta; k)$$

for $v = 0, 1, \ldots, d-1$, where $\tilde{H}^i(\Delta; k)$ denotes the reduced simplicial cohomology of Δ with coefficients in an arbitrary fixed field k.

Chapter 0

Some foundations of commutative and homological algebra

§ 1. Local algebra and homological algebra

Chapter 0 contains the fundamental tools needed for the following chapters. In § 1 we will assume familiarity with the basic techniques of local algebra and homological algebra. Since notation and terminology vary from one source to another, we will assemble in this paragraph (more or less without proofs) the basic definitions and results needed. More details can be found in standard sources such as: Atiyah-Macdonald [1], Matsumura [1], Zariski-Samuel [1], Cartan-Eilenberg [1], Grothendieck [3] and Serre [2].

All rings are tacitly assumed to be commutative and Noetherian with unit element. Let A be a ring. By an A-module we mean a unitary module over A. A Noetherian A-module is then a finitely generated A-module.

1. Associated primes

Let M be an A-module. We say that a prime ideal $\mathfrak{p}$ of A is an *associated prime* of M, if one of the following equivalent conditions holds

(i) there exists an element $x \in M$ with $\operatorname{Ann} xA = \mathfrak{p}$ where $\operatorname{Ann} N := \{a \in A \mid aN = 0\}$ is the *annihilator* of the A-module N.

(ii) M contains a submodule isomorphic to $A/\mathfrak{p}$.

The set of associated primes of M is denoted by $\operatorname{Ass}_A M$ or by $\operatorname{Ass} M$.

If M is an A-module, the *support* of M, written $\operatorname{Supp}_A M$ or $\operatorname{Supp} M$, is the set of prime ideals $\mathfrak{p}$ of A such that the localization $M_\mathfrak{p}$ of M at $\mathfrak{p}$ is $\neq O$.

The *Krull dimension* of M, written $\dim_A M$ or $\dim M$, is defined to be the supremum of length of chains of prime ideals of $\operatorname{Supp} M$ if it exists, and ∞ if it does not.

We have $\operatorname{Ass} M \subseteq \operatorname{Supp} M$, and any minimal element of $\operatorname{Supp} M$ is in $\operatorname{Ass} M$.

For example, let $\mathfrak{a}$ be an ideal of A. Then the minimal associated primes of the A-module $A/\mathfrak{a}$ are precisely the minimal prime over-ideals of $\mathfrak{a}$.

If M is a Noetherian A-module then $\operatorname{Ass} M$ is a finite set.

An A-module M is said to be *co-primary* if it has only one associated prime. A submodule N of M is said to be a primary submodule of M if M/N is co-primary. If $\operatorname{Ass} M/N = \{\mathfrak{p}\}$, we say N is $\mathfrak{p}$-primary or that N belongs to $\mathfrak{p}$ (as a submodule of M). The connection with the classical definition of a primary ideal $\mathfrak{q}$ of A is given by the following equivalence:

(i) the module M is co-primary,

(ii) $M \neq O$, and if $a \in A$ is a zero divisor for M then a is locally nilpotent on M; that is, for each $x \in M$ there is an integer $n > 0$ such that $a^n x = 0$.

Let N be a submodule of M. A *primary decomposition* of N is an equation $N = Q_1 \cap \ldots \cap Q_r$ with Q_i primary in M. Such a decomposition is said to be *irredundant* if no Q_i can be omitted and if the associated primes of M/Q_i $(1 \leq i \leq r)$ are all distinct. Any primary decomposition can be simplified to an irredundant one.

If $N = Q_1 \cap \ldots \cap Q_r$ is an irredundant primary decomposition and if Q_i belongs to $\mathfrak{p}_i$, then we have $\operatorname{Ass} M/N = \{\mathfrak{p}_1, \ldots, \mathfrak{p}_r\}$.

If $\mathfrak{p}_i$ is an embedded prime of M/N, that is $\mathfrak{p}_i$ is not minimal in $\operatorname{Ass} M/N$, then the corresponding primary component Q_i is not necessarily unique.

On the other hand if $\mathfrak{p}_j$ is minimal in $\operatorname{Ass} M/N$ then the primary component Q_j is uniquely determined by N and by $\mathfrak{p}_j$.

We have the following well-known main result:

If M is a Noetherian A-module then any submodule N of M has a primary decomposition.

Having a primary decomposition $N = Q_1 \cap \ldots \cap Q_r$ we put $U(N) = \bigcap Q_i$ such that $\dim M/Q_i = \dim M/N$.

Finally, we set $\operatorname{Spec} A = \{\text{prime ideals of } A\}$. $X := \operatorname{Spec} A$ is an affine scheme:

$$\operatorname{Spec} A = \begin{cases} \text{As a point set, the set of primes of } A. \\ \text{As a topological space a basis of open sets is given by the subsets } X_f \\ := \{\mathfrak{p} \in \operatorname{Spec} A \mid f \notin \mathfrak{p}\} \text{ for all } f \in A. \\ \text{As a locally ringed space, its structure sheaf is defined by } \Gamma(X_f, \mathcal{O}_X) := A_f \\ = \text{localization of } M \text{ at the multiplicatively closed set } \{1, f, f^2, \ldots\}. \end{cases}$$

2. *Systems of parameters and multiplicity*

A ring A which has only one maximal ideal $\mathfrak{m}$ is called a *local ring*, and $A/\mathfrak{m}$ is called the *residue field* of A. If A is a local ring then the Krull dimension of A is finite.

Let A be a local ring; $\mathfrak{m}$ its maximal ideal. Let M be a Noetherian A-module of dimension $d \geq 0$. A family $(x_1, \ldots, x_d)$ of elements $x_1, \ldots, x_d$ of $\mathfrak{m}$ is said to be a *system of parameters of* M if $\dim M/(x_1, \ldots, x_d)\, M = 0$. If no confusion is possible we denote a system of parameters $(x_1, \ldots, x_d)$ of M simply by its elements $x_1, \ldots, x_d$. We call an ideal $\mathfrak{q}$ of A a *parameter ideal* of M if there is a system of parameters $x_1, \ldots, x_d$ of M contained in $\mathfrak{q}$ such that $\mathfrak{q} \cdot M = (x_1, \ldots, x_d) \cdot M$.

If $d = 0$, any ideal contained in $\operatorname{Ann} M$ is a parameter ideal of M; especially, the zero ideal is a parameter ideal.

We have the existence of systems of parameters in any local ring and for every Noetherian A-module. Notice that if $x_1, \ldots, x_d$ is a system of parameters for M, then the dimension of $M/(x_1, \ldots, x_j)\, M$ is $d - j$ for all $j = 1, \ldots, d$.

A system of elements $x_1, \ldots, x_j$, $j \leq d$, of $\mathfrak{m}$ is said to be a *part* of a system of parameters for M if $\dim M/(x_1, \ldots, x_j)\, M = d - j$.

Let $x_1, \ldots, x_n$ be a sequence of elements of A. Recall that if M is an A-module, then $x_1, \ldots, x_n$ is an M-sequence if

1) x_{i+1} is a non-zero divisor on $M/(x_1, \ldots, x_i)\, M$ for $i = 0, \ldots, n-1$, and
2) $M \neq (x_1, \ldots, x_n)\, M$.

This property does not depend on the order of $x_1, \ldots, x_n$.

Every M-sequence is a part of a system of parameters.

Let M be a Noetherian A-module. The *depth* of M, denoted by depth M, is defined as the supremum of all integers r such that there exists an M-sequence $x_1, \ldots, x_r$.

If $x \in \mathfrak{m}$ is a non-zero divisor on M then

$$\operatorname{depth} M/xM = \operatorname{depth} M - 1.$$

Furthermore, depth $M \leq$ the infimum of dim $A/\mathfrak{p}$ as $\mathfrak{p}$ runs through Ass M.

In particular we get

$$\operatorname{depth} M \leq \dim M \quad \text{if} \quad M \neq 0.$$

M is said to be *Cohen-Macaulay* if depth $M = \dim M$. The ring A is said to be Cohen-Macaulay if it is a Cohen-Macaulay A-module.

A is a regular local ring if and only if $\mathfrak{m}$ is generated by an A-sequence. Hence a regular local ring is Cohen-Macaulay.

A local ring A is called a (*local*) *Gorenstein ring* if A is Cohen-Macaulay, and whenever $x_1, \ldots, x_n$ is a maximal A-sequence, then the ideal $(x_1, \ldots, x_n) A$ is irreducible.

We have the following useful

Lemma 1.1. *Let M be a Cohen-Macaulay A-module. Then:*

(i) *M is equidimensional (i.e.* dim $A/\mathfrak{p} = \dim M$ *for all minimal primes $\mathfrak{p}$ in* Supp M) *without embedded primes (i.e. $0 \subset M$ has no embedded primes).*

(ii) *Let x be an element of $\mathfrak{m}$ such that* dim $M/xM = \dim M - 1$. *Then x is a non-zero divisor on M and M/xM is Cohen-Macaulay.*

Because of our investigations in Chapter I we need to characterize the Cohen-Macaulay property by using local multiplicities. For this we examine the Hilbert-Samuel function.

Let M be a Noetherian A-module of dimension $d \geq 0$. Let $\mathfrak{q}$ be an ideal of A such that the length, $l_A(M/\mathfrak{q}M)$, of $M/\mathfrak{q}M$ over A is finite. Then we define the so-called *Hilbert-Samuel function*, denoted by $P_{\mathfrak{q},M}(n)$, as follows:

$$P_{\mathfrak{q},M}(n) = l_A(M/\mathfrak{q}^{n+1}M) \qquad \text{for all integers } n \geq 0.$$

It is well-known that there is a polynomial in n, denoted by $p_{\mathfrak{q},M}(n)$, such that $P_{\mathfrak{q},M}(n) = p_{\mathfrak{q},M}(n)$ for all large n. The polynomial $p_{\mathfrak{q},M}(n)$ is the so-called (*characteristic*) *Hilbert Samuel polynomial* of the ideal $\mathfrak{q}$ with respect to M. There exist integers $e_0 := e_0(\mathfrak{q}, M)$ (> 0), $e_1 := e_1(\mathfrak{q}, M), \ldots, e_d := e_d(\mathfrak{q}, M)$, where $d = \dim M$, such that

$$p_{\mathfrak{q},M}(n) = e_0(\mathfrak{q}, M)\binom{n+d}{d} + e_1(\mathfrak{q}, M)\binom{n+d-1}{d-1} + \ldots + e_d(\mathfrak{q}, M).$$

The leading coefficient $e_0(\mathfrak{q}, M)$ of $p_{\mathfrak{q},M}$ is called the *multiplicity of $\mathfrak{q}$ with respect to M.*

When $\mathfrak{q} = \mathfrak{m}$ and $M = A$, we simplify the notation as follows: $e_0(\mathfrak{q}, A) = e_0(A)$ and $e_0(A)$ is called the *multiplicity* of A.

For our purposes we obtain the following main result:

Theorem 1.2. *Let A be a local ring, let M be a Noetherian A-module. The following properties are equivalent:*

(a) *M is a Cohen-Macaulay module.*

(b) *There exists a parameter ideal $\mathfrak{q}$ of M in A such that* $e_0(\mathfrak{q}, M) = l(M/\mathfrak{q}M)$.

(c) *For every parameter ideal $\mathfrak{q}$ of M in A,*

$$e_0(\mathfrak{q}, M) = l(M/\mathfrak{q}M).$$

M. Auslander and D. A. Buchsbaum [1] used the methods of homological algebra to give an explicit expression for a general multiplicity in terms of the Euler-Poincaré characteristic of the graded homology module of a certain Koszul complex. In particular, they gave an axiomatic description of multiplicity. This development has opened the subject to a much simpler treatment. An example of this may be found in D. J. Wright's paper [1]. Here an inductive definition is given for the so-called general multiplicity symbol for n elements, relative to an arbitrary module over a commutative ring, which is suggested by Auslander-Buchsbaum [1], Theorem 3.3. We will have to use some well-known properties of this multiplicity symbol. It should be noted that the account of the general multiplicity theory in this section is restricted to what is required for our immediate application. However, the theory is more extensive than indicated here, and some of the results are valid under wider hypotheses, see, for instance, D. G. Northcott [1].

Let A be a local ring and let M be a Noetherian A-module. Let $x_1, \ldots, x_n$ be elements of A such that the A-module $M/(x_1, \ldots, x_n) M$ has finite length. When $n = 0$ this condition is to be understood as meaning that $l_A(M)$ is finite. The definition of the *multiplicity symbol of* $x_1, \ldots, x_n$ *with respect to* M, denoted by $e(x_1, \ldots, x_n|M)$, uses induction on n.

First suppose that $n = 0$. In this case, by our convention, $l_A(M)$ is finite. We may therefore put

$$e(\emptyset|M) := l_A(M).$$

Now assume that $n \geq 1$. We set $O :_M x_1 = \{m \in M \mid mx_1 = 0\}$.

Since $l_A\big(M/(x_1, \ldots, x_n) M\big) < \infty$ it follows that $l_A\big((O :_M x_1)/(x_2, \ldots, x_n) \cdot (O :_M x_1)\big) < \infty$. Accordingly, by our assumptions, $e(x_2, \ldots, x_n \mid M/x_1M)$ and $e(x_2, \ldots, x_n \mid O :_M x_1)$ are both defined and so we may put

$$e(x_1, \ldots, x_n|M) := e(x_2, \ldots, x_n|M/x_1M) - e(x_2, \ldots, x_n|O :_M x_1).$$

We collect some elementary properties that we need.

Lemma 1.3.

(i) *If* $x_1, \ldots, x_n$ *is a system of parameters for* M, *i.e.,* $n = \dim M$, *then we have for* $\mathfrak{q} := (x_1, \ldots, x_n) A$:

$$e_0(\mathfrak{q}, M) = e(x_1, \ldots, x_n|M),$$

(ii) $0 \leq e(x_1, \ldots, x_n|M) \leq l_A\big(M/(x_1, \ldots, x_n) M\big)$.

(iii) (*The additivity property*) *Let* $O \to M_p \to \ldots \to M_1 \to M_0 \to O$ *be an exact sequence of Noetherian* A*-modules and suppose that the* A*-modules* $M_i/(x_1, \ldots, x_n) M_i$ *have finite length for all* $i = 0, 1, \ldots, p$ *then we have*

$$\sum_{i=0}^{p} (-1)^i \cdot e(x_1, \ldots, x_n|M_i) = 0.$$

(iv) *Assume that for some particular value of* i *we have* $x_i^m M = 0$, *where* m *is some positive integer, then we get*

$$e(x_1, \ldots, x_n|M) = 0.$$

(v) *Let $r_1, r_2, \ldots, r_n$ be positive integers. Then*

$$e(x_1^{r_1}, x_2^{r_2}, \ldots, x_n^{r_n}|M) = r_1 \cdot r_2 \cdot \ldots \cdot r_n \cdot e(x_1, \ldots, x_n|M).$$

(vi) $l_A\big(M/(x_1, \ldots, x_n)M\big) - e(x_1, \ldots, x_n|M)$

$$= \sum_{i=1}^{n} e\big(x_{i+1}, \ldots, x_n \mid (x_1, \ldots, x_{i-1})\, M :_M x_i/(x_1, \ldots, x_{i-1})\, M\big).$$

The next lemma is useful, apart from its intrinsic interest. It enables us to prove some statements in Chapter I. It is precisely the property (i) of Lemma 1.3 and the following lemma which are of interest in the general multiplicity theory to our investigations in Chapter I. The point is to have a criterion for a multiplicity to be zero.

Lemma 1.4. *If M and $x_1, \ldots, x_n$ are as above, then $e(x_1, \ldots, x_n|M) = 0$ if and only if $n > \dim M$.*

Sketch of the proof (see also D. J. Wright [1], Lemma 9, or D. G. Northcott [1], Proposition 7 on p. 334): Note that we have always $n \geq \dim M$. Let $\dim M = n$. Then Lemma 1.3 (i) implies that $e(x_1, \ldots, x_n|M) > 0$. Let $n > \dim M$. We use induction on $\dim M$. If $\dim M = 0$, then x_1 is nilpotent with respect to M and so $x_1^m M = 0$ for a suitable integer m. Accordingly $e(x_1, \ldots, x_n|M) = 0$ by Lemma 1.3, (iv).

From now on assume that $\dim M \geq 1$. If x_1 is nilpotent, then we get the desired result by Lemma 1.3, (iv). We shall therefore suppose that x_1 is not nilpotent. Since M is a Noetherian A-module we can choose an integer m so that x_1 is not a zero divisor on $M' := M/(O :_M x_1^m)$. Now, by Lemma 1.3, (iv), $e(x_1, \ldots, x_n|O :_M x_1^m) = 0$. Consequently we have

$$e(x_1, \ldots, x_n|M) = e(x_1, \ldots, x_n|M'),$$

and x_1 is not a zero divisor of M'. This shows that for the remainder of the proof we may assume that x_1 is not a zero divisor of M. But then we get

$$e(x_1, \ldots, x_n|M) = e(x_2, \ldots, x_n|M/x_1M).$$

Since $\dim M/x_1M < \dim M$ the assertion follows by the inductive hypothesis, q.e.d.

As indicated, some important results in Chapter I will be established with this useful criterion for a multiplicity to be zero.

3. *Local cohomology theory and cohomology of the Koszul complex*

A recent addition to local algebra has been provided by Grothendieck's local cohomology theory; basic facts about this theory are available from Grothendieck [3] and Sharp [1]. If A is a local ring, then any ideal $\mathfrak{a}$ of A determines the following additive, A-linear, covariant, left exact functor $\Gamma_{\mathfrak{a}}^{\cdot}$ (on the category of all A-modules and homomorphisms) called the *local cohomology functor* with respect to $\mathfrak{a}$:

Let N be a submodule of an A-module M, let

$$N :_M \langle \mathfrak{a} \rangle := \{m \in M \mid \text{there exists an integer } n > 0 \text{ such that } \mathfrak{a}^n \cdot m \subseteq N\},$$

and

$$N :_M \mathfrak{a} := \{m \in M \mid \mathfrak{a} \cdot m \subseteq N\}.$$

For an A-module M, we define

$$\Gamma_{\mathfrak{a}}(M) := 0 :_M \langle \mathfrak{a} \rangle = \bigcup_{k=1}^{\infty} (0 :_M \mathfrak{a}^k).$$

If M is an A-module and i is an integer ≥ 0, then we denote by $H^i_{\mathfrak{a}}(M)$ the module obtained by applying to M the *right derived functor* of $\Gamma_{\mathfrak{a}}$ $\big(\Gamma_{\mathfrak{a}}(M) \cong H^0_{\mathfrak{a}}(M)\big)$.

Suppose again that A is a local ring having maximal ideal $\mathfrak{m}$. It is well-known that if M is a Noetherian A-module having Krull dimension $d \geq 0$ then all the modules $H^i_{\mathfrak{m}}(M)$ for $i \geq 0$ are Artinian A-modules and $H^d_{\mathfrak{m}}(M) \neq O$. Moreover, if $d > 0$ then $H^d_{\mathfrak{m}}(M)$ is not Noetherian.

Furthermore, $H^i_{\mathfrak{m}}(M) = O$ for all $i > \dim M$, and depth M is the least integer i for which $H^i_{\mathfrak{m}}(M) \neq O$.

We can give yet another description of local cohomology as follows:

The positive integers with the usual ordering form a directed set I. If $i, j \in I$ with $i \leq j$, then $\mathfrak{a}^j \subseteq \mathfrak{a}^i$, and the natural A-homomorphism $A/\mathfrak{a}^j \to A/\mathfrak{a}^i$ induces, for an arbitrary A-module M, an A-homomorphism $\pi_{ij}(M)\colon \operatorname{Ext}^n(A/\mathfrak{a}^i, M) \to \operatorname{Ext}^n(A/\mathfrak{a}^j, M)$. Also, the $\operatorname{Ext}^n(A/\mathfrak{a}^i, M)$ and $\pi_{ij}(M)$ form a direct system of A-modules and A-homomorphisms over I, and so the direct limit $\varinjlim_k \operatorname{Ext}^n(A/\mathfrak{a}^k, M)$ can be formed. It is now easy to check that $\varinjlim_k \operatorname{Ext}^n(A/\mathfrak{a}^k, \ )$ becomes a covariant, additive, A-linear functor on the category of all A-modules; one can show that the functors

$$H^n_{\mathfrak{a}}(\ ) \quad \text{and} \quad \varinjlim_k \operatorname{Ext}^n(A/\mathfrak{a}^k, \ )$$

are naturally equivalent for each $n \geq 0$. Therefore we get isomorphisms for all $n \geq 0$:

$$H^n_{\mathfrak{a}}(M) \cong \varinjlim_k \operatorname{Ext}^n(A/\mathfrak{a}^k, M).$$

In particular, we obtain canonical maps

$$\varphi^i_M\colon \operatorname{Ext}^i_A(A/\mathfrak{a}, M) \to H^i_{\mathfrak{a}}(M) \qquad \text{for each } i \geq 0.$$

Now, we want to examine how this relates to the cohomology of the Koszul complex. Let A be a local ring. For an element x of A we define the *Koszul complex* $K(x; A)$ generated over A by x as follows:

$$K_i(x; A) = 0 \qquad \text{for all } i \neq 0, 1,$$

$$K_0(x; A) \cong K_1(x; A) \cong A$$

and a map

$$d_1\colon K_1(x; A) \to K_0(x; A) \qquad \text{defined by } d_1(a) = xa \text{ for all } a \in A;$$

that is, $K(x; A)$ is the complex $0 \to A \xrightarrow{x} A \to 0$.

Let M be an A-module and let $x_1, \ldots, x_r$ be elements of A. We define the Koszul complex $K(x_1, \ldots, x_r; M)$ generated over A by $x_1, \ldots, x_r$ with respect to M by:

$$K(x_1; A) \otimes \ldots \otimes K(x_r; A) \otimes M.$$

Its homology we denote by $H_i(x_1, \ldots, x_r; M)$. We now put

$$K^i(x_1, \ldots, x_r; M) := K_{r-i}(x_1, \ldots, x_r; M)$$

and

$$H^i(x_1, \ldots, x_r; M) := H^i(K^{\cdot}(x_1, \ldots, x_r; M)) = H_{r-i}(x_1, \ldots, x_r; M).$$

Clearly, $K^i(x_1, \ldots, x_r; A)$ is a free A-module of rank $\binom{r}{i}$. We take free generators $e^{l_1 \ldots l_i}$, $1 \leq l_1 < \ldots < l_i \leq r$, of $K^i(x_1, \ldots, x_r; A)$ such that

$$d^i \left(\sum_{1 \leq l_1 < \ldots < l_i \leq r} a^{l_1 \ldots l_i} e^{l_1 \ldots l_i} \right)$$

$$= \sum_{1 \leq j_1 < \ldots < j_{i+1} \leq r} \left(\sum_{p=1}^{i+1} (-1)^{p-1} x_{j_p} a^{j_1 \ldots \hat{j}_p \ldots j_{i+1}} \right) e^{j_1 \ldots j_{i+1}},$$

where $d^{\cdot}$ denotes the differentiation in $K^{\cdot}(x_1, \ldots, x_r; A)$. Then similar formulas hold for $K^{\cdot}(x_1, \ldots, x_r; M)$.

Let $y_1, \ldots, y_s$ be elements of A such that $(y_1, \ldots, y_s) A \subseteq (x_1, \ldots, x_r) A$, i.e. there are $a_{ij} \in A$, $1 \leq i \leq s$, $1 \leq j \leq r$, with $y_i = \sum_{j=1}^{r} a_{ij} x_j$ for $i = 1, \ldots, s$. Let

$$\mathfrak{A} := \begin{pmatrix} a_{11} & \ldots & a_{1r} \\ \vdots & & \vdots \\ a_{s1} & \ldots & a_{sr} \end{pmatrix}.$$

If $1 \leq l_1 < \ldots < l_n \leq s$, $1 \leq j_1 < \ldots < j_n \leq r$, denote by $\Delta^{j_1 \ldots j_n}_{l_1 \ldots l_n}$ the minor of $\mathfrak{A}$ consisting of the elements of the l_1th, ..., l_nth row and the j_1th, ..., j_nth column (set $\Delta = 0$ if $n > r$ or $n > s$). Then it is easy to see (using well-known methods of linear algebra) that we have a homomorphism of complexes

$$\Phi : K^{\cdot}(x_1, \ldots, x_r; A) \to K^{\cdot}(y_1, \ldots, y_s; A)$$

defined by (denote by $\bar{e}^{l_1 \ldots l_i}$ the free generators of $K^i(y_1, \ldots, y_s; A)$):

$$\Phi^i(e^{l_1 \ldots l_i}) = \sum_{1 \leq j_1 < \ldots < j_i \leq s} \Delta^{l_1 \ldots l_i}_{j_1 \ldots j_i} \cdot \bar{e}^{j_1 \ldots j_i}.$$

Tensoring with M we find a homomorphism of complexes

$$\Psi : K^{\cdot}(x_1, \ldots, x_r; M) \to K^{\cdot}(y_1, \ldots, y_s; M).$$

Clearly, Ψ^0 is an isomorphism. We consider two special cases:

1. Let $r = s$ and assume that $\mathfrak{a} := (x_1, \ldots, x_r) A = (y_1, \ldots, y_r) A$ with $r = \operatorname{rank}_{A/\mathfrak{m}} \mathfrak{a}/\mathfrak{m}\mathfrak{a}$. Then $\mathfrak{A}$ is invertible over A since $\mathfrak{A}$ mod $\mathfrak{m}$ is invertible over $A/\mathfrak{m}$, i.e. $\det \mathfrak{A} \notin \mathfrak{m}$.

Therefore Φ and hence Ψ is an isomorphism of complexes. We define $K^{\cdot}(\mathfrak{a}; M) := K^{\cdot}(x_1, \ldots, x_r; M)$ and $H^i(\mathfrak{a}; M) := H^i(x_1, \ldots, x_r; M)$ for all A-modules M. If we choose another set of generators of $\mathfrak{a}$, the Koszul complex and hence its cohomology is unchanged (up to an isomorphism).

Since $H^0(\mathfrak{a}; M) \cong \operatorname{Hom}_A(A/\mathfrak{a}; M)$, we get from Cartan-Eilenberg [1], Chap. III, Proposition 5.2, canonical homomorphisms $\psi^i_M : \operatorname{Ext}^i_A(A/\mathfrak{a}; M) \to H^i(\mathfrak{a}; M)$ for all $i \geq 0$.

2. Consider the sequence of ideals

$$(x_1, \ldots, x_r)\,A \supseteq (x_1^2, \ldots, x_r^2)\,A \supseteq (x_1^3, \ldots, x_r^3)\,A \supseteq \ldots,$$

from which a direct system of complexes

$$K^{\cdot}(x_1, \ldots, x_r; M) \to K^{\cdot}(x_1^2, \ldots, x_r^2; M) \to \ldots \qquad (M \text{ an } A\text{-module})$$

is obtained. The direct limit of this system is denoted by

$$K^{\cdot}_{\infty}(x_1, \ldots, x_r; M).$$

Since direct limits commute with exact sequences, we get

$$H^i\big(K^{\cdot}_{\infty}(x_1, \ldots, x_r; M)\big) \cong \varinjlim_n H^i(x_1^n, \ldots, x_r^n; M),$$

in particular

$$H^r\big(K^{\cdot}_{\infty}(x_1, \ldots, x_r; M)\big) \cong \varinjlim_n M/(x_1^n, \ldots, x_r^n)\,M,$$

where the maps of the last right hand direct system are given by multiplication by $x_1 \cdot \ldots \cdot x_r$.

Furthermore, $H^0\big(K^{\cdot}_{\infty}(x_1, \ldots, x_r; M)\big) = \varinjlim_n \big(0 :_M (x_1^n, \ldots, x_r^n)\,A\big) = H^0_{\mathfrak{a}}(M)$, where $\mathfrak{a} := (x_1, \ldots, x_r)\,A$.

Now, for $r = 1$ it is easy to verify that $K^{\cdot}_{\infty}(x; M)$ is the complex

$$0 \to M \xrightarrow{h_x} M_x \to 0,$$

where M_x is the localization of M with respect to the set $\{1, x, x^2, \ldots\}$ and $h_x(m) = \dfrac{m}{1}$. (Observe $K^{\cdot}_{\infty}(x_1, \ldots, x_r; M) \cong K^{\cdot}_{\infty}(x_1; A) \otimes \ldots \otimes K^{\cdot}_{\infty}(x_r; A) \otimes M$, since tensor products commute with direct limits.)

If M is an injective module results of Matlis [1] show that h_x is always surjective, i.e. $H^1\big(K^{\cdot}_{\infty}(x; M)\big) = 0$. Using an exact sequence defined by Corollary 1.7 below an easy inductive argument (on r) shows that $H^i\big(K^{\cdot}_{\infty}(x_1, \ldots, x_r; M)\big) = 0$ whenever M is injective and $i \geq 1$. Hence by Cartan-Eilenberg [1], Chap. III, Proposition 5.2, and Chap. V, Proposition 4.4, we get:

$$H^i\big(K^{\cdot}_{\infty}(x_1, \ldots, x_r; M)\big) \cong H^i_{\mathfrak{a}}(M)$$

for every A-module M and with $\mathfrak{a} := (x_1, \ldots, x_r)\,A$.

Let us denote by λ^i_M the *canonical homomorphism* (defined by direct limit) $H^i(\mathfrak{a}; M) \to H^i_{\mathfrak{a}}(M)$, where $\mathfrak{a}$ is an ideal of A. Then we have by the above remarks

Lemma 1.5. *For all $i \geq 0$ we have commutative diagrams*

(i)

$$\begin{array}{ccc} \mathrm{Ext}^i_A(A/\mathfrak{a}; M) & \xrightarrow{\;\varphi^i_M\;} & \\ \Big\downarrow \psi^i_M & & H^i_{\mathfrak{a}}(M) \\ H^i(\mathfrak{a}; M) & \xrightarrow{\;\lambda^i_M\;} & \end{array}$$

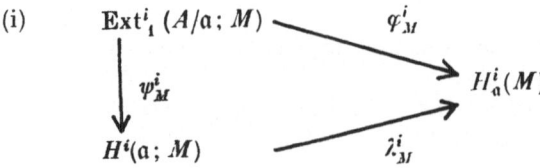

(ii) *If* $\mathfrak{b}$ *is another ideal of* A *with* $\mathfrak{a} \subseteq \mathfrak{b}$ *then there are for all* i *commutative diagrams*

$$\begin{array}{ccc} H^i(\mathfrak{b}; M) & \longrightarrow & H^i(\mathfrak{a}; M) \\ \downarrow & & \downarrow \\ H^i_{\mathfrak{b}}(M) & \longrightarrow & H^i_{\mathfrak{a}}(M) \end{array}$$

where the vertical maps are the canonical ones and the other homomorphisms are induced by ψ.

Next, analyzing the proof of Proposition 1 of Serre [2], IV-2, we get the following

Lemma 1.6. *Let* L *be a complex of* A*-modules and let* x *be an element of* A. *Then we have for all* $p \geq 1$ *commutative diagrams with exact rows:*

$$\begin{array}{ccccccccc} 0 & \to & H^1(x; H^{p-1}(L)) & \to & H^p(K^{\cdot}(x; A) \otimes L) & \to & H^0(x; H^p(L)) & \to & 0 \\ & & \alpha\downarrow & & \beta\downarrow & & \gamma\downarrow & & \\ 0 & \to & H^1_{xA}(H^{p-1}(L)) & \to & H^p(K^{\cdot}_{\infty}(x; A) \otimes L) & \to & H^0_{xA}(H^p(L)) & \to & 0 \end{array}$$

where α *and* γ *are induced by the canonical map* $K^{\cdot}(x;\ ) \to K^{\cdot}_{\infty}(x;\ )$ *which also defines a map* $K^{\cdot}(x; A) \otimes L \to K^{\cdot}_{\infty}(x; A) \otimes L$ *which specifies* β.

From Lemma 1.6 we obtain a very useful corollary.

Corollary 1.7. *Let* A *be a local ring and* M *an* A*-module. Let* $x_1, \ldots, x_r$, $r \geq 2$, *be elements of* A. *Then we have for all* $p \geq 1$ *commutative diagrams with exact rows:*

$$\begin{array}{ccccccccc} 0 & \to & H^1(x_1;\ H^{p-1}(x_2, \ldots, x_r; M)) & \to & H^p(x_1, \ldots, x_r; M) & \to & H^0(x_1; H^p(x_2, \ldots, x_r; M)) & \to & 0 \\ & & \downarrow & & \downarrow & & \downarrow & & \\ 0 & \to & H^1_{x_1A}(H^{p-1}_{(x_2\ldots x_r)A}(M)) & \to & H^p_{(x_1\ldots x_r)A}(M) & \to & H^0_{x_1A}(H^p_{(x_2\ldots x_r)A}(M)) & \to & 0. \end{array}$$

If $x_1 M = 0$, *we get the following commutative diagram*

$$\begin{array}{ccccccccc} 0 & \to & H^{p-1}(x_2, \ldots, x_r; M) & \to & H^p(x_1, \ldots, x_r;\ M) & \to & H^p(x_2,\ \ldots, x_r; M) & \to & 0 \\ & & \downarrow & & \downarrow & & \downarrow & & \\ 0 & \to & 0 & \to & H^p_{(x_1\ldots x_r)A}(M) & \to & H^p_{(x_2\ldots x_r)A}(M) & \to & 0. \end{array}$$

Proof: The second commutative diagram is a consequence of the first since for an A-module N with $x_1 N = 0$

$$H^1(x_1; N) = H^0(x_1; N) = N, \quad H^1_{x_1A}(N) = 0, \quad H^0_{x_1A}(N) = N$$

follows. Thus we need to prove the commutativity of the first diagram.

We set $L = K^{\cdot}(x_2, \ldots, x_r; M)$ and apply Lemma 1.6. Then the top row of the commutative diagram of Lemma 1.6 agrees with the top row of the diagram under consideration. Now we set for $n \geq 1$

$$L_n := K^{\cdot}(x_2^n, \ldots, x_r^n; M)$$

and

$$L_{\infty} := K^{\cdot}_{\infty}(x_2, \ldots, x_r; M)$$

and denote by (E_n) the bottom row of the commutative diagram of Lemma 1.6 if $L := L_n$ and by (E_∞) for $L := L_\infty$. Then we have a direct system of exact sequences $(E_1) \to (E_2) \to \ldots$ Since direct limits commute with exact sequences and with the local cohomology functors, we get $\varinjlim_n (E_n) = (E_\infty)$ and the map $(E_1) \to (E_\infty)$ (into the direct limit) describes a commutative diagram (with top row (E_1) and bottom row (E_∞)). But (E_∞) is just the bottom row of our commutative diagram. Thus, combining the commutative diagram constructed above and this commutative diagram we find a commutative diagram of the desired type, q.e.d.

Lemma 1.8. *Let $\mathfrak{a} \subset A$ be an ideal and $S \subset A$ a multiplicatively closed set with $1 \in S$, $\mathfrak{a} \cap S \neq \emptyset$. Let M be an A-module and for all $i \geq 0$, in abreviated notation, $H^i(M) := \varinjlim_n \mathrm{Ext}^i_A(\mathfrak{a}^n; M)$. Then we have:*

(i) *There is an exact sequence*

$$0 \to H^0_{\mathfrak{a}}(M) \to M \xrightarrow{f} H^0(M) \to H^1_{\mathfrak{a}}(M) \to 0$$

and for all $i \geq 1$ there are isomorphisms

$$H^i(M) \cong H^{i+1}_{\mathfrak{a}}(M).$$

(ii) *The following conditions are equivalent:*

(a) $\mathrm{Supp}\, M \subseteq V(\mathfrak{a})$, (b) $H^0(M) = 0$, (c) $H^i(M) = 0$ *for all* $i \geq 0$.

(iii) $$H^0_{\mathfrak{a}}(H^0(M)) = H^1_{\mathfrak{a}}(H^0(M)) = 0,$$

$$H^1_{\mathfrak{a}}(H^0(M)) \cong H^i_{\mathfrak{a}}(M) \quad \textit{for all } i \geq 2.$$

(iv) *There is a (natural) homomorphism (of A-modules)*

$$g: H^0(M) \to M_S$$

such that the composition gf is just the nautral map $M \to M_S$.

(v) *g is injective if and only if $S \cap \mathfrak{p} = \emptyset$ for all $\mathfrak{p} \in \mathrm{Ass}\, M \setminus V(\mathfrak{a})$.*

(vi) (Formula of Deligne) *There is a natural (A-)isomorphism*

$$H^0(M) \cong \varprojlim_{a \in \mathfrak{a}} M_a,$$

where M_a denotes the localization of M at the multiplicatively closed set $\{1, a, a^2, \ldots\}$. The (A-)modules M_a, $a \in \mathfrak{a}$, form an inverse system: We define for $a, b \in \mathfrak{a}$ a homomorphism

$$\varrho_{a,b}: M_b \to M_a$$

if and only if $a \in \sqrt{bA}$, i.e. if $a^t = bc$ for $t \in \mathbf{N}$, $c \in A$ by setting

$$\varrho_{a,b}\left(\frac{m}{b^n}\right) = \frac{mc^n}{a^{tn}} \quad (m \in M,\ n \in \mathbf{N}).$$

Proof: Apply the functors $\mathrm{Ext}^i_A(\ ; M)$ to the exact sequence $0 \to \mathfrak{a}^n \to A \to A/\mathfrak{a}^n \to O$ and take the direct limit of the resulting long exact cohomology sequence for $n = 1, 2, \ldots$ This is again an exact sequence and (i) follows by virtue of $\mathrm{Hom}_A(A; M) \cong M$ and $\mathrm{Ext}^i_A(A; M) = 0$ for $i \geq 1$.

(ii) (a) $\Rightarrow$ (b): Let $\operatorname{Supp} M \subseteq V(\mathfrak{a})$ and take $\sigma \in H^0(M)$. Choose $s \in \operatorname{Hom}(\mathfrak{a}^n; M)$ representing σ. Then for all $a \in \mathfrak{a}^n$ there is a $p \geq 0$ with $\mathfrak{a}^p s(a) = 0$. Since $\mathfrak{a}^n$ is finitely generated, there is a $q \geq 0$ with $\mathfrak{a}^q s(a) = 0$ for all $a \in \mathfrak{a}^n$. Therefore $s|_{\mathfrak{a}^{n+q}} = 0$, that is, $\sigma = 0$.

(a) $\Rightarrow$ (c): Take a minimal injective resolution of M

$$0 \to M \to I_0 \to I_1 \to \ldots$$

Since $\operatorname{Supp} M \subseteq V(\mathfrak{a})$, $\operatorname{Supp} I_j \subseteq V(\mathfrak{a})$ for all $i \geq 0$. Now $H^i(M)$ is the ith cohomology module of the complex

$$0 \to H^0(I_0) \to H^0(I_1) \to \ldots$$

But by the first part of the proof $H^0(I_j) = O$ for all $j \geq 0$ which proves this assertion.

(b) $\Rightarrow$ (a): If $H^0(M) = 0$ the exact sequence of (i) yields an isomorphism $H^0_{\mathfrak{a}}(M) \cong M$. Therefore each element of M is annihilated by some power of $\mathfrak{a}$, i.e. $\operatorname{Supp} M \subseteq V(\mathfrak{a})$. The implication (c) $\Rightarrow$ (b) is trivial and (ii) is therefore proven.

(iii) From (i) and (ii) we get:

$$\operatorname{Supp} M \subseteq V(\mathfrak{a}) \text{ if and only if } H^0_{\mathfrak{a}}(M) \cong M \text{ and } H^i_{\mathfrak{a}}(M) = 0 \quad \text{for all } i \geq 1.$$

We split the exact sequence of (i) into two exact sequences:

(a) $0 \to H^0_{\mathfrak{a}}(M) \to M \to M' \to 0$

and

(b) $0 \to M' \to H^0(M) \to H^1_{\mathfrak{a}}(M) \to 0$.

Since $\operatorname{Supp} H^0_{\mathfrak{a}}(M)$, $\operatorname{Supp} H^1_{\mathfrak{a}}(M) \subseteq V(\mathfrak{a})$ we get

$$H^i_{\mathfrak{a}}(M) \cong H^i_{\mathfrak{a}}(M') \qquad \text{for all } i \geq 1$$

and

$$H^i_{\mathfrak{a}}(M') \cong H^i_{\mathfrak{a}}\big(H^0(M)\big) \qquad \text{for all } i \geq 2,$$

hence

$$H^i_{\mathfrak{a}}(M) \cong H^i_{\mathfrak{a}}\big(H^0(M)\big) \qquad \text{for all } i \geq 2.$$

Also (a) yields $H^0_{\mathfrak{a}}(M') = 0$ and thus (b) gives rise to the following exact sequence $\big(H^0_{\mathfrak{a}}(H^1_{\mathfrak{a}}(M)) \cong H^1_{\mathfrak{a}}(M)\big)$:

$$0 \to H^0_{\mathfrak{a}}\big(H^0(M)\big) \to H^1_{\mathfrak{a}}(M) \to H^1_{\mathfrak{a}}(M') \to H^1_{\mathfrak{a}}\big(H^0(M)\big) \to 0.$$

Now it is not difficult to see that the middle homomorphism is nothing but the isomorphism $H^1_{\mathfrak{a}}(M) \cong H^1_{\mathfrak{a}}(M')$ obtained from (a) and this proves (iii).

(iv) Let $\sigma \in H^0(M)$. Take $n \geq 0$ and $s \in \operatorname{Hom}_A(\mathfrak{a}^n; M)$ representing σ. Choose $a \in \mathfrak{a} \cap S$. We define

$$g(\sigma) := \frac{s(a^n)}{a^n}.$$

Straightforeward calculations show that g is a well-defined A-homomorphism. Finally, for all $m \in M$ the homomorphism $h\colon \mathfrak{a} \to M$ defined by $h(a) = am$ for all $a \in \mathfrak{a}$ represents $f(m) \in H^0(M)$. Therefore, if $a \in \mathfrak{a} \cap S$, $gf(m) = \dfrac{h(a)}{a} = \dfrac{am}{a} = \dfrac{m}{1}$. This proves (iv).

(v) Assume g is injective. For $\operatorname{Ass} M \not\subseteq V(\mathfrak{a})$ let $\mathfrak{p} \in \operatorname{Ass} M \setminus V(\mathfrak{a})$. Then there is a monomorphism $A/\mathfrak{p} \to M$ giving rise to the following commutative diagram

$$\begin{array}{ccccc} A/\mathfrak{p} & \xrightarrow{f} & H^0(A/\mathfrak{p}) & \xrightarrow{g} & (A/\mathfrak{p})_S \\ & & \downarrow & & \downarrow \\ & & H^0(M) & \xrightarrow{g} & M_S \end{array}$$

where all homomorphisms are injective ($H^0(\ )$ is left exact and $H^0_{\mathfrak{a}}(A/\mathfrak{p}) = 0$ since $\mathfrak{p} \notin V(\mathfrak{a})$). Therefore $(A/\mathfrak{p})_S \neq 0$, i.e. $\mathfrak{p} \cap S = \emptyset$.

Assume now $\mathfrak{p} \cap S = \emptyset$ for all $\mathfrak{p} \in \operatorname{Ass} M \setminus V(\mathfrak{a})$. Let $\sigma \in \operatorname{Ker} g$ and choose $s \in \operatorname{Hom}(\mathfrak{a}^n; M)$ representing σ. Then for some $a \in \mathfrak{a} \cap S$ we get $\dfrac{s(a^n)}{a^n} = g(\sigma) = 0$, i.e. there is a $b \in S$ with $bs(a^n) = 0$. Then $b \in \mathfrak{p} \cap S$ for all $\mathfrak{p} \in \operatorname{Ass}(A/\operatorname{Ann} s(a^n)) \subseteq \operatorname{Ass} M$. Therefore $\operatorname{Ass}(A/\operatorname{Ann} s(a^n)) \subseteq V(\mathfrak{a})$, i.e. there is an $m \geq 0$ with $\mathfrak{a}^m s(a^n) = 0$. Assume without loss of genērality that $m \geq n$. Then $a^n s(c) = cs(a^n) = 0$ for all $c \in \mathfrak{a}^m$. Hence we have for all $c \in \mathfrak{a}^m$:

$$a \in \mathfrak{p} \cap S \quad \text{for all } \mathfrak{p} \in \operatorname{Ass}(A/\operatorname{Ann} s(c)) \subseteq \operatorname{Ass} M.$$

This implies $\operatorname{Ass}(A/\operatorname{Ann} s(c)) \subseteq V(\mathfrak{a})$, i.e. there is a $q \geq 0$ with $\mathfrak{a}^q s(c) = 0$. Since $\mathfrak{a}^m$ is finitely generated we can find an $r \geq 0$ with $\mathfrak{a}^r s(c) = 0$ for all $c \in \mathfrak{a}^m$. Therefore $s|_{\mathfrak{a}^{r+m}} = 0$, which means $\sigma = 0$, i.e. g is injective.

We now prove (vi). We see that the maps $H^0(M) \to M_a$ ($a \in \mathfrak{a}$) given in (iv) are compatible with the homomorphisms $\varrho_{a,b}$. Therefore we have a homomorphism $\psi: H^0(M) \to \varprojlim_{a \in \mathfrak{a}} M_a$.

If $\sigma \in \operatorname{Ker} \psi$, choose an $s \in \operatorname{Hom}(\mathfrak{a}^n, M)$ representing σ. Then $\dfrac{s(a^n)}{a^n} = 0$ (in M_a) for all $a \in \mathfrak{a}$, i.e. for fixed $a \in \mathfrak{a}$ there is an $m \in \mathbf{N}$ with $s(a^{m+n}) = a^m \cdot s(a^n) = 0$. Since $\mathfrak{a}$ is finitely generated, we find a $p \in \mathbf{N}$ with $s|_{\mathfrak{a}^p} = 0$, i.e. $\sigma = 0$ and ψ is injective.

Now let $\mu \in \varprojlim_{a \in \mathfrak{a}} M_a =: N$ and let $\varphi_a: N \to M_a$, $a \in \mathfrak{a}$, denote the canonical maps. If $\mathfrak{a} = (a_1, \ldots, a_t)\, A$, write $\varphi_{a_i}(\mu) = \dfrac{m_i'}{a_i^n} \in M_{a_i}$ with $m_i' \in M$, $n \in \mathbf{N}$. Then for all $i, j = 1, \ldots, t$:

$$\frac{m_i' a_j^n}{(a_i a_j)^n} = \varrho_{a_i a_j, a_i} \varphi_a(\mu) = \varrho_{a_i a_j, a_j} \varphi_{a_j}(\mu) = \frac{m_j' a_i^n}{(a_i a_j)^n},$$

i.e. there is an $l \in \mathbf{N}$ with $(a_i a_j)^l (m_i' a_j^n - m_j' a_i^n) = 0$.

We put $m_i := a_i^l m_i'$ and $p := n + l$. Then $\varphi_{a_i}(\mu) = \dfrac{m_i}{a_i^p}$ and $m_i a_j^p = m_j a_i^p$ for all $i, j = 1, \ldots, t$. Let L denote the submodule of M generated by $m_1, \ldots, m_t$. Assume $\sum_{i=1}^t r_i a_i^p = 0$ with $r_1, \ldots, r_t \in A$. Then for all $j = 1, \ldots, t$:

$$a_j^p \sum_{i=1}^t r_i m_i = \sum_i r_i m_i a_j^p = \sum_i r_i a_i^p m_j = 0,$$

i.e. $\sum r_i m_i \in 0 :_L \langle \mathfrak{a} \rangle$. Choose a $q \in \mathbf{N}$ with $\mathfrak{a}^q L \cap (0 :_L \langle \mathfrak{a} \rangle) = 0$ (which exists by the lemma of Artin-Rees since L is finitely generated). We define a map

$$s: (a_1^{p+q}, \ldots, a_t^{p+q})\, A \to M \quad \text{by} \quad s(a_i^{p+q}) := a_i^q m_i, \quad i = 1, \ldots, t.$$

If $\sum s_i a_i^{p+q} = 0$ with $s_1, \ldots, s_t \in A$ then by the preceding we get $\sum s_i a_i^q m_i \in (O :_L \langle\mathfrak{a}\rangle) \cap \mathfrak{a}^q L = 0$, i.e. s is a homomorphism. Choose $n \in \mathbf{N}$ with $\mathfrak{a}^n \subseteq (a_1^{p+q}, \ldots, a_t^{p+q}) A$. Then the restriction of s to $\mathfrak{a}^n$ represents an element $\sigma \in H^0(M)$ and it is easy to see that $\psi(\sigma) = \mu$, q.e.d.

Remarks.

1. The lemma is true for arbitrary Noetherian rings A.
2. The lemma also has a graded version: It remains true if A is a Noetherian graded ring and M a graded R-module. The homomorphisms occuring here are all of degree zero (compare with the notations of § 2).
3. Let $X := \operatorname{Spec} A$, $U := X \setminus V(\mathfrak{a})$. Then (vi) says that $H^0(M) \cong \Gamma(U, \tilde{M})$, where $\tilde{M}$ is the sheaf on X associated with M. This appears in the literature as "Formula of Deligne" (c.f. R. Hartshorne [4], Chap. III, Exercise 3.7).

§ 2. Graded modules and Künneth formulas

In this paragraph we will develop basic facts and notations concerning graded rings and modules which we will need in the sequel.

A *graded ring* R is always understood to be a commutative Noetherian ring with unit 1 which is (considered as an abelian group) a direct sum of subgroups R_i, $i \in \mathbf{Z}$, such that $R_i \cdot R_j \subseteq R_{i+j}$ for all $i, j \in \mathbf{Z}$.

A *graded R-module* M is a unitary R-module which is (also considered as an abelian group) a direct sum of subgroups M_i, $i \in \mathbf{Z}$, such that $R_i \cdot M_j \subseteq M_{i+j}$ for all $i, j \in \mathbf{Z}$.

An element $m \in M$ is called *homogeneous* if $m \in M_i$ for some $i \in \mathbf{Z}$. For $m \neq 0$ this i is uniquely determined. We call it the *degree* of m and write $i := \deg m$.

We say that a graded R-module $N = \bigoplus_{i \in \mathbf{Z}} N_i$ is a *submodule* of M if $N_i \subseteq M_i$ for all $i \in \mathbf{Z}$.

If M is a given graded R-module and if $n \in \mathbf{Z}$ let

$$[M]_n := \{m \in M \mid m = 0 \text{ or } \deg m = n\} \qquad (\text{i.e. } [M]_n = M_n \text{ if } M = \bigoplus_{i \in \mathbf{Z}} M_i).$$

Let $p \in \mathbf{Z}$. By $M(p)$ we denote the graded R-module which is given by $[M(p)]_n = [M]_{p+n}$ for all $n \in \mathbf{Z}$. We say that $M(p)$ is obtained from M by *shifting* of degrees.

By an ideal of R we always understand a graded R-submodule of R, i.e. an ideal (in the usual sense) which may be generated by homogeneous elements of R.

By a *Noetherian graded R-module* we mean a graded R-module satisfying the ascending chain condition for graded submodules. Since R is Noetherian itself, a graded R-module M is Noetherian if and only if M is finetely generated by homogeneous elements.

1. Associated primes

Let R be a graded ring and let M be a graded R-module. A homogeneous prime ideal $\mathfrak{p}$ of R is called an *associated prime* of M if one of the following equivalent conditions is fulfilled:

(i) There is an homogeneous element $x \in M$ with $\operatorname{Ann} xA = \mathfrak{p}$.

(ii) M contains a submodule isomorphic to $(R/\mathfrak{p})(n)$ for some $n \in \mathbf{Z}$.

The set of associated primes of M is denoted again by $\mathrm{Ass}_R M$ or $\mathrm{Ass}\, M$ if no confusion is possible.

Let T denote a multiplicatively closed set of homogeneous elements of R with $1 \in T$. Then the localization of M at T (denoted by $T^{-1}M$) is in a natural way a graded $T^{-1}R$-module where $T^{-1}R$ is a graded ring (let $\deg \frac{m}{t} := \deg m - \deg t$, where $m \in M$ is a homogeneous element and $t \in T$). If $f \in R$ is a homogeneous element let $T := \{1, f, f^2, \ldots\}$ and denote by R_f (M_f) the localization $T^{-1}R$ ($T^{-1}M$).

By $R_{(f)}$ ($M_{(f)}$) we denote the subring of R_f (submodule of M_f) consisting of homogeneous elements of degree zero. Clearly, $M_{(f)}$ is a (non-graded) $R_{(f)}$-module.

If $\mathfrak{p}$ is a homogeneous prime ideal of R then there are three different localizations of R and M at $\mathfrak{p}$:

1. $R_\mathfrak{p} := T^{-1}R$, $M_\mathfrak{p} := T^{-1}M$, where $T := R \setminus \mathfrak{p}$. This is the usual localization where we forget about the grading of R and M.
2. By $R_{\mathfrak{p},\mathrm{homog}}$ or $R_{\mathfrak{p},h}$ and $M_{\mathfrak{p},\mathrm{homog}}$ or $M_{\mathfrak{p},h}$ we denote the localization at the multiplicatively closed set of homogeneous elements of R not contained in $\mathfrak{p}$. The ring $R_{\mathfrak{p},h}$ is graded and $M_{\mathfrak{p},h}$ is a graded $R_{\mathfrak{p},h}$-module (see the above remarks).
3. $R_{(\mathfrak{p})}$ ($M_{(\mathfrak{p})}$) denotes the subring of $R_{\mathfrak{p},h}$ (submodule of $M_{\mathfrak{p},h}$) consisting of elements of degree zero.

We have the following inclusions $R_{(\mathfrak{p})} \subseteq R_{\mathfrak{p},h} \subseteq R_\mathfrak{p}$ and a natural homomorphism of graded rings $R \to R_{\mathfrak{p},h}$. It is clear that $R_{(\mathfrak{p})}$ and $R_\mathfrak{p}$ are local rings. $R_{\mathfrak{p},h}$ is a graded ring with exactly one maximal homogeneous ideal (the ideal $\mathfrak{p}R_{\mathfrak{p},h}$).

The following lemma relates these localizations:

Lemma 2.1. *Let $\mathfrak{p} \subset R$ be a homogeneous prime ideal. We have*

(i) $M_\mathfrak{p} \cong (M_{\mathfrak{p},h})_{\mathfrak{p}R_{\mathfrak{p},h}}$

and

(ii) *Assume that* $[R]_1 \not\subseteq \mathfrak{p}$. *Then for some* $f \in [R]_1 \setminus \mathfrak{p}$:

$$M_{\mathfrak{p},h} = M_{(\mathfrak{p})}[f]_f \cong M_{(\mathfrak{p})}[X]_X, \quad \text{where } X \text{ is an indeterminate.}$$

The proof is easy and we omit it.

Remarks.

1. Using this lemma we will show in Chapter I (Lemma 2.27): $R_\mathfrak{p} \cong R_{(\mathfrak{p})}[X]_{\mathfrak{m}[X]}$ for all homogeneous primes $\mathfrak{p} \subset R$ with $\mathfrak{p} \not\supseteq [R]_1$, where X is an indeterminate and $\mathfrak{m}$ denotes the maximal ideal of $R_{(\mathfrak{p})}$. (This implies $M_\mathfrak{p} \cong M_{(\mathfrak{p})}[X]_{\mathfrak{m}[X]}$.)
   Also a consequence of Corollary I.2.28 will be that for all homogeneous primes $\mathfrak{p}$ and R with $\mathfrak{p} \not\supseteq [R]_1$ we have:

   $M_\mathfrak{p}$ is a Cohen-Macaulay module (over $R_\mathfrak{p}$) if and only if
   $M_{(\mathfrak{p})}$ is a Cohen-Macaulay module (over $R_{(\mathfrak{p})}$).

2. Statement (ii) of Lemma 2.1 has a more general version:
   Assume $T \subset R$ is a multiplicatively closed set consisting of homogeneous elements with $1 \in T$ and $T \cap [R]_1 \neq \emptyset$. Let S denote the subring of $T^{-1}R$ of elements of degree zero (i.e. $S = [T^{-1}R]_0$).

If $t \in T \cap [R]_1$ and if X is an indeterminate then we have

$$T^{-1}R = S[t]_t \cong S[X]_X = S\left[X, \frac{1}{X}\right].$$

If M is a graded R-module and if we put $N := [T^{-1}M]_0$, then

$$T^{-1}M = N[t]_t \cong N[X]_X = N\left[X, \frac{1}{X}\right].$$

Set $\operatorname{Supp}_R M := \{\mathfrak{p} \subset R \mid \mathfrak{p} \text{ prime}, M_{\mathfrak{p}} \neq 0\}$. The *Krull dimension* of M, written $\dim_R M$ or for short $\dim M$, is defined to be the supremum of length of chains of prime ideals in Supp M if it exists and ∞ otherwise.
The same assertions made in § 1, 1. of this chapter on the existence of primary decompositions of submodules of Noetherian modules hold word for word in this situation. Therefore we omit an explicit discussion. Note that for any graded submodule N of a Noetherian graded R-module M there are graded primary submodules $Q_1, \ldots, Q_r$ of M belonging to homogeneous primes $\mathfrak{p}_1, \ldots, \mathfrak{p}_r$ of R where $\{\mathfrak{p}_1, \ldots, \mathfrak{p}_r\} = \operatorname{Ass} M/N$ such that $N = Q_1 \cap \ldots \cap Q_r$.

2. *Graded k-algebras, systems of parameters and Hilbert functions*

Let k be any field. By a *graded k-algebra* we understand a graded Noetherian ring R with $[R]_i = 0$ for all $i < 0$ and $[R]_0 = k$. The ring R has a unique homogeneous maximal ideal $\mathfrak{m}_R = \bigoplus_{i \geq 1} [R]_i$.

Let M be a Noetherian graded R-module. The same results as in § 1, 2. of this chapter on systems of parameters and M-sequence are true also in the graded case. Therefore we give only a short summary.

We set $d := \dim M$. Homogeneous elements $x_1, \ldots, x_d$ of $\mathfrak{m}_R$ are called a *system of parameters* of M if $\dim M/(x_1, \ldots, x_d) M = 0$. An ideal $\mathfrak{q}$ for which there is a (homogeneous) system of parameters $x_1, \ldots, x_d$ of M such that $\mathfrak{q} \cdot M = (x_1, \ldots, x_d) \cdot M$ is called a *parameter ideal* of M. Homogeneous elements $y_1, \ldots, y_r$ of $\mathfrak{m}_R$ are called an *M-sequence* if for $i = 0, 1, \ldots, r - 1$

$$(y_1, \ldots, y_i) M :_M y_{i+1} = (y_1, \ldots, y_i) M .$$

As before depth M is defined to be the supremum of all integers r such that there is an M-sequence consisting of r elements. Clearly any M-sequence is part of a system of parameters. Therefore we get

$$\operatorname{depth} M \leq \dim M.$$

M is called a *Cohen-Macaulay module* if depth $M = \dim M$.

Now we define the Hilbert function of a Noetherian module. Let M be a Noetherian graded R-module where R is a graded k-algebra. We set

$$H_M(n) := \operatorname{rank}_k([M]_n)$$

(note that for all $n \in \mathbf{Z}[M]_n$ is a k-vector space of finite rank). H_M is called the *Hilbert function* of M. There is a numerical polynomial h_M ($\in \mathbf{Q}[T]$, T an indeterminate) such that $H_M(n) = h_M(n)$ for all sufficiently large n. The polynomial h_M is called the

Hilbert polynomial of M. We write

$$h_M = h_0(M)\binom{T}{d} + h_1(M)\binom{T}{d-1} + \dots + h_d(M)$$

with $d := \deg h_M$ (considered as a polynomial in T), and where the integers $h_0(M) > 0$, $h_1(M), \dots, h_d(M)$ are called the *Hilbert coefficients* of M. It is well-known that $d = \deg h_M = \dim M - 1$.

If M is a Noetherian (Artinian) graded R-module then there is an $m \in \mathbf{Z}$ with $[M]_n = 0$ for all $n < m$ ($[M]_n = 0$ for all $n > m$). Therefore we can make the following definition:

Definition 2.2. Let M be a graded R-module. We define

(i) $a(M) := \inf\{n \in \mathbf{Z} \mid [M]_n \neq 0\}$,

(ii) $e(M) := \sup\{n \in \mathbf{Z} \mid [M]_n \neq 0\}$,

(iii) $r(M) := \inf\{n \in \mathbf{Z} \mid H_M(m) = h_M(m) \text{ for all } m \geq n\}$, if M is a Noetherian R-module. $r(M)$ is called the *index of regularity* of M.

If $M \neq O$ is Noetherian (Artinian) then $a(M)$ $\big(e(M)\big)$ is finite.

3. *Proj and cohomology*

Let R denote a graded k-algebra (k a field) with maximal homogeneous ideal $\mathfrak{m}_R$. By Proj R we denote the *projective spectrum* of R. It is the following scheme:

$$\operatorname{Proj} R = \begin{cases} \text{As a point set, the set of homogeneous primes } \mathfrak{p} \text{ of } R \text{ with } \mathfrak{m}_R \not\subseteq \mathfrak{p}. \\ \text{As a topological space, defined by a basis of open sets } X_f := \{\mathfrak{p} \in \operatorname{Proj} R \mid \\ f \not\subseteq \mathfrak{p}\} \text{ for all homogeneous } f \in \mathfrak{m}_R. \\ \text{As a locally ringed space with structure sheaf defined by } \Gamma(X_f, \mathcal{O}_{\operatorname{Proj} R}) \\ := R_{(f)}. \end{cases}$$

As is well known Proj R is a scheme. Note that in our notation

$$\dim R = 1 + \dim \operatorname{Proj} R,$$

where dim Proj R denotes the dimension of the topological space Proj R. Put $X := \operatorname{Proj} R$. There are two functors which relate graded modules over R with sheaves of modules over X. One is the "sheafification" functor, which associates to each graded R-module M a (quasi-coherent) sheaf of modules over X:

$$M \mapsto \tilde{M}.$$

This functor is exakt. For each point $\mathfrak{p}$ of X, the stalk $(\tilde{M})_\mathfrak{p}$ of M at $\mathfrak{p}$ is simply $M_{(\mathfrak{p})}$, the degree 0 localization of M by the homogeneous prime ideal $\mathfrak{p} \subset R$. If $M = R$, M is precisely the structure sheaf of X and is denoted by the symbol $\mathcal{O}_X$. More generally, the sheaf $\widetilde{R(v)}$ is denoted as $\mathcal{O}_X(v)$.

In the opposite direction we have the "twisted global sections" functor, which associates to each sheaf of modules $\mathcal{F}$ over X (quasi-coherent or not) a graded R-module:

$$\mathcal{F} \mapsto H^0_*(X, \mathcal{F}) = \bigoplus_{v \in \mathbf{Z}} \Gamma\big(X, \mathcal{F} \otimes \mathcal{O}_X(v)\big).$$

This functor is *not* exact. It is left exakt, however. Therefore it is natural to consider its right derived functors, which are the higher cohomology modules $H^1_*(X,\ )$ and $H^2_*(X,\ ), \ldots$

In contrast to the analogous affine case, the functors $\sim$ and H^0_* do not quite establish an equivalence of categories between graded R-modules and quasi-coherent sheaves of modules over X.

In one direction the two functors do compose to give the identity: if $\mathcal{F}$ is a quasi-coherent sheaf of modules over X, then the sheaf $\widetilde{H^0_*(X, \mathcal{F})}$ is canonically isomorphic to $\mathcal{F}$. However, in the other direction, if M is a graded R-module, the module $H^0_*(X, \tilde{M})$ is not isomorphic to M in general. In fact $H^0_*(X, \tilde{M})$ need not even be finitely generated if M is finitely generated. But there are important cases in which $H^0_*(X, \tilde{M}) \cong M$ namely for $R = k[X_0, \ldots, X_n]$, K a field, $X_0, \ldots, X_n$ indeterminates, $n \geq 1$ and

a) $M =$ any free graded R-module,

b) $M =$ any homogeneous ideal $I \subset R$ which is saturated (that is, $I :_R \mathfrak{m}_R = I$),

c) $M = H^0_*(X, \mathcal{F})$ where $\mathcal{F}$ is any quasi-coherent sheaf of modules over X.

In this situation (i.e. $R = k[X_0, \ldots, X_n]$) it is a basic fact that $\sim$ and H^0_* define a 1-1 correspondence:

$$\{\text{closed subschemes of } X\} \leftrightarrow \{\text{saturated homogeneous ideals of } R\},$$

$$V \mapsto H^0(X, \mathcal{J}_V),$$

where $\mathcal{J}_V$ is the ideal sheaf of V and

$$\operatorname{Proj}(R/I, \widetilde{R/I}) \leftarrow I.$$

If a homogeneous ideal $I \subset R$ is not saturated and not equal to an $\mathfrak{m}_R$-primary ideal, it is contained in a unique homogeneous ideal J of R which *is* saturated and which defines the same closed subscheme V as does I. This latter ideal is called the *total* ideal defining the closed subscheme V and will be denoted by I_V. We denote that

a) $I_V = H^0_*(X, \mathcal{J}_V)$, b) $\tilde{I}_V = \mathcal{J}_V$.

We observe also that b) is a consequence of a) since the ideal sheaf $\mathcal{J}_V$ of a closed subscheme V is always quasi-coherent.

We shall use the above 1-1 correspondence to translate all our statements about projective subschemes of X into ring-theoretic statements about R and its homogeneous ideals.

Let now R be again an arbitrarity graded k-algebra. It is a basic fact of commutative algebra that in many important respects R may be treated as a local ring with $\mathfrak{m}_R$ as its maximal ideal. For example, Nakayama's Lemma admits a graded version which holds for $(R, \mathfrak{m}_R)$.

Now we want to establish the connection between these concepts and our discussion of the local case. Let V denote a k-vector space. By $\underline{V}$ we denote the graded R-module given by $[\underline{V}]_n = O$ for all $n \neq 0$ and $[\underline{V}]_0 = V$.

Let M, N be graded R-modules. $\operatorname{Hom}_R(M, N)$ is the k-vector space of all R-homomorphisms $f\colon M \to N$ of degree zero, i.e. $f([M]_n) \subseteq [N]_n$ for all $n \in \mathbf{Z}$. We denote by $\underline{\operatorname{Hom}}_R(M, N)$ the graded R-module given by

$$[\underline{\operatorname{Hom}}_R(M, N)]_n = \operatorname{Hom}_R\big(M, N(n)\big) \cong \operatorname{Hom}_R\big(M(-n), N\big) \text{ for all } n \in \mathbf{Z}.$$

Note that $\mathrm{Hom}_R(\ ,\ )$ and $\underline{\mathrm{Hom}}_R(\ ,\ )$ are left exact functors from the abelian category of graded R-modules with R-homomorphisms of degree zero as morphisms to the category of k-vector spaces and graded R-modules, resp.

We denote by $\mathrm{Ext}^i_R(\ ,\ )$ and $\underline{\mathrm{Ext}}^i_R(\ ,\ )$ the right derived functors of $\mathrm{Hom}_R(\ ,\ )$ and $\underline{\mathrm{Hom}}_R(\ ,\ )$, resp. $\underline{\mathrm{Ext}}^i_R(M, N)$ again are graded R-modules with

$$[\underline{\mathrm{Ext}}^i_R(M, N)]_n \simeq \mathrm{Ext}^i_R(M, N(n)) \simeq \mathrm{Ext}^i_R(M(-n), N)$$

for all graded R-modules M, N and all $n \in \mathbf{Z}$.

Next, we define for a graded R-module M

$$\underline{H}^i_{\mathfrak{m}_R}(M) := \varinjlim_n \underline{\mathrm{Ext}}^i_R(R/\mathfrak{m}^n_R, M) \quad (\textit{local cohomology}),$$

$$\underline{H}^i(R, M) := \varinjlim_n \underline{\mathrm{Ext}}^i_R(\mathfrak{m}^n_R, M) \quad (\textit{Serre cohomology}).$$

If no ambiguity exists we will write $\underline{H}^i(M)$ for $\underline{H}^i(R, M)$.

$\underline{H}^i_{\mathfrak{m}_R}(M)$ and $\underline{H}^i(R, M)$ are related by the exact sequence

$$O \to \underline{H}^0_{\mathfrak{m}_R}(M) \to M \to \underline{H}^0(R, M) \to \underline{H}^1_{\mathfrak{m}_R}(M) \to 0$$

and by isomorphisms

$$\underline{H}^i(R, M) \simeq \underline{H}^{i+1}_{\mathfrak{m}_R}(M) \quad \text{for all } i \geq 1 \quad \text{(c.f. Lemma 1.8(i))}.$$

Notice that $\underline{H}^i_{\mathfrak{m}_R}(\ )$ and $\underline{H}^i(R,\ )$ are the right derived functors of the left exact functors $\underline{H}^0_{\mathfrak{m}_R}(\ )$ resp. $\underline{H}^0(R,\ )$. It is clear that $\underline{H}^0_{\mathfrak{m}_R}(M) = 0 :_M \langle\mathfrak{m}_R\rangle$ for any graded R-module M.

If M is Noetherian then $\underline{H}^i_{\mathfrak{m}_R}(M)$ is Artinian for all i.

Now we have

Proposition 2.3. *Let $X := \mathrm{Proj}\, R$ and assume that M is a graded R-module. Then for all i here are natural isomorphisms*

$$H^i_*(X, \tilde{M}) \simeq \underline{H}^i(R, M).$$

Proof: We prove $H^i(X, \tilde{M}) \simeq H^i(R, M)$ (then taking the direct sum over all shifts of degrees the result will follow). For $i = 0$ we get

$$H^0(X, M) = \varprojlim_{X_f \subseteq X} \Gamma(X_f, M) = \varprojlim_{f \in \mathfrak{m}_R} M_{(f)} \overset{\mu}{\simeq} \varinjlim_n \mathrm{Hom}_R(\mathfrak{m}^n_R, M) = H^0(R, M)$$

(μ is obtained from Lemma 1.8(vi), c.f. the remarks of Lemma 1.8). This isomorphism is a natural one. If M is an injective object in the category of graded R-modules, $\tilde{M}$ is flasque and hence $H^i(X, \tilde{M}) = 0$ and $H^i(R, M) = 0$ for all $i > 0$. Therefore we get the desired isomorphisms of all $i \geq 0$ using Cartan-Eilenberg [1], Chap. III, Proposition 5.2, and Chap. V, Proposition 4.4, q.e.d.

Like in the local case we have for a Noetherian graded R-module M

$$\dim M = \sup\{n \in \mathbf{N} \mid \underline{H}^i_{\mathfrak{m}_R}(M) \neq 0\}$$

and

$$\operatorname{depth} M = \inf\{n \in \mathbf{N} \mid \underline{H}^i_{\mathfrak{m}_R}(M) \neq 0\}.$$

The same relations between the Koszul cohomology and the local cohomology examined in § 1, 3. hold in the graded case. Therefore we mention only some differences in the definition of graded Koszul complexes. Let $x \in R$ and set $d := \deg x$. Then by $K.(x, R)$ we denote the complex

$$0 \to R \xrightarrow{x} R(d) \to 0.$$

For t homogeneous elements $x_1, \ldots, x_t \in R$ and a graded R-module M we set

$$K.(x_1, \ldots, x_t, M) := K.(x_1, R) \otimes \ldots \otimes K.(x_t, M) \otimes M.$$

We now define again the graded Koszul cohomology $\underline{H}^i(\mathfrak{m}_R, M)$ of M to be the $(t - i)$th homology of the Koszul complex $K.(x_1, \ldots, x_t, M)$ where $\{x_1, \ldots, x_t\}$ is a minimal basis of $\mathfrak{m}_R$ consisting of homogeneous elements, c.f. Lemma 4.2. Then we obtain the following (see also Lemma 1.5)

Lemma 2.4. *For all $i \geq 0$ we have a commutative diagram:*

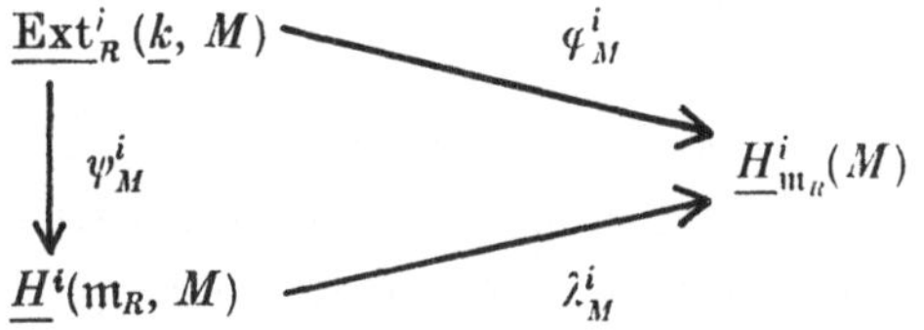

4. *Segre products*

Let k be a field and let R_1, R_2 be graded k-algebras with maximal ideals $\mathfrak{m}_1$ and $\mathfrak{m}_2$, resp.

Definition 2.5.

(i) A graded k-algebra R is called a *Segre product of R_1 and R_2 over k*, denoted by $R = \sigma_k(R_1, R_2)$ or $R = \sigma(R_1, R_2)$, if for every $p \in \mathbf{N}$

$$[R]_p = [R_1]_p \otimes_k [R_2]_p.$$

(ii) Let M_1, M_2 be graded R_1-, R_2-modules, resp. Let R be the Segre product of R_1 and R_2 (if it exists). A graded R-module M is called a *Segre product of M_1 and M_2 over k*, denoted by $M = \sigma_k(M_1, M_2)$ or $M = \sigma(M_1, M_2)$, if for all $p \in \mathbf{Z}$

$$[M]_p = [M_1]_p \otimes_k [M_2]_p.$$

As is known Segre products exist and are uniquely determined (up to isomorphisms) by definition.

Also since a sequence of graded modules over a graded k-algebra is exact if and only if the corresponding sequences for each grading are exact (as sequences of k-vector spaces), the Segre product is an exact covariant functor in each variable. Using this fact it is not difficult to show that the Segre product of Noetherian graded modules is again a Noetherian module. Note that the converse is generally not true. For example, if one takes an Artinian non-Noetherian graded R_1-module A_1 (for instance $A_1 = \underline{H}^{d_1}_{\mathfrak{m}_1}(R_1)$, $d_1 := \dim R_1$) and considers $\sigma(A_1, \underline{k}(d))$ where d is chosen such that $[A_1]_{-d} \neq 0$. Then $\sigma(A_1, k(d)) \cong \underline{[A_1]_{-d}}\,(d)$ and this is a finite-dimensional k-vector space and hence a Noetherian R-module.

If R_1 and R_2 for example are generated (over k) by $x_1, \ldots, x_n \in [R_1]_1, y_1, \ldots, y_m \in [R_2]_1$, resp., then $R := \sigma(R_1, R_2)$ is generated by $x_1 \otimes y_1, \ldots, x_n \otimes y_m$.

If M_1 and M_2 are finitely generated by $m_{11}, \ldots, m_{1r}$ and $m_{21}, \ldots, m_{2s}$ (over R_1 resp. R_2) then $M := \sigma(M_1, M_2)$ is generated (over R) by the following elements:

$m_{1i} \otimes \pi_2 m_{2j}$ for all $1 \leq i \leq r$, $1 \leq j \leq s$ with $\deg m_{1i} \geq \deg m_{2j}$ and π_2 running over a basis of the k-vector space $[R_2]_{\deg m_{1i} - \deg m_{2j}}$,

and

$\pi_1 m_{1l} \otimes m_{2n}$ for all $1 \leq l \leq r$, $1 \leq n \leq s$ with $\deg m_{2n} \geq \deg m_{1l}$ and π_1 running over a basis of the k-vector space $[R_1]_{\deg m_{2n} - \deg m_{1l}}$.

Hence M is again finitely generated, i.e. M is a Noetherian graded R-module. This explicit description is useful for studying some specific examples.

5. *Künneth relations for Segre products*

In this section we want to give some Künneth relations for Segre products. They allow us to "compare" the "cohomology of the Segre product" and the "Segre product of the cohomology".

The main point of interest in the following is that our Künneth relations are given by algebra homomorphisms and not only by k-linear maps (k denotes again the ground field) as would be the case using results of Grothendieck [2].

First we prove several lemmas. Let R_1, R_2 be graded k-algebras.

Lemma 2.6. *Let M_i and N_i be graded R_i-modules for $i = 1, 2$. Then there is a natural R-homomorphism*

$$\tau^0 : \sigma(\underline{\mathrm{Hom}}_{R_1}(N_1, M_1), \underline{\mathrm{Hom}}_{R_2}(N_2, M_2)) \to \underline{\mathrm{Hom}}_R(\sigma(N_1, N_2), \sigma(M_1, M_2))$$

where $R = \sigma_k(R_1, R_2)$.

Proof: We only give a short outline since all constructions are canonical. Let $p \in \mathbf{Z}$ and $f_i \in \mathrm{Hom}_{R_i}(N_i, M_i(p)) = [\underline{\mathrm{Hom}}_{R_i}(N_i, M_i)]_p$ for $i = 1, 2$. Then we define for all $n_i \in [N_i]_q$, $q \in \mathbf{Z}$ and $i = 1, 2$:

$$\tau^0(f_1 \otimes f_2)(n_1 \otimes n_2) = f_1(n_1) \otimes f_2(n_2).$$

It is clear that $\tau^0(f_1 \otimes f_2)$ is a well-defined R-homomorphism

$$\sigma(N_1, N_2) \to \sigma(M_1(p), M_2(p)) = \sigma(M_1, M_2)(p),$$

i.e. an element of $\underline{\mathrm{Hom}}_R(\sigma(N_1, N_2), \sigma(M_1, M_2))$ of degree p. Hence we have a k-linear map

$$\tau^0 : \sigma(\underline{\mathrm{Hom}}_{R_1}(N_1, M_1), \underline{\mathrm{Hom}}_{R_2}(N_2, M_2)) \to \underline{\mathrm{Hom}}_R(\sigma(N_1, N_2), \sigma(M_1, M_2))$$

preserving degrees.

It is not difficult to verify that τ^0 is even an R-homomorphism. To this end we choose $p, q \in \mathbf{Z}$, $f_i \in [\underline{\mathrm{Hom}}_{R_i}(N_i, M_i)]_p$, $r_i \in [R_i]_q$ for $i = 1, 2$ and prove

$$\tau^0((r_1 \otimes r_2)(f_1 \otimes f_2)) = (r_1 \otimes r_2)\, \tau^0(f_1 \otimes f_2).$$

Let $n_i \in [N_i]_s$, $s \in \mathbf{Z}$, $i = 1, 2$. Then we get:

$$\begin{aligned}
&\bigl(\tau^0((r_1 \otimes r_2)(f_1 \otimes f_2))\bigr)(n_1 \otimes n_2) \\
&= \bigl(\tau^0(r_1 f_1 \otimes r_2 f_2)\bigr)(n_1 \otimes n_2) = (r_1 f_1)(n_1) \otimes (r_2 f_2)(n_2) \\
&= r_1\bigl(f_1(n_1)\bigr) \otimes r_2\bigl(f_2(n_2)\bigr) = (r_1 \otimes r_2)\bigl(f_1(n_1) \otimes f_2(n_2)\bigr) \\
&= (r_1 \otimes r_2)\bigl(\tau^0(f_1 \otimes f_2)(n_1 \otimes n_2)\bigr) = \bigl((r_1 \otimes r_2)\,\tau^0(f_1 \otimes f_2)\bigr)(n_1 \otimes n_2),
\end{aligned}$$

which finishes the proof. Finally, the naturalness of τ^0 is evident, q.e.d.

Lemma 2.7. *Let τ^0 be as in Lemma 2.6.*

(i) *If for $i = 1, 2$ $N_i = R_i^{s_i}(p)$ with $s_1, s_2 \in \mathbf{N}$, $p \in \mathbf{Z}$, then τ^0 is an isomorphism.*

(ii) *If for $i = 1, 2$ N_i are finitely generated free R_i-modules and if $\underline{H}^0_{\mathfrak{m}_i}(M_i) = 0$ for $i = 1, 2$, then τ^0 is injective.*

Proof: By the naturalness of τ^0 we may assume that $N_i = R_i(p_i)$ with $p_i \in \mathbf{Z}$ for $i = 1, 2$ ($p = p_1 = p_2$ in case (i)). Now we have natural isomorphisms for $i = 1, 2$:

$$\xi_i : \underline{\mathrm{Hom}}_{R_i}\bigl(R_i(p_i), M_i\bigr) \cong M_i(-p_i)$$

sending each φ_i to $\varphi_i(1)$ where $1 \in [R_i(p_i)]_{-p_i}$.

Let $\mu := \tau^0\bigl(\sigma(\xi_1, \xi_2)\bigr)^{-1}$. Since $\sigma(\xi_1, \xi_2)$ is an isomorphism it is sufficient to prove (i) and (ii) for μ (instead of τ^0). We have for $m_1 \otimes m_2 \in \bigl[\sigma\bigl(M_1(-p_1), M_2(-p_2)\bigr)\bigr]_t$ and $r_1 \otimes r_2 \in \bigl[\sigma\bigl(R_1(p_1), R_2(p_2)\bigr)\bigr]_s$ with $s, t \in \mathbf{Z}$:

$$\bigl(\mu(m_1 \otimes m_2)\bigr)(r_1 \otimes r_2) = r_1 m_1 \otimes r_2 m_2 .$$

If now $p_1 = p_2 = p$ and if $\sum_i m_{1i} \otimes m_{2i}$ is a homogeneous element of $\operatorname{Ker} \mu$, then we obtain for $1 \otimes 1 \in \bigl[\sigma\bigl(R_1(p), R_2(p)\bigr)\bigr]_{-p}$:

$$0 = \bigl(\mu(\textstyle\sum m_{1i} \otimes m_{2i})\bigr)(1 \otimes 1) = \sum m_{1i} \otimes m_{2i}, \qquad \text{i.e.} \qquad \operatorname{Ker} \mu = 0.$$

If $f \in \underline{\mathrm{Hom}}_R\bigl(\sigma(R_1(p_1), R_2(p_2)), \sigma(M_1, M_2)\bigr)$, then $f = \mu\bigl(\overline{f(1 \otimes 1)}\bigr)$ where $\overline{f(1 \otimes 1)}$ denotes that we consider $f(1 \otimes 1)$ as an element of degree $\deg f$ in $\sigma\bigl(M_1(-p), M_2(-p)\bigr)$. Hence f is surjective and (i) is proven.

To prove (ii) we initially assume that M_1, M_2 are finitely generated. Let $\sum m_{1i} \otimes m_{2i}$ be a homogeneous element of $\operatorname{Ker} \mu$ where the elements m_{1i} are linearly independent. By our assumption we can find an $s \in \mathbf{N}$ (sufficiently large) such that there are $r_1 \in [R_1]_s$, $r_2 \in [R_2]_{s+p_2-p_1}$ with $0 :_{M_i} r_i = 0$ for $i = 1, 2$. Then the elements $r_i m_{1i}$ are again linearly independent and we have:

$$0 = \Bigl(\mu\Bigl(\sum_i m_{1i} \otimes m_{2i}\Bigr)\Bigr)(r_1 \otimes r_2) = \sum_i r_1 m_{1i} \otimes r_2 m_{2i} .$$

Therefore we get $r_2 m_{ii} = 0$ for all i, i.e. $m_{2i} \in 0 :_{M_2} r_2 = 0$. Hence $\operatorname{Ker} \mu = 0$, i.e. μ is injective.

If now M_1, M_2 are arbitrary modules then they are direct limits of their finitely generated graded submodules. Since all functors which occur commute with these limits the statement is true in general, q.e.d.

Lemma 2.8. *Let τ^0 be as in Lemma 2.6. Additionally suppose that N_1, N_2 are finitely generated. If there is a $p \in \mathbf{Z}$ such that $[N_i]_p$ generates N_i as an R_i-module for $i = 1, 2$ then τ^0 is injective. If also $\underline{H}^0_{\mathfrak{m}_i}(M_i) = 0$ for $i = 1, 2$ then τ^0 is an isomorphism.*

Proof: Take the exact sequence $G_i \to F_i \to N_i \to O$ with F_i, G_i free and $F_i = R_i^{n_i}(p)$, $n_i \in \mathbf{N}$, for $i = 1, 2$ which is possible by our first assumption. Set $H_i := \underline{\mathrm{Hom}}_{R_i}(N_i, M_i)$, $N := \sigma(N_1, N_2)$, $M := \sigma(M_1, M_2)$. Then we have exact sequences

(i) $0 \to H_i \to \underline{\mathrm{Hom}}_{R_i}(F_i, M_i) \to \underline{\mathrm{Hom}}_{R_i}(G_i, M_i)$ for $i = 1, 2$,

and

(ii) $\sigma(G_1, F_2) \oplus \sigma(F_1, G_2) \to \sigma(F_1, F_2) \to N \to 0$

(see Cartan-Eilenberg [1], Chap. II, Prop. 4.3. (c)). From this we obtain a commutative diagram with exact rows (the exactness of the top row follows from (i) and Cartan-Eilenberg [1], II, Prop. 4.3 (c) and the exactness of the bottom row by (ii)):

$$\begin{array}{ccccc}
O \to & \sigma(H_1, H_2) & \to & \sigma\big(\underline{\mathrm{Hom}}_{R_1}(F_1, M_1), \underline{\mathrm{Hom}}_{R_2}(F_2, M_2)\big) & \xrightarrow{g} \cdot \\
 & \downarrow{\scriptstyle \tau^0} & & \downarrow{\scriptstyle \bar{\tau}^0} & \\
0 \to & \underline{\mathrm{Hom}}_R(N, M) & \to & \underline{\mathrm{Hom}}_R\big(\sigma(F_1, F_2), M\big) & \xrightarrow{h}
\end{array}$$

$$\begin{array}{cc}
\xrightarrow{g} & \sigma\big(\underline{\mathrm{Hom}}_{R_1}(F_1, M_1), \underline{\mathrm{Hom}}_{R_2}(G_2, M_2)\big) \oplus \sigma\big(\underline{\mathrm{Hom}}_{R_1}(G_1, M_1), \underline{\mathrm{Hom}}_{R_2}(F_2, M_2)\big) \\
 & \downarrow{\scriptstyle \tilde{\tau}^0 \oplus \tilde{\tilde{\tau}}^0} \\
\xrightarrow{h} & \underline{\mathrm{Hom}}_R\big((\sigma(F_1, G_2) \oplus \sigma(G_1, F_2)), M\big)
\end{array}$$

where $\bar{\tau}^0$, $\tilde{\tau}^0$, $\tilde{\tilde{\tau}}^0$ denote the appropriate natural homomorphisms. By Lemma 2.7. $\bar{\tau}^0$ is an isomorphism and therefore τ^0 is injective. Furthermore, if $\underline{H}^0_{\mathfrak{m}_i}(M_i) = 0$ for $i = 1, 2$ then again by Lemma 2.7 $\tilde{\tau}^0 \oplus \tilde{\tilde{\tau}}^0$ is injective and therefore τ^0 is an isomorphism, q.e.d.

Lemma 2.9. *Let I_1, I_2 be injective graded R_1-, resp. R_2-modules. Then $\underline{H}^n_{\mathfrak{m}}\big(\sigma(I_1, I_2)\big)$ and $\underline{H}^n\big(\sigma(I_1, I_2)\big)$ are zero for all $n \geq 1$.*

Proof: Since for $n \geq 1$ $\underline{H}^n\big(\sigma(I_1, I_2)\big) \simeq \underline{H}^{n+1}_{\mathfrak{m}}\big(\sigma(I_1, I_2)\big)$ it is sufficient to prove $\underline{H}^n_{\mathfrak{m}}\big(\sigma(I_1, I_2)\big) = 0$ for $n \geq 1$. By results of Matlis [1] we obtain that I_j is a direct sum of modules of the form $I(R_j/\mathfrak{p}_j)\,(p_j)$ where $\mathfrak{p}_j \in \mathrm{Proj}\, R_j \cup \{\mathfrak{m}_j\}$, $p_j \in \mathbf{Z}$, $j = 1, 2$ and where $I(M_j)$ denotes the injective hull of M_j. Since σ and $\underline{H}^n_{\mathfrak{m}}$ commute with direct sums we may assume without loss of generality that $I_j = I(R_j/\mathfrak{p}_j)\,(p_j)$ for $j = 1, 2$. If $\mathfrak{p}_1 = \mathfrak{m}_1$ (or $\mathfrak{p}_2 = \mathfrak{m}_2$) then $\mathrm{Supp}\, I_1 = \{\mathfrak{m}_1\}$ ($\mathrm{Supp}\, I_2 = \{\mathfrak{m}_2\}$) and in each case $\mathrm{Supp}\, \sigma(I_1, I_2) = \{\mathfrak{m}\}$. But this implies $\underline{H}^n_{\mathfrak{m}}\big(\sigma(I_1, I_2)\big) = 0$ for all $n \geq 1$.

If $\mathfrak{p}_1 \neq \mathfrak{m}_1$ and $\mathfrak{p}_2 \neq \mathfrak{m}_2$ we choose for $i = 1, 2$ homogeneous elements of the same degree in $\mathfrak{m}_i \setminus \mathfrak{p}_i$, say $x_i \in [R_i]_p$ for some $p > 0$. Let $x := x_1 \otimes x_2 \in [R]_p$. Each element x_i gives rise to an isomorphism

$$I_i \xrightarrow[\sim]{x_i} I_i(p).$$

Therefore x defines an isomorphism $I \xrightarrow[\sim]{x} I(p)$ where $I := \sigma(I_1, I_2)$. Hence we have for all $n \geq 0$ isomorphisms

$$\underline{H}^n_{\mathfrak{m}}(I) \xrightarrow[\sim]{x} \underline{H}^n_{\mathfrak{m}}\big(I(p)\big) \simeq \underline{H}^n_{\mathfrak{m}}(I)\,(p).$$

Since $\mathrm{Supp}\, \underline{H}^i_{\mathfrak{m}}(I) \subseteq \{\mathfrak{m}\}$, this implies $\underline{H}^n_{\mathfrak{m}}(I) = 0$ for all $n \geq 0$, q.e.d.

It is not true in general that the Segre product of two injective modules is injective. A counter-example is in Chapter V, § 5, Example 5.6.

Now we can prove the following statement:

Proposition 2.10. *Let N_1, M_1 and N_2, M_2 be graded R_1-, resp. R_2-modules. Then for all $n \geq 0$ we get:*

(i) *There are natural R-homomorphisms*

$$\tau^n: \bigoplus_{p+q=n} \sigma\big(\underline{\mathrm{Ext}}^p_{R_1}(N_1, M_1), \underline{\mathrm{Ext}}^q_{R_2}(N_2, M_2)\big) \to \underline{\mathrm{Ext}}^n_R\big(\sigma(N_1, N_2), \sigma(M_1, M_2)\big)$$

(generalized Künneth relations).

(ii) *There are natural R-isomorphisms*

$$\varkappa^n: \bigoplus_{p+q=n} \sigma\big(\underline{H}^p(M_1), H^q(M_2)\big) \to \underline{H}^n\big(\sigma(M_1, M_2)\big)$$

(Künneth relations).

Proof: We first prove (i): For $i = 1, 2$ let $I_i^{\cdot}$ be injective resolutions of M_i and $K_i^{\cdot} := \underline{\mathrm{Hom}}_{R_i}(N_i, I_i^{\cdot})$. Since σ is an exact functor we have by Cartan-Eilenberg [1], Chap. IV, Theorem 7.2 (see also Chap. IV, Prop. 6.1):

$$\bigoplus_{p+q=n} \sigma\big(\underline{\mathrm{Ext}}^p_{R_1}(N_1, M_1), \underline{\mathrm{Ext}}^q_{R_2}(N_2, M_2)\big) = \bigoplus_{p+q=n} \sigma\big(H^p(K_1^{\cdot}), H^q(K_2^{\cdot})\big) \cong H^n\big(\sigma(K_1^{\cdot}, K_2^{\cdot})\big)$$

where H^j denotes the jth cohomology of the underlying complex. By Lemma 2.6 we have a natural homomorphism of complexes

$$\sigma(K_1^{\cdot}, K_2^{\cdot}) = \sigma\big(\underline{\mathrm{Hom}}_{R_1}(N_1, I_1^{\cdot}), \underline{\mathrm{Hom}}_{R_2}(N_2, I_2^{\cdot})\big) \to \underline{\mathrm{Hom}}_R\big(\sigma(N_1, N_2), \sigma(I_1^{\cdot}, I_2^{\cdot})\big).$$

Now $\sigma(I_1^{\cdot}, I_2^{\cdot})$ is a (not necessary injective) resolution of $\sigma(M_1, M_2)$ (see, for example Cartan-Eilenberg [1], Chap. IV, Theorem 7.2) and therefore we have natural homomorphisms

$$H^n\big(\sigma(K_1^{\cdot}, K_2^{\cdot})\big) \to H^n\big(\underline{\mathrm{Hom}}_R(\sigma(N_1, N_2), \sigma(I_1^{\cdot}, I_2^{\cdot}))\big) \to \underline{\mathrm{Ext}}^n_R\big(\sigma(N_1, N_2), \sigma(M_1, M_2)\big).$$

Putting together everything we obtain (i).

Now we prove (ii). There are an integer p and for $i = 1, 2$ $\mathfrak{m}_i$-primary ideals $\mathfrak{q}_i$ which are generated by their homogeneous elements of degree p. For example, take the ideals generated by suitable powers of the basis elements of $\mathfrak{m}_1$ and $\mathfrak{m}_2$, resp. Then $\mathfrak{q}_i^t$ will be generated by homogeneous elements of degree pt for $i = 1, 2$. Now we set in (i) $N_i = \mathfrak{q}_i^t$ for $i = 1, 2$, $t \geq 1$ and take the (direct) limit over all t. Since $\sigma(\mathfrak{q}_1^t, \mathfrak{q}_2^t) = \big(\sigma(\mathfrak{q}_1, \mathfrak{q}_2)\big)^t$ we get natural homomorphisms

$$\varkappa^n: \bigoplus_{p+q=n} \sigma\big(\underline{H}^p(M_1), \underline{H}^q(M_2)\big) \to \underline{H}^n\big(\sigma(M_1, M_2)\big).$$

To prove (ii) it is sufficient to show that $\varkappa^0$ is an isomorphism for any modules M_1, M_2 and that $\varkappa^n$ is an isomorphism for all $n \geq 1$ whenever M_1, M_2 are injective (see Cartan-Eilenberg [1], Chap. V, Prop. 4.4). But the last statement is true by Lemma 2.9 (both sides are zero). Therefore we have only to verify that $\varkappa^0$ is an isomorphism.

Assume first that $\underline{H}^0_{\mathfrak{m}_i}(M_i) = O$ for $i = 1, 2$. Then by Lemma 2.8 we have for each $t \geq 1$ isomorphisms $(\mathfrak{q} := \sigma(\mathfrak{q}_1, \mathfrak{q}_2))$:

$$\sigma(\underline{\mathrm{Hom}}_{R_1}(\mathfrak{q}_1^t, M_1), \underline{\mathrm{Hom}}_{R_2}(\mathfrak{q}_2^t, M_2)) \to \underline{\mathrm{Hom}}_R(\mathfrak{q}^t, \sigma(M_1, M_2)).$$

Therefore in this case $\varkappa^0$ is an isomorphism.

Let now M_1, M_2 be arbitrary modules. We put $M_i' := M_i/\underline{H}^0_{\mathfrak{m}_i}(M_i)$ for $i = 1, 2$. Then

$$\underline{H}^0_{\mathfrak{m}_i}(M_i') = 0 \text{ for } i = 1, 2 \text{ and since } \underline{H}^0(\underline{H}^0_{\mathfrak{m}_i}(M_i)) = O$$
$$(\text{note that Supp } \underline{H}^0_{\mathfrak{m}_i}(M_i) \subseteq \{m_i\})$$

we obtain isomorphisms

$$\underline{H}^0(M_i) \xrightarrow{\sim} \underline{H}^0(M_i') \quad \text{for } i = 1, 2.$$

Let λ denote the natural projection $\sigma(M_1, M_2) \to \sigma(M_1', M_2')$. Then we get the following commutative diagram:

$$\begin{array}{ccc} \sigma(\underline{H}^0(M_1), \underline{H}^0(M_2)) & \xrightarrow{\varkappa^0} & \underline{H}^0(\sigma(M_1, M_2)) \\ \downarrow\wr & & \downarrow \underline{H}^0(\lambda) \\ \sigma(\underline{H}^0(M_1'), \underline{H}^0(M_2')) & \xrightarrow{\varkappa'^0} & \underline{H}^0(\sigma(M_1', M_2')) \end{array}$$

We also have an exact sequence (see Cartan-Eilenberg [1], Chap. IV, Prop. 4.3(c)):

$$\sigma(\underline{H}^0_{\mathfrak{m}_1}(M_1), M_2) \oplus \sigma(M_1, \underline{H}^0_{\mathfrak{m}_2}(M_2)) \to \sigma(M_1, M_2) \xrightarrow{\lambda} \sigma(M_1', M_2') \to O.$$

Therefore Supp Ker $\lambda \subseteq \{\mathfrak{m}\}$, i.e. $\underline{H}^i(\operatorname{Ker} \lambda) = O$ for all $i \geq 0$. Hence $H^0(\lambda)$ is an isomorphism, i.e. $\varkappa^0$ is an isomorphism, q.e.d.

Using the natural maps

$$\bar{\varphi}^s_{M_i}\colon \underline{\mathrm{Ext}}^s_{R_i}(\mathfrak{m}_i, M_i) \to \underline{H}^s(M_i) \quad \text{for } i = 1, 2$$

and

$$\bar{\varphi}^s_{\sigma(M_1,M_2)}\colon \underline{\mathrm{Ext}}^s_R(\mathfrak{m}, \sigma(M_1, M_2)) \to \underline{H}^s(\sigma(M_1, M_2))$$

we get

Corollary 2.11. *Let M_1, M_2 be graded R_1-, resp. R_2-modules. Then there are for each $n \geq 0$ commutative diagrams*

$$\begin{array}{ccc} \bigoplus_{p+q=n} \sigma(\underline{\mathrm{Ext}}^p_{R_1}(\mathfrak{m}_1, M_1), \underline{\mathrm{Ext}}^q_{R_2}(\mathfrak{m}_2, M_2)) & \to & \underline{\mathrm{Ext}}^n_R(\mathfrak{m}, \sigma(M_1, M_2)) \\ \Big\downarrow \bigoplus_{p+q=n} \sigma(\bar{\varphi}^p_{M_1}, \bar{\varphi}^q_{M_2}) & & \Big\downarrow \bar{\varphi}^n_{\sigma(M_1,M_2)} \\ \bigoplus_{p+q=n} \sigma(\underline{H}^p(M_1), H^q(M_2)) & \xrightarrow{\varkappa^n} & \underline{H}^n(\sigma(M_1, M_2)) \end{array}$$

Finally we prove

Corollary 2.12. *Let M_1, M_2 be Noetherian graded R_1-, resp. R_2-modules. Then*

$$\operatorname{depth} \sigma(M_1, M_2) > \inf \{\operatorname{depth} M_1, \operatorname{depth} M_2\}.$$

Proof: Using the canonical maps $M_i \to \underline{H}^0(M_i)$ $(i = 1, 2)$ and $\varkappa^0$ we obtain a commutative diagram with exact bottom row:

$$\begin{array}{ccc} \sigma(M_1, M_2) & \xrightarrow{\mu} & \sigma(H^0(M_1), H^0(M_2)) \\ \downarrow \wr & & \downarrow \varkappa^0 \end{array}$$

$$O \to \underline{H}^0_{\mathfrak{m}}(\sigma(M_1, M_2)) \to \sigma(M_1, M_2) \to \underline{H}^0(\sigma(M_1, M_2)) \to \underline{H}'_{\mathfrak{m}}(\sigma(M_1, M_2)) \to 0.$$

If depth $M_1 = 0$ or depth $M_2 = 0$ there is nothing to prove. Hence we assume depth $M_i \geq 1$ for $i = 1, 2$. Then μ is injective and therefore $\underline{H}^0_{\mathfrak{m}}(\sigma(M_1, M_2)) = 0$, i.e. depth $\sigma(M_1, M_2) \geq 1$. Therefore if depth $M_1 = 1$ or depth $M_2 = 1$ the conclusion follows. Assume now depth $M_i \geq 2$ for $i = 1, 2$. Then μ is even an isomorphism since $M_i \xrightarrow{\sim} \underline{H}^0(M_i)$ for $i = 1, 2$.

Therefore $\underline{H}^0_{\mathfrak{m}}(\sigma(M_1, M_2)) = \underline{H}^1_{\mathfrak{m}}(\sigma(M_1, M_2)) = 0$ and we get:

$$\begin{aligned} \operatorname{depth} \sigma(M_1, M_2) &= 1 + \inf\{n \in \mathbf{N} \mid \underline{H}^n(\sigma(M_1, M_2)) \neq 0\} \\ &= 1 + \inf\{n \in \mathbf{N} \mid \bigoplus_{p+q=n} \sigma(\underline{H}^p(M_1), \underline{H}^q(M_2)) \neq 0\} \\ &\geq 1 + \inf\{\inf\{n \in \mathbf{N} \mid \underline{H}^n(M_1) \neq 0\}, \inf\{m \in \mathbf{N} \mid \underline{H}^m(M_2) \neq 0\}\} \\ &= \inf\{\operatorname{depth} M_1, \operatorname{depth} M_2\}, \end{aligned}$$

q.e.d.

§ 3. Local duality

In this paragraph we shall review further basic facts on homological algebra. In particular we shall introduce the notion of the dualizing complex which is a very useful concept of homological algebra. For our purposes here we shall need the dualizing complex to give a homological description of Buchsbaum modules. First of all we need to define some terminology. By a complex we shall understand a complex of modules over a fixed (commutative and Noetherian) ring A. Let

$$X^\cdot : \ldots \to X^n \to X^{n+1} \to \ldots$$

denote a complex. We write X^n for its nth cochain module and $\partial^n : X^n \to X^{n+1}$ for its nth differential. If $X^\cdot$ and $Y^\cdot$ are complexes we define complexes

$$\operatorname{Hom}^\cdot(X^\cdot, Y^\cdot) \quad \text{resp.} \quad X^\cdot \otimes Y^\cdot$$

given by

$$\operatorname{Hom}^n(X^\cdot, Y^\cdot) = \prod_{i \in \mathbf{Z}} \operatorname{Hom}(X^i, Y^{i+n})$$

with differential $\partial^n(f) = \partial f - (-1)^n f \partial$,

resp.

$$(X^\cdot \otimes Y^\cdot)^n = \bigoplus_{i+j=n} X^i \otimes Y^j$$

with differential $\partial^n(x^i \otimes y^j) = \partial^i(x^i) \otimes y^j + (-1)^i x^i \otimes \partial^j(y^j)$,

$n \in \mathbf{Z}$.

Recall that $X^i \otimes Y^j \to (-1)^{ij} Y^j \otimes X^i$ induces an isomorphism

$$X^\cdot \otimes Y^\cdot \xrightarrow{\sim} Y^\cdot \otimes X^\cdot.$$

Also, for $n \in \mathbf{Z}$

$$Y^{\cdot}[n]$$

denotes the complex whose ith cochain module is Y^{i+n} and whose ith differential is given by $(-1)^n \partial^{i+n}$, that is $Y^{\cdot}[n]$ denotes the complex $Y^{\cdot}$ shifted n places to the left.

A morphism of complexes $f^{\cdot}: X^{\cdot} \to Y^{\cdot}$ is called a *quasi-isomorphism* if the induced homomorphism on the cohomology

$$H^i(f^{\cdot}): H^i(X^{\cdot}) \to H^i(Y^{\cdot})$$

is an isomorphism for all $i \in \mathbf{Z}$.

A complex $X^{\cdot}$ is called *bounded below* (resp. *bounded above*, resp. *bounded*) if $X^n = 0$ for all $n \ll 0$ (resp. $n \gg 0$, resp. $n \ll 0$ and $n \gg 0$).

All results which we list in the following are well-known. For proofs we refer to R. Hartshorne's lecture notes [2], R. Y. Sharp's more elementary exposition in [1, 3, 4], or B. Iversen's preprint [1].

Let $E^{\cdot}$ be a complex of injective modules which is bounded below. Then for any quasi-isomorphism $f^{\cdot}: X^{\cdot} \to Y^{\cdot}$

$$\operatorname{Hom}^{\cdot}(f^{\cdot}, 1): \operatorname{Hom}^{\cdot}(Y^{\cdot}, E^{\cdot}) \to \operatorname{Hom}^{\cdot}(X^{\cdot}, E^{\cdot})$$

is a quasi-isomorphism. This may be reformulated as follows: Let $f^{\cdot}: X^{\cdot} \to Y^{\cdot}$ be a quasi-isomorphism and $g^{\cdot}: X^{\cdot} \to E^{\cdot}$ a morphism into a bounded below complex of injective modules. Then there exists $h^{\cdot}: Y^{\cdot} \to E^{\cdot}$ such that the following diagram is homotopy commutative

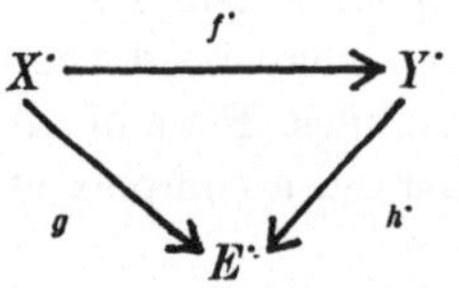

Moreover, $h^{\cdot}$ is unique up to homotopy. From this it follows that any quasi-isomorphism between bounded below complexes of injective modules is a homotopy equivalence. For any bounded below complex $Z^{\cdot}$ there exists a bounded below complex of injective modules $E^{\cdot}$ and a quasi-isomorphism $f^{\cdot}: Z^{\cdot} \to E^{\cdot}$. Furthermore, $E^{\cdot}$ may be chosen as a minimal injective complex, i.e.

$$\operatorname{Ker}(\partial^n) \to E^n$$

is an essential extension.

The above-mentioned results have the following "dual" form. Let $P^{\cdot}$ be a bounded above complex of projective modules. A quasi-isomorphism $X^{\cdot} \to Y^{\cdot}$ induces a quasi-isomorphism

$$\operatorname{Hom}^{\cdot}(P^{\cdot}, X^{\cdot}) \to \operatorname{Hom}^{\cdot}(P^{\cdot}, Y^{\cdot}).$$

A quasi-isomorphism between bounded above complexes of projective modules is a homotopy equivalence. For bounded above complexes $X^{\cdot}$ there exists a quasi-isomorphism $P^{\cdot} \to X^{\cdot}$ where $P^{\cdot}$ is a bounded above complex of projective modules. In the case $X^{\cdot}$ has finitely generated cohomology, $P^{\cdot}$ can be chosen as a bounded above complex of finitely generated projective modules. Let $F^\cdot$ be a bounded above complex of flat modules. Then for any quasi-isomorphism $f^\cdot: X^\cdot \to Y^\cdot$ of bounded above complexes,

$$f^\cdot \otimes 1: X^\cdot \otimes F^\cdot \to Y^\cdot \otimes F^\cdot$$

is a quasi-isomorphism.

Here we are mainly interested in complexes $X^\cdot$ such that their cohomology modules $H^i(X^\cdot)$, $i \in \mathbf{Z}$, are finitely generated A-modules. Regarding this we remark that if $X^\cdot$ is a bounded above complex and $E^\cdot$ is a bounded below complex of injective modules, and assuming that both complexes have finitely generated cohomology modules, then $\operatorname{Hom}^\cdot(X^\cdot, E^\cdot)$ has finitely generated cohomology modules, too.

Given a complex $E^\cdot$ of injective A-modules and an A-module M, we set for $i \in \mathbf{Z}$

$$\operatorname{Ext}^i(M, E^\cdot) = H^i(\operatorname{Hom}(M, E^\cdot)).$$

Let $X^\cdot$, $E^\cdot$ be complexes of modules over a fixed ring A. For $n \in \mathbf{Z}$ consider the map

$$X^n \to \operatorname{Hom}^n(\operatorname{Hom}^\cdot(X^\cdot, E^\cdot), E^\cdot)$$

which assigns to $x_n \in X^n$ the element $(f_i)_{i\in\mathbf{Z}} \in \operatorname{Hom}^n(\operatorname{Hom}^\cdot(X^\cdot, E^\cdot), E^\cdot)$ where

$$f_i: \prod_{j\in\mathbf{Z}} \operatorname{Hom}(X^j, E^{i+j}) = \operatorname{Hom}^i(X^\cdot, E^\cdot) \to E^{n+i}$$

is defined by

$$f_i((g_j)_{j\in\mathbf{Z}}) = (-1)^{in} g_n(x_n) \qquad \text{for all } (g_j)_{j\in\mathbf{Z}} \in \prod_{j\in\mathbf{Z}} \operatorname{Hom}(X^j, E^{i+j}).$$

In fact, this defines a map of complexes

$$e: X^\cdot \to \operatorname{Hom}^\cdot(\operatorname{Hom}^\cdot(X^\cdot, E^\cdot), E^\cdot)$$

which is called the *evaluation map*. In particular, if $X^\cdot$ is the complex $X^i = O$ for $i \neq 0$ and $X^0 = A$ we get a map

$$e: A \to \operatorname{Hom}^\cdot(E^\cdot, E^\cdot).$$

Now we shall state the definition of a dualizing complex.

Definition and Theorem 3.1. *Let $D^\cdot$ be a bounded complex of injective modules with finitely generated cohomology modules. Then the following conditions are equivalent:*

(i) *The evaluation map*

$$X^\cdot \to \operatorname{Hom}^\cdot(\operatorname{Hom}^\cdot(X^\cdot, D^\cdot), D^\cdot)$$

is a quasi-isomorphism for any bounded complex $X^\cdot$ with finitely generated cohomology.

(ii) *The canonical homomorphism*

$$A \to \operatorname{Hom}^\cdot(D^\cdot, D^\cdot)$$

is a quasi-isomorphism.

If $D^\cdot$ satisfies one of these equivalent conditions we call it a *dualizing complex* for A.

In particular it follows that if A has a finite injective resolution $E^\cdot$ then $E^\cdot$ is a dualizing complex. This means that if A is a Gorenstein ring then it possesses a dualizing complex since A has finite injective dimension.

For the proof of this result see Hartshorne [2], Chap. V, § 2, or Sharp [3].

Let A be a ring and $D^\cdot$ a dualizing complex for A. For any ideal $\mathfrak{a}$ in A we have that $\operatorname{Hom}^\cdot_A(A/\mathfrak{a}, D^\cdot)$ is a dualizing complex for $A/\mathfrak{a}$. Let A be a local ring which is a quotient of a local Gorenstein ring. Then A possesses a dualizing complex. By the theorem of Cohen we know that any complete local ring is a quotient of a regular ring. Hence it follows that any complete local ring admits a dualizing complex. On the other hand there are local rings which don't have a dualizing complex, compare Hartshorne [2], Chap. V, Proposition 10.1.

Let A be a local ring with maximal ideal $\mathfrak{m}$. If $D^\cdot$ and D'' are dualizing complexes then there exists an $n \in \mathbf{Z}$ such that D and $D''[n]$ are homotopy equivalent, see Hartshorne [2], Chap. V, Theorem 3.1, or Sharp [3]. Since $k = A/\mathfrak{m}$ itself is a dualizing complex there exists an integer $d \in \mathbf{Z}$ such that

$$\operatorname{Hom}^\cdot(k, D^\cdot) = k[-d]$$

for the local ring A admitting a dualizing complex $D^\cdot$. We call $D^\cdot$ *normalized* if the integer $d = 0$. Since the translate of a dualizing complex is again a dualizing complex we can normalize by translation. In the sequel a dualizing complex is assumed to be normalized.

Proposition 3.2. *Let A be a ring admitting a dualizing complex $D^\cdot$.*

(a) *For any prime ideal $\mathfrak{p}$ the localization $A_\mathfrak{p}$ has a dualizing complex.*

(b) *Let $D^\cdot_{A_\mathfrak{p}}$ denote the normalized dualizing complex of $A_\mathfrak{p}$. Then there is a homotopy equivalence*

$$D^\cdot \otimes_A A_\mathfrak{p} \to D^\cdot_{A_\mathfrak{p}}[\dim A/\mathfrak{p}].$$

Proof: Let $X^\cdot$ be a bounded above complex and $E^\cdot$ a bounded below complex of injective modules and suppose that both complexes have finitely generated cohomology modules. Then the canonical map

$$\operatorname{Hom}^\cdot(X^\cdot, E^\cdot) \otimes_A A_\mathfrak{p} \to \operatorname{Hom}^\cdot_{A_\mathfrak{p}}(X^\cdot \otimes A_\mathfrak{p}, E^\cdot \otimes A_\mathfrak{p}) \qquad (*)$$

is a quasi-isomorphism. Furthermore, $D^\cdot \otimes_A A_\mathfrak{p}$ is a complex of injective $A_\mathfrak{p}$-modules whose cohomology modules are finitely generated over $A_\mathfrak{p}$. The quasi-isomorphism $(*)$ and the definition of the dualizing complex imply that $D^\cdot \otimes_A A_\mathfrak{p}$ is a dualizing complex for $A_\mathfrak{p}$. Hence we have proved (a). Since $D^\cdot \otimes_A A_\mathfrak{p}$ is a dualizing complex there exists an integer $s \in \mathbf{Z}$ such that

$$\operatorname{Hom}^\cdot_{A_\mathfrak{p}}\big(k(\mathfrak{p}), D^\cdot \otimes_A A_\mathfrak{p}\big) = k(\mathfrak{p})\,[s]$$

where $k(\mathfrak{p}) = A_\mathfrak{p}/\mathfrak{p}A_\mathfrak{p}$. The assertion (b) follows if we can show that $s = \dim A/\mathfrak{p}$. Our last condition gives

$$\operatorname{Ext}^{-s}_{A_\mathfrak{p}}\big(k(\mathfrak{p}), D^\cdot \otimes_A A_\mathfrak{p}\big) \neq 0.$$

By slightly modifying results of Bass [1] to the case of a bounded below complex of injective modules having finitely generated cohomology we get

$$\operatorname{Ext}_A^{-s+\dim A/\mathfrak{p}}(k^\cdot, D^\cdot) \neq 0.$$

Since $D^\cdot$ is a dualizing complex of A we have $\operatorname{Hom}^\cdot(k, D^\cdot) \simeq k$ and we therefore obtain $-s + \dim A/\mathfrak{p} = 0$. This concludes the proof, q.e.d.

Corollary 3.3. *Suppose that A has a dualizing complex. Then A admits a dualizing complex $D^\cdot$ with*

$$D^i = \bigoplus_{\substack{\mathfrak{p} \in \operatorname{Spec} A \\ \dim A/\mathfrak{p} = -i}} E(A/\mathfrak{p}).$$

Proof: We take a minimal injective complex which is dualizing and normalize it. Since $D^\cdot \otimes_A A_\mathfrak{p}[-\dim A/\mathfrak{p}]$ is homotopy equivalent to the normalized dualizing complex of $A_\mathfrak{p}$ it suffices to show for a local ring A with maximal ideal $\mathfrak{m}$ the following: $D^0 = E(A/\mathfrak{m})$ and $E(A/\mathfrak{m})$ does not occur in D^i for $i \neq 0$. Since $D^\cdot$ is chosen to be minimal

$$H^i(\operatorname{Hom}^\cdot(k, D^\cdot)) = \operatorname{Hom}(k, D^i)$$

follows. Hence $\operatorname{Hom}(k, D^i) = k$ if $i = 0$ and 0 otherwise. Since D^i is isomorphic to a direct sum of $E(A/\mathfrak{p})$, $\mathfrak{p} \in \operatorname{Spec} A$, we obtain the required result, q.e.d.

In the sequel we will assume a dualizing complex $D^\cdot$ of this form.

For a bounded below complex $X^\cdot$ we denote by $R\Gamma_\mathfrak{m}(X^\cdot)$ the complex obtained as follows: Choose a quasi-isomorphism $X^\cdot \to E^\cdot$ where $E^\cdot$ is a bounded below complex of injective modules, and let

$$R\Gamma_\mathfrak{m}(X^\cdot) = \Gamma_\mathfrak{m}(E^\cdot).$$

This complex is unique up to homotopy and we define

$$H^i_\mathfrak{m}(X^\cdot) = H^i(R\Gamma_\mathfrak{m}(X^\cdot)), \quad i \in \mathbf{Z}.$$

In particular, for a module M, considered as a complex, we recover the local cohomology modules considered in § 1. Now we state and prove the main result of this paragraph.

Local Duality Theorem 3.4. *Let A be a local ring with maximal ideal $\mathfrak{m}$. Suppose that A has a dualizing complex $D^\cdot$. Let $X^\cdot$ be a bounded below complex with finitely generated cohomology modules. Then there exists a quasi-isomorphism*

$$R\Gamma_\mathfrak{m}(X^\cdot) \to \operatorname{Hom}(\operatorname{Hom}^\cdot(X^\cdot, D^\cdot), E),$$

where $E = E(k)$ denotes the injective hull of the residue field k.

Proof: First we remark that for any bounded above complex $L^\cdot$ of finitely generated free modules and any bounded below complex of injective modules $E^\cdot$ we have an isomorphism of complexes

$$\Gamma_\mathfrak{m}(\operatorname{Hom}^\cdot(L^\cdot, E^\cdot)) \cong \operatorname{Hom}^\cdot(L^\cdot, \Gamma_\mathfrak{m}(E^\cdot)).$$

If we now start with a bounded above complex $Y^\cdot$ with finitely generated cohomology modules we can choose a quasi-isomorphism $L^\cdot \to Y^\cdot$ where $L^\cdot$ is as above. This induces a quasi-isomorphism

$$\operatorname{Hom}^\cdot(Y^\cdot, E^\cdot) \to \operatorname{Hom}^\cdot(L^\cdot, E^\cdot).$$

We observe that the complex on the right consists of injective modules whence

$$R\Gamma_\mathfrak{m}(\operatorname{Hom}^\cdot(Y^\cdot, E^\cdot)) \to \Gamma_\mathfrak{m}(\operatorname{Hom}^\cdot(L^\cdot, E^\cdot)).$$

Note also that the quasi-isomorphism $L^{\cdot} \to Y^{\cdot}$ induces a quasi-isomorphism

$$\operatorname{Hom}^{\cdot}(Y^{\cdot}, \Gamma_{\mathfrak{m}}(E^{\cdot})) \to \operatorname{Hom}^{\cdot}(L^{\cdot}, \Gamma_{\mathfrak{m}}(E^{\cdot}))$$

by using that $\Gamma_{\mathfrak{m}}(E^{\cdot})$ consists of injective modules. Putting this together we obtain a quasi-isomorphism

$$R\Gamma_{\mathfrak{m}}(\operatorname{Hom}^{\cdot}(Y^{\cdot}, E^{\cdot})) \to \operatorname{Hom}^{\cdot}(Y^{\cdot}, \Gamma_{\mathfrak{m}}(E^{\cdot})).$$

For the normalized dualizing complex $D^{\cdot}$ we obtain $\Gamma_{\mathfrak{m}}(D^{\cdot}) = E$ (see H.-B. Foxby [1]) and therefore we have

$$R\Gamma_{\mathfrak{m}}(\operatorname{Hom}^{\cdot}(Y^{\cdot}, D^{\cdot})) \to \operatorname{Hom}^{\cdot}(Y^{\cdot}, E).$$

If now $X^{\cdot}$ is a bounded below complex with finitely generated cohomology modules we get for $Y^{\cdot} := \operatorname{Hom}^{\cdot}(X^{\cdot}, D^{\cdot})$

$$R\Gamma_{\mathfrak{m}}(\operatorname{Hom}^{\cdot}(\operatorname{Hom}^{\cdot}(X^{\cdot}, D^{\cdot}), D^{\cdot})) \to \operatorname{Hom}(\operatorname{Hom}^{\cdot}(X^{\cdot}, D^{\cdot}), E).$$

By using the quasi-isomorphism

$$e\colon X^{\cdot} \to \operatorname{Hom}^{\cdot}(\operatorname{Hom}^{\cdot}(X^{\cdot}, D^{\cdot}), D^{\cdot})$$

we have proven the statement, q.e.d.

The particular case of a local Gorenstein ring is of some interest since in this situation the local duality is of quite a simple form.

Corollary 3.5. *Let A be a factor of the local Gorenstein ring B with* $\dim B =: n$. *Then we have for all $i \in \mathbf{Z}$ natural isomorphisms*

$$H^i_{\mathfrak{m}_A}(M) \cong \operatorname{Hom}_A(\operatorname{Ext}^{n-i}_B(M, B), E),$$

where E denotes the injective hull of $A/\mathfrak{m}_A$ (consider M as a B-module and $\operatorname{Ext}^{n-i}_B(M, B)$ *as an A-module).*

Proof: Since B is a Gorenstein ring the minimal injective resolution $E^{\cdot}$ of B is a dualizing complex. Hence

$$B[-n] \to E^{\cdot}[-n]$$

is a normalized dualizing complex. Then $D^{\cdot} := \operatorname{Hom}_B(A, E^{\cdot})$ is a dualizing complex of A. Taking the cohomology in

$$R\Gamma_{\mathfrak{m}_A}(M) \to \operatorname{Hom}_A(\operatorname{Hom}^{\cdot}_B(M, D^{\cdot}), E)$$

and taking into account that

$$\operatorname{Ext}^{n-i}_B(M, B) \cong H^{-i}(\operatorname{Hom}^{\cdot}_B(M, E^{\cdot}[-n])) \cong H^{-i}(\operatorname{Hom}^{\cdot}_A(M, D^{\cdot}[-n]))$$

we get the corollary, q.e.d.

Let M be a finitely generated A-module. Then we obtain by the Local Duality Theorem

$$H^i(\operatorname{Hom}^{\cdot}(M, D^{\cdot})) \neq 0 \quad \text{if } -\dim_A M = i,$$

and

$$H^i(\operatorname{Hom}^{\cdot}(M, D^{\cdot})) = 0 \quad \text{if } i \neq -\dim_A M.$$

This follows since $i = \dim_A M$ is the highest non-vanishing local cohomology module of M. We call

$$K_M := H^{-\dim M}(\operatorname{Hom}^{\cdot}(M, D^{\cdot}))$$

the *canonical module* of the A-module M.

Next we examine the complex $\operatorname{Hom}^{\cdot}(M, D^{\cdot})$. Since

$$D^i = \bigoplus_{\substack{\mathfrak{p} \in \operatorname{Spec} A \\ \dim A/\mathfrak{p} = -i}} E(A/\mathfrak{p}) \quad \text{for } 0 \leq i \leq \dim A$$

we have

$$\operatorname{Hom}^i(M, D^{\cdot}) = \bigoplus_{\substack{\mathfrak{p} \in \operatorname{Spec} A \\ \dim A/\mathfrak{p} = -i}} \operatorname{Hom}_A(M, E(A/\mathfrak{p})).$$

Furthermore, we have

$$\operatorname{Ass} \operatorname{Hom}_A(M, E(A/\mathfrak{p})) = \operatorname{Supp} M \cap \{\mathfrak{p}\}.$$

From this we obtain that $\operatorname{Hom}^{\cdot}(M, D^{\cdot})$ is a bounded complex with finitely generated cohomology modules such that

$$\operatorname{Hom}^i(M, D^{\cdot}) = 0 \quad \text{for } i > 0 \text{ and } i < -\dim_A M.$$

Therefore there is an injection of complexes

$$0 \to K_M[\dim_A M] \to \operatorname{Hom}^{\cdot}(M, D^{\cdot})$$

where we regard K_M as a complex concentrated in degree zero.

Next we recall some basic facts about K_M and the cohomology modules $\operatorname{Ext}_A^i(M, D^{\cdot})$ for $-\dim_A M < i \leq 0$. For this we introduce the following notation:

Let Z be a subset of $\operatorname{Spec} A$. Then we denote by Z_i, $i \in \mathbf{Z}$, the set of primes in Z of dimension i. We collect some useful data which we need later.

Proposition 3.6. *Let M be a finitely generated A-module. Then we get the following properties:*

(a) $\operatorname{Ass}_A K_M = (\operatorname{Ass}_A M)_d$, *where* $d = \dim_A M$,

(b) $(\operatorname{Ass}_A M)_i = (\operatorname{Ass}_A \operatorname{Ext}_A^{-i}(M, D^{\cdot}))_i, \quad 0 \leq i < \dim_A M$,

(c) $\dim_A \operatorname{Ext}_A^{-i}(M, D^{\cdot}) \leq i, \quad 0 \leq i < \dim_A M$,

(d) $\operatorname{depth}_A K_M \geq \inf\{2, \dim_A K_M\}$.

Proof: The assertions (a) and (c) follow immediately from the definition of the dualizing complex $D^{\cdot}$. Next we shall prove (b). For this let $\mathfrak{p} \in (\operatorname{Ass}_A M)_i$. It follows that

$$\mathfrak{p}A_{\mathfrak{p}} \in \operatorname{Ass}_{A_{\mathfrak{p}}} M_{\mathfrak{p}} \quad \text{and} \quad H^0_{\mathfrak{p}A_{\mathfrak{p}}}(M_{\mathfrak{p}}) \neq 0.$$

The Local Duality Theorem for $A_{\mathfrak{p}}$ implies

$$H^0(\operatorname{Hom}_{A_{\mathfrak{p}}}(M_{\mathfrak{p}}, D^{\cdot}_{A_{\mathfrak{p}}})) \neq 0.$$

Since $\operatorname{Hom}_{A_\mathfrak{p}}(M_\mathfrak{p}, D^{\cdot}_{A_\mathfrak{p}}) \xrightarrow{\sim} (\operatorname{Hom}_A(M, D^{\cdot})[-\dim A/\mathfrak{p}]) \otimes_A A_\mathfrak{p}$ we get

$$H^{-\dim A/\mathfrak{p}}(\operatorname{Hom}_A(M, D^{\cdot})) \otimes_A A_\mathfrak{p} \neq O,$$

that is $\mathfrak{p} \in \operatorname{Supp} \operatorname{Ext}_A^{-\dim A/\mathfrak{p}}(M, D^{\cdot})$. From (c) we get the inclusion

$$(\operatorname{Ass}_A M)_i \subseteq (\operatorname{Ass}_A \operatorname{Ext}_A^{-i}(M, D^{\cdot}))_i.$$

The other inclusion can be proved similarly.

It follows by (a) that $\dim_A M = \dim_A K_M$ and that $\operatorname{depth}_A K_M \geq 1$.

Without loss of generality we can assume the existence of an M-regular element x which is therefore also K_M-regular. The short exact sequence

$$0 \to M \xrightarrow{x} M \to M/xM \to 0$$

induces an exact sequence

$$0 \to K_M \xrightarrow{x} K_M \to K_{M/xM} \to \dots$$

by applying the functor $\operatorname{Hom}(\ , D^{\cdot})$. Therefore K_M/xK_M is isomorphic to a submodule of $K_{M/xM}$. Since $K_{M/xM}$ has depth ≥ 1 we obtain the property (d), q.e.d.

Corollary 3.7. *Let A be a local complete ring. Let M be a Noetherian A-module. Then we have*

$$(\operatorname{Ass}_A M)_i = (\operatorname{Ass}_A \operatorname{Hom}_A(H^i_\mathfrak{m}(M), E))_i$$

for $i = 0, \dots, \dim M - 1$.

Proof: Using local duality we get the corollary from Proposition 3.6(b), q.é.d.

§ 4. Resolutions and duality

Let R denote a Noetherian graded k-algebra (k a field) with maximal ideal $\mathfrak{m}$, i.e. we have $[R]_i = O$ for all $i < 0$, $[R]_0 = k$ and $\mathfrak{m} = \bigoplus_{i \geq 1} [R]_i$ (althought many of the following facts remain true in more general situations). For the notation compare § 2.

Our aim is to recall basic facts and some applications regarding the following topics:

— free resolutions of (Noetherian) graded R-modules;

— duality similar to the local duality studied in the previous section.

For the proof of the graded version of the Local Duality Theorem 4.14 we use the methods of R. Y. Sharp [2]. This may be considered as an alternative for proving these results without using dualizing complexes.

1. *Graded modules of finite length*

We shall often be concerned with graded R-modules of finite length. Recall that a module M has finite length if it has a finite composition series; then the length $l_R(M)$ is the common length of all composition series of M. (If M is a graded module then for any composition series there is a composition series of the same length consisting of graded submodules of M.) For graded R-modules we have the following characterizations of finite length:

Lemma 4.1. *Let M be a graded R-module. Then the following conditions are equivalent:*

(i) *M has finite length.*

(ii) *M is finite-dimensional as a k-vector space.*

(iii) *M is Noetherian and $\mathfrak{m}^n M = 0$ for some n.*

(iv) *M is Noetherian and* $\operatorname{Ass} M \subseteq \{\mathfrak{m}\}$.

(v) *M is Noetherian and* $\operatorname{Supp} M \subseteq \{\mathfrak{m}\}$.

(vi) *M is Noetherian and* Supp *M contains no homogeneous primes other than* $\mathfrak{m}$.

(vii) *M is Noetherian and $M_{(\mathfrak{p})} = O$ for all $\mathfrak{p} \in$* Proj R.

We note that modules of finite length represent a somewhat pathological case, for if M has finite length then $\tilde{M} = O$ on Proj R. This is because the stalks of $\tilde{M}$ are the $M_{(\mathfrak{p})}$, $\mathfrak{p} \neq \mathfrak{m}$, which are zero by (vii) of Lemma 4.1.

2. *Minimal sets of generators*

Let M be a Noetherian graded R-module. Then $\overline{M} := M/\mathfrak{m} \cdot M$ is a finite-dimensional vector space over the residue field $k \simeq R/\mathfrak{m}$. A set $\{m_1, \ldots, m_t\}$ of elements of M is called a *minimal set of generators* for M if the residues $\overline{m}_1, \ldots, \overline{m}_t$ in $\overline{M}$ form a vector space basis of $\overline{M}$. The graded version of Nakayama's Lemma guarantees that a minimal set of generators is in fact a set of generators for M as an R-module. We have the following basic lemma:

Lemma 4.2. *Let M be a Noetherian graded R-module. Then every set of generators for M as an R-module contains a minimal set of generators for M; the number of elements in a minimal set of generators is uniquely determined; it is possible to choose a set of homogeneous generators for M which is a minimal set of generators for M, and the degrees of such minimal set of generators are uniquely determined by M (up to order).*

If R is a graded or a local ring and M a Noetherian (graded) R-module then we denote by $\mu_R(M)$ the number of elements of a minimal set of generators of M. (We note that $\mu_R(\mathfrak{m})$ is called the *embedding dimension* of R, denoted by emb R, if R is local with maximal ideal $\mathfrak{m}$.) Using Lemma 2.1 we obtain:

Lemma 4.3. *Let M be a Noetherian graded R-module and $\mathfrak{p} \neq \mathfrak{m}$ a homogeneous prime ideal of R. Then*

$$\mu_{R_{(\mathfrak{p})}}(M_{(\mathfrak{p})}) = \mu_{R_{\mathfrak{p},h}}(M_{\mathfrak{p},h}) = \mu_{R_{\mathfrak{p}}}(M_{\mathfrak{p}}) \leq \mu_R(M).$$

This will allow us to show that a projective subscheme is a local complete intersection, or a generic complete intersection, by checking ordinary localizations instead of degree 0 localizations.

3. *Minimal free resolutions*

Let M be a graded R-module. Then there is a set B of homogeneous elements of M generating M (as an R-module). Therefore we can define an epimorphism $\varphi: \bigoplus_{b \in B} R(-\deg b) \to M$

sending $1 \in R(-\deg b)$ to b. The module $\bigoplus_{b \in B} R(-\deg b)$ is even a free graded R-module and $\operatorname{Ker} \varphi$ is again a graded R-module. So we can repeat this procedure using $\operatorname{Ker} \varphi$ instead of M and obtain an exact sequence (with degree zero homomorphisms)

$$\ldots \to F_i \xrightarrow{\varphi_i} F_{i-1} \to \ldots \to F_0 \to M \to 0,$$

where all F_i, $i \geq 0$, are free graded R-modules. This exact sequence is called a *free resolution* of M.

We note that we work in the category of graded R-modules with degree zero homomorphisms as morphisms. This is a complete (and cocomplete) Grothendieck category, (c.f. Schubert [1], Def. 14.6.1.). Direct and inverse limits are formed "degreewise", i.e. in the category of k-vector spaces. This category possesses enough projectives (as we have seen before) and therefore enough injectives. This fact we will use in Section 4 of this paragraph.

For any graded R-module M we have projective and injective resolutions (even free resolutions). Therefore we define the *projective* and the *injective dimension* of M:

$$\operatorname{pd}_R M := \inf\{n \in \mathbf{N} \mid \text{there is a projective resolution } 0 \to P_n \to \ldots \to P_0 \to M \to 0 \text{ of } M\}$$

$$\operatorname{inj\,dim}_R M := \inf\{n \in \mathbf{N} \mid \text{there is an injective resolution } 0 \to M \to I^0 \to \ldots \to I^n \to 0 \text{ of } M\}.$$

Assume now that M is a Noetherian graded R-module and let $\{m_1, \ldots, m_t\}$ be a minimal set of homogeneous generators for M. Let $d_i := \deg m_i$ for $i = 1, \ldots, t$. Using the above we obtain an epimorphism $\varphi_0 \colon \bigoplus_{i=1}^{t} R(-d_i) \to M$, where $F_0 := \bigoplus_{i=1}^{t} R(-d_i)$ is a finitely generated free graded R-module. Hence $\operatorname{Ker} \varphi_0$ is again finitely generated, i.e. Noetherian. Moreover, by the Lemma of Nakayama, $\operatorname{Ker} \varphi_0 \subseteq \mathfrak{m} \cdot F_0$. Therefore there is a graded free resolution

$$\mathbf{F}\colon \ldots \to F_i \xrightarrow{\varphi_i} F_{i-1} \to \ldots \xrightarrow{\varphi_1} F_0 \xrightarrow{\varphi_0} M \to 0$$

of M such that all F_i, $i \geq 0$, are finitely generated free graded R-modules and $\operatorname{Im} \varphi_{i+1} = \operatorname{Ker} \varphi_i \subseteq \mathfrak{m} \cdot F_i$ for all $i \geq 0$. Such a resolution we call a *minimal graded free resolution* of M. Using standard techniques (first of all Nakayama's Lemma) we see that two minimal free resolutions $\mathbf{F}$, $\mathbf{F}'$ of M (M finitely generated) are isomorphic, i.e. there are for all $i \geq 0$ isomorphisms $\psi_i \colon F_i \to F'_i$ such that $\varphi'_i \psi_i = \psi_{i-1} \varphi_i$ for all $i \geq 1$ and $\varphi'_0 \psi_0 = \varphi_0$.

Especially, any two minimal free resolutions of M have the same number of terms, thus $\operatorname{pd}_R M$ may be calculated by computing some minimal free resolution.

We recall that by the famous Syzygy Theorem of Hilbert $\operatorname{pd}_R M$ is always finite if R is a polynomial ring over k.

Using the above remarks we get the following well-known result:

Lemma 4.4. (Criterion of minimality) *Let M be a finitely generated graded R-module and let*

$$\mathbf{G}\colon \ldots \to G_i \xrightarrow{\varphi_i} G_{i-1} \to \ldots \xrightarrow{\varphi_1} G_0 \xrightarrow{\varphi_0} M \to 0$$

be a graded free resolution of M (not necessarily minimal) where all G_i, $i \geq 0$, are finitely generated. Assume we are given bases of G_i, $i \geq 0$, and matrices representing φ_i, $i \geq 1$. Then $\mathbf{G}$ is minimal if and only if all the matrix entries lie in $\mathfrak{m}$.

Corollary 4.5. *Let M be a Noetherian graded R-module with $p := \operatorname{pd}_R M < \infty$. Then $\underline{\operatorname{Ext}}^i_R(M, R) = O$ for all $i > p$ and $\underline{\operatorname{Ext}}^p_R(M, R) \neq 0$.*

Proof: Let

$$\mathbf{F}: O \to F_p \xrightarrow{\varphi_p} F_{p-1} \to \dots \xrightarrow{\varphi_1} F_0 \xrightarrow{\varphi_0} M \to O$$

be a minimal free resolution of M. Let $G^i := \underline{\operatorname{Hom}}_R(F_i, R)$ and $\psi^i := \underline{\operatorname{Hom}}_R(\varphi_{i+1}, R)$: $G^i \to G^{i+1}$ for all $i \geq 0$. Then $\underline{\operatorname{Ext}}^i_R(M, R)$ is the ith cohomology of the complex

$$\mathbf{G}: O \to G^0 \xrightarrow{\psi^0} G^1 \to \dots \xrightarrow{\psi^{p-1}} G^p \to O,$$

i.e. $\underline{\operatorname{Ext}}^i_R(M, R) = O$ for all $i > p$. Since $\operatorname{Im} \varphi_p \subseteq \mathfrak{m} \cdot F_{p-1}$, $\operatorname{Im} \psi^{p-1} \subseteq \mathfrak{m} \cdot G^p$. But $G^p \neq O$ and we get an epimorphism

$$\underline{\operatorname{Ext}}^p_R(M, R) \cong G^p / \operatorname{Im} \psi^{p-1} \twoheadrightarrow G^p / \mathfrak{m} \cdot G^p \neq O,$$

by the Lemma of Nakayama, q.e.d.

We now want to describe an application of Proposition 2.3 for curves in $\mathbf{P}^3_k$.

Definition 4.6. By a *curve* $C \subset \mathbf{P}^3_k$ we mean a closed subscheme of $\mathbf{P}^3_k$, given by the homogeneous ideal $I(C)$ of $S := k[X_0, \dots, X_3]$, satisfying any of the following equivalent conditions:

(i) C is a one-dimensional scheme,

(ii) $S/I(C)$ is a two-dimensional graded ring,

(iii) $I(C)$ has height 2.

Furthermore, we will assume that C is locally Cohen-Macaulay and equidimensional.

Let C be a curve in $\mathbf{P}^3_k$. Let A be the local ring of the vertex of the affine cone over C with maximal ideal $\mathfrak{m}$; that is $A = \big(S/I(C)\big)_{\mathfrak{n}}$, where $\mathfrak{n} := (X_0, \dots, X_3)\, S$ denotes the maximal homogeneous ideal of S. Then we get the following corollary:

Corollary 4.7. *Let C be a curve in $\mathbf{P}^3_k$. The following conditions are equivalent:*

(i) *The local cohomology module $H^1_{\mathfrak{m}}(A)$ is annihilated by $\mathfrak{m}$.*

(ii) *The graded S-module $\bigoplus_{m \in \mathbf{Z}} H^1\big(\mathbf{P}^3_k, \widetilde{I(C)}\,(m)\big)$ is annihilated by $\mathfrak{n}$.*

Proof: We have by Proposition 2.3 (and the isomorphisms given there):

$$\bigoplus_{m \in \mathbf{Z}} H^1\big(\mathbf{P}^3_k, \widetilde{I(C)}\,(m)\big) \cong \underline{H}^2_{\mathfrak{n}}\big(I(C)\big).$$

It follows immediately from the exact sequence $O \to I(C) \to S \to S/I(C) \to O$ that

$$\underline{H}^2_{\mathfrak{n}}\big(I(C)\big) \cong \underline{H}^1_{\mathfrak{n}}\big(S/I(C)\big).$$

Let M be a graded S-module with $\operatorname{Supp} M \subseteq \{\mathfrak{n}\}$. Then for any $m \in M$ there is an n with $\mathfrak{n}^n m = 0$. Hence we obtain an isomorphism between (non graded) S-modules:

$M \cong M_{\mathfrak{n}}$. Therefore we have:

$$H^1_{\mathfrak{m}}(A) \cong (\underline{H}^1_{\mathfrak{n}}(S/I(C)))_{\mathfrak{n}} \cong \underline{H}^1_{\mathfrak{n}}(S/I(C)) \cong \bigoplus_{m \in \mathbf{Z}} H^1(\mathbf{P}^3_k, \widetilde{I(C)}\,(m))$$

and this proves our Corollary 4.7. q.e.d.

Using the notation from Chapter III we therefore have given a new proof of the following corollary (see our Theorem III.1.2 and the note in the preface):

Corollary 4.8. *The arithmetical Buchsbaum property for curves in* $\mathbf{P}^3_k$ *is preserved under liaison.*

Proof: The property of Corollary 4.7(ii) is preserved by shifting degrees and dualization. Therefore we get our assertion from our observations on liaison among curves in $\mathbf{P}^3_k$ of Chapter III, Proposition I.2.12 and Corollary 4.7, q.e.d.

4. *Dualization*

If M is a graded R-module, the k-vector space $M^v := \bigoplus_{n \in \mathbf{Z}} \operatorname{Hom}_k([M]_{-n}, k)$ is in a natural way a graded R-module with $[M^v]_n := \operatorname{Hom}_k([M]_{-n}, k)$. We say that M^v is the *dual* (or sometimes the *k-dual*) of M. If $f\colon M \to N$ is an R-homomorphism (of degree zero) of graded R-modules, then f defines an R-homomorphism $f^v\colon N^v \to M^v$. The functor $M \mapsto M^v$, $f \mapsto f^v$ is an exact contravariant functor from the category of graded R-modules to itself. We have:

Lemma 4.9. *Let M, N be graded R-modules. Then*

(i) *In the subcategory of all graded R-modules with* $\operatorname{rank}_k[M]_n < \infty$ *for all* $n \in \mathbf{Z}$ *(especially, in the subcategory of Noetherian or Artinian graded R-modules) the duality is perfect: There is a canonical isomorphism* $M^{vv} \cong M$.

(ii) *There is a natural isomorphism of graded R-modules*

$$\lambda_{M,N}\colon \underline{\operatorname{Hom}}_R(M, N^v) \cong (M \otimes_R N)^v.$$

Proof: (i) and the construction of $\lambda_{M,N}$ in (ii) follows from standard techniques of linear algebra. (For $p \in \mathbf{Z}$ the k-vector space $[M \otimes_R N]_p$ has a basis of elements of the form $m \otimes n$ with $m \in [M]_i$, $n \in [N]_j$, $i + j = p$ and therefore we can define for all $p \in \mathbf{Z}$ k-linear maps

$$\lambda_p\colon \operatorname{Hom}_R(M, N^v(p)) \to [(M \otimes_R N)^v]_p = \operatorname{Hom}_k([M \otimes_R N]_{-p}, k)$$

in the usual manner. Some straightforeward calculations show that $\lambda_{M,N} := \bigoplus_{p \in \mathbf{Z}} \lambda_p$ is a natural R-homomorphism.) It is easy to see that $\lambda_{M,N}$ is an isomorphism whenever M is a graded free R-module, i.e. if $M \cong \bigoplus_{i \in I} R(n_i)$ where I is a set (of indices) and $n_i \in \mathbf{Z}$. Then consider an exact sequence $F \to G \to M \to 0$, where F, G are free graded R-modules. This gives rise to a commutative diagram with exact rows

$$\begin{array}{ccccccc} O \to & \underline{\operatorname{Hom}}_R(M, N^v) & \to & \underline{\operatorname{Hom}}_R(G, N^v) & \to & \underline{\operatorname{Hom}}_R(F, N^v) \\ & \downarrow{\scriptstyle \lambda_{M,N}} & & \downarrow{\scriptstyle \lambda_{G,N}} & & \downarrow{\scriptstyle \lambda_{F,N}} \\ 0 \to & (M \otimes_R N)^v & \to & (G \otimes_R N)^v & \to & (F \otimes_R N)^v. \end{array}$$

Since $\lambda_{G,N}$, $\lambda_{F,N}$ are isomorphisms, $\lambda_{M,N}$ is an isomorphism too, q.e.d.

Corollary 4.10. *R^v is an injective object in the category of graded R-modules. More precisely, R^v is the injective hull of $\underline{k}$ (or better, of $\underline{k}^v \cong \underline{k}$).*

Proof: By Lemma 4.9(ii) we have for all graded R-modules M: $\underline{\mathrm{Hom}}_R(M, R^v) \cong M^v$, i.e. the functor $\underline{\mathrm{Hom}}_R(\ , R^v)$ is exact and therefore R^v is injective.

The epimorphism $R \to \underline{k}$ gives rise to a monomorphism $\underline{k}^v \to R^v$ and it is easy to check that this is an essential extension, q.e.d.

Corollary 4.11. *For all graded R-modules M, N and all $i \geq 0$ there are natural isomorphisms (of connected sequences of derived functors):*

$$\underline{\mathrm{Ext}}^i_R(M, N^v) \cong \underline{\mathrm{Tor}}^R_i(M, N)^v.$$

Proof: Note that $\{\underline{\mathrm{Tor}}^R_i(M, N)^v\}_{i \in \mathbf{N}}$ are the right derived functors of the left exact contravariant functor $(\ \otimes_R N)^v$, N fixed. For $i = 0$ we take the isomorphism of Lemma 4.9(ii). Since $\underline{\mathrm{Tor}}^R_i(M, N)^v = O$, $\underline{\mathrm{Ext}}^i_R(M, N^v) = O$ for all $i > 0$ whenever M is projective, the result follows from Cartan-Eilenberg [1], Chap. III, Prop. 5.2, and Chap. V, Prop. 4.4, q.e.d.

The local version of the following very useful lemma can be found in R. Y. Sharp [2], Lemma 3.2:

Lemma 4.12. *Let T be a right exact covariant additive R-linear functor from the category of graded R-modules to itself which respects shifts of degrees. Then there is for every graded R-module M a natural homomorphism*

$$\xi_M : M \otimes_R T(R) \to T(M)$$

which is an isomorphism if M is finitely generated. If T commutes with direct limits, ξ_M is always an isomorphism.

Proof: Let M be a graded R-module. For any homogeneous element $m \in M$ let $h_m : R \to M(\deg m)$ denote the homomorphism defined by $h_m(r) := rm$ for all $r \in R$. Then $T(h_m): T(R) \to T(M(\deg m)) = T(M)(\deg m)$ is an R-homomorphism. Define $\xi_M : M \otimes_R T(R) \to T(M)$ by $\xi_M(m \otimes \varrho) := T(h_m)(\varrho)$ for all $m \in [M]_i$, $\varrho \in [T(R)]_j$. It is easy to verify that ξ_M is a natural R-homomorphism.

Now ξ_M is an isomorphism if $M = R(p)$ with $p \in \mathbf{Z}$. Therefore ξ_M is an isomorphism if M is a finitely generated free graded R-module, since T commutes with finite direct sums. If T commutes with direct limits, ξ_M is an isomorphism if M is an arbitrary free graded R-module, since tensor products also commute with direct limits.

Now we can take an exact sequence $F \xrightarrow{f} G \xrightarrow{g} M \to 0$, where F, G are (finitely generated) free graded R-modules. Then we obtain a commutative diagram with exact rows

$$\begin{array}{ccccccc} F \otimes_R T(R) & \xrightarrow{f \otimes \mathrm{id}} & G \otimes_R T(R) & \xrightarrow{g \otimes \mathrm{id}} & M \otimes_R T(R) & \to & O \\ \downarrow \xi_F & & \downarrow \xi_G & & \downarrow \xi_M & & \\ T(F) & \xrightarrow{T(f)} & T(G) & \xrightarrow{T(g)} & T(M) & \to & O. \end{array}$$

Since ξ_F, ξ_G are isomorphisms, ξ_M is an isomorphism by the 5-Lemma, q.e.d.

Using the definitions of Section 3 of this paragraph we define:

Definition 4.13. A graded k-algebra R is called a *graded Gorenstein* k-algebra, if $\operatorname{inj}\dim_R R < \infty$.

The graded versions of the result of H. Bass [1] yield that $\operatorname{inj}\dim R = \dim R$ and that R is a graded Cohen-Macaulay algebra, i.e. $\operatorname{depth} R = \dim R$. Now we are able to state and prove the following.

Theorem 4.14. (Duality Theorem).

(i) *Suppose that R is a graded Gorenstein k-algebra of dimension n. For any graded R-module M and all $i \in \mathbf{Z}$ there are natural R-isomorphisms*

$$\underline{H}^i_{\mathfrak{m}}(M)^v \cong \underline{\operatorname{Ext}}^{n-i}_R(M, R)\,(r-1),$$

where $r := r(R)$ is the index of regularity of R (c.f. Def. 2.2).

(ii) *If R is arbitrary and if S is a graded Gorenstein k-algebra of dimension n such that R is a factor of S, then we have for all graded R-modules M and all $i \in \mathbf{Z}$ natural isomorphisms*

$$\underline{H}^i_{\mathfrak{m}}(M)^v \cong \underline{\operatorname{Ext}}^{n-i}_S(M, S)\,(s-1)$$

where $s := r(S)$, M is considered as an S-module and $\underline{\operatorname{Ext}}^{n-i}_S(M, S)$ as an R-module.

(iii) *If under the assumptions of* (ii) *M is Noetherian, we get*

$$\underline{H}^i_{\mathfrak{m}}(M) \cong \underline{\operatorname{Ext}}^{n-i}_S(M, S)^v\,(1-s).$$

Proof: By Corollary 4.11 we get with $N = R^v$ ($R^{vv} \cong R$ by Lemma 4.9(i)): $\underline{\operatorname{Ext}}^j_R(M, R) \cong \underline{\operatorname{Tor}}^R_j(M, R^v)^v$ for all j. To prove (i) we establish the following claim:

Claim. Let R be a graded Gorenstein k-algebra with $n := \dim R$. Then we have $R^v \cong \underline{H}^n_{\mathfrak{m}}(R)\,(r-1)$, where $r := r(R)$.

We use induction on n. If $n = 0$, $\underline{H}^0_{\mathfrak{m}}(R) = R$ and therefore this is an injective graded R-module. Thus we have to show that $R^v \cong R(r-1)$. Now $r = 1 + e(R)$ (c.f. Def. 2.2). Take $a \in [R]_{r-1} \setminus \{0\}$. Then $\mathfrak{m} \cdot a = 0$, i.e. we can find a monomorphism $\underline{k} \to R(r-1)$. Since $R(r-1)$ is injective, this induces a monomorphism $R^v \to R(r-1)$ (R^v is the injective hull of $\underline{k}$). Therefore we obtain for all $i \in \mathbf{Z}$:

$$\operatorname{rank}_k [R]_{-i} = \operatorname{rank}_k [R^v]_i \leq \operatorname{rank}_k [R]_{r-1+i},$$

and replacing i by $-r+1-i$ we find $\operatorname{rank}_k [R]_{r-i+1} \leq \operatorname{rank}_k [R]_{-i}$, i.e. $\operatorname{rank}_k [R^v]_i = \operatorname{rank}_k [R(r-1)]_i$ for all $i \in \mathbf{Z}$ and this gives

$$R^v \cong R(r-1).$$

If $n > 0$, we take a homogeneous non-zero divisor $x \in \mathfrak{m}$ with $d := \deg x$. Since $\bar{R} := R/xR$ is again a graded Gorenstein k-algebra of dimension $n-1$ with $r(\bar{R}) = r(R) + d$, we have

$$\underline{H}^{n-1}_{\mathfrak{m}}(\bar{R})\,(r+d-1) \cong \underline{H}^{n-1}_{\mathfrak{m}\cdot\bar{R}}(\bar{R})\,(r+d-1) \cong \bar{R}^v.$$

The exact sequence $0 \to R(-d) \xrightarrow{x} R \to \bar{R} \to 0$ induces a diagram with exact rows

$$\begin{array}{ccccccc} O \to & \underline{H}^{n-1}_{\mathfrak{m}}(\bar{R})\,(r+d-1) & \to & \underline{H}^n_{\mathfrak{m}}(R)\,(r-1) & \xrightarrow{x} & \underline{H}^n_{\mathfrak{m}}(R)\,(r+d-1) & \to 0 \\ & \downarrow\wr & & & & & \\ 0 \to & \bar{R}^v & \to & R^v & \xrightarrow{x} & R^v(d) & \to O. \end{array}$$

Since R^v is injective there is a homomorphism $f\colon \underline{H}^n_{\mathfrak m}(R)\,(r-1) \to R^v$ such that the diagram is commutative. Now $f(d)\colon \underline{H}^n_{\mathfrak m}(R)\,(r+d-1) \to R^v(d)$ makes the whole diagram commutative. Therefore $\operatorname{Ker} f \cong (\operatorname{Ker} f)\,(d)$ and $\operatorname{Coker} f \cong (\operatorname{Coker} f)\,(d)$. Both $\operatorname{Ker} f$ and $\operatorname{Coker} f$ are Artinian graded R-modules and hence $\operatorname{Ker} f = \operatorname{Coker} f = 0$ by these isomorphisms, i.e. f is an isomorphism and our claim has been proven.

Now we have with Lemma 4.12 ($\underline{H}^n_{\mathfrak m}(\ )$ is right exact):

$$(M \otimes R^v)^v \cong (M \otimes \underline{H}^n_{\mathfrak m}(R)\,(r-1))^v = (M \otimes \underline{H}^n_{\mathfrak m}(R))^v\,(1-r) \cong \underline{H}^n_{\mathfrak m}(M)^v\,(1-r),$$

hence

$$\underline{H}^n_{\mathfrak m}(M)^v \cong \underline{\operatorname{Hom}}_R(M, R)\,(r-1).$$

The right derived functors of the left exact contravariant functor $\underline{H}^n_{\mathfrak m}(\ )^v$ are just the functors $\underline{H}^{n-i}_{\mathfrak m}(\ )^v$, $i > 0$. If F is a free graded R-module then $\underline{H}^{n-i}_{\mathfrak m}(F) = O$ for all $i > 0$ since F is a direct sum of the Cohen-Macaulay modules $R(m)$, $m \in \mathbf{Z}$.

If P is a projective graded R-module then there is a free graded R-module F such that P is a direct summand of F, hence $\underline{H}^{n-i}_{\mathfrak m}(P) = O$ for all $i > 0$. Since $\underline{\operatorname{Ext}}^i_R(P, R) = O$ for all $i > 0$, we find by Cartan-Eilenberg [1], Chap. III, Prop. 5.2, and Chap. V, Prop. 4.4, for all $i \geq 0$ natural isomorphisms $\underline{H}^{n-i}_{\mathfrak m}(M)^v \cong \underline{\operatorname{Ext}}^i_R(M, R)\,(r-1)$ which proves (i).

To prove (ii), let $\mathfrak n$ denote the maximal homogeneous ideal of S. Then $\underline{H}^i_{\mathfrak n}(M)^v \cong \underline{\operatorname{Ext}}^{n-i}_S(M, S)\,(s-1)$ where M is considered as an S-module. But $\underline{H}^i_{\mathfrak n}(M)^v \cong \underline{H}^i_{\mathfrak m}(M)^v$, considered as S-modules.

(iii) follows from (ii) by dualization and Lemma 4.9(i) since $\underline{H}^i_{\mathfrak m}(M)$ is an Artinian graded R-module for all i, q.e.d.

Corollary 4.15. *Let M be a Noetherian graded R-module. Then $\underline{H}^i_{\mathfrak m}(M)$ is of finite length for all $i \neq \dim M$ if and only if M is locally Cohen-Macaulay and equidimensional, i.e. $M_{(\mathfrak p)}$ is a Cohen-Macaulay $R_{(\mathfrak p)}$-module for all $\mathfrak p \in \operatorname{Proj} R$ and $\dim M = \dim R/\mathfrak p$ for all minimal primes $\mathfrak p \in \operatorname{Supp} M$.*

Proof. Let S be a graded Gorenstein k-algebra such that R is an epimorphic image of S. If M is locally Cohen-Macaulay then we have for all $\mathfrak p \in \operatorname{Proj} S$ with $M_{(\mathfrak p)} \neq 0$:

$$0 = \operatorname{Ext}^i_{S_{(\mathfrak p)}}(M_{(\mathfrak p)}, S_{(\mathfrak p)}) \cong (\underline{\operatorname{Ext}}^i_S(M, S))_{(\mathfrak p)}$$

$$\text{for all } i \neq \dim S_{(\mathfrak p)} - \dim M_{(\mathfrak p)} = \dim S - \dim M.$$

Therefore $\underline{\operatorname{Ext}}^i_S(M, S)$ is of finite length for all $i \neq \dim S - \dim M$ by Lemma 4.1.

Thus $\underline{H}^i_{\mathfrak m}(M)^v$ and hence $\underline{H}^i_{\mathfrak m}(M)$ is of finite length for all $i \neq \dim M$ by Theorem 4.14. The converse is also true, q.e.d.

For another important consequence of Theorem 4.14 we need the following notation:

Let M be a Noetherian graded non-free R-module. Then take a finitely generated free graded R-module F and an epimorphism $\pi\colon F \to M$ such that $\operatorname{Ker} \pi \subseteq \mathfrak m \cdot F$ (F is up to an isomorphism uniquely determined by M, see our investigations in 3.). Then let

$$M^\times := \operatorname{Coker}(\underline{\operatorname{Hom}}_R(M, R) \to \underline{\operatorname{Hom}}_R(F, R)) = \operatorname{Coker} \underline{\operatorname{Hom}}_R(\pi, R).$$

$M^\times$ is uniquely determined by M up to an isomorphism and $M^\times \neq O$.

Corollary 4.16. *Assume that* R *is a graded Gorenstein* k*-algebra with* $r := r(R)$ *and* $n := \dim R$. *Let* $M \neq O$ *be a Noetherian graded* R*-module with* $\dim M = \dim R = n$ *such that* $\underline{H}^i_{\mathfrak{m}}(M)$ *is of finite length for all* $i \neq n$. *Then we have for all* $i \neq 0$, n *isomorphisms*

$$\underline{H}^i_{\mathfrak{m}}(M^{\times}) \simeq \underline{H}^{n-i}_{\mathfrak{m}}(M)^v (1 - r).$$

Proof: By Theorem 4.14 we have to show that $\underline{H}^i_{\mathfrak{m}}(M^{\times}) \simeq \underline{\mathrm{Ext}}^i_R(M, R)$ for all $i = 1, \ldots, n-1$. By our assumption the $\underline{\mathrm{Ext}}^i_R(M, R)$ are modules of finite length for all $i \geq 1$ (use Theorem 4.14 and Corollary 4.15). Take a minimal free resolution $\ldots \to F_i \to F_{i-1} \to \ldots \to F_0 \to M \to O$ of M. Applying the functor $\underline{\mathrm{Hom}}_R(\ ,R)$ we obtain a complex

$$\mathbf{F}^{\cdot} : O \to M^{\times} \xrightarrow{d^0} F^1 \xrightarrow{d^1} F^2 \to \ldots,$$

where $F^i := \underline{\mathrm{Hom}}_R(F_i, R)$ for all $i \geq 1$. Then

$$H^i(\mathbf{F}^{\cdot}) = \begin{cases} O & \text{for } i = 0, \\ \underline{\mathrm{Ext}}^i_R(M, R) & \text{for } i \geq 1. \end{cases}$$

By using for all $i \geq 1$ the exact sequences

$$0 \to \mathrm{Im}\, d^{i-1} \to \mathrm{Ker}\, d^i \to \underline{\mathrm{Ext}}^i_R(M, R) \to 0,$$

$$0 \to \mathrm{Ker}\, d^i \to F^i \to \mathrm{Im}\, d^i \to 0$$

and $\underline{H}^j_{\mathfrak{m}}(F^i) = 0$ for all $j \neq n$, $\underline{H}^j_{\mathfrak{m}}\big(\underline{\mathrm{Ext}}^i_R(M, R)\big) = O$ for all $j \neq 0$, $i \geq 1$, we obtain for all $i = 1, \ldots, n-1$:

$$\underline{\mathrm{Ext}}^i_R(M, R) = \underline{H}^0_{\mathfrak{m}}\big(\underline{\mathrm{Ext}}^i_R(M, R)\big) \simeq \underline{H}^1_{\mathfrak{m}}(\mathrm{Im}\, d^{i-1})$$
$$\simeq \underline{H}^2_{\mathfrak{m}}(\mathrm{Ker}\, d^{i-1}) \simeq \ldots \simeq \underline{H}^i_{\mathfrak{m}}(\mathrm{Im}\, d^0) \simeq H^i_{\mathfrak{m}}(M^{\times}),$$

q.e.d.

If R is a graded Gorenstein k-algebra and $\mathfrak{a} \subset R$ is a homogeneous ideal with $\dim R/\mathfrak{a} = \dim R$, such that $R_{(\mathfrak{p})}/\mathfrak{a}_{(\mathfrak{p})}$ is a Cohen-Macaulay ring for all $\mathfrak{p} \in \mathrm{Proj}\, R/\mathfrak{a}$, then we have for all $i = 1, \ldots, n-1$

$$\underline{H}^i_{\mathfrak{m}}(R/O :_R \mathfrak{a}) \simeq \underline{H}^{n-i}_{\mathfrak{m}}(R/\mathfrak{a})^v (1 - r),$$

since $(R/\mathfrak{a})^{\times} = R/O :_R \mathfrak{a}$.

If $R = k[X_0, \ldots, X_m]/(F_1, \ldots, F_s)$, where $X_0, \ldots, X_m$ are variables (of degree 1) and $F_1, \ldots, F_s$ are forms such that $n := \dim R = m - s + 1$ then R is a graded complete intersection, hence a Gorenstein k-algebra with $r(R) = -m + \deg F_1 + \ldots + \deg F_s$. Using again the notation of Chapter III we get a further application of Theorem 4.14:

Corollary 4.17. *Let* $V \subset \mathbf{P}^m_k$ *be a pure* n*-dimensional,* $n < m$, *locally Cohen-Macaulay subscheme which is linked by a complete intersection* C *to a subscheme* $W \subset \mathbf{P}^m_k$. *If* $C = H_1 \cap \ldots \cap H_s$, $s = \mathrm{codim}_{\mathbf{P}^n_k} C$, $H_1, \ldots, H_s$ *hypersurfaces of* $\mathbf{P}^m_k$ *of degree* $d_1, \ldots, d_s$ *(thus* $n = m - s$*) we have for all* $i = 1, \ldots, n$ *isomorphisms*

$$\bigoplus_{p \in \mathbf{Z}} H^i\big(\mathbf{P}^m_k, \mathcal{J}_V(p)\big) \simeq \Big(\bigoplus_{p \in \mathbf{Z}} H^{n-i+1}\big(\mathbf{P}^m_k, \mathcal{J}_W(p)\big)\Big)^v (m + 1 - d_1 - . \quad - d_s),$$

where $\mathcal{J}_V$, $\mathcal{J}_W$ *denote the corresponding sheafs of ideals.*

Proof: Let $S := k[X_0, \ldots, X_m]$, $\mathfrak{n} := (X_0, \ldots, X_m)\,S$, $R := S/(F_1, \ldots, F_s)$, $\mathfrak{m} := (X_0, \ldots, X_m)\,R$, $F_i = O$ the equation of H_i for $i = 1, \ldots, s$. Then by Proposition 2.3 and the isomorphisms given there:

$$\bigoplus_{p\in\mathbf{Z}} H^i\big(\mathbf{P}^m_k, \mathcal{J}_V(p)\big) \cong \underline{H}^{i+1}_{\mathfrak{n}}\big(I(V)\big) \cong \underline{H}^i_{\mathfrak{n}}\big(S/I(V)\big) \cong \underline{H}^i_{\mathfrak{m}}\big(R/I(V)\,R\big)$$

$$\cong \underline{H}^{n-i+1}_{\mathfrak{m}}\big(R/I(W)\,R\big)^v\,(m + 1 - d_1 - \ldots - d_s) \cong \ldots$$

$$\cong \Big(\bigoplus_{p\in\mathbf{Z}} H^{n-i+1}\big(\mathbf{P}^m_k, \mathcal{J}_W(p)\big)\Big)^v\,(m + 1 - d_1 - \ldots - d_s),$$

q.e.d.

Chapter I
Characterizations of Buchsbaum modules

§ 1. Characterization of Buchsbaum modules by systems of parameters

We will always denote by A a local ring with (unique) maximal ideal $\mathfrak{m}$.

First we will recall the problem of D. A. Buchsbaum expressed in the language of local algebra (see Introduction).

To do this let M be a Noetherian A-module and $\mathfrak{q}$ a parameter ideal of M. We will use the notion of the Hilbert-Samuel polynomial $p_{\mathfrak{q},M}$ and its leading coefficient $e_0(\mathfrak{q}, M)$, the multiplicity of $\mathfrak{q}$ with respect to M (see Chap. 0, § 1, 2.).

D. A. Buchsbaum's original questions was as follows:

Does there exist a natural number $I(M)$, such that the difference of "length" and "multiplicity" of every parameter ideal $\mathfrak{q}$ of M is equal to $I(M)$, i.e., it is true that

$$l_A(M/\mathfrak{q}M) - e_0(\mathfrak{q}, M) = I(M) \quad \text{for all parameter ideals } \mathfrak{q} \text{ of } M?$$

Already in the Introduction we gave examples which showed that this will not be true in general (see Examples 1 and 5). But we have also seen (see Examples 2—7 of the Introduction) that there are a lot of in this sense "good" examples, see also Chapter V.

The main purpose of this paragraph is to give a first characterization of those A-modules M, for which the above question has a positive answer. We start with some definitions and easy lemmas.

Definition 1.1. Let M be a Noetherian A-module. A system of elements $x_1, \ldots, x_r \in \mathfrak{m}$ is called a *weak M-sequence,* if for each $i = 1, \ldots, r$

$$(x_1, \ldots, x_{i-1}) \cdot M : x_i = (x_1, \ldots, x_{i-1}) \cdot M : \mathfrak{m}.$$

Remark 1.2. For $r = 1$ we have

$$0 :_M x_1 = 0 :_M \mathfrak{m}$$

or, equivalently,

$$\mathfrak{m} \cdot (0 :_M x_1) = 0 \quad (\text{since } x_1 \in \mathfrak{m}).$$

Consequently, if x_1 is a non-zero divisor of M, x_1 also forms a weak M-sequence (consisting of one element). We therefore see that the weak M-sequences are a direct generalization of the well-known M-sequences (see Chap. 0, § 1, 2.).

We know that every M-sequence forms a part of a system of parameters of M. For weak M-sequences we have:

Lemma 1.3. *Let M be a Noetherian A-module of positive dimension d. Then every weak M-sequence $x_1, \ldots, x_r$ with $r \leq d$ is a part of a system of parameters of M.*

Proof: Clearly, $x_2, \ldots, x_r$ forms a weak M/x_1M-sequence. Therefore, by an easy induction argument, it is sufficient to prove the lemma for $r = 1$. Let $x \in \mathfrak{m}$ be an element with $0 :_M x = 0 :_M \mathfrak{m}$. Then $x \notin \mathfrak{p}$ for all $\mathfrak{p} \in \operatorname{Ass} M \setminus \{\mathfrak{m}\}$, particularly, $x \notin \mathfrak{p}$ for all $\mathfrak{p} \in \operatorname{Supp} M$ with $\dim A/\mathfrak{p} = \dim M$ (> 0). But this means $\dim M/x \cdot M = d - 1$, i.e. x forms a part of a system of parameters of M, q.e.d.

Remark 1.4. It was stated in Stückrad-Vogel [2], Corollary 4, that $r \leq d$ for every weak M-sequence $x_1, \ldots, x_r$. This is not the case, however, as was pointed out by Balwant Singh with the following example: Let k be a field and set $A := k \oplus kx$ with $x^2 = 0$. Then x is a weak A-sequence but $\dim A = 0$.

Also, we note that two maximal weak M-sequences (i.e. the number of elements in such a sequence is equal to or less then $\dim M$ and it is not possible to lengthen it) have not always the same number of elements. To see this, we refer to Proposition 2.1 of this chapter and Example I.2.5 or Example V.5.4. These examples render modules M fulfilling statement (iii) of Proposition 2.1 which are not Buchsbaum modules. By Proposition 2.1 every system of parameters contained in $\mathfrak{m}^2$ is a weak M-sequence. On the other hand it is easy to see that every element x with $\dim M/x \cdot M = \dim M - 1$ forms a weak M-sequence, i.e. every weak M-sequence of maximal length consists of at least one element. But not every system of parameters is a weak M-sequence since M need not be a Buchsbaum module, see our next definition.

Now we are able to define Buchsbaum rings, resp. modules.

Definition 1.5. A Noetherian A-module M is called a *Buchsbaum module* if every system of parameters of M is a weak M-sequence. A is called a *Buchsbaum ring* if it is a Buchsbaum module as a module over itself.

From the Introduction we already have quite a number of examples of Buchsbaum modules. Another easy consequence which enhances our knowledge of Buchsbaum modules is the following statement.

Lemma 1.6. *Assume that A is an epimorphic image of the local ring B. An A-module M is a Buchsbaum module over A if and only if it is a Buchsbaum module considered as a B-module by "restricting scalars".*

Proof: Clearly, $\dim_B M = \dim_A M$ and every system of parameters of M in A may be obtained by restricting a system of parameters of M as a B-module. Conversely, the restriction of any system of parameters of M in B to A is again a system of parameters of M as an A-module. Therefore the validity of the statement of our lemma becomes clear by the definition of weak M-sequences, q.e.d.

Before continuing we state a seemingly technical yet in the sequel very useful result. To this end we define:

Definition 1.7. Let $\mathfrak{a} \subset A$ be an ideal and M a Noetherian A-module with $\dim M/\mathfrak{a} \cdot M = 0$. Let $d := \dim M$. A system of elements $x_1, \ldots, x_t$ of A is called an *M-basis* of $\mathfrak{a}$ if the following conditions are fulfilled:

(i) $x_1, \ldots, x_t$ form a minimal basis of $\mathfrak{a}$.

(ii) For every system $i_1, \ldots, i_d$ of integers with $1 \leq i_1 < \ldots < i_d \leq t$ the elements $x_{i_1}, \ldots, x_{i_d}$ form a system of parameters of M.

Remark 1.8. 1. Since $0 = \dim M/\mathfrak{a} \cdot M = \dim M/(x_1, \ldots, x_t) \cdot M$ it follows that $t \geq \dim M = d$.
2. If $\dim M = 0$ any minimal basis of $\mathfrak{a}$ is an M-basis.

We now prove:

Proposition 1.9. *Let $\mathfrak{a} \subset A$ be an ideal and $M_1, \ldots, M_n$ Noetherian A-modules with $\dim_A M_i/\mathfrak{a}M_i = 0$ for all $i = 1, \ldots, n$. Then there are $a_1, \ldots, a_t \in \mathfrak{a}$ forming an M_i-basis of $\mathfrak{a}$ for all $i = 1, \ldots, n$.*

Proof: Clearly, we can omit zero-dimensional modules from our collection, i.e., we may assume $d_i := \dim M_i > 0$ for all $i = 1, \ldots, n$. We now prove by induction on m, $0 \leq m \leq t$, $t := \operatorname{rank}_{A/\mathfrak{m}} \mathfrak{a}/\mathfrak{m}\mathfrak{a}$, $\mathfrak{a} = (b_1, \ldots, b_t) A$:

There are elements $a_1, \ldots, a_m$ of $\mathfrak{a}$ with

(i) $\mathfrak{a} = (a_1, \ldots, a_m, b_{m+1}, \ldots, b_t) A$ and

(ii) for all $l = 1, \ldots, n$, all $j = 0, \ldots, \min(m, d_l)$ and all $1 \leq i_1 < \ldots < i_j \leq m$, $a_{i_1}, \ldots, a_{i_j}$ is a part of a system of parameters of M_l.

This is trivial for $m = 0$. Let therefore be $0 < m \leq t$ and assume that $a_1, \ldots, a_{m-1}$ ($\in \mathfrak{a}$) fulfil (i) and (ii). Define

$$\begin{aligned} L := \{\mathfrak{p} \in \operatorname{Spec} A \mid & \text{ there are } l, j, i_1, \ldots, i_j \in \mathbf{N},\ 1 \leq l \leq n, \\ & 0 \leq j < \min(m, d_l),\ 1 \leq i_1 < \ldots < i_j < m \\ & \text{with } \mathfrak{p} \in \operatorname{Supp} M_l/(a_{i_1}, \ldots, a_{i_j}) M_l, \dim A/\mathfrak{p} = d_l - j\}, \end{aligned}$$

$$L_1 := \{\mathfrak{p} \in L \mid b_m \notin \mathfrak{q} \text{ for all } \mathfrak{q} \in L \text{ with } \mathfrak{p} \subseteq \mathfrak{q}\} \quad \text{and} \quad L_2 := L \setminus L_1.$$

If $L_2 = \emptyset$, set $a_m := b_m$. If $L_2 \neq \emptyset$, we have

$$(a_1, \ldots, a_{m-1}, b_{m+1}, \ldots, b_t) A \cap \bigcap_{\mathfrak{p} \in L_1} \mathfrak{p} \nsubseteq \mathfrak{q} \quad \text{for all} \quad \mathfrak{q} \in L_2.$$

Therefore we can find an

$$x \in \Big[(a_1, \ldots, a_{m-1}, b_{m+1}, \ldots, b_t) A \cap \bigcap_{\mathfrak{p} \in L_1} \mathfrak{p}\Big] \setminus \Big[\bigcup_{\mathfrak{q} \in L_2} \mathfrak{q}\Big]$$

and we define $a_m := x + b_m$. Then $a_1, \ldots, a_{m-1}, a_m$ satisfy (i) and (ii), q.e.d.

Now we state a proposition which enables us to examine the Buchsbaum property in several different ways.

Proposition 1.10. *Let M be a Noetherian A-module with $d := \dim M > 0$. The following properties are equivalent:*

(i) *M is a Buchsbaum module, i.e., every system of parameters is a weak M-sequence.*

(i)′ *For every system of parameters $x_1, \ldots, x_d$ of M we have*

$$(x_1, \ldots, x_{d-1}) \cdot M : x_d = (x_1, \ldots, x_{d-1}) \cdot M : \mathfrak{m}.$$

(ii) *For every system of parameters $x_1, \ldots, x_d$ of M and all $i = 0, \ldots, d-1$ we have*

$$(x_1, \ldots, x_i) \cdot M : x_{i+1} = (x_1, \ldots, x_i) \cdot M : x \quad \textit{for all } x \in \mathfrak{m}$$

such that $x_1, \ldots, x_i, x$ form a part of a system of parameters of M.

(ii)′ *For every system of parameters $x_1, \ldots, x_d$ of M we have*

$$(x_1, \ldots, x_{d-1}) \cdot M : x_d = (x_1 \ldots, x_{d-1}) \cdot M : x \quad \textit{for all } x \in \mathfrak{m}$$

such that $x_1, \ldots, x_{d-1}, x$ form again a system of parameters of M.

(iii) *For every system of parameters* $x_1, \ldots, x_d$ *of* M *we have for all* $i = 0, \ldots, d-1$:

$$(x_1, \ldots, x_i) \cdot M : x_{i+1} = (x_1, \ldots, x_i) \cdot M : x_{i+1}^2.$$

(iii)′ *For every system of parameters* $x_1, \ldots, x_d$ *of* M *we have*

$$(x_1, \ldots, x_{d-1}) \cdot M : x_d = (x_1, \ldots, x_{d-1}) \cdot M : x_d^2.$$

(iv) *For every part of a system of parameters* $x_1, \ldots, x_i$ *of* M *with* $i < d$ *we have*

$$U\big((x_1, \ldots, x_i) \cdot M\big) = (x_1, \ldots, x_i) \cdot M : \mathfrak{m}.$$

Proof: The implications (i) ⇒ (i)′, (i) ⇒ (ii), (i)′ ⇒ (ii)′, (ii) ⇒ (iii), (ii)′ ⇒ (iii)′, (iii) ⇒ (iii)′ are trivial. It remains to prove (iii)′ ⇒ (iv) and (iv) ⇒ (i).

To prove (iii)′ ⇒ (iv) we show (iii)′ ⇒ (ii)′ and (ii)′ ⇒ (iv).

Let $x_1, \ldots, x_d$ be a system of parameters. (iii)′ implies that

$$(x_1, \ldots, x_{d-1}) \cdot M : x_d = (x_1, \ldots, x_{d-1}) \cdot M : x_d^2 = (x_1, \ldots, x_{d-1}) \cdot M : x_d^3 = \ldots,$$

i.e. $(x_1, \ldots, x_{d-1}) \cdot M : x_d = U\big((x_1, \ldots, x_{d-1}) \cdot M\big)$ since $\dim M/(x_1, \ldots, x_{d-1}) \cdot M = 1$. Consequently, $(x_1, \ldots, x_{d-1}) \cdot M : x_d$ does not depend on x_d and this implies (ii)′.

Assume now that (ii)′ is satisfied. Let $x_1, \ldots, x_i$ be any part of a system of parameters of M, $i < d$. Then we choose $x_{i+1}, \ldots, x_d$ such that $x_1, \ldots, x_d$ form a system of parameters. For all integers $n \geq 1$ the sequence $x_1, \ldots, x_i, x_{i+1}^n, \ldots, x_{d-1}^n, x_d$ is also a system of parameters. By Krull's Intersection Theorem we obtain for all $x \in \mathfrak{m}$ such that $x_1, \ldots, x_{d-1}, x$ is again a system of parameters of M the following:

$$\begin{aligned}(x_1, \ldots, x_i) \cdot M : x_d &= \Big(\bigcap_{n \geq 1} (x_1, \ldots, x_i, x_{i+1}^n, \ldots, x_{d-1}^n) \cdot M\Big) : x_d \\ &= \bigcap_{n \geq 1} \big((x_1, \ldots, x_i, x_{i+1}^n, \ldots, x_{d-1}^n) \cdot M : x_d\big) \\ &= \bigcap_{n \geq 1} \big((x_1, \ldots, x_i, x_{i+1}^n, \ldots, x_{d-1}^n) \cdot M : x\big) = (x_1, \ldots, x_i) \cdot M : x.\end{aligned}$$

Amongst the elements x we find an $\big(M/(x_1, \ldots, x_{d-1}) \cdot M\big)$-basis of the maximal ideal $\mathfrak{m}$ of A (see Proposition 1.9) and this gives

$$(x_1, \ldots, x_i) \cdot M : x_d = (x_1, \ldots, x_i) \cdot M : \mathfrak{m}.$$

On the other hand it is easy to see that x_d can be chosen in such a way that we have $(x_1, \ldots, x_i) \cdot M : x_d = U\big((x_1, \ldots, x_i) \cdot M\big)$ and therefore we obtain (iv).

Finally, we verify (iv) ⇒ (i). To do this let $x_1, \ldots, x_d$ be an arbitrary system of parameters for M. Then we have for all $i = 0, \ldots, d-1$:

$$\begin{aligned}(x_1, \ldots, x_i) \cdot M : \mathfrak{m} \subseteq (x_1, \ldots, x_i) \cdot M : x_{i+1} &\subseteq U\big((x_1, \ldots, x_i) \cdot M\big) \\ &= (x_1, \ldots, x_i) \cdot M : \mathfrak{m},\end{aligned}$$

i.e. $x_1, \ldots, x_d$ is a weak M-sequence, q.e.d.

Corollary 1.11. *Let* M *be a Buchsbaum module. Assume* $x_1, \ldots, x_r$ *is a part of a system of parameters of* M *with* $r < \dim M$. *Then*

(i) $M/(x_1, \ldots, x_r) \cdot M$ *and* (ii) $M/U\big((x_1, \ldots, x_r) \cdot M\big)$

are Buchsbaum modules (of dimension $\dim M - r$*).*

Furthermore, the localizations $M_{\mathfrak{p}}$ *are Cohen-Macaulay modules for all prime ideals* $\mathfrak{p} \neq \mathfrak{m}$ *of* Supp M.

Proof: (i) is a direct consequence of the definition of weak M-sequences.

Since $M/U\big((x_1, \ldots, x_r) \cdot M\big) = \big(M/(x_1, \ldots, x_r) \cdot M\big)/U(0)$ ($U(0)$ in $M/(x_1, \ldots, x_r) \cdot M$) it is sufficient to prove (ii) for $r = 0$. Let $y_1, \ldots, y_i$, $i < \dim M/U(0) = \dim M$ be a part of a system of parameters of $M/U(0)$. Then it is a part of a system of parameters of M as well and we have by Proposition 1.10(iv)

$$\begin{aligned}\big(U(0) + (y_1, \ldots, y_i) \cdot M\big):\mathfrak{m} &\subseteq U\big(U(0) + (y_1, \ldots, y_i) \cdot M\big)\\ &= U\big((y_1, \ldots, y_i) \cdot M\big) = (y_1, \ldots, y_i) \cdot M:\mathfrak{m}\\ &\subseteq \big(U(0) + (y_1, \ldots, y_i) \cdot M\big):\mathfrak{m}.\end{aligned}$$

Therefore $M/U(0)$ is a Buchsbaum module by Proposition 1.10(iv).

The last statement follows from Proposition 1.10(iv), q.e.d.

We note that the converse of (i) or (ii) of Corollary 1.11 is not true in general. We will return to this topic later when we discuss the so-called "lifting property", see Proposition 2.19 and Proposition 2.23.

Now we come to the main result of this paragraph. It shows the connection between the Buchsbaum modules defined above and the original problem of D. A. Buchsbaum (see Theorem 2 of the Introduction).

Theorem 1.12. *Let M be a Noetherian A-module with $d := \dim M > 0$. Then M is a Buchsbaum module if and only if there is an integer $I(M) \geq 0$ such that*

$$l(M/\mathfrak{q} \cdot M) - e_0(\mathfrak{q}, M) = I(M) \qquad \textit{for all parameter ideals } \mathfrak{q} \textit{ of } M.$$

Proof: For every parameter ideal $\mathfrak{q}$ of M we set

$$c(\mathfrak{q}, M) := l(M/\mathfrak{q} \cdot M) - e_0(\mathfrak{q}, M) \geq 0 \qquad \text{(see Lemma 0.1.3(ii))}.$$

Assume first that there is an integer $I(M)$ such that $c(\mathfrak{q}, M) = I(M)$ for all parameter ideals $\mathfrak{q}$ of M.

Let $x_1, \ldots, x_d$ be an arbitrary system of parameters for M, $\mathfrak{q} := (x_1, \ldots, x_d) \cdot A$. For $i = 1, \ldots, d$ we let

$$M_i := (x_1, \ldots, x_{i-1}) \cdot M:x_i/(x_1, \ldots, x_{i-1}) \cdot M$$

and

$$e_i := e(x_{i+1}, \ldots, x_d|M_i).$$

Then by Lemma 0.1.3(vi)

$$c(\mathfrak{q}, M) = \sum_{i=1}^{d} e_i.$$

Now we replace x_d by x_d^2 and we let $(1 \leq i \leq d-1)$

$$\mathfrak{q}' := (x_1, \ldots, x_{d-1}, x_d^2) \cdot A, \qquad e_i' := e(x_{i+1}, \ldots, x_{d-1}, x_d^2|M_i).$$

Then by Lemma 0.1.3(v) we have for $i = 1, \ldots, d-1$:

$$e_i' = 2e_i$$

and this results in

$$l\big((x_1, \ldots, x_{d-1}) \cdot M : x_d^2/(x_1, \ldots, x_{d-1}) \cdot M : x_d\big) + \sum_{i=1}^{d-1} e_i$$

$$= l\big((x_1, \ldots, x_{d-1}) \cdot M : x_d^2/(x_1, \ldots, x_{d-1}) \cdot M : x_d\big) + \sum_{i=1}^{d-1} e_i' - l(M_d) - \sum_{i=1}^{d-1} e_i$$

$$= c(\mathfrak{q}', M) - c(\mathfrak{q}, M) = I(M) - I(M) = 0.$$

Since all terms of the left sum are non-negative integers,

$$l\big((x_1, \ldots, x_{d-1}) \cdot M : x_d^2/(x_1, \ldots, x_{d-1}) \cdot M : x_d\big) = 0,$$

i.e.

$$(x_1, \ldots, x_{d-1}) \cdot M : x_d = (x_1, \ldots, x_{d-1}) \cdot M : x_d^2.$$

Hence, by Proposition 1.10 (iii)', M is a Buchsbaum module.

Now assume that M is a Buchsbaum module. Let $x_1, \ldots, x_d$ be a system of parameters for M. If we use the same notations as defined above, we have for $i = 1, \ldots, d$:

$$\mathfrak{m} \cdot M_i = 0, \quad \text{especially,} \quad \dim M_i = 0.$$

Therefore by Lemma 0.1.4

$$e(x_{i+1}, \ldots, x_d | M_i) = 0 \quad \text{for } i = 1, \ldots, d-1$$

and we have

$$c(\mathfrak{q}, M) = l(M_d).$$

We notice that the left part of this equation does not depend on the order of the elements $x_1, \ldots, x_d$ and, consequently, the right-hand term does not depend on this order.

Let now $\mathfrak{q}' := (y_1, \ldots, y_d) \cdot A$ be another parameter ideal of M. We show by induction on d that $c(\mathfrak{q}, M) = c(\mathfrak{q}', M)$.

If $d = 1$, $c(\mathfrak{q}, M) = l(0 :_M x_1) = l(0 :_M \mathfrak{m}) = l(0 :_M y_1) = c(\mathfrak{q}', M)$ and we are done.

Suppose now that $d \geq 2$. We choose an element $z \in \mathfrak{m}$ such that $x_1, \ldots, x_{d-1}, z$ and $y_1, \ldots, y_{d-1}, z$ are again systems of parameters with respect to M. Then by Proposition 1.10 (ii)' and the induction hypothesis we have with $\bar{\mathfrak{q}} := (x_1, \ldots, x_{d-1}) \cdot A$, $\bar{\mathfrak{q}}' := (y_1, \ldots, y_{d-1}) \cdot A$, $\bar{M} := M/z \cdot M$:

$$\begin{aligned} c(\mathfrak{q}, M) &= l(\bar{\mathfrak{q}} \cdot M : x_d/\bar{\mathfrak{q}} \cdot M) = l(\bar{\mathfrak{q}} \cdot M : z/\bar{\mathfrak{q}} \cdot M) \\ &= l\big((z, x_1, \ldots, x_{d-2}) \cdot M : x_{d-1}/(z, x_1, \ldots, x_{d-2}) \cdot M\big) \\ &= l\big((x_1, \ldots, x_{d-2}) \cdot \bar{M} : x_{d-1}/(x_1, \ldots, x_{d-2}) \cdot \bar{M}\big) \\ &= c(\bar{\mathfrak{q}}, \bar{M}) = c(\bar{\mathfrak{q}}', \bar{M}) = \ldots = l(\bar{\mathfrak{q}}' \cdot M : z/\bar{\mathfrak{q}}' \cdot M) \\ &= l(\bar{\mathfrak{q}}' \cdot M : y_d/\bar{\mathfrak{q}}' \cdot M) = c(\mathfrak{q}', M), \end{aligned}$$

since $\bar{M}$ is a Buchsbaum module of dimension $d - 1$ by Corollary 1.11 and $\bar{\mathfrak{q}}$ resp. $\bar{\mathfrak{q}}'$ are parameter ideals for $\bar{M}$, q.e.d.

Now we are able to state the following

Lemma 1.13. *Let M be a Noetherian A-module of positive dimension. M is a Buchsbaum module if and only if the $\mathfrak{m}$-adic completion $\hat{M}$ of M is a Buchsbaum module over $\hat{A}$. In this case $I(M) = I(\hat{M})$.*

Proof: Let M be a Buchsbaum module and let $\mathfrak{v}$ denote a parameter ideal of $\hat{M}$. Then there is a parameter ideal $\mathfrak{q}$ of M with $\mathfrak{v} \cdot \hat{M} = \hat{\mathfrak{q}} \cdot \hat{M}$ and we have

$$l_{\hat{A}}(\hat{M}/\mathfrak{v} \cdot \hat{M}) - e_0(\mathfrak{v}, \hat{M}) = l_{\hat{A}}(\hat{M}/\hat{\mathfrak{q}} \cdot \hat{M}) - e_0(\hat{\mathfrak{q}}, \hat{M}) = l_{\hat{A}}(\widehat{M/\mathfrak{q}M}) - e_0(\hat{\mathfrak{q}}, \hat{M})$$
$$= l_A(M/\mathfrak{q} \cdot M) - e_0(\mathfrak{q}, M) = I(M).$$

This is independent of the choice of $\mathfrak{v}$, i.e., $\hat{M}$ is a Buchsbaum module with $I(\hat{M}) = I(M)$ by Theorem 1.12.

Conversely, if $\mathfrak{q}$ is a parameter ideal with respect to M then $\hat{\mathfrak{q}}$ is a parameter ideal of $\hat{M}$ and the same reasoning as before provides our statement, q.e.d.

Next we prove two lemmas needed in the sequel.

Lemma 1.14. *Let M be a Noetherian A-module and $\mathfrak{a} \subset A$ be an ideal such that $M/\mathfrak{a} \cdot M$ is a Buchsbaum module of positive dimension. Then for every part of a system of parameters $x_1, \ldots, x_r$ of $M/\mathfrak{a} \cdot M$ we have with $\mathfrak{b} := (x_1, \ldots, x_r) \cdot A$:*

$$(\mathfrak{a} \cdot M : \mathfrak{m}) \cap \mathfrak{b}^k \cdot M \subseteq \mathfrak{a} \cdot \mathfrak{b}^{k-1} \cdot M \quad \textit{for all } k \geq 1 \quad (\mathfrak{b}^0 := A).$$

Proof: Assume the statement of the lemma is false. Then there is a maximal member among all ideals $\mathfrak{a}$ for which the assumptions are fulfilled but the statement is not true. Let it be $\mathfrak{a}_0$. Then there is an $r \geq 0$, a part of a system of parameters $x_1, \ldots, x_r$ of $M/\mathfrak{a}_0 \cdot M$ and an integer $k \geq 1$ with

$$(\mathfrak{a}_0 \cdot M : \mathfrak{m}) \cap \mathfrak{b}^k \cdot M \nsubseteq \mathfrak{a}_0 \cdot \mathfrak{b}^{k-1} \cdot M \quad \big(\mathfrak{b} = (x_1, \ldots, x_r) \cdot A\big).$$

We also may assume that k is minimal with respect to this property. We set

$$U := \mathfrak{a}_0 \cdot M : \mathfrak{m} = U(\mathfrak{a}_0 \cdot M) \quad \text{(see Proposition 1.10(iv))}.$$

Clearly, $r = 0$ is impossible (since then $\mathfrak{b} = 0$). If $r = 1$,

$$U \cap x_1^k \cdot M = x_1^k \cdot (U :_M x_1^k) = x_1^k \cdot U = x_1^{k-1} \cdot x_1 \cdot U \subseteq x_1^{k-1} \cdot \mathfrak{a}_0 \cdot M,$$

a contradiction. Therefore $r \geq 2$.

Choose $a \in (U \cap \mathfrak{b}^k \cdot M) \setminus \mathfrak{a}_0 \cdot \mathfrak{b}^{k-1} \cdot M$. Then we can write $a = \sum_{i=1}^{r} x_i \cdot u_i$ with $u_i \in (x_1, \ldots, x_i)^{k-1} \cdot M$. Notice that $\mathfrak{m} \cdot a \subseteq \mathfrak{a}_0 \cdot M$. Set

$$b := \sum_{i=1}^{r-1} x_i \cdot u_i \in \mathfrak{b}'^k \cdot M, \quad \text{where } \mathfrak{b}' := (x_1, \ldots, x_{r-1}) \cdot A.$$

Now, for all $x \in \mathfrak{m}$ we have $x \cdot b + x \cdot x_r \cdot u_r = x \cdot a \in \mathfrak{a}_0 \cdot M$, i.e.

$$b \in \big((\mathfrak{a}_0 + x_r \cdot A) \cdot M : \mathfrak{m}\big) \cap \mathfrak{b}'^k \cdot M.$$

Since $M/(\mathfrak{a}_0 + x_r \cdot A) \cdot M \cong (M/\mathfrak{a}_0 \cdot M)/x_r \cdot (M/\mathfrak{a}_0 \cdot M)$ is a Buchsbaum module of positive dimension ($r \geq 2$) by Corollary 1.11(i) and since $\mathfrak{b}'$ is generated by a part of a system of parameters with respect to $M/(\mathfrak{a}_0 + x_r \cdot A) \cdot M$, we have by the maximality of $\mathfrak{a}_0$ that

$$b \in (\mathfrak{a}_0 + x_r \cdot A) \cdot \mathfrak{b}'^{k-1} \cdot M.$$

Hence

$$\begin{aligned} a = b + x_r \cdot u_r \in (\mathfrak{a}_0 \cdot \mathfrak{b}'^{k-1} + x_r \cdot \mathfrak{b}'^{k-1} + x_r \cdot \mathfrak{b}^{k-1}) \cdot M \cap U \\ = \mathfrak{a}_0 \cdot \mathfrak{b}'^{k-1} \cdot M + (x_r \cdot \mathfrak{b}^{k-1} \cdot M \cap U) \\ = \mathfrak{a}_0 \cdot \mathfrak{b}'^{k-1} \cdot M + x_r \cdot (\mathfrak{b}^{k-1} \cdot M \cap U :_M x_r) \\ = \mathfrak{a}_0 \cdot \mathfrak{b}'^{k-1} \cdot M + x_r \cdot (\mathfrak{b}^{k-1} \cdot M \cap U). \end{aligned}$$

If $k \geq 2$, by the minimality of k, we obtain

$$a \in \mathfrak{a}_0 \cdot \mathfrak{b}'^{k-1} \cdot M + x_r \cdot \mathfrak{a}_0 \cdot \mathfrak{b}^{k-2} \cdot M = \mathfrak{a}_0 \cdot \mathfrak{b}^{k-1} \cdot M$$

which is not possible. Hence $k = 1$ and this implies $a \in \mathfrak{a}_0 \cdot M + x_r \cdot (M \cap U) = \mathfrak{a}_0 \cdot M$ which is also impossible. This contradiction proves the lemma, q.e.d.

An easy consequence of this is

Lemma 1.15. *Let M be a Buchsbaum module over A of positive dimension. Then for every system of parameters $x_1, \ldots, x_d$ of M we have for $\mathfrak{q} := (x_1, \ldots, x_d) \cdot A$:*

$$(\mathfrak{q}^{k+1} \cdot M : x_d) \cap \mathfrak{q} \cdot M = \mathfrak{q}^k \cdot M \qquad \text{for all } k \geq 1.$$

If depth $M \geq 1$ *then*

$$\mathfrak{q}^{k+1} \cdot M : x_d = \mathfrak{q}^k \cdot M \qquad \text{for all } k \geq 1.$$

Proof: Set $\mathfrak{q}' := (x_1, \ldots, x_{d-1}) \cdot A$. Then we have by Lemma 1.14 ($\mathfrak{a} = x_d \cdot A$):

$$\begin{aligned} \mathfrak{q}^{k+1} \cdot M : x_d = (\mathfrak{q}^k \cdot x_d \cdot M + \mathfrak{q}'^{k+1} \cdot M) : x_d = \mathfrak{q}^k \cdot M + (\mathfrak{q}'^{k+1} \cdot M : x_d) \\ = \mathfrak{q}^k \cdot M + \big((x_d \cdot M : \mathfrak{m}) \cap q'^{k+1} \cdot M\big) : x_d = \mathfrak{q}^k \cdot M + x_d \cdot \mathfrak{q}'^k \cdot M : x_d \\ = \mathfrak{q}^k \cdot M + 0 :_M x_d. \end{aligned}$$

If depth $M \geq 1$, $0 :_M x_d = 0 :_M \mathfrak{m} = 0$ and we have obtained the required equality. If depth $M = 0$, we have by Lemma 1.14:

$$\begin{aligned} (\mathfrak{q}^{k+1} \cdot M : x_d) \cap \mathfrak{q} \cdot M = (\mathfrak{q}^k \cdot M + 0 :_M \mathfrak{m}) \cap \mathfrak{q} \cdot M \\ = \mathfrak{q}^k \cdot M + (0 :_M \mathfrak{m} \cap \mathfrak{q} \cdot M) = \mathfrak{q}^k \cdot M, \end{aligned}$$

since $0 :_M \mathfrak{m} \cap \mathfrak{q} \cdot M = 0$ ($\mathfrak{a} = 0$ in Lemma 1.14), q.e.d.

Remark. Our characterization of Buchsbaum local rings A resulted in the notion of weak A-sequences. Several authors have studied further generalizations of a regular sequence (for example, M. Fiorentini [1], C. Huneke [1], N. V. Trung [10], or P. Schenzel [3]). We will show that these generalizations coincide in a Buchsbaum local ring. First we recall some definitions:

Definition 1.16. Let A be a local ring of dimension $n > 0$. Suppose that $x_1, \ldots, x_n$ is a system of parameters of A. Then:

1. $x_1, \ldots, x_n$ is a *weak A-sequence* if

$$(x_1, \ldots, x_i) \cdot A : x_{i+1} = (x_1, \ldots, x_i) \cdot A : \mathfrak{m} \qquad \text{for every } i = 0, \ldots, n-1.$$

2. $x_1, \ldots, x_n$ is said to be a *d-sequence* if for each subset $\{i_1, \ldots, i_j\}$ (possibly $\emptyset$) of $\{1, \ldots, n\}$ and all $k, m \in \{1, \ldots, n\} \setminus \{i_1, \ldots, i_j\}$ we have

$$\big((x_{i_1}, \ldots, x_{i_j}) \cdot A : x_k \cdot x_m\big) = (x_{i_1}, \ldots, x_{i_j}) \cdot A : x_k.$$

3. $x_1, \ldots, x_n$ is a *relative regular sequence* if for every integer $i = 1, \ldots, n$ we have

$$\big((x_1, \ldots, x_{i-1}, x_{i+1}, \ldots, x_n) \cdot A : x_i\big) \cap (x_1, \ldots, x_n) \cdot A = (x_1, \ldots, x_{i-1}, x_{i+1}, \ldots, x_n) \cdot A .$$

4. An element x in an $\mathfrak{m}$-primary ideal $\mathfrak{q}$ is an *absolutely superficial element for* $\mathfrak{q}$ *in* A if

$$(\mathfrak{q}^{k+1} : x) \cap \mathfrak{q} = \mathfrak{q}^k \quad \text{for all integers } k \geq 1,$$

$x_1, \ldots, x_n$ is said to be an *absolutely superficial system of parameters* if the element x_i is an absolutely superficial element for the image of $(x_1, \ldots, x_n) \cdot A$ in $A/(x_1, \ldots, x_{i-1}) \cdot A$ for all integers $i = 1, \ldots, n$.

5. $x_1, \ldots, x_n$ has property (F), if

$$\big((x_1, \ldots, x_{i-1}) \cdot A : x_i\big) \cap (x_1, \ldots, x_n) \cdot A = (x_1, \ldots, x_{i-1}) \cdot A$$

for every integer i with $1 \leq i \leq n$.
(Note for $i = 1$ we obtain $(0 : x_1) \cap (x_1, \ldots, x_n) \cdot A = 0$.)

Proposition 1.17. *A is a Buchsbaum ring if and only if one of the five conditions of Definition* 1.16 *holds for all systems of parameters of A. In this case all five conditions are equivalent.*

Proof: By definition of course A is a Buchsbaum ring if and only if 1. holds for every system of parameters of A.

The equivalence of 1. and 2. follows from Proposition 1.10(ii) and (iii). Now we prove the implications 1. $\Rightarrow$ 4. $\Rightarrow$ 5. $\Rightarrow$ 3. $\Rightarrow$ 1. (for all systems of parameters of A).

1. $\Rightarrow$ 4. follows from Lemma 1.15.

4. $\Rightarrow$ 5.: Let $B := A/(x_1, \ldots, x_{i-1}) \cdot A$ $(1 \leq i \leq n)$. Then

$$\begin{aligned}(0 :_B x_i) \cap (x_i, \ldots, x_n) \cdot B &\subseteq \big((x_i, \ldots, x_n)^{k+1} \cdot B : x_i\big) \cap (x_i, \ldots, x_n) \cdot B \\ &= (x_i, \ldots, x_n)^k \cdot B \quad \text{for all } k \geq 1\end{aligned}$$

by 4. Since $\bigcap_{k \geq 1} (x_i, \ldots, x_n)^k \cdot B = 0$, we get for A:

$$\big((x_1, \ldots, x_{i-1}) \cdot A : x_i\big) \cap (x_1, \ldots, x_n) \cdot A = (x_1, \ldots, x_{i-1}) \cdot A .$$

5. $\Rightarrow$ 3. is trivial.

3. $\Rightarrow$ 1.: For $i = n$ we obtain from 3.:

$$\begin{aligned}(x_1, \ldots, x_{n-1}) \cdot A &= \big((x_1, \ldots, x_{n-1}) \cdot A : x_n\big) \cap (x_1, \ldots, x_n) \cdot A \\ &= (x_1, \ldots, x_{n-1}) \cdot A + x_n \cdot A \cap \big((x_1, \ldots, x_{n-1}) \cdot A : x_n\big) \\ &= (x_1, \ldots, x_{n-1}) \cdot A + x_n \cdot \big((x_1, \ldots, x_{n-1}) \cdot A : x_n^2\big),\end{aligned}$$

hence $(x_1, \ldots, x_{n-1}) \cdot A : x_n^2 = (x_1, \ldots, x_{n-1}) \cdot A : x_n$ and 1. now follows by Proposition 1.10(iii)′, q.e.d.

§ 2. Cohomological characterization of Buchsbaum modules

The aim of this paragraph is to establish "parameter-free" cohomological criteria for the Buchsbaum property of modules over a local ring. The main result describes a cohomological characterization of Buchsbaum modules without the use of systems of parameters.

An essential role in our investigations will be played by the local cohomology modules with support in the maximal ideal $\mathfrak{m}$ of A, i.e. the modules $H^i_{\mathfrak{m}}(M)$ (see Chap. 0, § 1, 3.).

Let A denote again a local ring with maximal ideal $\mathfrak{m}$ and residue field $k := A/\mathfrak{m}$. We know that a Noetherian A-module M is a Cohen-Macaulay module if and only if $H^i_{\mathfrak{m}}(M) = O$ for all i with $0 \leq i < \dim M$ (see Chap. 0, § 1, 3.). Related to this is our next result which gives a first indication that local cohomology is an appropriate tool for studying Buchsbaum modules.

Proposition 2.1. *Let M be a Noetherian A-module of positive dimension. The following conditions are equivalent:*

(i) *There is a system of parameters for M contained in $\mathfrak{m}^2$ which is a weak M-sequence.*

(ii) *Every system of parameters for M in $\mathfrak{m}^2$ is a weak M-sequence.*

(iii) $\mathfrak{m} \cdot H^i_{\mathfrak{m}}(M) = O$ *for all i with* $0 \leq i < \dim M$.

Furthermore, if one of these conditions is fulfilled, we have

(iv) $l(M/\mathfrak{q} \cdot M) - e_0(\mathfrak{q}, M)$ *is independent of $\mathfrak{q}$ for all parameter ideals* $\mathfrak{q} \subseteq \mathfrak{m}^2$.

Proof: The implication (ii) $\Rightarrow$ (i) is trivial.

First we prove (i) $\Rightarrow$ (iii). Let $d := \dim M \geq 1$.

Let $x_1, \ldots, x_d$ be a weak M-sequence in $\mathfrak{m}^2$. Since $x_1 \in \mathfrak{m}^2$ we have $O :_M \mathfrak{m}^2 \subseteq O :_M x_1 = O :_M \mathfrak{m} \subseteq O :_M \mathfrak{m}^2$, i.e. $O :_M \mathfrak{m} = O :_M \mathfrak{m}^2$. This implies $O :_M \mathfrak{m} = O :_M \mathfrak{m}^n$ for all $n \geq 1$ and consequently $H^0_{\mathfrak{m}}(M) = \bigcup_{n \geq 1} O :_M \mathfrak{m}^n = O :_M \mathfrak{m}$, i.e. $\mathfrak{m} \cdot H^0_{\mathfrak{m}}(M) = O$.

Now we use induction on d. For $d = 1$ the proof has been given. Suppose $d \geq 2$. We have two exact sequences

$$0 \to O :_M x_1 \to M \xrightarrow{p} x_1 \cdot M \to 0 \tag{E$_1$}$$

and

$$0 \to x_1 \cdot M \xrightarrow{i} M \to M' \to 0 \quad \text{with } M' := M/x_1 \cdot M. \tag{E$_2$}$$

Since $\operatorname{Supp} O :_M x_1 = \operatorname{Supp} H^0_{\mathfrak{m}}(M) \subseteq \{\mathfrak{m}\}$, $H^j_{\mathfrak{m}}(O :_M x_1) = O$ for all $j \geq 1$. Therefore $H^j_{\mathfrak{m}}(p)$ is an isomorphism for all $j \geq 1$ and we have

$$\operatorname{Ker} H^j_{\mathfrak{m}}(i) \cong \operatorname{Ker} \left(H^j_{\mathfrak{m}}(p) \cdot H^j_{\mathfrak{m}}(i)\right) = \operatorname{Ker} H^j_{\mathfrak{m}}(p \cdot i) \cong O :_{H^j_{\mathfrak{m}}(M)} x_1 .$$

The long exact cohomology sequence obtained from (E$_2$) yields epimorphisms

$$H^{j-1}_{\mathfrak{m}}(M') \to \operatorname{Ker} H^j_{\mathfrak{m}}(i) \cong O :_{H^j_{\mathfrak{m}}(M)} x_1 .$$

By the induction hypothesis ($\dim M' = d - 1$ and $x_2, \ldots, x_d$ form a weak M'-sequence in $\mathfrak{m}^2$) $\mathfrak{m} \cdot H^{j-1}_{\mathfrak{m}}(M') = O$ for all j with $1 \leq j \leq d - 1$. Thus $\mathfrak{m} \cdot (O :_{H^j_{\mathfrak{m}}(M)} x_1) = O$ for all $j = 1, \ldots, d - 1$. Since $x_1 \in \mathfrak{m}^2$ we again obtain

$$H^j_{\mathfrak{m}}(M) = H^0_{\mathfrak{m}}\left(H^j_{\mathfrak{m}}(M)\right) = \bigcup_{n \geq 1} O :_{H^j_{\mathfrak{m}}(M)} \mathfrak{m}^n = O :_{H^j_{\mathfrak{m}}(M)} \mathfrak{m},$$

i.e. $\mathfrak{m} \cdot H^j_{\mathfrak{m}}(M) = O$ for all $j = 1, \ldots, d - 1$. Since the case $j = 0$ has already been settled, (iii) is proven.

To prove the implication (iii) $\Rightarrow$ (ii) we need the following

Lemma 2.2. *Let M be a Noetherian A-module of positive dimension such that $H^i_{\mathfrak{m}}(M)$ are modules of finite length for all $i < \dim M$. Then for every part of a system of parameters*

$x_1, \ldots, x_r$ with respect to M the submodule $(x_1, \ldots, x_r) \cdot M$ of M is unmixed up to $\mathfrak{m}$-primary components, i.e., if $\mathfrak{p} \in \operatorname{Ass} M/(x_1, \ldots, x_r) \cdot M$, then either $\mathfrak{p} = \mathfrak{m}$ or $\dim A/\mathfrak{p} = \dim M - r$.

Proof: Let $d := \dim M$. We use induction on r. If $r = 0$ we have to show that $(\operatorname{Ass} M)_i := \{\mathfrak{p} \in \operatorname{Ass} M \mid \dim A/\mathfrak{p} = i\} = \emptyset$ for all i, $0 < i < d$. Since $l_{\hat{A}}(H^i_{\hat{\mathfrak{m}}}(\hat{M})) = l_{\hat{A}}(\widehat{H^i_{\mathfrak{m}}(M)}) = l_A(H^i_{\mathfrak{m}}(M)) < \infty$ for all $i < d$ we can assume (using Proposition 3 and Lemma 8(i) of the Appendix) that A is complete. Then Corollary 0.3.7 yields

$$(\operatorname{Ass} M)_i = \big(\operatorname{Ass} \operatorname{Hom}(H^i_{\mathfrak{m}}(M), E)\big)_i \quad \text{for all } i < d.$$

Since $H^i_{\mathfrak{m}}(M)$ are modules of finite length for each $i < d$, the same holds for the module $\operatorname{Hom}(H^i_{\mathfrak{m}}(M), E)$, i.e., $\operatorname{Ass} \operatorname{Hom}(H^i_{\mathfrak{m}}(M), E) \subseteq \{\mathfrak{m}\}$ for all $i < d$ and the lemma is therefore proven for this case.

Suppose now $r \geq 1$. Since O is unmixed up to $\mathfrak{m}$-primary components, $\operatorname{Supp} O :_M x_1 \subseteq \{\mathfrak{m}\}$, i.e., $H^j_{\mathfrak{m}}(O :_M x_1) = O$ for all $j \geq 1$. Therefore the two exact sequences (E_1) and (E_2) of the first part of the proof of Proposition 2.1 yield for each $j \geq 0$ an exact sequence

$$H^j_{\mathfrak{m}}(M) \to H^j_{\mathfrak{m}}(M/x_1 \cdot M) \to H^{j+1}_{\mathfrak{m}}(M).$$

This shows that $H^j_{\mathfrak{m}}(M/x_1 \cdot M)$ is of finite length for all $j < d - 1 = \dim M/x_1 \cdot M$. By our induction hypothesis the lemma is therefore proven, q.e.d.

Continuation of the proof of Proposition 2.1. Since (iii) is satisfied, $H^i_{\mathfrak{m}}(M)$ is a module of finite length for all $i = 0, \ldots, d - 1$. Thus we can apply Lemma 2.2.

Let $x_1, \ldots, x_d$ denote an arbitrary system of parameters of M in $\mathfrak{m}^2$. Then $\mathfrak{m} \cdot (O :_M x_1) \subseteq \mathfrak{m} \cdot H^0_{\mathfrak{m}}(M) = O$, i.e. $O :_M x_1 = O :_M \mathfrak{m}$ and x_1 is a weak M-sequence.

Now we use induction on d. If $d = 1$ we are already done.

Therefore let $d \geq 2$ and $1 \leq i < d$. We need to show

$$(x_1, \ldots, x_i) \cdot M : x_{i+1} = (x_1, \ldots, x_i) \cdot M : \mathfrak{m}.$$

We proceed in two steps:

(a) $x_1 = x^n$ with $x \in \mathfrak{m}$, $n \geq 2$.

We have $O :_M \mathfrak{m} \subseteq O :_M x \subseteq O :_M x^2 \subseteq \ldots \subseteq H^0_{\mathfrak{m}}(M) = O :_M \mathfrak{m}$ (since $\mathfrak{m} \cdot H^0_{\mathfrak{m}}(M) = O$) and this results in

$$H^0_{\mathfrak{m}}(M) = O :_M \mathfrak{m} = O :_M x^t \quad \text{for all } t \geq 1.$$

Now let $M' := M/H^0_{\mathfrak{m}}(M)$, $f_n \colon M' \to M$ defined by $f_n(\overline{m}) := x^n \cdot m$ for all $\overline{m} \in M'$ and all $n \geq 1$.

If $\pi \colon M \to M'$ denotes the canonical projection,

$$(f_n \cdot \pi)(m) = x^n \cdot m \quad \text{for all } n \geq 1.$$

Since $H^i_{\mathfrak{m}}(H^0_{\mathfrak{m}}(M)) = O$ for all $i \geq 1$, $H^i_{\mathfrak{m}}(\pi)$ is an isomorphism for all $i \geq 1$. Furthermore, $x^n \cdot H^i_{\mathfrak{m}}(M) = O$ for all $n \geq 1$ and $i < d$. Hence $H^i_{\mathfrak{m}}(f_n) = O$ for all $n \geq 1$ and all $i < d$.

Thus the commutative diagram

$$\begin{array}{ccccccccc}
0 & \to & M' & \xrightarrow{f_n} & M & \xrightarrow{g_n} & M/x^n \cdot M & \to & 0 \\
 & & \downarrow{\scriptstyle x^{n-1}} & & \| & & \downarrow{\scriptstyle h_n} & & \\
0 & \to & M' & \xrightarrow{f_1} & M & \xrightarrow{g_1} & M/x \cdot M & \to & 0
\end{array}$$

(g_i denotes for each $i \geq 1$ the canonical projection and h_i is defined by the inclusion $x^i \cdot M \subseteq x \cdot M$) gives us for all i, $1 \leq i \leq d-1$, commutative diagrams

$$\begin{array}{ccccccc} O \to H^{i-1}_{\mathfrak{m}}(M) & \xrightarrow{H^{i-1}_{\mathfrak{m}}(g_n)} & H^{i-1}_{\mathfrak{m}}(M/x^n \cdot M) & \to & H^i_{\mathfrak{m}}(M') & \to O \\ \| & & \downarrow{\scriptstyle H^{i-1}_{\mathfrak{m}}(h_n)} & & \downarrow{\scriptstyle x^{n-1}} & \\ 0 \to H^{i-1}_{\mathfrak{m}}(M) & \xrightarrow{H^{i-1}_{\mathfrak{m}}(g_1)} & H^{i-1}_{\mathfrak{m}}(M/x \cdot M) & \to & H^i_{\mathfrak{m}}(M') & \to O. \end{array}$$

Since $x^j \cdot H^i_{\mathfrak{m}}(M') = x^j \cdot H^i_{\mathfrak{m}}(M) = O$ for all $i \leq d-1$ and all $j \geq 1$, we have for all $n \geq 2$ and all $i \leq d-2$ uniquely determined homomorphisms

$$g_{i,n}: H^i_{\mathfrak{m}}(M/x^n \cdot M) \to H^i_{\mathfrak{m}}(M) \quad \text{with} \quad H^i_{\mathfrak{m}}(g_1) \cdot g_{i,n} = H^i_{\mathfrak{m}}(h_n).$$

It is not difficult to verify that with $g_{i-1,n}$ the top row of our last commutative diagram becomes a "split sequence", i.e.

$$H^i_{\mathfrak{m}}(M/x^n \cdot M) \simeq H^i_{\mathfrak{m}}(M) \oplus H^{i+1}_{\mathfrak{m}}(M') \simeq H^i_{\mathfrak{m}}(M) \oplus H^{i+1}_{\mathfrak{m}}(M)$$

for all $i \leq d-2$ and all $n \geq 2$. This means

$$\mathfrak{m} \cdot H^i_{\mathfrak{m}}(M/x^n \cdot M) = 0 \quad \text{for all } i \leq d-2 \quad (n \geq 2).$$

Since $\dim M/x^n \cdot M = d-1$, the induction hypothesis proves our statement for this case.

(b) x_1 is an arbitrary element (in $\mathfrak{m}^2$).

By an easy induction argument (induction on i) we may assume that $x_1, \ldots, x_i$ is already a weak M-sequence, in particular,

$$(x_1, \ldots, x_{i-1}) \cdot M : x_i = (x_1, \ldots, x_{i-1}) \cdot M : \mathfrak{m} = (x_1, \ldots, x_{i-1}) \cdot M : \langle \mathfrak{m} \rangle,$$

where the last equation follows since $x_i \in \mathfrak{m}^2$.

Now, let $y \in \mathfrak{m}$, $z \in \mathfrak{m}^2$ be elements such that $x_1, \ldots, x_{i-1}, y, z$ is again a part of a system of parameters of M. Set

$$N := (x_1, \ldots, x_{i-1}) \cdot M.$$

Since $N : \mathfrak{m} \subseteq N : y \subseteq N : \langle \mathfrak{m} \rangle$ and $N : \mathfrak{m} \subseteq N : z \subseteq N : \langle \mathfrak{m} \rangle$ (Lemma 2.2) we get

(1) $\quad N : \mathfrak{m} = N : y = N : z = N : \langle \mathfrak{m} \rangle.$

From this we obtain

$$\begin{aligned} (y \cdot M + N) : \langle \mathfrak{m} \rangle &= (y \cdot M + N : y) : \langle \mathfrak{m} \rangle = \big((y^2 \cdot M + N) : y\big) : \langle \mathfrak{m} \rangle \\ &= \big((y^2 \cdot M + N) : \langle \mathfrak{m} \rangle\big) : y = \big((y^2 \cdot M + N) : \mathfrak{m}\big) : y \quad \text{(by step (a))} \\ &= \big((y^2 \cdot M + N) : y\big) : \mathfrak{m} = (y \cdot M + N : y) : \mathfrak{m} = (y \cdot M + N : \mathfrak{m}) : \mathfrak{m}. \end{aligned}$$

Furthermore, we have ($z \in \mathfrak{m}^2$)

$$(y \cdot M + N : \mathfrak{m}) : \mathfrak{m} \subseteq (y \cdot M + N) : \mathfrak{m}^2 \subseteq (y \cdot M + N) : z \subseteq (y \cdot M + N) : \langle \mathfrak{m} \rangle.$$

From all this we obtain

(2) $\quad (y \cdot M + N) : z = (y \cdot M + N) : \langle \mathfrak{m} \rangle = (y \cdot M + N : \mathfrak{m}) : \mathfrak{m}.$

Now, let $m \in (x_1, \ldots, x_i) \cdot M : x_{i+1} = \big((x_1, \ldots, x_{i-1}) \cdot M : \mathfrak{m} + x_i \cdot M\big) : \mathfrak{m}$ (using equation (2) with $y = x_i$, $z = x_{i+1}$). Choose an $\big(M/(x_1, \ldots, x_i) \cdot M\big)$-basis $y_1, \ldots, y_t$ of the maximal ideal $\mathfrak{m}$ of A.

Let $y \in \{y_1, \ldots, y_t\}$. Then

$$y \cdot m = u + x_i m' \quad \text{with} \quad u \in (x_1, \ldots, x_{i-1}) \cdot M : \mathfrak{m}, \quad m' \in M$$

and

$$y^2 \cdot m = y \cdot u + x_i \cdot y \cdot m', \quad y \cdot u \in (x_1, \ldots, x_{i-1}) \cdot M.$$

Hence, by step (a) and equations (1) and (2) with $z = x_i$ we get

$$\begin{aligned} m' &\in \big((y^2, x_1, \ldots, x_{i-1}) \cdot M : x_i\big) : y = \big((y^2, x_1, \ldots, x_{i-1}) \cdot M : \mathfrak{m}\big) : y \\ &= \big((y^2, x_1, \ldots, x_{i-1}) \cdot M : y\big) : \mathfrak{m} = \big(y \cdot M + (x_1, \ldots, x_{i-1}) \cdot M : \mathfrak{m}\big) : \mathfrak{m} \\ &= (y, x_1, \ldots, x_{i-1}) \cdot M : x_i . \end{aligned}$$

Therefore

$$\begin{aligned} u = y \cdot m - x_i \cdot m' &\in (y, x_1, \ldots, x_{i-1}) \cdot M \cap \big((x_1, \ldots, x_{i-1}) \cdot M : \mathfrak{m}\big) \\ &= (x_1, \ldots, x_{i-1}) \cdot M + y \cdot M \cap \big((x_1, \ldots, x_{i-1}) \cdot M : \mathfrak{m}\big) \\ &= (x_1, \ldots, x_{i-1}) \cdot M + y \cdot \big((x_1, \ldots, x_{i-1}) \cdot M : y \cdot \mathfrak{m}\big) \\ &= (x_1, \ldots, x_{i-1}) \cdot M + y \cdot \big((x_1, \ldots, x_{i-1}) \cdot M : \mathfrak{m}\big) \\ &= (x_1, \ldots, x_{i-1}) \cdot M . \end{aligned}$$

Thus $y \cdot m = u + x_i \cdot m' \in (x_1, \ldots, x_i) \cdot M$, and consequently

$$m \in (x_1, \ldots, x_i) \cdot M : y \quad \text{for all } y \in \{y_1, \ldots, y_t\}.$$

But this implies $m \in (x_1, \ldots, x_i) \cdot M : \mathfrak{m}$ and $x_1, \ldots, x_{i+1}$ is therefore a weak M-sequence. This proves (ii).

Finally, the implication (ii) $\Rightarrow$ (iv) is obtained using similar arguments (with some obvious modifications) as was in the corresponding part of the proof of Theorem **1.12**, q.e.d.

Remark 2.3. Condition (iv) of Proposition 2.1 does not imply (i) or (ii) or (iii), in general.

For example, let k be any field and X, Y indeterminates. Set $R := k[\![X, Y]\!]$,

$$A := k[\![X, Y]\!]/X \cdot R \cap (X^3, Y) \cdot R, \quad \mathfrak{m} = (X, Y) \cdot A.$$

Then $H^0_\mathfrak{m}(A) = X \cdot R/X \cdot R \cap (X^3, Y) \cdot R$ and therefore $\mathfrak{m} \cdot H^0_\mathfrak{m}(A) \neq 0$. But for all parameters z of A in $\mathfrak{m}^2$ we have (where z is the picture of $Z \in R$ in A)

$$\begin{aligned} l(A/z \cdot A) - e_0(z \cdot A, A) = l(0 :_A z) &= l\big((X \cdot R \cap (X^3, Y) \cdot R) : Z/X \cdot R \cap (X^3, Y) \cdot R\big) \\ &= l\big(X \cdot R/X \cdot R \cap (X^3, Y) \cdot R\big) \end{aligned}$$

and this last number does not depend on z.

Also, we note that N. Suzuki [5] and S. Goto [10] have studied the class of local rings for which condition (i) of Proposition 2.1 is satisfied. These rings are called *quasi-Buchsbaum rings*. S. Goto [10] has established the ubiquity of quasi-Buchsbaum rings which are not Buchsbaum rings.

Corollary 2.4. *If M is a Buchsbaum module then $\mathfrak{m} \cdot H^i_{\mathfrak{m}}(M) = 0$ for all $i \neq \dim M$. In particular, the local cohomology modules are modules of finite length in this case.*

Unfortunately, the following example will show that the converse of this statement is false. Nevertheless Corollary 2.4 gives a first necessary condition for Buchsbaum modules independent of systems of parameters.

Example 2.5. Let k be a field and X_1, X_2, X_3, X_4 indeterminates. Take

$$A := k[\![X_1, \ldots, X_4]\!]/(X_1, X_2) \cap (X_3, X_4) \cap (X_1^2, X_2, X_3, X_4^2).$$

Then $U(0) = X_1 \cdot X_4 \cdot A$ and therefore $\mathfrak{m}_A \cdot H^0_{\mathfrak{m}_A}(A) = \mathfrak{m}_A \cdot U(0) = 0$. The exact sequence

$$0 \to H^0_{\mathfrak{m}_A}(A) \to A \to B \to 0$$

$$\text{with } B := A/U(0) = k[\![X_1, \ldots, X_4]\!]/(X_1, X_2) \cap (X_3, X_4)$$

results for all $i \geq 1$ in isomorphisms $\left(H^i_{\mathfrak{m}_A}(H^0_{\mathfrak{m}_A}(A)\right) = 0$ for $i \geq 1)$:

$$H^i_{\mathfrak{m}_A}(A) \cong H^i_{\mathfrak{m}_A}(B).$$

Therefore (see Example 6 of the Introduction or Proposition 2.25 which show that B is a Buchsbaum ring and hence a Buchsbaum module over A by Lemma 1.6) $\mathfrak{m}_A \cdot H^1_{\mathfrak{m}_A}(A) = 0$. Notice that $\dim A = 2$.

We next show that A is not a Buchsbaum ring. Take $z := X_1 + X_4 \bmod A$. Clearly, $\dim A/z \cdot A = 1$ and we have

$$U(z \cdot A) = (X_1, X_4) \cdot A.$$

Therefore $X_1 \cdot U(z \cdot A) \not\subseteq z \cdot A$, i.e. $\mathfrak{m}_A \cdot U(z \cdot A) \not\subseteq z \cdot A$. Thus by Proposition 1.10(iv) we find that A is not a Buchsbaum ring.

It is possible to construct similar examples which will have depth zero. In Chapter V (see § 5, 3.) we will give an example with depth greater then zero (Example V.5.4).

The usefulness of local cohomology for our purposes will next be demonstrated by the following result which enables us to give a "parameter-free" expression for the invariant $I(M)$ of a Buchsbaum module M.

Proposition 2.6. *For any Buchsbaum module M with $d := \dim M$ we have*

$$I(M) = \sum_{i=0}^{d-1} \binom{d-1}{i} \cdot l\big(H^i_{\mathfrak{m}}(M)\big).$$

Proof: We already know from the proof of Theorem 1.12 that

$$\begin{aligned} I(M) &= l\big((x_1, \ldots, x_{d-1}) \cdot M : x_d/(x_1, \ldots, x_{d-1}) \cdot M\big) \\ &= l\big((x_1, \ldots, x_{d-1}) \cdot M : \mathfrak{m}/(x_1, \ldots, x_{d-1}) \cdot M\big), \end{aligned}$$

where $x_1, \ldots, x_d$ is some system of parameters of M. Let us write $x_1' := x_1^2$, $M' := M/x_1' \cdot M$.

We use induction on d. If $d = 1$, $I(M) = l(0 :_M \mathfrak{m}) = l\big(H^0_{\mathfrak{m}}(M)\big)$, since $\mathfrak{m} \cdot H^0_{\mathfrak{m}}(M) = 0$ by Corollary 2.4 which finishes the proof. If $d \geq 2$ we have already seen in the proof of

Proposition 2.1 (implication (iii) $\Rightarrow$ (ii), step (a)) that

$$H^i_{\mathfrak{m}}(M') \simeq H^i_{\mathfrak{m}}(M) \oplus H^{i+1}_{\mathfrak{m}}(M)$$

and we obtain with the induction hypothesis ($\dim M' = d - 1$):

$$\begin{aligned} I(M) &= l\big((x_1', x_2, \ldots, x_{d-1}) \cdot M : \mathfrak{m}/(x_1', x_2, \ldots, x_{d-1}) \cdot M\big) \\ &= l\big((x_2, \ldots, x_{d-1}) \cdot M' : \mathfrak{m}/(x_2, \ldots, x_{d-1}) \cdot M'\big) = I(M') \\ &= \sum_{i=0}^{d-2} \binom{d-2}{i} \cdot l\big(H^i_{\mathfrak{m}}(M')\big) = \sum_{i=0}^{d-2} \binom{d-2}{i} \cdot \Big(l\big(H^i_{\mathfrak{m}}(M)\big) + l\big(H^{i+1}_{\mathfrak{m}}(M)\big)\Big) \\ &= \sum_{i=0}^{d-1} \left(\binom{d-2}{i-1} + \binom{d-2}{i}\right) \cdot l\big(H^i_{\mathfrak{m}}(M)\big) \\ &= \sum_{i=0}^{d-1} \binom{d-1}{i} \cdot l\big(H^i_{\mathfrak{m}}(M)\big), \end{aligned}$$

q.e.d.

Now we want to state a result which shows first that for a Buchsbaum module M the Hilbert-Samuel function $P_{\mathfrak{q},M}(n)$ and the Hilbert-Samuel polynomial $p_{\mathfrak{q},M}(n)$ coincide for each parameter ideal $\mathfrak{q}$ of M and all $n \geq 0$.

Secondly, the Hilbert-Samuel coefficients $e_i(\mathfrak{q}, M)$ are shown to be independent of $\mathfrak{q}$ for all $i \geq 1$. We obtain an expression for $I(M)$ using these $e_i(\mathfrak{q}, M)$.

Finally we find for each $t \geq 0$ non-negative integers $I_t(M)$ such that

$$l(M/\mathfrak{q}^{t+1} \cdot M) - \binom{t+d}{d} \cdot e_0(\mathfrak{q}, M) = I_t(M)$$

for all $t \geq 0$ and all parameter ideals $\mathfrak{q}$ of M, where $d := \dim M > 0$.

Proposition 2.7. *Let M be a Buchsbaum module with $d := \dim M \geq 0$. Then for every parameter ideal $\mathfrak{q}$ of M*

(i) $$l(M/\mathfrak{q}^{t+1} \cdot M) = \sum_{i=0}^{d} \binom{t+d-i}{d-i} \cdot e_i(\mathfrak{q}, M) \quad \text{for all } t \geq 0.$$

(ii) $$e_i(\mathfrak{q}, M) = \sum_{j=0}^{d-i} \binom{d-i-1}{j-1} \cdot l\big(H^j_{\mathfrak{m}}(M)\big) \quad \text{for all } i = 1, \ldots, d,$$

$$\text{where } \binom{p}{-1} := \begin{cases} 0 & \text{for } p \neq -1, \\ 1 & \text{for } p = -1. \end{cases}$$

(iii) $$I(M) = \sum_{i=1}^{d} e_i(\mathfrak{q}, M).$$

Proof: (iii) is a consequence of (i) if we set $t = 0$.

We prove (i) and (ii) by induction on d. For $d = 0$ there is nothing to prove ($\mathfrak{q} = 0$). Assume $d \geq 1$. Let $\mathfrak{q} = (x_1, \ldots, x_d) \cdot A$ be any parameter ideal of M and let $\mathfrak{q}'$

$:= (x_1, \ldots, x_{d-1}) \cdot A$, $M' := M/x_d \cdot M$. Then we have an exact sequence for each $t \geq 1$:

$$0 \to \mathfrak{q}^{t+1} \cdot M : x_d/\mathfrak{q}^t \cdot M \to M/\mathfrak{q}^t \cdot M \xrightarrow{f} M/\mathfrak{q}^{t+1} \cdot M \to M'/\mathfrak{q}'^{t+1} \cdot M' \to 0$$

where f is obtained from multiplication by x_d.

As in the proof of Lemma 1.15, we obtain

$$\mathfrak{q}^{t+1} \cdot M : x_d = \mathfrak{q}^t \cdot M + 0 :_M x_d = \mathfrak{q}^t \cdot M + 0 :_M \mathfrak{m}$$

and therefore

$$\mathfrak{q}^{t+1} \cdot M : x_d/\mathfrak{q}^t \cdot M \cong 0 :_M \mathfrak{m}/0 :_M \mathfrak{m} \cap \mathfrak{q}^t \cdot M = 0 :_M \mathfrak{m} \cong H^0_{\mathfrak{m}}(M)$$

since $0 :_M \mathfrak{m} \cap \mathfrak{q}^t \cdot M = 0$ by Lemma 1.14.

Also

$$\begin{aligned} l(\mathfrak{q}^t \cdot M/\mathfrak{q}^{t+1} \cdot M) &= l(M/\mathfrak{q}^{t+1} \cdot M) - l(M/\mathfrak{q}^t \cdot M) \\ &= -l\big(H^0_{\mathfrak{m}}(M)\big) + l(M'/\mathfrak{q}'^{t+1} \cdot M'). \end{aligned}$$

By our induction hypothesis we have

$$l(M'/\mathfrak{q}'^{t+1} \cdot M') = \sum_{i=0}^{d-1} \binom{t+d-i-1}{d-i-1} \cdot e_i(\mathfrak{q}', M')$$

and

$$e_i(\mathfrak{q}', M') = \sum_{j=0}^{d-i-1} \binom{d-i-2}{j-1} \cdot l\big(H^j_{\mathfrak{m}}(M')\big) \quad \text{for all } i = 1, \ldots, d-1.$$

From the exact sequence

$$0 \to M/0 :_M \mathfrak{m} \xrightarrow{g} M \to M' \to 0$$

where g is induced by multiplying the cosets by x_d, we find for all $j \leq d-2$ short exact sequences (see also the proof of Proposition 2.1(iii) $\Rightarrow$ (ii), step (a)):

$$0 \to H^j_{\mathfrak{m}}(M) \to H^j_{\mathfrak{m}}(M') \to H^{j+1}_{\mathfrak{m}}(M) \to 0.$$

Therefore for $i = 1, \ldots, d-1$

$$\begin{aligned} e_i(\mathfrak{q}', M') &= \sum_{j=0}^{d-i-1} \binom{d-i-2}{j-1} \cdot l\big(H^j_{\mathfrak{m}}(M)\big) + \sum_{j=0}^{d-i-1} \binom{d-i-2}{j-1} \cdot l\big(H^{j+1}_{\mathfrak{m}}(M)\big) \\ &= \sum_{j=0}^{d-i-1} \binom{d-i-2}{j-1} \cdot l\big(H^j_{\mathfrak{m}}(M)\big) + \sum_{j=1}^{d-i} \binom{d-i-2}{j-2} \cdot l\big(H^j_{\mathfrak{m}}(M)\big) \\ &= \sum_{j=0}^{d-i} \left(\binom{d-i-2}{j-1} + \binom{d-i-2}{j-2}\right) \cdot l\big(H^j_{\mathfrak{m}}(M)\big) \\ &= \begin{cases} \displaystyle\sum_{j=0}^{d-i} \binom{d-i-1}{j-1} \cdot l\big(H^j_{\mathfrak{m}}(M)\big) & \text{for } i \leq d-2, \\ l\big(H^0_{\mathfrak{m}}(M)\big) + l\big(H^1_{\mathfrak{m}}(M)\big) & \text{for } i = d-1. \end{cases} \end{aligned}$$

Now for all $t \geq 0$

$$
\begin{aligned}
l(M/\mathfrak{q}^{t+1} \cdot M) &= l(M/\mathfrak{q} \cdot M) + \sum_{i=1}^{t} l(\mathfrak{q}^i \cdot M/\mathfrak{q}^{i+1} \cdot M) \\
&= l(M/\mathfrak{q} \cdot M) + \sum_{i=1}^{t} l(M'/\mathfrak{q}'^{i+1} \cdot M') - t \cdot l(H^0_\mathfrak{m}(M)) \\
&= l(M/\mathfrak{q} \cdot M) + \sum_{i=1}^{t} \sum_{j=0}^{d-1} \binom{d+i-j-1}{d-j-1} \cdot e_j(\mathfrak{q}', M') - t \cdot l\big(H^0_\mathfrak{m}(M)\big) \\
&= l(M/\mathfrak{q} \cdot M) + \sum_{j=0}^{d-1} \left(\binom{t+d-j}{d-j} - 1\right) \cdot e_j(\mathfrak{q}', M') - t \cdot l\big(H^0_\mathfrak{m}(M)\big) \\
&= l(M/\mathfrak{q} \cdot M) - \sum_{j=0}^{d-1} e_j(\mathfrak{q}', M') + l\big(H^0_\mathfrak{m}(M)\big) + \sum_{j=0}^{d-2} \binom{t+d-j}{d-j} \cdot e_j(\mathfrak{q}', M') \\
&\quad + \binom{t+1}{1} \cdot \big(e_{d-1}(\mathfrak{q}', M') - l\big(H^0_\mathfrak{m}(M)\big)\big).
\end{aligned}
$$

For sufficiently large t this polynomial in t coincides with $p_{\mathfrak{q},M}(t)$ and comparing coefficients we obtain

$$
\begin{aligned}
&e_j(\mathfrak{q}, M) = e_j(\mathfrak{q}', M') \quad \text{for } j = 0, \ldots, d-2, \\
&e_{d-1}(\mathfrak{q}, M) = e_{d-1}(\mathfrak{q}', M') - l\big(H^0_\mathfrak{m}(M)\big), \\
&e_d(\mathfrak{q}, M) = l(M/\mathfrak{q} \cdot M) - \sum_{j=0}^{d-1} e_j(\mathfrak{q}', M') + l\big(H^0_\mathfrak{m}(M)\big).
\end{aligned}
$$

But this proves (i) since the above equation is true for all $t \geq 0$. From our expressions for $e_i(\mathfrak{q}', M')$ and

$$l(M/\mathfrak{q} \cdot M) = l(M'/\mathfrak{q}' \cdot M') = \sum_{j=0}^{d-1} e_j(\mathfrak{q}', M')$$

(see (i) applied to M' and $\mathfrak{q}'$, $t = 0$) we also obtain (ii), q.e.d.

Corollary 2.8. *Let M be a Buchsbaum module of dimension $d > 0$. Then for every $t \geq 0$ there is a natural number $I_t(M)$ such that for every parameter ideal $\mathfrak{q}$ of M*

$$l(M/\mathfrak{q}^{t+1} \cdot M) - \binom{t+d}{d} \cdot e_0(\mathfrak{q}, M) = I_t(M).$$

This follows immediately from Proposition 2.7(i) and (ii).

Next we will prove a first sufficient "parameter-free" criterion for the Buchsbaum property which allows us to find many examples for Buchsbaum modules and which is also a necessary condition if A is a regular local ring. First we need the following

Lemma 2.9. *Let $0 \to M' \to M \xrightarrow{f} M'' \to 0$ be an exact sequence of A-modules with $\mathfrak{m} \cdot M'' = 0$. Then the sequences*

$$0 \to \operatorname{Ext}^i_A(k, M') \to \operatorname{Ext}^i_A(k, M) \xrightarrow{f_i} \operatorname{Ext}^i_A(k, M'') \to 0 \qquad (k = A/\mathfrak{m})$$

with $f_i := \operatorname{Ext}^i_A(k, f)$ *are exact for each* $i \geq 0$ *if and only if we have exactness for* $i = 0$, *i.e. the sequence*

$$0 \to \operatorname{Hom}_A(k, M') \to \operatorname{Hom}_A(k, M) \to M'' \to 0$$

is exact.

Proof: The only if part is trivial.

Assume now that $0 \to \operatorname{Hom}_A(k, M') \to \operatorname{Hom}_A(k, M) \xrightarrow{f_0} M'' \to 0$ is exact (note: $\operatorname{Hom}_A(k, M'') \cong 0 :_{M''} \mathfrak{m} = M''$ since $\mathfrak{m} \cdot M'' = 0$). We apply $\operatorname{Hom}_A(k, \ )$ to $M \xrightarrow{f} M'' \to 0$ and get a commutative diagram

$$\begin{array}{ccc} \operatorname{Hom}_A(k, M) & \xrightarrow{f_0} & M'' \to 0 \\ \cap\!| & & \| \\ M & \longrightarrow & M'' \to 0. \end{array}$$

Next, by using $\operatorname{Ext}^i_A(k, \ )$, we obtain commutative diagrams

$$\begin{array}{ccc} \operatorname{Ext}^i_A(k, \operatorname{Hom}_A(k, M)) & \longrightarrow & \operatorname{Ext}^i_A(k, M'') \to 0 \\ \downarrow & & \| \\ \operatorname{Ext}^i_A(k, M) & \xrightarrow{f_i} & \operatorname{Ext}^i_A(k, M'') \to 0. \end{array}$$

Since M'' is a direct summand of $\operatorname{Hom}_A(k, M)$ (by virtue of f_0), the top row is exact. Therefore the bottom row is also exact, q.e.d.

Theorem 2.10. *Let* M *be a Noetherian* A*-module with* $d := \dim M \geq 1$. *If the canonical maps* (see Chap. 0, § 1, 3.)

$$\varphi^i_M : \operatorname{Ext}^i_A(k, M) \to H^i_{\mathfrak{m}}(M)$$

are surjective for all $i \neq d$ *then* M *is a Buchsbaum module.*

Proof: By the surjectivity of φ^i_M the $H^i_{\mathfrak{m}}(M)$ are modules of finite length and we conclude by Lemma 2.2 that for every part of a system of parameters $x_1, \ldots, x_r$ of M with $r < d$ the submodule $(x_1, \ldots, x_r) \cdot M$ of M is unmixed up to $\mathfrak{m}$-primary components.

Now, φ^0_M is the inclusion $\operatorname{Hom}_A(k, M) \cong 0 :_M \mathfrak{m} \subseteq 0 :_M \langle \mathfrak{m} \rangle = H^0_{\mathfrak{m}}(M)$ and therefore we have for every parameter x (i.e. $\dim M/x \cdot M = d - 1$)

$$\mathfrak{m} \cdot (0 :_M x) \subseteq \mathfrak{m} \cdot (0 :_M \langle \mathfrak{m} \rangle) = \mathfrak{m} \cdot H^0_{\mathfrak{m}}(M) \cong \mathfrak{m} \cdot \operatorname{Hom}_A(k, M) = 0,$$

i.e., x is a weak M-sequence.

We use induction on d. If $d = 1$, we are already done.

Assume $d \geq 2$ and let $x_1, \ldots, x_d$ be an arbitrary system of parameters for M. Suppose first that depth $M > 0$. In this case, since $0 :_M x_1 = 0$, we have an exact sequence

$$0 \to M \xrightarrow{x_1} M \to M/x_1 \cdot M \to 0.$$

This gives for $i = 0, \ldots, d - 2$ commutative diagrams with exact rows (set $M' := M/x_1 \cdot M$):

$$\begin{array}{ccccccccc} 0 \to & \operatorname{Ext}^i_A(k, M) & \to & \operatorname{Ext}^i_A(k, M') & \to & \operatorname{Ext}^{i+1}_A(k, M) & \to O \\ & \downarrow{\scriptstyle \varphi^i_M} & & \downarrow{\scriptstyle \varphi^i_{M'}} & & \downarrow{\scriptstyle \varphi^{i+1}_M} & \\ O \to & H^i_{\mathfrak{m}}(M) & \to & H^i_{\mathfrak{m}}(M') & \to & H^{i+1}_{\mathfrak{m}}(M) & \to O. \end{array}$$

By hypothesis φ_M^i, φ_M^{i+1} are surjective and so $\varphi_{M'}^i$ is surjective by the 4-Lemma. By the inductive hypothesis ($\dim M' = d - 1$) M' is a Buchsbaum module and therefore $x_1, \ldots, x_d$ is a weak M-sequence.

Let now depth $M = 0$. We know already that $O :_M x_1 = O :_M \mathfrak{m}$.

Let $M' := M/x_1 \cdot M$. Consider the exact sequence

$$\begin{array}{ccccccccccc} 0 & \to & O :_M \mathfrak{m} & \xrightarrow{\alpha} & M & & \xrightarrow{x_1} & & M & \to M' \to & 0 \\ & & & & & \searrow^{\pi} & & \nearrow^{\iota} & & & \\ & & & & & & x_1 \cdot M & & & & \\ & & & & & \nearrow & & \searrow & & & \\ & & & & 0 & & & & 0 & & \end{array}$$

where $M \xrightarrow{\pi} x_1 \cdot M \xrightarrow{\iota} M$ is the factorization of x_1 through its image. Since $H_\mathfrak{m}^i(O :_M \mathfrak{m}) = O$ for all $i \geq 1$, $H_\mathfrak{m}^i(\pi)$ is an isomorphism for each $i \geq 1$. We now prove that $\operatorname{Ext}_A^i(k, \pi)$ is surjective for all $i \leq d - 1$. To establish this we prove the following

Claim. The homomorphisms $\operatorname{Ext}_A^i(k, \alpha)$ are injective for all $i \leq d$. Let

$$\begin{array}{cccccccc} 0 \to & O :_M \mathfrak{m} & \to & I_0 & \to & I_1 & \to & \ldots \\ & \downarrow^{\alpha} & & \downarrow & & \downarrow & & \\ 0 \to & M & \to & I_0' & \to & I_1' & \to & \ldots \end{array}$$

be minimal injective resolutions of $O :_M \mathfrak{m}$ and M, respectively.

Since $\operatorname{Supp} O :_M \mathfrak{m} \subseteq \{\mathfrak{m}\}$, we have $I_i = E^{m_i}$, where E denotes the injective hull of k and $m_i = l\big(\operatorname{Ext}_A^i(k, O :_M \mathfrak{m})\big)$.

It is well known that $I_i' = J_i \oplus E^{l_i}$, where J_i is a direct sum of copies of injective hulls of $A/\mathfrak{p}$, $\mathfrak{p} \in \operatorname{Supp} M \setminus \{\mathfrak{m}\}$, and $l_i = l\big(\operatorname{Ext}_A^i(k, M)\big)$. The homomorphisms $I_i \to I_i'$ are induced by α. Since $\operatorname{Hom}_A\big(k, H_\mathfrak{m}^0(\ )\big) = \operatorname{Hom}_A(k,\ )$, we have

$$\operatorname{Ext}_A^i(k, M) = H^i\big(\operatorname{Hom}_A(k, I_\cdot')\big) = H^i\big(\operatorname{Hom}_A\big(k, H_\mathfrak{m}^0(I_\cdot')\big)\big) = H^i\big(\operatorname{Hom}_A(k, E^{l_\cdot})\big)$$

(H^i denotes the ith cohomology of the underlying complex).

We write the complexes $E^{m_\cdot}$ and $E^{l_\cdot}$ in the following manner:

$$\begin{array}{ccccccccccccccccc} 0 \to & O :_M \mathfrak{m} & \to & E^{m_0} & & \to & Z_0^1 & & \to & E^{m_1} \to \ldots \to & E^{m_{i-1}} & & \to & Z_0^i & & \to & E^{m_i} \to \ldots \\ & \downarrow & & \downarrow & & \swarrow^{\beta^1} & & \searrow^{\alpha^1} & & \downarrow & \downarrow & & \swarrow^{\beta^i} & & \searrow^{\alpha^i} & & \downarrow \\ 0 \to & O :_M \mathfrak{m} & \to & E^{l_0} & \to & B_1^0 & \xrightarrow{\iota^1} & Z_1^1 & \to & E^{l_1} \to \ldots \to & E^{l_{i-1}} & \to & B_1^{i-1} & \xrightarrow{\iota^i} & Z_1^i & \to & E^{l_i} \to \ldots \end{array}$$

where $E^{m_i} \to E^{l_i} \doteq H_\mathfrak{m}^0(I_i \to I_i')$ which induces the other homomorphisms. Since the resolutions were minimal, we have

$$\operatorname{Hom}_A(k, Z_0^i) \simeq \operatorname{Ext}_A^i(k, O :_M \mathfrak{m}), \qquad \operatorname{Hom}_A(k, Z_1^i) \simeq \operatorname{Ext}_A^i(k, M),$$

$$\operatorname{Hom}_A(k, \alpha^i) \simeq \operatorname{Ext}_A^i(k, \alpha).$$

Next we show that $\operatorname{Ext}_A^j(k, \alpha^i)$ is injective for all $j \geq 0$ and all $i = 0, \ldots, d - 1$.

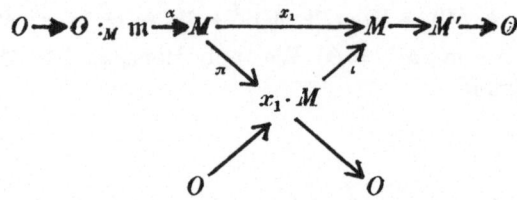

The following commutative diagrams with exact rows ($i \geq 1$)

$$\begin{array}{ccccccccc} O & \to & Z_0^{i-1} & \to & E^{m_{i-1}} & \to & Z_0^i & \to & 0 \\ & & \downarrow{\scriptstyle \alpha^{i-1}} & & \downarrow & & \downarrow{\scriptstyle \beta^i} & & \\ 0 & \to & Z_1^{i-1} & \to & E^{l_{i-1}} & \to & B_1^{i-1} & \to & O \end{array}$$

show that $\operatorname{Ext}_A^j(k, \beta^i) = \operatorname{Ext}_A^{j+1}(k, \alpha^{i-1})$ for all $j \geq 0$ and $i \geq 1$. For each $i \leq d-1$ the induced homomorphisms

$$\operatorname{Ext}_A^i(k, M) \cong \operatorname{Hom}_A(k, Z_1^i) \to \operatorname{Hom}_A\big(k, H_{\mathfrak{m}}^i(M)\big) \cong H_{\mathfrak{m}}^i(M)$$

in the commutative diagram

$$\begin{array}{ccccccccc} O & \to & Z_0^i & = & Z_0^i & \longrightarrow & 0 & \longrightarrow & 0 \\ & & \downarrow{\scriptstyle \beta^i} & & \downarrow{\scriptstyle \alpha^i} & & \downarrow & & \\ 0 & \to & B_1^{i-1} & \xrightarrow{\iota^i} & Z_1^i & \to & H_{\mathfrak{m}}^i(M) & \to & O \end{array}$$

are just the canonical maps φ_M^i and thus they are surjective. Therefore by Lemma 2.9 the sequences

$$0 \to \operatorname{Ext}_A^j(k, B_1^{i-1}) \to \operatorname{Ext}_A^j(k, Z_1^i) \to \operatorname{Ext}_A^j\big(k, H_{\mathfrak{m}}^i(M)\big) \to 0$$

are exact for each $j \geq 0$ and $i \leq d-1$. In particular, $\operatorname{Ext}_A^j(k, \iota^i)$ is injective for all $j \geq 0$ and $i \leq d-1$. Also, we have

$$\operatorname{Ext}_A^j(k, \alpha^i) = \operatorname{Ext}_A^j(k, \iota^i) \cdot \operatorname{Ext}_A^j(k, \beta^i) = \operatorname{Ext}_A^j(k, \iota^i) \cdot \operatorname{Ext}_A^{j+1}(k, \alpha^{i-1}).$$

Since α^0 is the identy map (of $0 :_M \mathfrak{m}$) an easy inductive argument (induction on i) now shows that $\operatorname{Ext}_A^j(k, \alpha^i)$ is injective for all $j \geq 0$ and $i \leq d-1$. In particular, for $j = 0$ we have that $\operatorname{Hom}_A(k, \alpha^i)$ is injective. But since

$$\operatorname{Hom}_A(k, \alpha^d) = \operatorname{Hom}_A(k, \iota^d) \cdot \operatorname{Hom}_A(k, \beta^d) = \operatorname{Hom}_A(k, \iota^d) \cdot \operatorname{Ext}_A^1(k, \alpha^{d-1})$$

is also injective, our claim is therefore proven.

Now we continue with the proof of Theorem 2.10.

Since $0 = \operatorname{Ext}_A^i(k, x_1) = \operatorname{Ext}_A^i(k, \iota) \cdot \operatorname{Ext}_A^i(k, \pi)$ and $\operatorname{Ext}_A^i(k, \pi)$ is surjective for each $i \leq d-1$, we obtain $\operatorname{Ext}_A^i(k, \iota) = 0$ for all $i \leq d-1$. Analogously $H_{\mathfrak{m}}^i(\iota) = 0$ (since $H_{\mathfrak{m}}^i(\pi)$ is an isomorphism for $i \geq 1$) for all $i \leq d-1$.

Therefore we have for $i = 0, \ldots, d-2$ commutative diagrams with exact rows

$$\begin{array}{ccccccccc} O & \to & \operatorname{Ext}_A^i(k, M) & \to & \operatorname{Ext}_A^i(k, M') & \to & \operatorname{Ext}_A^{i+1}(k, x_1 \cdot M) & \to & 0 \\ & & \downarrow{\scriptstyle \varphi_M^i} & & \downarrow{\scriptstyle \varphi_{M'}^i} & & \downarrow{\scriptstyle \varphi_{x_1 \cdot M}^{i+1}} & & \\ 0 & \to & H_{\mathfrak{m}}^i(M) & \to & H_{\mathfrak{m}}^i(M') & \to & H_{\mathfrak{m}}^{i+1}(x_1 \cdot M) & \to & O. \end{array}$$

Furthermore, we have commutative diagrams

$$\begin{array}{ccc} \operatorname{Ext}_A^i(k, M) & \xrightarrow{\operatorname{Ext}_A^i(k, \pi)} & \operatorname{Ext}_A^i(k, x_1 \cdot M) \\ \downarrow{\scriptstyle \varphi_M^i} & & \downarrow{\scriptstyle \varphi_{x_1 \cdot M}^i} \\ H_{\mathfrak{m}}^i(M) & \xrightarrow{\sim} & H_{\mathfrak{m}}^i(x_1 \cdot M) \end{array}$$

Thus the surjectivity of φ_M^i implies the surjectivity of $\varphi_{x_1 \cdot M}^i$ for all $i \leq d-1$ and this means the surjectivity of $\varphi_{M'}^i$ (use the 4-Lemma), i.e. by the induction hypothesis M' is a Buchsbaum module. Since $\mathfrak{m} \cdot (0 :_M x_1) = 0$, $x_1, \ldots, x_d$ is a weak M-sequence, q.e.d.

Remark 2.11. It is not true in general that this sufficient criterion for Buchsbaum modules is also a necessary condition. This is shown by the Example V.5.7. But as mentioned already, this criterion is necessary if A is a regular local ring, see Corollary 2.16.

We now give the following important consequence of this criterion.

Proposition 2.12. *Let M be a Noetherian A-module with $r :=$ depth $M <$ dim $M =: d$. and $H_{\mathfrak{m}}^i(M) = 0$ for all $i \neq r, d$. The following properties are equivalent:*

(i) *M is a Buchsbaum module.*

(ii) $\mathfrak{m} \cdot H_{\mathfrak{m}}^r(M) = 0$.

(iii) *Let $x_1, \ldots, x_r$ be an M-sequence in $\mathfrak{m}^2$. Then*

$$(x_1, \ldots, x_r) \cdot M : \langle \mathfrak{m} \rangle = (x_1, \ldots, x_r) \cdot M : \mathfrak{m}.$$

Proof: (i) $\Rightarrow$ (ii) follows from Corollary 2.4 and (ii) $\Rightarrow$ (iii) is a consequence of Proposition 2.1.

To prove (iii) $\Rightarrow$ (i) it is sufficient to verify the surjectivity of $\varphi_M^r : \mathrm{Ext}_A^r(k, M) \to H_{\mathfrak{m}}^r(M)$ since the other φ_M^i's, $i \neq d$ are obviously surjective and therefore Theorem 2.10 implies (i).

To this end we prove the more general statement:

Claim. Let M be a Noetherian A-module with $r :=$ depth $M <$ dim M. If there is an M-sequence $x_1, \ldots, x_r$ in $\mathfrak{m}^2$ such that

$$(x_1, \ldots, x_r) \cdot M : \langle \mathfrak{m} \rangle = (x_1, \ldots, x_r) \cdot M : \mathfrak{m}$$

then φ_M^r is an isomorphism.

For the proof we use induction on r. If $r = 0$, φ_M^0 is the embedding $0 :_M \mathfrak{m} \subseteq 0 :_M \langle \mathfrak{m} \rangle$. Also $0 :_M \langle \mathfrak{m} \rangle = 0 :_M \mathfrak{m}$ by (iii) and φ_M^0 is therefore an isomorphism.

If $r > 0$ we have an exact sequence

$$0 \to M \xrightarrow{x_1} M \to M' \to 0 \quad \text{with } M' := M/x_1 \cdot M.$$

This gives rise to a commutative diagram

$$\begin{array}{ccccccc} O \to & \mathrm{Ext}_A^{r-1}(k, M') & \xrightarrow{\sim} & \mathrm{Ext}_A^r(k, M) & \longrightarrow 0 & & \\ & \downarrow \varphi_{M'}^{r-1} & & \downarrow \varphi_M^r & & & \\ 0 \to & H_{\mathfrak{m}}^{r-1}(M') & \longrightarrow & H_{\mathfrak{m}}^r(M) & \xrightarrow{x_1} & H_{\mathfrak{m}}^r(M). & \end{array}$$

By the induction hypothesis $\varphi_{M'}^{r-1}$ is an isomorphism and thus φ_M^r splits into an isomorphism $\tilde{\varphi}_M^r : \mathrm{Ext}_A^r(k, M) \to 0 :_{H_{\mathfrak{m}}^r(M)} x_1 \cong H_{\mathfrak{m}}^{r-1}(M')$ and the embedding $\iota : 0 :_{H_{\mathfrak{m}}^r(M)} x_1 \subseteq H_{\mathfrak{m}}^r(M)$.

From $\tilde{\varphi}_M^r$ we obtain $\mathfrak{m} \cdot \left(0 :_{H^r_{\mathfrak{m}}(M)} x_1\right) = 0$. Since $x_1 \in \mathfrak{m}^2$ this results in

$$0 :_{H^r_{\mathfrak{m}}(M)} \mathfrak{m}^2 \subseteq 0 :_{H^r_{\mathfrak{m}}(M)} x_1 \subseteq 0 :_{H^r_{\mathfrak{m}}(M)} \mathfrak{m}$$

and consequently

$$0 :_{H^r_{\mathfrak{m}}(M)} x_1 = 0 :_{H^r_{\mathfrak{m}}(M)} \mathfrak{m} = O :_{H^r_{\mathfrak{m}}(M)} \langle \mathfrak{m} \rangle = H^r_{\mathfrak{m}}(M)$$

which shows that ι is the identy map. Hence φ_M^r is an isomorphism, q.e.d.

We are now able to give our first class of examples for Buchsbaum modules. To do this we define:

Definition 2.13. Let r, d be integers with $1 \leq r < d$. Further let k denote a field and $X_1, \ldots, X_d, Y_1, \ldots, Y_d$ indeterminates. We set

$$R_d := k[X_1, \ldots, X_d, Y_1, \ldots, Y_d]_{\mathfrak{m}_d}$$

where $\mathfrak{m}_d$ is the ideal generated by the indeterminates or

$$R_d := k[\![X_1, \ldots, X_d, Y_1, \ldots, Y_d]\!].$$

We define by induction on r what is meant for an ideal $\mathfrak{a} \subset R_d$ to be of type (r, d):
1. $\mathfrak{a}$ is said to be of *type* $(1, d)$ if $\mathfrak{a} = (X_1, \ldots, X_d) R_d \cap (Y_1, \ldots, Y_d) R_d$;
2. $\mathfrak{a}$ is said to be of *type* $(r + 1, d)$ with $r + 1 < d$, if $\mathfrak{a} = \mathfrak{a}_1 \cap \mathfrak{a}_2$ and
   a) $R_d/\mathfrak{a}_1$ a Cohen-Macaulay ring with $\dim R_d/\mathfrak{a}_1 = d$,
   b) the automorphism of R_d given by exchanging the indeterminates $(X_i \leftrightarrow Y_i)$ carries $\mathfrak{a}_1$ into $\mathfrak{a}_2$,
   c) $\mathfrak{a}_1 + \mathfrak{a}_2 = (X_d, Y_d) \cdot R_d + \mathfrak{b} \cdot R_d$, where $\mathfrak{b} \subset R_{d-1}$ is an ideal of type $(r, d - 1)$.

By Definition 2.13 it is not difficult to give an explicit description of ideals of type (r, d) for arbitrary integers r, d with $1 \leq r < d$ (c.f. Proposition V.2.7).

For these ideals the following statement is true:

Lemma 2.14. *Let* $\mathfrak{a} \subset R_d$ (see Definition 2.13) *be an ideal of type* (r, d) *with* $1 \leq r < d$. *Then*

$$H^i_{\mathfrak{m}_d}(R_d/\mathfrak{a}) = 0 \quad \text{for all } i \neq r, d \quad \text{and} \quad H^r_{\mathfrak{m}_d}(R_d/\mathfrak{a}) \cong k.$$

Therefore $R_d/\mathfrak{a}$ *is a Buchsbaum module (over* R_d*) of* depth r *and dimension* d.

Proof: We have an exact sequence (recall: $\mathfrak{a} = \mathfrak{a}_1 \cap \mathfrak{a}_2$, see Definition 2.13)

$$0 \to R_d/\mathfrak{a} \xrightarrow{f} (R_d/\mathfrak{a}_1) \oplus (R_d/\mathfrak{a}_2) \xrightarrow{g} R_d/(\mathfrak{a}_1 + \mathfrak{a}_2) \to 0,$$

where

$$f(a \bmod \mathfrak{a}) = (a \bmod \mathfrak{a}_1, a \bmod \mathfrak{a}_2), \qquad a \in R_d,$$

$$g\big((a_1 \bmod \mathfrak{a}_1, a_2 \bmod \mathfrak{a}_2)\big) = (a_1 - a_2) \bmod (\mathfrak{a}_1 + \mathfrak{a}_2), \qquad a_1, a_2 \in R_d.$$

But $R_d/\mathfrak{a}_i$, $i = 1, 2$ are Cohen-Macaulay rings (and therefore Cohen-Macaulay modules over R_d) and $R_d/(\mathfrak{a}_1 + \mathfrak{a}_2) \cong R_{d-1}/\mathfrak{b}$. Hence

$$H^i_{\mathfrak{m}_d}(R_d/\mathfrak{a}_1 \oplus R_d/\mathfrak{a}_2) \cong H^i_{\mathfrak{m}_d}(R_d/\mathfrak{a}_1) \oplus H^i_{\mathfrak{m}_d}(R_d/\mathfrak{a}_2) = O$$

for all $i \leq d - 1$ and

$$H^i_{\mathfrak{m}_d}(R_d/(\mathfrak{a}_1 + \mathfrak{a}_2)) \simeq H^i_{\mathfrak{m}_{d-1}}(R_{d-1}/\mathfrak{b})$$

for all i. Thus the long exact cohomology sequence renders for $i \leq d - 1$ isomorphisms

$$H^i_{\mathfrak{m}_d}(R_d/\mathfrak{a}) \simeq H^{i-1}_{\mathfrak{m}_{d-1}}(R_{d-1}/\mathfrak{b}).$$

Now, induction on d proves the lemma, q.e.d.

In order to obtain a necessary and sufficient characterization of Buchsbaum modules one has to look for "better" criteria. We will show that this is possible if we replace the "Ext"-functors by the "cohomology modules" of the Koszul complex in the sense which was made precise in Chapter 0, § 1, 3.

The following theorem is the main result of this chapter. First it gives a complete cohomological characterization of Buchsbaum modules. Secondly it shows how to verify the Buchsbaum property by considering only a finite set of weak M-sequences. We note that Theorem 20 of the Appendix gives another approach for proving this Theorem 2.15.

Theorem 2.15. *Let M be a Noetherian A-module of positive dimension d. The following properties are equivalent:*

(i) *M is a Buchsbaum module.*

(ii) *The canonical maps*

$$\lambda^i_M : H^i(\mathfrak{m}, M) \to H^i_{\mathfrak{m}}(M) \qquad \text{(cf. Lemma 0.1.5)}$$

are surjective for all $i < d$.

(iii) *Let $x_1, \ldots, x_t$ be an M-basis of the maximal ideal $\mathfrak{m}$ of A* (c.f. Definition 1.7 and Proposition 1.9). *For every system $i_1, \ldots, i_d$ of integers with $1 \leq i_1 < \ldots < i_d \leq t$ the sequence $x^{r_1}_{i_1}, \ldots, x^{r_d}_{i_d}$ is a weak M-sequence for all $r_1, \ldots, r_d \in \{1, 2\}$.*

Proof: (i) $\Rightarrow$ (iii) follows from the definition of Buchsbaum modules (Definition 1.5) and M-bases (Definition 1.7).

We now prove (iii) $\Rightarrow$ (ii). λ^0_M is (up to a natural equivalence) the embedding $0 :_M \mathfrak{m} \subseteq 0 :_M \langle\mathfrak{m}\rangle$. But we have $0 :_M \mathfrak{m}^2 \subseteq 0 :_M x_1^2 = 0 :_M \mathfrak{m} \subseteq 0 :_M \mathfrak{m}^2$ and, consequently, $0 :_M \mathfrak{m} = 0 :_M \mathfrak{m}^2 = \ldots = 0 :_M \langle\mathfrak{m}\rangle$, i.e. λ^0_M is an isomorphism.

Now we use induction on d. If $d = 1$ we are done. Assume $d > 1$. Since $x_1^2, \ldots, x_d^2$ is a weak M-sequence in $\mathfrak{m}^2$, $\mathfrak{m} \cdot H^i_{\mathfrak{m}}(M) = 0$ for all $i < d$ by Proposition 2.1.

Assume first depth $M > 0$. By Lemma 2.2, (0) is unmixed up to $\mathfrak{m}$-primary components and hence unmixed. Therefore we have an exact sequence $0 \to M \xrightarrow{x_1} M \to M' \to 0$ with $M' := M/x_1 \cdot M$. This gives rise to commutative diagrams with exact rows

$$(\mathrm{D}_i) \qquad \begin{array}{ccccccccc} 0 & \to & H^{i-1}(\mathfrak{m}, M) & \to & H^{i-1}(\mathfrak{m}, M') & \to & H^i(\mathfrak{m}, M) & \to & 0 \\ & & \downarrow{\scriptstyle \lambda^{i-1}_M} & & \downarrow{\scriptstyle \lambda^{i-1}_{M'}} & & \downarrow{\scriptstyle \lambda^i_M} & & \\ 0 & \to & H^{i-1}_{\mathfrak{m}}(M) & \to & H^{i-1}_{\mathfrak{m}}(M') & \to & H^i_{\mathfrak{m}}(M) & \to & 0 \end{array}$$

for all $i < d$. (Notice $x_1 \cdot H^i(\mathfrak{m}, M) = 0$ for all i and $x_1 \cdot H^i_{\mathfrak{m}}(M) \subseteq \mathfrak{m} \cdot H^i_{\mathfrak{m}}(M) = 0$ for all $i < d$.)

Let $A' := A/x_1 \cdot A$, $\mathfrak{m}' := \mathfrak{m} \cdot A$. Then $\bar{x}_2, \ldots, \bar{x}_t$ is an M'-basis of $\mathfrak{m}'$ (considered as A'-modules; we write $\bar{x}$ for the image of $x \in A$ in A') satisfying the hypothesis of (iii) (with respect to A' and M'). Therefore the natural homomorphisms

$$\lambda'^{i}_{M'}: H^i(\mathfrak{m}, M') \to H^i_{\mathfrak{m}'}(M')$$

are surjective for all $i < d - 1$ by the induction hypothesis.

Let $\mathfrak{n} := (x_2, \ldots, x_t) \cdot A$. Then obviously as A-modules

$$H^i(\mathfrak{m}', M') \cong H^i(\mathfrak{n}, M')$$

and (for instance by R. Y. Sharp [1], Theorem 4.3)

$$H^i_{\mathfrak{m}'}(M') \cong H^i_{\mathfrak{n}}(M').$$

Hence the corresponding natural homomorphisms

$$\tilde{\lambda}^i_{M'}: H^i(\mathfrak{n}, M') \to H^i_{\mathfrak{n}}(M')$$

are surjective for all $i < d - 1$.

By use of the commutative diagrams of Corollary 0.1.7 we get:

$$\lambda^i_{M'}: H^i(\mathfrak{m}, M') \to H^i_{\mathfrak{m}}(M')$$

are surjective for all $i < d - 1$.

Now the commutative diagrams (D_i) imply (ii) in this case.

Assume next depth $M = 0$. The exact sequence $0 \to 0 :_M \mathfrak{m} \to M \to N \to 0$ with $N := M/0 :_M \mathfrak{m} = M/0 :_M \langle\mathfrak{m}\rangle$ gives for each $i > 0$ rise to a commutative diagram $\left(H^i_{\mathfrak{m}}(0 :_M \mathfrak{m}) = 0 \text{ for all } i > 0\right)$:

$$(D'_i) \qquad \begin{array}{ccccc} H^i(\mathfrak{m}, 0 :_M \mathfrak{m}) & \xrightarrow{f^i} & H^i(\mathfrak{m}, M) & \xrightarrow{g^i} & H^i(\mathfrak{m}, N) \\ & & \downarrow{\scriptstyle \lambda^i_M} & & \downarrow{\scriptstyle \lambda^i_N} \\ & & H^i_{\mathfrak{m}}(M) & \xrightarrow{\sim} & H^i_{\mathfrak{m}}(N). \end{array}$$

We have depth $N > 0$. Further N also fulfils the assumptions of (iii) too, since $x_1, \ldots, x_t$ is of course also an N-basis of $\mathfrak{m}$. For each $1 \leq i_1 < \ldots < i_d \leq t$, $r_1, \ldots, r_d \in \{1, 2\}$, $0 \leq j < d$ we have

$$\begin{aligned} &\left((O :_M \mathfrak{m}) + (x_{i_1}^{r_1}, \ldots, x_{i_j}^{r_j}) \cdot M\right) : x_{i_{j+1}}^{r_{j+1}} \\ &\subseteq \left((0 :_M \mathfrak{m}) + (x_{i_1}^{r_1}, \ldots, x_{i_j}^{r_j}) \cdot M\right) : \langle\mathfrak{m}\rangle = (x_{i_1}^{r_1}, \ldots, x_{i_j}^{r_j}) \cdot M : \langle\mathfrak{m}\rangle \\ &= (x_{i_1}^{r_1}, \ldots, x_{i_j}^{r_j}) \cdot M : \mathfrak{m} \subseteq \left((0 :_M \mathfrak{m}) + (x_{i_1}^{r_1}, \ldots, x_{i_j}^{r_j}) \cdot M\right) : \mathfrak{m} \end{aligned}$$

(notice $(x_{i_1}^{r_1}, \ldots, x_{i_j}^{r_j}) \cdot M : \mathfrak{m}^2 \subseteq (x_{i_1}^{r_1}, \ldots, x_{i_j}^{r_j}) \cdot M : x_{i_{j+1}}^2 = (x_{i_1}^{r_1}, \ldots, x_{i_j}^{r}) \cdot M : \mathfrak{m}$ implies

$$(x_{i_1}^{r_1}, \ldots, x_{i_j}^{r_j}) \cdot M : \langle\mathfrak{m}\rangle = (x_{i_1}^{r_1}, \ldots, x_{i_j}^{r_j}) \cdot M : \mathfrak{m}).$$

Therefore λ^i_N is surjective for all $i < d$. We now prove that also g^i (in the diagram (D'_i)) is surjective for all $i < d$. Then the proof of (iii) $\Rightarrow$ (ii) will be complete by the diagrams (D'_i).

To this end we prove the injectivity of f^i for all $i \leq d$. We use the notation introduced in Chapter 0, § 1, 3.

Let $K^i := K^i(x_1, \ldots, x_t; M)$, $\bar{K}^i := K^i(x_1, \ldots, x_t; O :_M \mathfrak{m})$ and let H^i, $\bar{H}^i$ denote the cohomology modules of these complexes. Since $(x_1, \ldots, x_t)(O :_M \mathfrak{m}) = 0$, the differentiation of $\bar{K}^i$ is zero, i.e. $\bar{H}^i = \bar{K}^i$ for all i. Let d^i denote the differentiation of $K^\cdot$. We have to show that $\bar{K}^i \cap \operatorname{Im} d^{i-1} = 0$ for all $i \leq d$ (considered as submodules of K^i). Let $x \in \bar{K}^i \cap \operatorname{Im} d^{i-1}$, i.e. $x = d^{i-1}(y)$ with $y \in K^{i-1}$: We write

$$y = \sum_{1 \leq l_1 < \ldots < l_{i-1} \leq t} m^{l_1 \ldots l_{i-1}} e^{l_1 \ldots l_{i-1}} \quad (m^{l_1 \ldots l_{i-1}} \in M),$$

$$x = \sum_{1 \leq j_1 < \ldots < j_i \leq t} u^{j_1 \ldots j_i} e^{j_1 \ldots j_i} \quad (u^{j_1 \ldots j_i} \in O :_M \mathfrak{m}).$$

Then we obtain

$$x = d^{i-1}(y) = \sum_{1 \leq n_1 < \ldots < n_i \leq t} \left(\sum_{p=1}^{i} (-1)^{p-1} x_{n_p} m^{n_1 \ldots \hat{n}_p \ldots n_i} \right) e^{n_1 \ldots n_i},$$

hence $u^{n_1 \ldots n_i} \in (O :_M \mathfrak{m}) \cap (x_{n_1}, \ldots, x_{n_i}) M$ for all $1 \leq n_1 < \ldots < n_i \leq t$.

Let $u := u^{n_1 \ldots n_i}$ ($n_1, \ldots, n_i$ fixed) and choose s minimal with $u \in (x_{n_1}, \ldots, x_{n_s}) M$, $0 \leq s \leq i$. Since $i \leq d$, $x_{n_1}, \ldots, x_{n_{s-1}}, x_{n_s}^2$ forms a weak M-sequence. We write

$$u = x_{n_1} m_1 + \ldots + x_{n_s} m_s, \quad m_i \in M.$$

If $s \geq 1$, multiply by x_{n_s} and obtain $(x_{n_s} u = 0)$:

$$x_{n_s}^2 m_s \in (x_{n_1}, \ldots, x_{n_{s-1}}) M \quad \text{and thus} \quad m_s \in (x_{n_1}, \ldots, x_{n_{s-1}}) M : \mathfrak{m}$$

by our assumption on $x_1, \ldots, x_t$. Therefore $u \in (x_{n_1}, \ldots, x_{n_{s-1}}) \cdot M$ which contradicts the minimality of s. Hence $s = 0$, i.e. $u = 0$.

Since this is true for all integers $j_1, \ldots, j_i$ with $1 \leq j_1 < \ldots < j_i \leq t$, we get $x = 0$ and (ii) is proven.

Finally, we show (ii) $\Rightarrow$ (i). If $\hat{A}$ denotes the $\mathfrak{m}$-adic completion of A, $\hat{M} := M \otimes_A \hat{A}$ then

$$H^i(\mathfrak{m}, M) \otimes_A \hat{A} \simeq H^i(\hat{\mathfrak{m}}, \hat{M}) \quad \text{and} \quad H^i_{\mathfrak{m}}(M) \otimes_A \hat{A} \simeq H^i_{\hat{\mathfrak{m}}}(\hat{M}),$$

i.e. the canonical maps

$$\hat{\lambda}^i_{\hat{M}} : H^i(\hat{\mathfrak{m}}, \hat{M}) \to H^i_{\hat{\mathfrak{m}}}(\hat{M})$$

are surjective for all $i < d = \dim_{\hat{A}} \hat{M}$.

By Cohen's structure theorem for complete local rings there is a regular local ring R with maximal ideal $\mathfrak{n}$ such that $\hat{A}$ is an epimorphic image of R. Without loss of generality we may assume that

$$\dim R = \operatorname{rank}_k(\hat{\mathfrak{m}}/\hat{\mathfrak{m}}^2) = \operatorname{rank}_k(\mathfrak{m}/\mathfrak{m}^2) \quad (k = A/\mathfrak{m} = \hat{A}/\hat{\mathfrak{m}}).$$

Then $H^i(\hat{\mathfrak{m}}, \hat{M}) \simeq H^i(\mathfrak{n}, \hat{M})$ and $H^i_{\hat{\mathfrak{m}}}(\hat{M}) \simeq H^i_{\mathfrak{n}}(\hat{M})$ as R-modules. The corresponding natural maps

$$\lambda^i_{\hat{M}} : H^i(\mathfrak{n}, \hat{M}) \to H^i_{\mathfrak{n}}(\hat{M})$$

therefore are surjective for all $i < d$ ($= \dim_R \hat{M}$).

Using the commutative diagrams of Lemma 0.1.5 we see that $\varphi^i_{\hat{M}} : \operatorname{Ext}^i_R(k, \hat{M}) \to H^i_{\mathfrak{n}}(\hat{M})$ are surjective for all $i < d$ since $\mathfrak{n}$ is generated by an R-sequence and the canonical maps $\psi^i_{\hat{M}}$ are therefore isomorphisms. Thus $\hat{M}$ is a Buchsbaum module over R by

Theorem 2.10 and by Lemma 1.6 a Buchsbaum module over $\hat{A}$. Now M is a Buchsbaum module over A by Lemma 1.13, q.e.d.

Corollary 2.16. *Let A be a regular local ring and M a Noetherian A-module of positive dimension. M is a Buchsbaum module if and only if the canonical maps*

$$\varphi_M^i\colon \operatorname{Ext}_A^i(k, M) \to H_{\mathfrak{m}}^i(M)$$

are surjective for all $i <$ dim M.

Remark 2.17. Our initial investigations of Buchsbaum modules made it necessary to study all systems of parameters (see, for example, Proposition 1.10 and Theorem 1.12). Using the cohomological characterization of Buchsbaum modules we are able to reduce these considerations to a well-defined finite set of systems of parameters, see Theorem 2.15. In special situations Proposition 2.12 enables us to work with only one system of parameters. Therefore we would like to pose the following

Problem. Is there a criterion for Buchsbaum modules which uses only one fixed system of parameters?

One might be tempted to try to answer the following

Question. Let M be a Noetherian A-module of dimension $d \geq 1$. Suppose that there are elements $x_1, \ldots, x_d$ of $\mathfrak{m}$ such that

$$x_{s(1)}^{n_1}, \ldots, x_{s(d)}^{n_d}$$

is a weak M-sequence for every permutation s on $\{1, \ldots, d\}$ and for all integers $n_1, \ldots, n_d > 0$.

Is M then a Buchsbaum module?

The following (unpublished) example due to S. Goto shows that this question has a negative answer.

Example. Let $R := k[\![X_1, \ldots, X_d, Y_1, \ldots, Y_d]\!]$, $d \geq 3$, be the formal power series ring in the indeterminates $X_1, \ldots, X_d, Y_1, \ldots, Y_d$ over an arbitrary field k.

Put

$$\mathfrak{a} := (X_1, \ldots, X_d)\, R \cap (Y_1, \ldots, Y_d) \cdot R,$$

$$\mathfrak{q} := (X_1^2, X_2, \ldots, X_d, Y_1^2, Y_2, \ldots, Y_d) \cdot R,$$

$$F_i := X_i + Y_i \quad \text{for } i = 1, \ldots, d,$$

$$A := R/\big((\mathfrak{a} \cap \mathfrak{q}) + F_1^n \cdot R\big) \quad \text{with } n \geq 3.$$

Then dim $A = d - 1$ and A is not a Buchsbaum ring since $\mathfrak{m} \cdot U(O) \neq O$ in A. It is now easy to see that the images of $F_2, \ldots, F_d$ in A form a system of parameters for A and have the required property.

We note that the images of $F_2, \ldots, F_d$ are even a part of an A-basis of the maximal ideal $\mathfrak{m}$ of A since they form a part of a minimal basis of $\mathfrak{m}$.

A first and important application of Theorem 2.15 is the solution of the so-called lifting problem for Buchsbaum modules, i.e. the possibility of lifting the Buchsbaum property by a non-zero divisor. M. Hochster asked the following related question:

Let $A = R/\mathfrak{a}$ be a local ring where R is regular and $\mathfrak{a}$ is an ideal of R. Suppose that:
(i) $A_\mathfrak{p}$ is a Cohen-Macaulay ring for all $\mathfrak{p} \in \operatorname{Spec} A \setminus \{\mathfrak{m}\}$.
(ii) there exists a non-zero divisor x of A such that $A/x \cdot A$ is a Buchsbaum ring.
Is it true that then A is a Buchsbaum ring?

The following example shows that the answer to this question is negative.

Example 2.18. Take $A := k[\![X_1, X_2, X_3, X_4]\!]/(X_1^2, X_2) \cap (X_3, X_4)$ where k is an arbitrary field and $X_1, \ldots, X_4$ are indeterminates. Then we get the following:
(i) $A_\mathfrak{p}$ is a Cohen-Macaulay ring for all $\mathfrak{p} \in \operatorname{Spec} A \setminus \{\mathfrak{m}\}$.
(ii) $A/(X_1 + X_3) \cdot A$ is a Buchsbaum ring (of dimension one).
(iii) A is not a Buchsbaum ring.
(iv) $\mathfrak{m} \cdot H^1_\mathfrak{m}(A) \neq 0$.

The statements (i), (iii), (iv) are clear (see e.g. Proposition 2.25).
To prove statement (ii) it is sufficient to show that

$$(X_1, \ldots, X_4) \cdot U(\mathfrak{a} + F \cdot R) \subseteq \mathfrak{a} + F \cdot R$$

(by Proposition 1.10) where

$$\mathfrak{a} = (X_1^2, X_2) \cdot R \cap (X_3, X_4) \cdot R = (X_1^2 \cdot X_3, X_1^2 \cdot X_4, X_2 \cdot X_3, X_2 \cdot X_4) \cdot R,$$
$$F := X_1 + X_3, \qquad R := k[\![X_1, X_2, X_3, X_4]\!].$$

But this follows from

$$U(\mathfrak{a} + F \cdot R) = (X_1, X_3, X_4) \cdot R \cap (X_1^2, X_2, X_3^2, F) \cdot R$$
$$= (X_1^2, X_3^2, X_1 X_2, X_2 X_4, F) \cdot R.$$

What we can prove is the following:

Proposition 2.19. *Let M be a Noetherian A-module with* depth $M > 0$. *The following conditions are equivalent:*
(i) *M is a Buchsbaum module.*
(ii) *There is a non-zero divisor $x \in \mathfrak{m}^2$ of M such that $M/x \cdot M$ is a Buchsbaum module.*
(ii') *$M/x \cdot M$ is a Buchsbaum module for every non-zero divisor $x \in \mathfrak{m}^2$ of M.*
(iii) *There is a non-zero divisor $x \in \mathfrak{m}$ of M such that:*
a) *$M/x \cdot M$ is a Buchsbaum module.*
b) *$x \cdot H^i_\mathfrak{m}(M) = O$ for all $i <$* dim M.
(iii') *For all non-zero divisors $x \in \mathfrak{m}$ of M* a) *and* b) *of* (iii) *are true.*
(iv) *There is a non-zero divisor $x \in \mathfrak{m}$ of M such that:*
c) *$M/x \cdot M$ is a Buchsbaum module.*
d) *$x \cdot H^i_\mathfrak{m}(M/x^2 \cdot M) = 0$ for all $i <$* dim $M - 1$.
(iv') *For all non-zero divisors $x \in \mathfrak{m}$ of M* c) *and* d) *of* (iv) *are true.*

Proof: The necessity of the conditions (ii), (ii'), (iii), ... is obvious by Corollary 1.11(i) and Corollary 2.4. Therefore it remains to prove the implications (ii) $\Rightarrow$ (i), (iii) $\Rightarrow$ (i), (iv) $\Rightarrow$ (i).

We start with the exact sequence

$$0 \to M \xrightarrow{x} M \to M' \to 0, \qquad \text{where } M' := M/x \cdot M.$$

From it one obtains a commutative diagram with exact rows

$$(\mathrm{D}_i) \qquad \begin{array}{ccccccccc} & & O \to & H^i(\mathfrak{m}, M) & \to & H^i(\mathfrak{m}, M') & \to & H^{i+1}(\mathfrak{m}, M) & \to O \\ & & & \downarrow \lambda^i_M & & \downarrow \lambda^i_{M'} & & \downarrow \lambda^{i+1}_M & \\ \ldots \to & H^i_{\mathfrak{m}}(M) & \xrightarrow{x} & H^i_{\mathfrak{m}}(M) & \to & H^i_{\mathfrak{m}}(M') & \to & H^{i+1}_{\mathfrak{m}}(M) & \xrightarrow{x} H^{i+1}_{\mathfrak{m}}(M) \to \ldots \end{array}$$

Now in each case $\lambda^i_{M'}$ is surjective for all $i < \dim M' = \dim M - 1$ by Theorem 2.15.

We want to show that we have always $x \cdot H^j_{\mathfrak{m}}(M) = O$ for all $j < \dim M$. Then by the commutative diagrams (D_{j-1}) λ^i_M is surjective for all $j < \dim M$ and (i) follows by Theorem 2.15.

In case (iii) this is clear. In case (ii) we obtain by the exactness of the bottom row of (D_{j-1}) an epimorphism

$$H^{j-1}_{\mathfrak{m}}(M') \to O :_{H^j_{\mathfrak{m}}(M)} x .$$

Since $\mathfrak{m} \cdot H^{j-1}_{\mathfrak{m}}(M') = O$ for all $j < \dim M$ (Corollary 2.4), we get

$$\mathfrak{m} \cdot \left(0 :_{H^j_{\mathfrak{m}}(M)} x\right) = 0 \qquad \text{for all } j < \dim M .$$

But $x \in \mathfrak{m}^2$ yields $0 :_{H^j_{\mathfrak{m}}(M)} x = O :_{H^j_{\mathfrak{m}}(M)} \mathfrak{m} = O :_{H^j_{\mathfrak{m}}(M)} \langle \mathfrak{m} \rangle = H^j_{\mathfrak{m}}(M)$ and we are done.

If (iv) holds we obtain for all j by the exact sequence

$$0 \to M \xrightarrow{x^2} M \to M/x^2 \cdot M \to 0$$

epimorphisms

$$H^{j-1}_{\mathfrak{m}}(M/x^2 \cdot M) \to 0 :_{H^j_{\mathfrak{m}}(M)} x^2 .$$

Hence, by d) $x \cdot \left(0 :_{H^j_{\mathfrak{m}}(M)} x^2\right) = 0$ and this implies

$$0 :_{H^j_{\mathfrak{m}}(M)} x = 0 :_{H^j_{\mathfrak{m}}(M)} x^2 = \ldots = H^j_{\mathfrak{m}}(M),$$

since every element of $H^j_{\mathfrak{m}}(M)$ is annihilated by some power of x.

But this also implies $x \cdot H^j_{\mathfrak{m}}(M) = O$ for all $j < \dim M$, q.e.d.

Remark 2.20. We can apply this lifting property to the classical investigations on hyperplane sections. This is connected with Bertini's Theorems on linear systems. In 1977 H. Flenner [1] proved local analogues of these theorems which answer a conjecture of A. Grothendieck. These striking results also improve the observations on the hyperplane sections of normal varieties (see, e.g., Seidenberg [1, 2] and Kuan [2]). Let us consider a hyperplane section through a rational normal point of an algebraic variety. To do this let V be an irreducible algebraic variety of dimension ≥ 3 defined over an infinite field k in affine n-space over k, and let H be the generic hyperplane defined by $u_0 + u_1 \cdot x_1 + \ldots + u_n \cdot x_n = 0$, where $u_0, \ldots, u_n$ are indeterminates over k. Let $P \in V$ be a rational point, that is the coordinates of P are elements of k. Then one is interested in the intersection of V with H. If V is normal at P over k then it is not true in general that $V \cap H$ is normal at P over $k(u_0, \ldots, u_n)$; however, $V \cap H$ is normal at P if the local ring of V at P is a Cohen-Macaulay ring (see Kuan [2]). Therefore Trung [1, 3] has studied the hyperplane section through a Buchsbaum (and Cohen-Macaulay) point of an algebraic variety and the theory of specializations of Buchsbaum points. From the view-point of local algebra these results follow from our Buchsbaum lifting property, Corollary 1.11 and Lemma 2.26 below.

Using geometric language we will discuss one result of a generic hyperplane section through a Buchsbaum point. Let Ω be a universal domain and $k \subseteq \Omega$, that is Ω has infinite degree of transcendence over k. By *affine n-space* $\mathbf{A}^n$ we mean the n-fold cartesian product of Ω:

$$\mathbf{A}^n = \Omega \times \ldots \times \Omega.$$

A point in n-space, or briefly a point, is an n-tuple $(a_1, \ldots, a_n)$ with components $a_i \in \Omega$. If k is our fixed ground field we denote the ring of polynomials in n indeterminates by $k[X] := k[X_1, \ldots, X_n]$.

Let $V \subset \mathbf{A}^n$ be an algebraic variety over k denoted by V/k; that is, the defining ideal of V is an ideal in $k[X]$. Let $u_1, \ldots, u_r$ be elements of Ω algebraically independent over k. Let $f_1, \ldots, f_r$ be polynomials of $k[X]$. We consider the hypersurface F_u over $k(u) = k(u_1, \ldots, u_r)$ in $\mathbf{A}^n$ defined by the equation $u_1 \cdot f_1 + \ldots + u_r \cdot f_r = 0$. We want to examine the general hypersurface section $V \cap F_u$ over $k(u)$. Let P be a point of V/k with defining ideal $\mathfrak{p} \subset k[X]$. We denote by A the local ring of V/k at P. We state without proof:

Proposition 2.21. *Assume $P \in V \cap F_u$ and height $(f_1, \ldots, f_r) \cdot A > 1$. If P is a Buchsbaum point of V/k then P is a Buchsbaum point of $V \cap F_u/k(u)$.*

The converse is true if grade $(f_1, \ldots, f_r) \cdot A > 1$ and $(f_1, \ldots, f_r) \cdot A \subseteq \mathfrak{p}^2 \cdot A$.

The next result was proven by N. V. Trung, see [2], Theorem 4. It gives informations about the lifting property in case depth $M = 0$.

Proposition 2.22. *Let M be a Noetherian A-module of positive dimension d and* depth $M = 0$. *M is a Buchsbaum module if and only if the following conditions are satisfied:*

(i) $\mathfrak{m} \cdot H^0_\mathfrak{m}(M) = 0$.

(ii) *$M/H^0_\mathfrak{m}(M)$ is a Buchsbaum module.*

(iii) *There is an M-basis $x_1, \ldots, x_t$ of $\mathfrak{m}$ such that*

$$H^0_\mathfrak{m}(M) \cap (x_{i_1}, \ldots, x_{i_d}) \cdot M = 0 \quad \textit{for all } 1 \leq i_1 < \ldots < i_d \leq t.$$

Proof: Let M be a Buchsbaum module. Then (i) follows from Corollary 2.4, (ii) from Corollary 1.11 and (iii) from the proof of Theorem 2.15(iii) $\Rightarrow$ (ii).

Conversely, the proof of Theorem 2.15(iii) $\Rightarrow$ (ii) shows that the canonical maps $\lambda^i_M: H^i(\mathfrak{m}, M) \to H^i_\mathfrak{m}(M)$ are surjective for all $i < \dim M$, i.e. M is a Buchsbaum module by Theorem 2.15, q.e.d.

The next proposition which has its origin with M. Brodmann [4], gives a further criterion for "lifting" the Buchsbaum property. It is closely related to the statements of Proposition 2.19.

Proposition 2.23. *Let M be a Noetherian A-module with $d := \dim M \geq 2$ and* depth $M > 0$. *M is a Buchsbaum module if and only if $\mathfrak{m} H^1_\mathfrak{m}(M) = 0$ and $M/xM : \langle \mathfrak{m} \rangle$ is a Buchsbaum module for all $x \in \mathfrak{m}$ with* $\dim M/xM = d - 1$.

Proof: If M is a Buchsbaum module then $\mathfrak{m}H^1_{\mathfrak{m}}(M) = O$ by Corollary 2.4 and $M/xM:\langle\mathfrak{m}\rangle = M/U(xM)$ is a Buchsbaum module by Corollary 1.11.

Assume $\mathfrak{m}H^1_{\mathfrak{m}}(M) = O$ and that $M/xM:\langle\mathfrak{m}\rangle$ is a Buchsbaum module for all $x \in \mathfrak{m}$ with $\dim M/xM = \dim M/xM:\langle\mathfrak{m}\rangle = d - 1$. If $d = 2$, there is nothing else to prove. Let $d \geq 3$. Choose $x \in \mathfrak{m}^2$ with $x \notin \mathfrak{p}$ for all $\mathfrak{p} \in \operatorname{Ass} M$. The exact sequence $O \to M \xrightarrow{x} M \to M/xM \to O$ induces for all i epimorphisms

$$H^{i-1}_{\mathfrak{m}}(M/xM) \to O:_{H^i_{\mathfrak{m}}(M)} x.$$

Since $\mathfrak{m}H^{i-1}_{\mathfrak{m}}(M/xM) \cong \mathfrak{m}H^{i-1}_{\mathfrak{m}}(M/xM:\langle\mathfrak{m}\rangle) = O$ for all i with $2 \leq i < d$ (this follows from $M/xM:\langle\mathfrak{m}\rangle$ being a Buchsbaum module, compare Corollary 2.4) we have $\mathfrak{m}(O:_{H^i_{\mathfrak{m}}(M)} x) = O$ for $2 \leq i < d$. Since $x \in \mathfrak{m}^2$, this implies $O:_{H^i_{\mathfrak{m}}(M)} \mathfrak{m}^2 \subseteq O:_{H^i_{\mathfrak{m}}(M)} x \subseteq O:_{H^i_{\mathfrak{m}}(M)} \mathfrak{m} \subseteq O:_{H^i_{\mathfrak{m}}(M)} \mathfrak{m}^2$. Therefore

$$O:_{H^i_{\mathfrak{m}}(M)} \mathfrak{m} = O:_{H^i_{\mathfrak{m}}(M)} \mathfrak{m}^2 = \ldots = O:_{H^i_{\mathfrak{m}}(M)} \langle\mathfrak{m}\rangle = H^i_{\mathfrak{m}}(M)$$

and this shows that $H^i_{\mathfrak{m}}(M)$ is annihilated by $\mathfrak{m}$ for $i = 2, \ldots, d - 1$. Since $\mathfrak{m}H^1_{\mathfrak{m}}(M) = O$ by our assumption (and $H^0_{\mathfrak{m}}(M) = O$), we have by Lemma 2.2 that for every system of parameters $x_1, \ldots, x_d$ of M the submodule $(x_1, \ldots, x_i)M$ is unmixed (in M) up to $\mathfrak{m}$-primary components for all $i = 0, \ldots, d - 1$.

If $x \in \mathfrak{m}$ with $\dim M/xM = d - 1$, x is a non-zero divisor of M (since O is unmixed in M). Therefore the exact sequence $O \to M \xrightarrow{x} M \to M/xM \to O$ induces an isomorphism $H^0_{\mathfrak{m}}(M/xM) \cong H^1_{\mathfrak{m}}(M)$, i.e. we have $\mathfrak{m}H^0_{\mathfrak{m}}(M/xM) = O$. But this means $\mathfrak{m}(xM:\langle\mathfrak{m}\rangle) \subseteq xM$ or $xM:\langle\mathfrak{m}\rangle = xM:\mathfrak{m}$. In particular for every $y \in \mathfrak{m}$ with $\dim M/(x, y)\, M = d - 2$ we have $xM:y = xM:\mathfrak{m}$.

Let now $x_1, \ldots, x_d$ be an arbitrary system of parameters of M.

We choose an $x \in \mathfrak{m}$ such that $x, x_1, \ldots, x_{d-1}$ is again a system of parameters of M. Since $M/x^nM:\langle\mathfrak{m}\rangle$ is a Buchsbaum module for all $n \geq 1$, it follows for all $j = 1, \ldots, d - 1$ and all $n \geq 1$ that

$$\begin{aligned}(x^n, x_1, \ldots, x_{j-1})M:x_j &\subseteq \big(x^nM:\langle\mathfrak{m}\rangle + (x_1, \ldots, x_{j-1})\, M\big):x_j\\ &= \big(x^nM:\langle\mathfrak{m}\rangle + (x_1, \ldots, x_{j-1})\, M\big):\mathfrak{m}.\end{aligned}$$

If we take the intersection over all $n \geq 1$ we get by Krull's Intersection Theorem

$$(x_1, \ldots, x_{j-1})M:x_j \subseteq (x_1, \ldots, x_{j-1})M:\mathfrak{m},$$

in particular

$$(x_1, \ldots, x_{j-1})\, M:x_j = (x_1, \ldots, x_{j-1})\, M:\mathfrak{m} = (x_1, \ldots, x_{j-1})\, M:\langle\mathfrak{m}\rangle.$$

Claim. $x_1M:\mathfrak{m} \cap \big(x_2M:\mathfrak{m} + (x_3, \ldots, x_j)\, M\big) \subseteq x_1M$ for all $j = 2, \ldots, d - 1$.

For $j = 2$ we have to prove: $x_1M:\mathfrak{m} \cap x_2M:\mathfrak{m} \subseteq x_1M$.

Now $x_1x_2M:\mathfrak{m} \subseteq x_1x_2M:x_2 = x_1M$, hence

$$\begin{aligned}x_1x_2M:\mathfrak{m} &= (x_1x_2M:\mathfrak{m}) \cap x_1M = x_1\big((x_1x_2M:\mathfrak{m}):x_1\big)\\ &= x_1\big((x_1x_2M:x_1):\mathfrak{m}\big) = x_1(x_2M:\mathfrak{m}).\end{aligned}$$

Since $\dim M/x_1x_2M = d - 1$, we conclude

$$\begin{aligned} x_1M:\mathfrak{m} \cap x_2M:\mathfrak{m} &= x_1M:\mathfrak{m} \cap x_2M:\mathfrak{m}^2 = \big(x_1M \cap (x_2M:\mathfrak{m})\big):\mathfrak{m} \\ &= \big(x_1(x_2M:\mathfrak{m}x_1)\big):\mathfrak{m} = \big(x_1(x_2M:\mathfrak{m})\big):\mathfrak{m} = (x_1x_2M:\mathfrak{m}):\mathfrak{m} \\ &= x_1x_2M:\mathfrak{m} = x_1(x_2M:\mathfrak{m}) \subseteq x_1M. \end{aligned}$$

Let $3 \leq j \leq d-1$ and $m \in (x_1M:\mathfrak{m}) \cap \big(x_2M:\mathfrak{m} + (x_3, \ldots, x_j)\, M\big)$. Then $m = u_2 + x_3m_3 + \ldots + x_jm_j$ with $u_2 \in x_2M:\mathfrak{m}$, $m_3, \ldots, m_j \in M$. Therefore

$$\begin{aligned} m_j &\in \big(x_1M:\mathfrak{m} + x_2M:\mathfrak{m} + (x_3, \ldots, x_{j-1})\, M\big):x_j \\ &\subseteq (x_1, x_2, \ldots, x_{j-1})\, M:\langle\mathfrak{m}\rangle = (x_1, \ldots, x_{j-1})\, M:\mathfrak{m}, \end{aligned}$$

i.e. we have $x_jm_j = m_1'x_1 + \ldots + m_{j-1}'x_{j-1}$. But this implies

$$\begin{aligned} m - m_1'x_1 &= u_2 + m_2x_2 + (m_3 + m_3')x_3 + \ldots + (m_{j-1} + m_{j-1}')x_{j-1} \\ &\in x_1M:\mathfrak{m} \cap \big(x_2M:\mathfrak{m} + (x_3, \ldots, x_{j-1})\, M\big) \subseteq x_1M, \end{aligned}$$

i.e. $m \in x_1M$ which proves the contention.

Now we have ($M/x_1M:\mathfrak{m}$ is a Buchsbaum module):

$$\begin{aligned} \mathfrak{m}\big((x_1, \ldots, x_{d-1})\, M:x_d\big) &\subseteq \mathfrak{m}\big((x_1M:\mathfrak{m} + (x_2, \ldots, x_{d-1})\, M):x_d\big) \\ &\subseteq x_1M:\mathfrak{m} + (x_2, \ldots, x_{d-1})\, M \end{aligned}$$

and exchanging x_1 and x_2

$$\mathfrak{m}\big((x_1, \ldots, x_{d-1})\, M:x_d\big) \subseteq x_2M:\mathfrak{m} + (x_1, x_3, \ldots, x_{d-1})\, M.$$

Hence

$$\begin{aligned} &\mathfrak{m}\big((x_1, \ldots, x_{d-1})\, M:x_d\big) \\ &\subseteq \big(x_1M:\mathfrak{m} + (x_2, \ldots, x_{d-1})\, M\big) \cap \big(x_2M:\mathfrak{m} + (x_1, x_3, \ldots, x_{d-1})\, M\big) \\ &= (x_2, x_3, \ldots, x_{d-1})\, M + \big((x_1M:\mathfrak{m}) \cap (x_2M:\mathfrak{m} + (x_1, x_3, \ldots, x_{d-1}M))\big) \\ &= (x_1, x_2, \ldots, x_{d-1})\, M + \big((x_1M:\mathfrak{m}) \cap (x_2M:\mathfrak{m} + (x_3, \ldots, x_{d-1})\, M)\big) \\ &= (x_1, \ldots, x_{d-1})\, M. \end{aligned}$$

Therefore by Proposition 1.10 M is a Buchsbaum module, q.e.d.

Corollary 2.24. *Let M be a Noetherian A-module with $d := \dim M \geq 3$ and $\operatorname{depth} M > 0$. Assume furthermore that either A is an epimorphic image of a Gorenstein ring or $H^1_\mathfrak{m}(M)$ is a Noetherian module. Then M is a Buchsbaum module if and only if $M/xM:\langle\mathfrak{m}\rangle$ is a Buchsbaum module for all $x \in \mathfrak{m}$ with $\dim M/xM = d - 1$.*

Proof: The "only-if-part" needs no further elaboration. Assume that $M/xM:\langle\mathfrak{m}\rangle$ is a Buchsbaum module for all $x \in \mathfrak{m}$ with $\dim M/xM = d - 1$. Suppose A is an epimorphic image of a local Gorenstein ring. If there is a $\mathfrak{p} \in \operatorname{Ass} M$ with $1 \leq \dim A/\mathfrak{p} \leq d - 1$ then $(0:_M \mathfrak{p})_\mathfrak{p} = 0:_{M_\mathfrak{p}} \mathfrak{p}A_\mathfrak{p} \neq 0$.

We choose an $x \in \mathfrak{q}$ with $x \notin \mathfrak{q}$ for all $\mathfrak{q} \in \operatorname{Ass} M$ with $\dim A/\mathfrak{q} = d$ and $xM_\mathfrak{p} \cap (0:_{M_\mathfrak{p}} \mathfrak{p}A_\mathfrak{p}) = 0$. Then

$$\operatorname{Hom}(A_\mathfrak{p}/\mathfrak{p}A_\mathfrak{p}, xM_\mathfrak{p}) \cong 0:_{xM_\mathfrak{p}} \mathfrak{p}A_\mathfrak{p} = xM_\mathfrak{p} \cap (0:_{M_\mathfrak{p}} \mathfrak{p}A_\mathfrak{p}) = 0$$

and the exact sequence $0 \to xM_{\mathfrak{p}} \to M_{\mathfrak{p}} \to (M/xM)_{\mathfrak{p}} \to 0$ induces a monomorphism

$$\mathrm{Hom}(A_{\mathfrak{p}}/\mathfrak{p}A_{\mathfrak{p}}, M_{\mathfrak{p}}) \to \mathrm{Hom}\big(A_{\mathfrak{p}}/\mathfrak{p}A_{\mathfrak{p}}, (M/xM)_{\mathfrak{p}}\big).$$

Since $\mathrm{Hom}(A_{\mathfrak{p}}/\mathfrak{p}A_{\mathfrak{p}}, M_{\mathfrak{p}}) \cong 0:_{M_{\mathfrak{p}}}\mathfrak{p}A_{\mathfrak{p}} \neq 0$, it follows that $\mathrm{Hom}\big(A_{\mathfrak{p}}/\mathfrak{p}A_{\mathfrak{p}}, (M/xM)_{\mathfrak{p}}\big) \neq 0$, i.e. $\mathfrak{p} \in \mathrm{Ass}\, M/xM$. Since $\mathrm{Ass}\, M/xM:\langle\mathfrak{m}\rangle = \mathrm{Ass}\, M/xM \setminus \{\mathfrak{m}\}$, we obtain $\mathfrak{p} \in \mathrm{Ass}\, M/xM:\langle\mathfrak{m}\rangle$. Also since $\dim M/xM = d-1$, $M/xM:\langle\mathfrak{m}\rangle$ is a Buchsbaum module, hence $\dim A/\mathfrak{p} = d - 1 \geq 2$. Let A be an epimorphic image of the local Gorenstein ring B. Then by the local duality theorem (cf. Corollary 0.3.5), $H^1_{\mathfrak{m}}(M) \cong \mathrm{Hom}_A\big(\mathrm{Ext}^{n-1}_B(M, B), E\big)$, where E denotes the injective envelope of $A/\mathfrak{m}$ (as an A-module) and $n := \dim B$. Let $H^1_{\mathfrak{m}}(M) \neq 0$ and $\mathfrak{p} \in \mathrm{Ass}_A \mathrm{Ext}^{n-1}_B(M, B)$. Then $\dim A/\mathfrak{p} \leq 1$ by Sharp [2], Proposition (3.8) and Theorem (2.3), and we have with $\mathfrak{q}$ denoting the inverse image of $\mathfrak{p}$ in B):

$$\mathrm{Ext}^{n-1}_B(M, B)_{\mathfrak{p}} \cong \mathrm{Ext}^{n-1}_{B_{\mathfrak{q}}}(M_{\mathfrak{p}}, B_{\mathfrak{q}})$$

where $B_{\mathfrak{q}}$ is a Gorenstein ring of dimension $n-1$ and $A_{\mathfrak{p}}$ is an epimorphic image of $B_{\mathfrak{q}}$. If I denotes the injective envelope of the $A_{\mathfrak{p}}$-module $A_{\mathfrak{p}}/\mathfrak{p}A_{\mathfrak{p}}$, then by local duality

$$\mathrm{Hom}_{A_{\mathfrak{p}}}\big(\mathrm{Ext}^{n-1}_{B_{\mathfrak{q}}}(M_{\mathfrak{p}}, B_{\mathfrak{q}}), I\big) \cong H^0_{\mathfrak{p}A_{\mathfrak{p}}}(M_{\mathfrak{p}}) = 0 \quad \text{if } \dim A/\mathfrak{p} = 1$$

as was shown previously. Hence $\dim A/\mathfrak{p} = 0$, i.e. $\mathfrak{p} = \mathfrak{m}$. Thus $\mathrm{Ext}^{n-1}_B(M, B)$ is a module of finite length and therefore $H^1_{\mathfrak{m}}(M) \cong \mathrm{Hom}_A\big(\mathrm{Ext}^{n-1}_B(M, B), E\big)$ is a module of finite length, i.e. a Noetherian A-module. Next assume that $H^1_{\mathfrak{m}}(M)$ is Noetherian. Take an element $x \in \mathfrak{m}$ with $\dim M/xM = d - 1$ and $xH^1_{\mathfrak{m}}(M) = 0$. Then from the exact sequence $0 \to M \xrightarrow{x} M \to M/xM \to 0$ we get a monomorphism $H^1_{\mathfrak{m}}(M) \to H^1_{\mathfrak{m}}(M/xM) \cong H^1_{\mathfrak{m}}(M/xM:\langle\mathfrak{m}\rangle)$. The last module is annihilated by $\mathfrak{m}$ since $M/xM:\langle\mathfrak{m}\rangle$ is Buchsbaum and $1 < d - 1 = \dim M/xM:\langle\mathfrak{m}\rangle$. Hence $\mathfrak{m}H^1_{\mathfrak{m}}(M) = 0$ and the Corollary follows from Proposition 2.23, q.e.d.

Another application of Theorem 2.15 gives informations on the Buchsbaum property of a local ring whose zero ideal is the intersection of two "perfect" ideals. We note that the ideals "of type (r, d)" defined above (Definition 2.13) belong to this setting, see also Lemma 2.14. Additionally the following statement has some useful applications with respect to liaison among arithmetically Buchsbaum curves in $\mathbf{P}^3$.

Proposition 2.25. *Let A be a local ring with $d := \dim A \geq 2$ and $\mathfrak{a}$, $\mathfrak{b}$ ideals of A with $\mathfrak{a} \cap \mathfrak{b} = 0$, $\dim A/\mathfrak{a} + \mathfrak{b} < d$. Assume $A/\mathfrak{a}$ and $A/\mathfrak{b}$ to be Cohen-Macaulay rings of dimension d. Then A is a Buchsbaum ring if and only if either $\mathfrak{a} + \mathfrak{b} = \mathfrak{m}$ or $B := A/(\mathfrak{a} + \mathfrak{b})$ is a Buchsbaum ring of dimension $d - 1$.*

Proof: As in the first part of the proof of Lemma 2.14 we have an exact sequence:

$$0 \to A \to A/\mathfrak{a} \oplus A/\mathfrak{b} \to B \to 0.$$

Since depth $A/\mathfrak{a}$ = depth $A/\mathfrak{b} = d$, we have for all $i < d$ commutative diagrams:

$$\begin{array}{ccc} H^{i-1}(\mathfrak{m}, B) & \xrightarrow{\sim} & H^i(\mathfrak{m}, A) \\ \downarrow{\scriptstyle \lambda_B^{i-1}} & & \downarrow{\scriptstyle \lambda_A^i} \\ H^{i-1}_{\mathfrak{m}}(B) & \xrightarrow{\sim} & H^i_{\mathfrak{m}}(A). \end{array}$$

By Theorem 2.15 A is a Buchsbaum ring if and only if λ_B^j is surjective for all $j \leq d - 2$.

If $\dim B = 0$, i.e. $H^j_{\mathfrak{m}}(B) = O$ for all $j \geq 1$ this is equivalent to the surjectivity of λ^0_B. But $H^0(\mathfrak{m}, B) \cong O :_B \mathfrak{m} \cong (\mathfrak{a} + \mathfrak{b}) : \mathfrak{m}/(\mathfrak{a} + \mathfrak{b})$ and $H^0_{\mathfrak{m}}(B) \cong B = A/(\mathfrak{a} + \mathfrak{b})$ and thus A is a Buchsbaum ring if and only if $\mathfrak{a} + \mathfrak{b} = \mathfrak{m}$.

If $\dim B > 0$, $\lambda^{\dim B}_B$ cannot be surjective since $H^{\dim B}_{\mathfrak{m}}(B)$ is not a module of finite length. Since $\dim B \leq d - 1$, A is a Buchsbaum ring if and only if $\dim B = d - 1$ and λ^j_B are surjective for all $j < d - 1$, i.e. if and only if B is a Buchsbaum module over A by Theorem 2.15 and hence a Buchsbaum ring itself by Lemma 1.6, q.e.d.

Another application of Theorem 2.15 was proven by U. Daepp and A. Evans in [1]. In order to formulate this result we need to introduce some further notions.

Let M be an A-module and X an indeterminate. Then we set

$$M^* := M[X]_{\mathfrak{m}[X]},$$

where $\mathfrak{m}[X]$ denotes the kernel of the map $A[X] \to (A/\mathfrak{m})[X]$ given by the canonical projection $A \to A/\mathfrak{m}$, i.e. $\mathfrak{m}[X]$ is a prime ideal of $A[X]$.

Now, it is clear that $M^* = M \otimes_A A^*$ and that the natural map $A \to A^*$ is a local flat homomorphism.

For every A-module N of finite length $l_{A^*}(N^*) = l_A(N)$. This implies for example $e_0(\mathfrak{q}^*, M^*) = e_0(\mathfrak{q}, M)$ for every Noetherian A-module M and every ideal $\mathfrak{q} \subset A$ with $l(M/\mathfrak{q} \cdot M) < \infty$.

For each ideal $\mathfrak{a} \subset A$ we have $\mathfrak{a}^* = \mathfrak{a} \otimes A^* = \mathfrak{a} \cdot A^*$ and thus for the Koszul complex

$$K_.(\mathfrak{a} \cdot A^*, M^*) \cong K_.(\mathfrak{a}, M) \otimes_A A^* \quad \text{(as complexes)}.$$

Therefore $H^i(\mathfrak{a}^*, M^*) \cong H^i(\mathfrak{a}, M) \otimes_A A^*$ and $H^i_{\mathfrak{a}^*}(M^*) = H^i_{\mathfrak{a}}(M) \otimes_A A^*$ since tensor products commute with direct limits. Consequently, ($\mathfrak{m}^*$ is the maximal ideal of A^*) $\dim_A M = \dim_{A^*} M^*$.

Now, we are in position to prove.

Lemma 2.26. *Let M be a Noetherian A-module of positive dimension. M is a Buchsbaum module over A if and only if M^* is a Buchsbaum module over A. Moreover, $I(M^*) = I(M)$.*

Proof: We have $\lambda^i_{M^*} \cong \lambda^i_M \otimes_A \mathrm{id}_{A^*}$ and since $\otimes_A A^*$ is an exact functor, $\lambda^i_{M^*}$ is surjective if and only if λ^i_M is surjective which proves (using Theorem 2.15) the first statement.

Next, let M (and therefore M^*) be a Buchsbaum module. Let $\mathfrak{q}$ be a parameter ideal of M. Then $\mathfrak{q}^*$ is a parameter ideal of M^* and

$$\begin{aligned} I(M^*) &= l(M^*/\mathfrak{q}^* \cdot M^*) - e_0(\mathfrak{q}^*, M^*) = l\big((M/\mathfrak{q} \cdot M)^*\big) - e_0(\mathfrak{q}^*, M^*) \\ &= l(M/\mathfrak{q} \cdot M) - e_0(\mathfrak{q}, M) = I(M), \end{aligned}$$

q.e.d.

The previous lemma has an interesting consequence which was first proven for a special case by Daepp and Evans in [1]. To establish this we need a lemma (for notations see Chap. 0, § 2, 1.):

Lemma 2.27. *Let R be a graded ring and assume that $\mathfrak{p} \subset R$ is a homogeneous prime ideal with $[R]_1 \not\subseteq \mathfrak{p}$. Then*

$$R_{\mathfrak{p}} \cong R^*_{(\mathfrak{p})}$$

and hence we have for every graded R-module M:

$$M_{\mathfrak{p}} \cong M^*_{(\mathfrak{p})}.$$

Proof: This follows immediately from Lemma 0.2.1. (It is sufficient to prove $R_{\mathfrak{p}} \cong R^*_{(\mathfrak{p})}$. From Lemma 0.2.1 we obtain $R_{\mathfrak{p}} \cong (R_{\mathfrak{p},h})_{\mathfrak{p}\cdot R_{\mathfrak{p},h}} \cong (R_{(\mathfrak{p})}[X]_X)_{\mathfrak{m}[X]_X} \cong R_{(\mathfrak{p})}[X]_{\mathfrak{m}[X]} = R^*_{(\mathfrak{p})}$, where $\mathfrak{m}$ is the maximal ideal of $R_{(\mathfrak{p})}$.)

Corollary 2.28. *Let R be a Noetherian graded ring and let M denote a Noetherian graded R-module. Then we have for every homogeneous prime ideal* $\mathfrak{p} \subset R$ *with* $[R]_1 \not\subseteq \mathfrak{p}$: $M_{\mathfrak{p}}$ *is a Buchsbaum module if and only if* $M_{(\mathfrak{p})}$ *is a Buchsbaum module. In this case,* $I(M_{\mathfrak{p}}) = I(M_{(\mathfrak{p})})$.

§ 3. Graded Buchsbaum modules

The goal of this paragraph is to expand the results of both previous paragraphs to graded modules. The geometric background and the motivation for this is to get information about the Buchsbaum property of the local ring at the vertex of the affine cone over a projective variety.

Throughout we use the concepts introduced in Chapter 0, § 2. In addition we always suppose that our graded k-algebras (k a field) are generated by their homogeneous elements of degree one, i.e. they are of the form $k[X_0, \ldots, X_n]/\mathfrak{a}$ where $X_0, \ldots, X_n$ are indeterminates (of degree 1) and $\mathfrak{a}$ is a homogeneous ideal of $k[X_0, \ldots, X_n]$.

Let R always denote such a graded k-algebra with the maximal (homogeneous) ideal $\mathfrak{m} = \bigoplus_{n\geq 1} [R]_n$.

Definition 3.1. Let M be a Noetherian graded R-module of positive dimension. M is called a *Buchsbaum module* if $M_{\mathfrak{m}}$ is a Buchsbaum module (over $R_{\mathfrak{m}}$). M is called an *h-Buchsbaum module* if every homogeneous system of parameters with respect to M is a weak M-sequence.

Thereby weak M-sequences are defined analogously to the local case.

It is clear that M is an h-Buchsbaum module if it is a Buchsbaum module. Our goal is to study the converse of this statement. We are able to prove it if k is an infinite field. This means, geometrically speaking, that the Buchsbaum property of the local ring of the affine cone at the vertex over a projective variety may be verified by regarding only homogeneous systems of parameters. In a similar way as in § 1 M-bases consisting of homogeneous elements (M a Noetherian graded R-module) of homogeneous ideals $\mathfrak{a}$ with $\dim M/\mathfrak{a}\cdot M$ are defined. Therefore we omit an additional definition. But in contrast to the local case such bases may not exist. We give two examples:

1. In a homogeneous basis of $\mathfrak{a}$ appear elements of different degrees: We choose $R = k[X, Y]$, $\mathfrak{a} = (X, Y^2)\cdot R$, $M = R/X\cdot R$ (X, Y indeterminates).
2. The field k is finite: We choose $\mathfrak{a} = \mathfrak{m}$ and $M := R/p\cdot R$ where p denotes the product of all elements of degree one in R.

If we exclude these two cases we are able to prove the existence of M-bases. The proof is essentially the same as in the local case (Proposition 1.9). In addition we need here the following easy

Lemma 3.2. *Assume that the ground field k is infinite. If $y_1, \ldots, y_t$ are elements of $[R]_l$, $l \geq 0$, and if there are homogeneous ideals $\mathfrak{b}_1, \ldots, \mathfrak{b}_s$ in R with $(y_1, \ldots, y_t) R \not\subseteq \mathfrak{b}_i$ for all $i = 1, \ldots, s$, then there are elements $\alpha_1, \ldots, \alpha_t$ of k with*

$$\alpha_1 \cdot y_1 + \ldots + \alpha_t \cdot y_t \notin \mathfrak{b}_1 \cup \ldots \cup \mathfrak{b}_s.$$

Proof: Let $V \subseteq [R]_l$ be the vector space spanned by the y_i's. We put $V_i := V \cap \mathfrak{b}_i$, $i = 1, \ldots, s$. Since $(y_1, \ldots, y_t) \cdot R \not\subseteq \mathfrak{b}_i$ for all i, V_i is a proper subspace of V for each i. Since k is infinite, this implies $V \cap (\mathfrak{b}_1 \cup \ldots \cup \mathfrak{b}_s) = V_1 \cup \ldots \cup V_s \subsetneqq V$, hence we get $V \not\subseteq \mathfrak{b}_1 \cup \ldots \cup \mathfrak{b}_s$, q.e.d.

Now we can prove:

Proposition 3.3. *Assume that the ground field k is infinite. Let $\mathfrak{a} \subset R$ be an ideal possessing a basis which consists of homogeneous elements of the same degree r and $M_1, \ldots, M_n$ Noetherian graded R-modules with* $\dim_R M_i/\mathfrak{a}M_i = 0$ *for all $i = 1, \ldots, n$. Then there are $a_1, \ldots, a_t \in \mathfrak{a}$ forming a homogeneous M_i-basis of $\mathfrak{a}$ for all $i = 1, \ldots, n$.*

Proof: We can use the proof of Proposition 1.9 word for word with only one modification: when constructing a_m we simply choose an element of degree r not contained in $(a_1, \ldots, a_{m-1}) R \cup \bigcup_{\mathfrak{p} \in L} \mathfrak{p}$ (apply Lemma 3.2), q.e.d.

Also we have:

Proposition 3.4. *Let M be a Noetherian graded R-module with $d := \dim M > 0$. The following conditions are equivalent:*

(i) *There is a homogeneous system of parameters of M contained in $\mathfrak{m}^2$ which is a weak M-sequence.*

(ii) *Every homogeneous system of parameters of M contained in $\mathfrak{m}^2$ is a weak M-sequence.*

(iii) $\mathfrak{m} \cdot \underline{H}^i_{\mathfrak{m}}(M) = 0$ *for all $i \neq d$.*

Proof: (ii) $\Rightarrow$ (i) is trivial and (i) $\Rightarrow$ (iii) may be verified as in the proof of Proposition 2.1. We only have to pay attention to the necessary shifting of degrees. Finally, we obtain the implication (iii) $\Rightarrow$ (ii) by localizing at $\mathfrak{m}$ and applying Proposition 2.1, q.e.d.

Likewise by localizing at $\mathfrak{m}$ and applying Theorem 2.10 we obtain

Theorem 3.5. *Let M be a Noetherian graded R-module with* $\dim M > 0$. *If the natural maps ($\underline{k} = R/\mathfrak{m}$)*

$$\varphi^i_M : \underline{\mathrm{Ext}}^i_R(\underline{k}, M) \to \underline{H}^i_{\mathfrak{m}}(M)$$

are surjective for all $i <$ $\dim M$ *then M is a Buchsbaum module.*

Corollary 3.6. *Let M be as in Theorem 3.5. If in addition $r :=$* $\operatorname{depth} M < \dim M =: d$ *and $\underline{H}^i_{\mathfrak{m}}(M) = 0$ for all $i \neq r, d$ then the following conditions are equivalent:*

(i) *M is a Buchsbaum module.*

(ii) *M is an h-Buchsbaum module.*

(iii) $\mathfrak{m} \cdot \underline{H}^r_{\mathfrak{m}}(M) = 0$.

Proof: (i) $\Rightarrow$ (ii) is clear, (ii) $\Rightarrow$ (iii) follows from Proposition 3.4 and (iii) $\Rightarrow$ (i) we obtain if we localize at $\mathfrak{m}$ and apply Proposition 2.12, q.e.d.

Now we prove the main result of this paragraph:

Theorem 3.7. *Assume that k is an infinite field. If M is a Noetherian graded R-module with $d := \dim M > 0$, the following conditions are equivalent:*

(i) *M is a Buchsbaum module.*

(ii) *M is an h-Buchsbaum module.*

(iii) *Take a homogeneous M-basis $x_1, \ldots, x_t$ of $\mathfrak{m}$. Then for each system $i_1, \ldots, i_d$ of integers with $1 \leq i_1 < \ldots < i_d \leq t$ the sequence $x_{i_1}^{r_1}, \ldots, x_{i_d}^{r_d}$ is a weak M-sequence for all $r_1, \ldots, r_d \in \{1, 2\}$.*

(iv) *The natural maps*

$$\lambda_M^i : \underline{H}^i(\mathfrak{m}, M) \to \underline{H}_{\mathfrak{m}}^i(M)$$

are surjective for all $i < d$.

If R is a free k-algebra then (i)—(iv) *are equivalent to*

(v) *The natural maps*

$$\varphi_M^i : \underline{\mathrm{Ext}}_R^i(\underline{k}, M) \to \underline{H}_{\mathfrak{m}}^i(M)$$

are surjective for all $i < d$.

The proof is not difficult. The implications are either self evident or follow by localizing at $\mathfrak{m}$ and applying Theorem 2.15 and Corollary 2.16.

The following example shows that our assumption on k is necessary. It implies that (for finite k) the conditions (i) and (ii) of Theorem 3.7 are not equivalent.

Example 3.8. Let k be a finite field. Choose a polynomial ring $R = k[X_1, \ldots, X_n]$ over k and a Noetherian graded R-module M such that:

(i) M is not a Buchsbaum module, $\dim M \geq 3$ and depth $M > 0$.

(ii) $\mathfrak{m} \cdot \underline{H}_{\mathfrak{m}}^i(M) = 0$ for all $i < \dim M$.

We note that such R and M exist (see Example V.5.4). We have to distinguish the following cases:

1. Each linear form of R is not a part of a system of parameters of M, i.e. all homo geneous systems of parameters of M are contained in $\mathfrak{m}^2$. Then M itself is an h Buchsbaum module by applying Proposition 3.4.
2. There is (at least) one linear form l of R with $\dim M/l \cdot M < \dim M$. Let p be the square of the product of all these linear forms and set $N := M/p \cdot M$. We note that p is a non-zero divisor of M contained in $\mathfrak{m}^2$.

Let $x_1, \ldots, x_d$ be a homogeneous system of parameters of N. Then $x_1, \ldots, x_d \in \mathfrak{m}^2$ and by Proposition 3.4 $p, x_1, \ldots, x_d$ is a weak M-sequence. Therefore $x_1, \ldots, x_d$ is a weak N-sequence and N is an h-Buchsbaum module. If N would be a Buchsbaum module, i.e. $N_{\mathfrak{m}}$ is a Buchsbaum module over $R_{\mathfrak{m}}$, the lifting property (Proposition 2.19) implies that $M_{\mathfrak{m}}$ is a Buchsbaum module. This is a contradiction and we have found an h-Buchsbaum module which is not a Buchsbaum module.

As an application of our last theorem we will state a new sufficient criterion for graded Buchsbaum modules using only local cohomology modules. To this end we need:

Lemma 3.9. *Let* $R := k[X_1, \ldots, X_n]$ *($X_1, \ldots, X_n$ indeterminates) and let* H *be a graded* R*-module with* $\mathfrak{m} \cdot H = O$. *Then*

$$\underline{\mathrm{Ext}}^i_R(\underline{k}, H) \cong \underline{\mathrm{Hom}}_R\left(R^{\binom{n}{i}}(-i), H\right) \cong H^{\binom{n}{i}}(i) \quad \text{for all } i \geq 0.$$

Proof: The graded Koszul complex $K_.(X_1, \ldots, X_n; R)$:

$$0 \to R^{\binom{n}{n}}(-n) \to \ldots \to R^{\binom{n}{2}}(-2) \to R^{\binom{n}{1}}(-1) \to R \to O$$

provides a free resolution of $\underline{k} = R/\mathfrak{m}$. Applying $\underline{\mathrm{Hom}}_R(\ , H)$ we get for all $i \geq 0$ $\left(\mathfrak{m} \cdot \underline{\mathrm{Hom}}_R(\ , H) = O\right)$:

$$\underline{\mathrm{Ext}}^i_R(\underline{k}, H) \cong \underline{\mathrm{Hom}}_R\left(R^{\binom{n}{i}}(-i), H\right) \cong \underline{\mathrm{Hom}}_R\left(R^{\binom{n}{i}}, H(i)\right) \cong H^{\binom{n}{i}}(i),$$

q.e.d.

For abbreviated notation we define for each graded R-module M the following set of integers:

$$g(M) := \{i \in \mathbf{Z} \mid [M]_i \neq O\}.$$

Proposition 3.10. *Let* M *be a Noetherian graded* R*-module with* $d := \dim M > 0$ *and* $\mathfrak{m} \cdot \underline{H}^i_\mathfrak{m}(M) = O$ *for all* $i < d$. *If for each pair of integers* i, j *with* $0 \leq i < j < d$ *and all* $p \in g\left(\underline{H}^i_\mathfrak{m}(M)\right)$, $q \in g\left(\underline{H}^j_\mathfrak{m}(M)\right)$,

$$(i + p) - (j + q) \neq 1$$

then M *is a Buchsbaum module.*

Proof: First assume $R = k[X_1, \ldots, X_n]$, $X_1, \ldots, X_n$ indeterminates. If $0 \leq i < j < d$ then for $q \in g\left(\underline{H}^j_\mathfrak{m}(M)\right)$ (by Lemma 3.9):

$$\left[\underline{\mathrm{Ext}}^{j-i+1}_R\left(\underline{k}, \underline{H}^i_\mathfrak{m}(M)\right)\right]_q \cong \left[\underline{H}^i_\mathfrak{m}(M)^{\binom{n}{j-i+1}}\right]_{q+j-i+1} = O,$$

since by our assumption $q + j - i + 1 \notin g\left(\underline{H}^i_\mathfrak{m}(M)\right)$.

Now let $0 \to I^0 \to I^1 \to \ldots$ be a minimal graded injective resolution of M and set $J^i := \underline{H}^0_\mathfrak{m}(I^i)$. We examine the corresponding complex

$$0 \to J^0 \xrightarrow{d^0} J^1 \xrightarrow{d^1} \ldots$$

Set $B^i := \mathrm{Im}\, d^i$, $Z^i := \mathrm{Ker}\, d^i$. Then there are exact sequences

$$0 \to B^{j-1} \to Z^j \to \underline{H}^j_\mathfrak{m}(M) \to O \quad \text{(for all } j \geq 0)$$

and isomorphisms

$$\underline{\mathrm{Hom}}_R(\underline{k}, Z^j) \cong \underline{\mathrm{Ext}}^j_R(\underline{k}, M) \quad \text{(since the resolution was minimal).}$$

The map $Z^j \to \underline{H}^j_\mathfrak{m}(M)$ induces a homomorphism π^j:

$$\underline{\mathrm{Ext}}^j_R(\underline{k}, M) \cong \underline{\mathrm{Hom}}_R(\underline{k}, Z^j) \xrightarrow{\pi^j} \underline{\mathrm{Hom}}_R\left(\underline{k}, \underline{H}^j_\mathfrak{m}(M)\right) \cong \underline{H}^j_\mathfrak{m}(M).$$

The composition of these maps is just the natural map φ^j_M.

By induction on j we show the surjectivity of π^j. Since $\mathfrak{m} \cdot \underline{H}^0_\mathfrak{m}(M) = O$, π^0 is even an isomorphism. Let $0 < j < d$. Then we have for each $q \in g\left(\underline{H}^j_\mathfrak{m}(M)\right)$ an exact sequence

$$0 \to [\underline{\mathrm{Hom}}_R(\underline{k}, B^{j-1})]_q \to [\underline{\mathrm{Hom}}_R(\underline{k}, Z^j)]_q \xrightarrow{[\pi^j]_q} \left[\underline{\mathrm{Hom}}_R\left(\underline{k}, \underline{H}^j_\mathfrak{m}(M)\right)\right]_q$$
$$\to [\underline{\mathrm{Ext}}^1_R(\underline{k}, B^{j-1})]_q.$$

For $i < j$ we have by the induction hypothesis and by Lemma 2.9 for each $l \geq 0$ exact sequences

$$0 \to \underline{\mathrm{Ext}}^l_R(\underline{k}, B^{i-1}) \to \underline{\mathrm{Ext}}^l_R(\underline{k}, Z^i) \to \underline{\mathrm{Ext}}^l_R(\underline{k}, \underline{H}^i_{\mathfrak{m}}(M)) \to O.$$

Since $\underline{\mathrm{Ext}}^l_R(\underline{k}, B^{i-1}) \cong \underline{\mathrm{Ext}}^{l+1}_R(\underline{k}, Z^{i-1})$ for all $l \geq 0$, $i \geq 1$ we get

$$\begin{aligned}[\underline{\mathrm{Ext}}^1_R(\underline{k}, B^{j-1})]_q &\cong [\underline{\mathrm{Ext}}^2_R(\underline{k}, Z^{j-1})]_q \cong [\underline{\mathrm{Ext}}^2_R(\underline{k}, B^{j-2})]_q \\ &\cong [\underline{\mathrm{Ext}}^3_R(\underline{k}, Z^{j-2})]_q \cong \dots \cong [\underline{\mathrm{Ext}}^j_R(\underline{k}, B^0)]_q \\ &\cong [\underline{\mathrm{Ext}}^{j+1}_R(\underline{k}, \underline{H}^0_{\mathfrak{m}}(M))]_q = O.\end{aligned}$$

Moreover, for $r \notin g(\underline{H}^j_{\mathfrak{m}}(M))$ we have $[\underline{\mathrm{Hom}}_R(\underline{k}, \underline{H}^j_{\mathfrak{m}}(M))]_r = [\underline{H}^j_{\mathfrak{m}}(M)]_r = 0$ and therefore $[\pi^j]_i$ is surjective for all $i \in \mathbf{Z}$. Hence π^j is surjective and M is a Buchsbaum module by Theorem 3.5.

Now let R be an arbitrary graded k-algebra which is an epimorphic image of $S := k[X_1, \dots, X_n]$. Let $\mathfrak{n} := (X_1, \dots, X_n) \cdot S$. Then $\underline{H}^i_{\mathfrak{m}}(M)$ and $\underline{H}^i_{\mathfrak{n}}(M)$ are isomorphic as S-modules. Hence M is a Buchsbaum module over S and consequently also over R by Lemma 1.6, q.e.d.

We note that this criterion is not necessary. To show this we will provide an example using Segre products of graded modules, see Example V.5.5.

§ 4. Segre products of graded Cohen-Macaulay modules

In this paragraph we will investigate Segre products of graded Cohen-Macaulay modules in order to get additional examples of Buchsbaum modules. In this context we obtain as an easy consequence the results of W. L. Chow on the Cohen-Macaulay property of such Segre products (for our special graded k-algebras), see Chow [1]. Throughout this paragraph we make the following assumptions:

1. Our ground field k is infinite.
2. Cohen-Macaulay module always means Noetherian graded Cohen-Macaulay module.
3. Our graded k-algebras are finitely generated by their homogeneous elements of degree one.

Always let R_1, R_2 denote two graded k-algebras with maximal homogeneous ideals $\mathfrak{m}_1$, resp. $\mathfrak{m}_2$. We set $R := \sigma(R_1, R_2)$ and $\mathfrak{m} := \sigma(\mathfrak{m}_1, \mathfrak{m}_2)$. $\mathfrak{m}$ is then the maximal homogeneous ideal of the graded k-algebra R, see also Chapter 0, § 2, 4.

First we state and prove some preliminary results:

Lemma 4.1. *Let R be a graded k-algebra. Suppose that M is a Noetherian graded R-module with $d := \dim M > 0$. We have:*

(i) *If* depth $M \geq 1$, *then* $[M]_n \neq 0$ *for all* $n \geq a(M)$.

(ii) $\mathfrak{m} \cdot [\underline{H}^d_{\mathfrak{m}}(M)]_n \neq 0$ *for all* $n < e(\underline{H}^d_{\mathfrak{m}}(M))$

($a(M)$, $e(\underline{H}^d_{\mathfrak{m}}(M))$ are defined in Chapter 0, Definition 2.2).

Proof:

(i) If $[M]_p = 0$, then we have for all $q < p$:

$$[R]_{p-q} \cdot [M]_q \subseteq [M]_p = 0, \quad \text{i.e.} \quad [M]_q \subseteq 0 :_M [R]_{p-q} = 0 :_M \mathfrak{m}^{p-q} \subseteq \underline{H}^0_{\mathfrak{m}}(M) = O,$$

since depth $M \geq 1$. Hence $p < a(M)$.

(ii) Since $d \geq 1$, $\underline{H}^d_{\mathfrak{m}}(M) \simeq \underline{H}^d_{\mathfrak{m}}(M/\underline{H}^0_{\mathfrak{m}}(M))$, i.e. we may assume depth $M \geq 1$. Let $x \in [R]_1$ be a non-zero divisor of M (k is infinite!). For each $t \geq 1$ let $M_t := M/x^t \cdot M$. Then $\dim M_t = d - 1$ and the exact sequence $0 \to M \xrightarrow{x^t} M(t) \to M_t(t) \to 0$ yields an epimorphism

$$\underline{H}^d_{\mathfrak{m}}(M) \xrightarrow{x^t} \underline{H}^d_{\mathfrak{m}}(M(t)) \simeq \underline{H}^d_{\mathfrak{m}}(M)\,(t).$$

From this we obtain for each $n \in \mathbf{Z}$ epimorphisms

$$[\underline{H}^d_{\mathfrak{m}}(M)]_n \xrightarrow{x^t} [\underline{H}^d_{\mathfrak{m}}(M)]_{n+t}.$$

If $\mathfrak{m} \cdot [\underline{H}^d_{\mathfrak{m}}(M)]_p = 0$, then we have for all $q > p$:

$$[\underline{H}^d_{\mathfrak{m}}(M)]_q = 0, \quad \text{i.e.} \quad p \geq e(\underline{H}^d_{\mathfrak{m}}(M)),$$

q.e.d.

Corollary 4.2. *Let M_1, M_2 be Noetherian graded R_1-, resp. R_2-modules with* $d_i := \dim M_i > 0$ *for* $i = 1, 2$. *Then*

$$\dim \sigma(M_1, M_2) = d_1 + d_2 - 1.$$

Proof: Let $M := \sigma(M_1, M_2)$, $d := d_1 + d_2 - 1$. Then by the Künneth formulas (see Proposition 0.2.1(ii))

$$\underline{H}^n(M) = \bigoplus_{p+q=n} \sigma(\underline{H}^p(M_1), \underline{H}^q(M_2)) \qquad \text{for all } n.$$

Since $\underline{H}^p(M_1) = 0$ for all $p > d_1 - 1$, $\underline{H}^q(M_2) = 0$ for all $q > d_2 - 1$, we get $\underline{H}^n(M) = 0$ for all $n > d_1 + d_2 - 2$, i.e. $\dim M \leq d$.

Now by Lemma 4.1 there are $e_1, e_2 \in \mathbf{Z}$ such that $[\underline{H}^{d_i-1}(M_i)]_{n_i} \neq 0$ for all $n_i \leq e_i$, $i = 1, 2$. Hence

$$\underline{H}^{d-1}(M) \simeq \sigma(\underline{H}^{d_1-1}(M_1), \underline{H}^{d_2-1}(M_2)) \neq 0.$$

If $d > 1$ this implies $\underline{H}^d_{\mathfrak{m}}(M) \neq 0$, i.e. $\dim M = d$. Let $d = 1$. If $\dim M = 0$, then $\underline{H}^0_{\mathfrak{m}}(M) \simeq M$ and $\underline{H}^1_{\mathfrak{m}}(M) = 0$. Therefore $\underline{H}^0(M) = 0$ which is impossible, q.e.d.

Lemma 4.3. *Let R be a graded k-algebra and M a Noetherian graded R-module with* $d := \dim M > 0$. *If M is a Cohen-Macaulay module then*

$$H_M(n) - h_M(n) = (-1)^d \operatorname{rank}_k[\underline{H}^d_{\mathfrak{m}}(M)]_n.$$

Proof: By Serre [1], Nr. 79, we have for arbitrary Noetherian graded R-modules M:

$$h_{\tilde{M}}(n) = \sum_{i \geq 0} (-1)^i \operatorname{rank}_k(H^i(X, \tilde{M}(n)) \qquad \text{for all } n \in \mathbf{Z}$$

where $X := \operatorname{Proj} R$ and $\tilde{M}$ denotes the sheaf associated to M.

For all $i \geq 0$ and $n \in \mathbf{Z}$ one has (see Proposition 0.2.3)

$$H^i(X, \tilde{M}(n)) \simeq H^i(M(n)) = [\underline{H}^i(M)]_n$$

and therefore

$$\begin{aligned} h_M(n) = h_{\tilde{M}}(n) &= \operatorname{rank}_k[\underline{H}^0(M)]_n + \sum_{i \geq 1} (-1)^i \operatorname{rank}_k[\underline{H}^{i+1}_{\mathfrak{m}}(M)]_n \\ &= \operatorname{rank}_k[M]_n - \sum_{i \geq 0} (-1)^i \operatorname{rank}_k[\underline{H}^i_{\mathfrak{m}}(M)]_n \end{aligned}$$

is obtained. Since $\operatorname{rank}_k[M]_n = H_M(n)$ and since by our assumption $\underline{H}^i_{\mathfrak{m}}(M) = 0$ for all $i \neq d$, the statement now follows, q.e.d.

Together with Lemma 4.1(ii) this implies

Corollary 4.4. *Let R, M be as in Lemma 4.3. Then* (see Definition 0.2.2):
(i) $r(M) = 1 + e(\underline{H}^d_{\mathfrak{m}}(M))$.
(ii) $r(M) = \inf\{n \in \mathbf{Z} \mid h_M(n) = H_M(n)\}$.

Our next result enables us to calculate the index of regularity $r(R)$ of a graded Cohen-Macaulay algebra R without knowledge of the Hilbert function $H_R(n)$ of R.

Lemma 4.5. *Let R denote a graded Cohen-Macaulay algebra over k with $d := \dim R > 0$. Then we have for every system of parameters $x_1, \ldots, x_d$ with respect to R which is contained in $[R]_1$:*

$$d + r(R) = \inf\{t \in \mathbf{N} \mid \mathfrak{m}^t \subseteq (x_1, \ldots, x_d) \cdot R\}.$$

Proof: Let $t_0 := \inf\{t \in \mathbf{N} \mid \mathfrak{m}^t \subseteq (x_1, \ldots, x_d) \cdot R\}$, $r := r(R)$ and $S := R/(x_1, \ldots, x_d) \cdot R$. Since R is a Cohen-Macaulay algebra we have for all $t \geq 0$:

$$H_R(t) = \sum_{i=0}^{t} \binom{t-i+d-1}{d-1} H_S(i)$$

and therefore

$$h_R(t) = \sum_{i\geq 0} \binom{t-i+d-1}{d-1} H_S(i).$$

The validity of the first equation follows by an easy induction argument (induction on d).

Now, $H_R(r) = h_R(r)$ and $\binom{r-i+d-1}{d-1} = 0$ for all $i = r+1, \ldots, r+d-1$. Therefore $\sum_{i\geq r+d} \binom{r-i+d-1}{d-1} H_S(i) = 0$. All binomial coefficients occuring in this sum have the same sign. Hence $H_S(i) = 0$ for all $i \geq d+r$, i.e. $\mathfrak{m}^{d+r} \subseteq (x_1, \ldots, x_d) \cdot R$. Thus $t_0 \leq d + r$. On the other hand $\mathfrak{m}^{t_0} \subseteq (x_1, \ldots, x_d) \cdot R$, i.e. $H_S(i) = 0$ for all $i \geq t_0$. Since $\binom{t_0-i-1}{d-1} = 0$ for all $i = t_0-d-1, \ldots, t_0-1$, we have

$$h_R(t_0-d) = \sum_{i=0}^{t_0-d} \binom{t_0-d-i+d-1}{d-1} H_S(i) = H_R(i).$$

Now Corollary 4.4(ii) implies $r \leq t_0 - d$, i.e. $t_0 \geq d + r$, q.e.d.

We are now able to prove the main result of this paragraph:

Theorem 4.6. *Let M_1, M_2 be Cohen-Macaulay modules over R_1, resp. R_2 with $d_i :=$* dim M *≥ 2 for $i = 1, 2$. Then*
(i) *$\sigma(M_1, M_2)$ is a Cohen-Macaulay module if and only if*

$$r(M_1) \leq a(M_2) \quad \text{and} \quad r(M_2) \leq a(M_1).$$

(ii) *$\sigma(M_1, M_2)$ is a Buchsbaum module if and only if*

$$r(M_1) \leq 1 + a(M_2) \quad \text{and} \quad r(M_2) \leq 1 + a(M_1).$$

Proof: Let $M := \sigma(M_1, M_2)$. By Corollary 0.2.12 we have depth $M \geq 2$, i.e. $\underline{H}^0_{\mathfrak{m}}(M) = \underline{H}^1_{\mathfrak{m}}(M) = O$, $\underline{H}^0(M) \simeq M$. Also by our Künneth formulas Proposition 0.2.10 we have

$$\underline{H}^n(M) = 0 \quad \text{for all } n \neq d_1 - 1,\ d_2 - 1,\ d_1 + d_2 - 2$$

and

$$\left.\begin{aligned} \underline{H}^{d_1-1}(M) &\simeq \sigma\big(\underline{H}^{d_1-1}(M_1), \underline{H}^0(M_2)\big) \simeq \sigma\big(\underline{H}^{d_1}_{\mathfrak{m}_1}(M_1), M_2\big), \\ \underline{H}^{d_2-1}(M) &\simeq \sigma\big(\underline{H}^0(M_1), \underline{H}^{d_2-1}(M_2)\big) \simeq \sigma\big(M_1, \underline{H}^{d_2}_{\mathfrak{m}_2}(M_2)\big) \end{aligned}\right\} \text{ if } d_1 \neq d_2,$$

$$\begin{aligned} \underline{H}^{d-1}(M) &\simeq \sigma\big(\underline{H}^{d-1}(M_1), \underline{H}^0(M_2)\big) \oplus \sigma\big(\underline{H}^0(M_1), \underline{H}^{d-1}(M_2)\big) \\ &\simeq \sigma\big(\underline{H}^d_{\mathfrak{m}_1}(M_1), M_2\big) \oplus \sigma\big(M_1, \underline{H}^d_{\mathfrak{m}_2}(M_2)\big) \quad \text{if } d_1 = d_2 =: d, \text{ resp.} \end{aligned}$$

By the commutative diagrams of Corollary 0.2.11 we obtain commutative diagrams:

$$\begin{array}{ccc} \sigma\big(\underline{\mathrm{Ext}}^{d_1}_{R_1}(\underline{k}, M_1), M_2\big) & \to & \underline{\mathrm{Ext}}^{d_1}_R(\underline{k}, M) \\ \downarrow{\scriptstyle \sigma(\varphi^{d_1}_{M_1}, \mathrm{id}_{M_2})} & & \downarrow{\scriptstyle \varphi^{d_1}_M} \\ \sigma\big(\underline{H}^{d_1}_{\mathfrak{m}_1}(M_1), M_2\big) & \to & \underline{H}^{d_1}_{\mathfrak{m}}(M) \end{array} \qquad \begin{array}{ccc} \sigma\big(M_1, \underline{\mathrm{Ext}}^{d_2}_{R_2}(\underline{k}, M_2)\big) & \to & \underline{\mathrm{Ext}}^{d_2}_R(\underline{k}, M) \\ \downarrow{\scriptstyle \sigma(\mathrm{id}_{M_1}, \varphi^{d_2}_{M_2})} & & \downarrow{\scriptstyle \varphi^{d_2}_M} \\ \sigma\big(M_1, \underline{H}^{d_2}_{\mathfrak{m}_2}(M_2)\big) & \to & \underline{H}^{d_2}_{\mathfrak{m}}(M) \end{array}$$

if $d_1 \neq d_2$ and a similar diagram if $d_1 = d_2$.

Now, M is a Cohen-Macaulay module if and only if $\underline{H}^i_{\mathfrak{m}}(M) = O$ for all $i < d_1 + d_2 - 1$, i.e. if and only if $\sigma\big(\underline{H}^{d_1}_{\mathfrak{m}_1}(M_1), M_2\big) = \sigma\big(M_1, \underline{H}^{d_2}_{\mathfrak{m}_2}(M_2)\big) = 0$. But this is equivalent to (Lemma 4.1(i), (ii) and Corollary 4.4(i))

$$r(M_1) = 1 + e\big(\underline{H}^{d_1}_{\mathfrak{m}_1}(M_1)\big) \leq a(M_2)$$

and

$$r(M_2) = 1 + e\big(\underline{H}^{d_2}_{\mathfrak{m}_2}(M_2)\big) \leq a(M_1).$$

Next assume that M is a Buchsbaum module. Then $\mathfrak{m} \cdot \underline{H}^i_{\mathfrak{m}}(M) = O$ for all $i < d_1 + d_2 - 1$ by Proposition 3.4 and, consequently,

$$0 = \mathfrak{m} \cdot \sigma\big(\underline{H}^{d_1}_{\mathfrak{m}_1}(M_1), M_2\big) = \sigma\big(\mathfrak{m}_1 \cdot \underline{H}^{d_1}_{\mathfrak{m}_1}(M_1), \mathfrak{m}_2 \cdot M_2\big)$$

and

$$0 = \mathfrak{m} \cdot \sigma\big(M_1, \underline{H}^{d_2}_{\mathfrak{m}_2}(M_2)\big) = \sigma\big(\mathfrak{m}_1 \cdot M_1, \mathfrak{m}_2 \cdot \underline{H}^{d_2}_{\mathfrak{m}_2}(M_2)\big).$$

But depth $\mathfrak{m}_i \cdot M_i \geq 1$ for $i = 1, 2$ and by Lemma 4.1

$$r(M_1) = 1 + e\big(\underline{H}^{d_1}_{\mathfrak{m}_1}(M_1)\big) = 1 + e\big(\mathfrak{m}_1 \cdot \underline{H}^{d_1}_{\mathfrak{m}_1}(M_1)\big) \leq a(\mathfrak{m}_2 \cdot M_2) = 1 + a(M_2)$$

and, by exchanging M_1 and M_2

$$r(M_2) \leq 1 + a(M_1).$$

Conversely suppose that these both relations hold. We want to show that the canonical maps φ^i_M are surjective for all $i < d_1 + d_2 - 1$. By Theorem 3.5 this will prove our statement. According to the above commutative diagrams it will be sufficient to prove the surjectivity of $\sigma(\varphi^{d_1}_{M_1}, \mathrm{id}_{M_2})$ and $\sigma(\mathrm{id}_{M_1}, \varphi^{d_2}_{M_2})$.

If $a(M_2) > e\big(\underline{H}^{d_1}_{\mathfrak{m}_1}(M_1)\big)$, $\sigma\big(\underline{H}^{d_1}_{\mathfrak{m}_1}(M_1), M_2\big) = 0$ and there is nothing to prove. Assume therefore $a(M_2) = e\big(\underline{H}^{d_1}_{\mathfrak{m}_1}(M_1)\big) =: e$.

Now,

$$[\sigma(\underline{H}^{d_1}_{\mathfrak{m}_1}(M_1), M_2)]_p = [\underline{H}^{d_1}_{\mathfrak{m}_1}(M_1)]_p \otimes_k [M_2]_p = 0 \quad \text{for all } p \neq e,$$

i.e. $[\sigma(\varphi^{d_1}_{M_1}, \mathrm{id}_{M_2})]_p$ is surjective.

It is also easy to see that $\underline{\mathrm{Ext}}^{d_1}_{R_1}(\underline{k}, M_1) \simeq \underline{\mathrm{Hom}}_{R_1}(\underline{k}, \underline{H}^{d_1}_{\mathfrak{m}_1}(M_1)) \cong 0 :_{\underline{H}^{d_1}_{\mathfrak{m}_1}(M_1)} \mathfrak{m}_1$ and that $\varphi^{d_1}_{M_1}$ is the corresponding embedding. On the other hand $\mathfrak{m}_1 \cdot [\underline{H}^{d_1}_{\mathfrak{m}_1}(M_1)]_e = 0$, i.e. $[\underline{H}^{d_1}_{\mathfrak{m}_1}(M_1)]_e \subseteq [0 :_{\underline{H}^{d_1}_{\mathfrak{m}_1}(M_1)} \mathfrak{m}_1]_e$. Hence $[\varphi^{d_1}_{M_1}]_e$ is even an isomorphism and therefore $[\sigma(\varphi^{d_1}_{M_1}, \mathrm{id}_{M_2})]_e$ is also an isomorphism. Thus $\sigma(\varphi^{d_1}_{M_1}, \mathrm{id}_{M_2})$ is surjective. Exchanging M_1 and M_2 we obtain the same for $\sigma(\mathrm{id}_{M_1}, \varphi^{d_2}_{M_2})$ and the proof is finished, q.e.d.

For $\dim M_1 = 1$ or $\dim M_2 = 1$ we need another formulation of our statement. We do this in the following

Lemma 4.7. *Let M_1, M_2 be as in Theorem 4.6, but $\dim M_1 = 1$, $\dim M_2 \geq 1$. Then we have for $M := \sigma(M_1, M_2)$:*

(i) *If $\dim M_2 = 1$, M is a Cohen-Macaulay module.*

(ii) *If $\dim M_2 \geq 2$, M is a Cohen-Macaulay module if and only if $r(M_1) \leq a(M_2)$ and M is a Buchsbaum module if and only if $r(M_1) \leq 1 + a(M_2)$.*

Proof: (i) By Corollary 4.2 we get $\dim M = 1$ and by Corollary 0.2.12: depth $M \geq 1$.

(ii) is obtained by the same methods used in the proof of Theorem 4.6, q.e.d.

Next we state two corollaries of Theorem 4.6. The first is a main result of Chow [1] for our graded k-algebras and the second gives a very easy method for calculating the Cohen-Macaulay resp. Buchsbaum property of graded complete intersections.

Corollary 4.8. *Let R_1, R_2 be graded Cohen-Macaulay algebras over k with $d_i := \dim R_i \geq 2$ for $i = 1, 2$. $\sigma(R_1, R_2)$ is a Cohen-Macaulay algebra if and only if R_1 and R_2 are proper k-algebras, i.e. there are systems of parameters $x_1, \ldots, x_{d_1}$ of R_1 and $y_1, \ldots, y_{d_2}$ of R_2 consisting of homogeneous elements of degree one such that $\mathfrak{m}_1^{d_1} \subseteq (x_1, \ldots, x_{d_1}) \cdot R_1$ and $\mathfrak{m}_2^{d_2} \subseteq (y_1, \ldots, y_{d_2}) \cdot R_2$.*

Proof: By Lemma 4.5 we have for $i = 1, 2$: $d_i + r(R_i) \leq d_i$. Since $a(R_i) = 0$, the proof is finished, q.e.d.

Corollary 4.9. *Let $R_1 := k[X_1, \ldots, X_n]$, $R_2 := k[Y_1, \ldots, Y_m]$ ($X_1, \ldots, X_n, Y_1, \ldots, Y_m$ indeterminates) and let $\mathfrak{a}_1 = (f_1, \ldots, f_r) \cdot R_1 \subset R_1$, $\mathfrak{a}_2 = (g_1, \ldots, g_s) \cdot R_2 \subset R_2$ be ideals of the principal class r, resp. s with $n - r$, $m - s \geq 2$. Then we have:*

(i) *$\sigma(R_1/\mathfrak{a}_1, R_2/\mathfrak{a}_2)$ is a Cohen-Macaulay algebra if and only if*

$$\sum_{i=1}^{r} \deg f_i \leq n \quad \textit{and} \quad \sum_{j=1}^{s} \deg g_j \leq m.$$

(ii) *$\sigma(R_1/\mathfrak{a}_1, R_2/\mathfrak{a}_2)$ is a Buchsbaum algebra if and only if*

$$\sum_{i=1}^{r} \deg f_i \leq n + 1 \quad \textit{and} \quad \sum_{j=1}^{s} \deg g_j \leq m + 1.$$

Proof: We apply again Theorem 4.6 and notice that for instance

$$r(R_1/\mathfrak{a}_1) = -n + \sum_{i=1}^{r} \deg f_i \qquad \text{(see Gröbner [1], 142.)}$$

The modifications of these corollaries to the case $\dim R_1 = 1$ or $\dim R_2 = 1$ (Corollary 4.8) or $n - r = 1$ or $m - s = 1$ (Corollary 4.9) are easy to obtain (use Lemma 4.7) and are left to the reader.

By reason of the following geometrical discussions we still prove the following

Proposition 4.10. *Let M_1, M_2 be Noetherian graded R_1- resp. R_2-modules with $d_i := \dim M_i \geq 2$ for $i = 1, 2$ and let $M := \sigma(M_1, M_2)$. M is locally Cohen-Macaulay and equidimensional if and only if M_1 and M_2 are locally Cohen-Macaulay and equidimensional.*

Proof: By Corollary 0.4.15 and the existing relations between the local cohomology modules $\underline{H}^i_{\mathfrak{m}}(M)$ and the cohomology modules $\underline{H}^i(M) := \underline{H}^i(R, M) = \varinjlim_n \underline{\mathrm{Ext}}^i_R(\mathfrak{m}^n, M)$ (c.f. Chapter 0, § 2, 3.) we have that M is a locally Cohen-Macaulay module if and only if $\underline{H}^i(M)$ are Noetherian R-modules for all $i < d_1 + d_2 - 2$ $(= \dim M - 1)$ since $R_{\mathfrak{m}}$ is an epimorphic image of a local Gorenstein ring. Further, we recall that the Segre product of two Noetherian graded modules is again a Noetherian graded module and that the Segre product of a Noetherian and an Artinian graded module is a Noetherian (and Artinian) graded module.

If M_1 and M_2 are locally Cohen-Macaulay modules these remarks and our Künneth formulas (Proposition 0.2.10(ii)) imply that $\underline{H}^n(M)$ is Noetherian for all $n < d_1 + d_2 - 2$, i.e. M is a locally Cohen-Macaulay module.

Conversely, assume that M is a locally Cohen-Macaulay module. Let $0 \leq i < d_1 - 1$. Then again by our Künneth formulas $\sigma\big(\underline{H}^i(M_1), \underline{H}^{d_2-1}(M_2)\big)$ is a direct summand of $\underline{H}^{i+d_2-1}(M)$ and hence Noetherian. But there is an integer e such that $[\underline{H}^{d_2-1}(M_2)]_p \neq 0$ for all $p \leq e$ (see Lemma 4.1(ii)). Therefore there must be an integer a with $[\underline{H}^i(M_1)]_q = 0$ for all $q < a$, i.e. $\underline{H}^i(M_1)$ is Noetherian. Hence M_1 is a locally Cohen-Macaulay module. Exchanging M_1 and M_2 we obtain that M_2 is also a locally Cohen-Macaulay module, q.e.d.

We next state some corollaries and make some comments in a geometric context.

Corollary 4.11. *Let $V \subseteq \mathbf{P}^n$ and $W \subseteq \mathbf{P}^m$ be arithmetically Cohen-Macaulay varieties with positive dimensions in projective n-space resp. m-space. Then the Segre embedding $S(V \times W)$ of V and W in $\mathbf{P}^{n+m+n\cdot m}$ is an arithmetically Cohen-Macaulay variety if and only if the arithmetic genus of V and W is 0.*

We note that we mean by the arithmetic genus $p_a(X)$ of a projective variety X the following number:

$$p_a(X) = (-1)^{\dim X}\big(h_X(0) - 1\big)$$

where $h_X(0)$ is the constant term of the Hilbert polynomial of $\tilde{\mathcal{O}}_X$ (see Chap. 0, § 2, 3. and 2.). Now the corollary follows immediately from Theorem 4.6 and Corollary 4.4(ii).

Since for non-singular curves the arithmetic genus agrees with the usual (geometric) genus, we obtain from this corollary better information than does A. Seidenberg in [2] for the construction of arithmetically normal irregular surfaces free of singularities. First, we prove the following result of A. Seidenberg [2].

Corollary 4.12. *Let G, H be plane curves without singularities and such that at least one of them has positive genus. The Segre embedding of $G \times H$ is than arithmetically normal non-Cohen-Macaulay surface F free of singularities.*

Proof: Corollary 4.11 shows that F is an arithmetically non-Cohen-Macaulay surface. Let A be the local ring at the vertex of the affine cone over F. Corollary 0.2.12 implies depth $A \geq 2$.

Clearly F is free of singularities. Therefore Serre's characterization of normal rings (see, e.g., Matsumura [1], p. 125) shows that F is an arithmetically normal surface, q.e.d.

From the foregoing we obtain immediately an example:

Example 4.13. Let G be the cubic curve defined by the equation $X_1^2 \cdot X_2 - X_0^3 + X_0 \cdot X_2^2 = 0$ in $\mathbf{P}^2$ and let $H = \mathbf{P}^1$. Let F be the Segre embedding of $G \times H$ in $\mathbf{P}^5$. Let A be the local ring at the vertex of the affine cone over F. By Corollary 4.12, A is a normal non-Cohen-Macaulay ring. By Corollary 4.9(ii) we obtain that A is a Buchsbaum ring.

These observations render moreover for every given dimension $d \geq 3$ and depth $t \geq 2$, $d > t$, normal (local) Buchsbaum rings A such that $\dim A = d$ and depth $A = t$. This is shown by the following example. Another class of such projective varieties is given in Chapter V, § 2.

Example 4.14. Let $X \subset \mathbf{P}_k^n$, $n \geq 2$, char $k \nmid n + 1$, be the variety defined by the equation $X_0^{n+1} + \ldots + X_n^{n+1} = 0$. Let Y be the Segre embedding of $X \times \mathbf{P}^m$ in $\mathbf{P}^{n+m+n\cdot m}$, $m \geq 1$.

Let A be the local ring at the vertex of the affine cone over Y. By Corollary 4.9(ii) A is a Buchsbaum ring. Since Y is free of singularities we have that A is a normal ring. By Corollary 4.2 $\dim A = n + m$. From our Künneth formulas (Proposition 0.2.10) we get depth $A = n$. Choosing $n = t$, char $k \nmid t + 1$, $m = d - t$ we obtain what was claimed before.

Furthermore, we note that the Segre embedding $S(V \times W)$ of two projective varieties is a locally Cohen-Macaulay variety if and only if V and W are locally Cohen-Macaulay varieties. Clearly this statement is true because the local behavior of $V \times W$ does not depend on the embedding, thus the theorem that the tensor product of two k-algebras is Cohen-Macaulay if and only if the factors are can be applied. Using Proposition 4.10 we have a direct proof:

Corollary 4.15. *Let $V \subseteq \mathbf{P}^n$ and $W \subseteq \mathbf{P}^m$ be varieties of positive dimension. The Segre embedding $S(V \times W)$ is a locally Cohen-Macaulay variety if and only if V and W are locally Cohen-Macaulay varieties.*

In analogy to Corollary 4.15 one might be tempted to ask whether $S(V \times W)$ is a locally Buchsbaum variety, when V and W are locally Buchsbaum. However, this is not the case as we will show by Proposition V.5.1.

We note that the investigations in Chapter II, § 2, also permit us to work with arithmetically non-Cohen-Macaulay varieties. For instance, the Segre embedding $S(X \times X)$ is an arithmetically Buchsbaum (non-Cohen-Macaulay) variety where X is our well-known curve in $\mathbf{P}^3$ given parametrically by $(s^4, s^3 \cdot t, s \cdot t^3, t^4)$. The proof of this statement follows from Proposition II.2.10, see also Chapter V, § 5, 2. for generalizations.

Chapter II
Hochster-Reisner theory for monomial ideals. An interaction between algebraic geometry, algebraic topology and combinatorics

We shall in this chapter develop a new topic, that of face rings of simplicial complexes. The final paragraph of this chapter will be a brief introduction to Buchsbaum complexes. Following P. Schenzel [1] we will study an expression which can be interpreted as measuring the error when we no longer have the Cohen-Macaulay case of simplicial complexes. Hence we will get an inequality for the so-called h-vector of a Buchsbaum complex which generalizes the inequality for a Cohen-Macaulay complex which implies the Upper Bound Conjecture. The motivation for §§ 1—4 is provided by examples of Chapter III. We will discover through liaison addition that the ideals

$$(x_0, x_1)^n \cap (x_1, x_2)^{n-1} \cap (x_2, x_3)^n \cap (x_3, x_0)^{n-1}$$

never define projectively Cohen-Macaulay but projectively Buchsbaum curves in $\mathbf{P}^3_K$. Further, the ideals

$$(x_0, x_1)^n \cap (x_1, x_2)^n \cap (x_2, x_3)^n \cap (x_3, x_0)^{n-1}$$

define projectively Cohen-Macaulay curves for all $n \geq 1$. In this chapter we would like to understand the reason for this independent of liaison addition. Each of these ideals it generated by monomials.

Thus we may try to apply again the Hochster-Reisner theory of monomial ideals, which associates to each such ideal I a simplicial complex Σ_I. In particular we have the Reisner homology criterion, which relates the Cohen-Macaulay property of the quotient $K[x_0, \ldots, x_n]/I$ to the homology of Σ_I and its link subcomplexes. There are two problems with this approach. One is that the Hochster-Reisner theory was developed only for ideals generated by square-free monomials, whereas the above-mentioned ideals are not square-free in general. The other problem is that the computation of simplicial homology is a laborious process, making the Reisner homology criterion difficult to apply in practice.

The first problem is solved by Schwartau's polarization, which extends the Hochster-Reisner theory to the non-square-free case. This will be explained in § 1. Following Schwartau [1] we deal with the second problem by developing new Cohen-Macaulay criteria for monomial ideals of height 2. We discover that in the orientable case it is possible to replace the Reisner criterion altogether by conditions on Σ_I which have nothing to do with its homology but rather its singularities. Also we show that these new criteria may be given in purely algebraic terms, as a condition on the primary decomposition of the polarization of I. This in turn can be easily measured by means of a certain graph we associate to any monomial ideal of height 2.

The main results of this chapter rest at crucial points on some completely concrete construction in which algebra, geometry, topology and combinatorics are significantly intertwined.

§ 1. Foundations

We will assume some basic facts about monomial ideals. For example, we will use the so-called splitting lemma:

If $\lambda, \tau, m_1, \ldots, m_n$ are monomials and λ, τ are relatively prime, then

$$(\lambda \cdot \tau, m_1, \ldots, m_n) = (\lambda, m_1, \ldots, m_n) \cap (\tau, m_1, \ldots, m_n).$$

First we want to discuss Schwartau's polarization, which extends the Hochster-Reisner theory to the non-square-free case of ideals generated by monomials. Throughout this chapter K will denote an arbitrary algebraically closed field. We shall always write S for $K[x_0, \ldots, x_n]$, the ring of polynomials on $\mathbf{P}^n_K$.

Definition 1.1. Let m be a monomial $x_0^{a_0} \cdots x_n^{a_n}$ of S. Then we define the *polar* m to be the square-free monomial obtained by replacing all repeated variables in m by new variables:

$$\text{Polar } m = x_0^{(1)} \cdots x_0^{(a_0)} x_1^{(1)} \cdots x_1^{(a_1)} \cdots x_n^{(1)} \cdots x_n^{(a_n)}.$$

Let $I = (m_1, \ldots, m_r)$ be a monomial ideal of S, where $(m_1, \ldots, m_r)$ is a minimal set of generators of I consisting of monomials. Then define the "polarization of I" to be the ideal

$$\text{Polar } I = (\text{Polar } m_1, \ldots, \text{Polar } m_r).$$

Note that if I is generated by square-free monomials Polar $I = I$; whereas in general the square-free ideal Polar I lies in a higher dimensional polynomial ring than the original ring S.

We collect some properties on Polar I. These assertions are immediate consequences from basic facts on monomial ideals.

Theorem 1.2.

(i) *Let $I^{(1)} = (m_1^{(1)}, \ldots, m_{n(1)}^{(1)}), \ldots, I^{(e)} = (m_1^{(e)}, \ldots, m_{n(e)}^{(e)})$ be monomial ideals of S. Then*

$$\text{Polar } (I^{(1)} \cap \ldots \cap I^{(e)}) = \text{Polar } I^{(1)} \cap \ldots \cap \text{Polar } I^{(e)}.$$

(ii) *Let $I \subset S$ be a monomial ideal. Then if I is equidimensional and unmixed, so is* Polar I *and both ideals have the same height.*

(iii) *For a primary decomposition we have*

$$\text{Polar } \left((x_a, x_b)^m\right) = \bigcap_{\substack{i+j \leq m+1 \\ 1 \leq i,j \leq m}} (x_a^{(i)}, x_b^{(j)}).$$

We will review the correspondence between monomial ideals of S and finite simplicial complexes. Our treatment is based on M. Hochster [4] the reader is referred to this paper and to G. Reisner [1] for more details. Also, the reader may consult to Hilton-Wylie [1], Spanier [1] and McMullen-Shephard [1] for basic facts of algebraic topology.

First we want to recall some basic definitions and results from algebraic topology; for the most part we follow Stanley [1] and Baclawski [3].

An (abstract) *simplicial complex* Δ on a vertex set V is a collection of subsets F of V satisfying:

(a) if $x \in V$ then $\{x\} \in \Delta$,

(b) if $F \in \Delta$ and $G \subset F$, then $G \in \Delta$.

Elements of Δ are called *faces* or *simplices*. If $F \in \Delta$, then define $\dim F = \# F - 1$ and the *dimension* of Δ, $\dim \Delta = \max\limits_{F \in \Delta} (\dim F)$.

If $\# F = q + 1$, then F is a *q-face* or *q-simplex*. We frequently identify the vertex x with the face $\{x\}$.

Suppose V is finite, say $V = \{x_1, \ldots, x_n\}$. Let e_i be the ith unit coordinate vector in $\mathbf{R}^n$. Given a subset $F \subseteq V$, define

$$|F| = \operatorname{cx}\{e_i \mid x_i \in F\},$$

where cx denotes convex hull (see, e.g., McMullen-Shephard [1]).

Thus if F is an (abstract) q-simplex, then $|F|$ is a geometric q-simplex in $\mathbf{R}^n$. Define the *geometric realization* $|\Delta|$ of the simplicial complex Δ by

$$|\Delta| = \bigcup_{F \in \Delta} |F|.$$

Thus $|\Delta|$ inherits from the usual topology on $\mathbf{R}^n$ the structure of a topological space. If X is a topological space homeomorphic to $|\Delta|$, then we (somewhat inaccurately) call Δ a *triangulation* of X.

An *oriented q-simplex* of Δ is a q-simplex F together with an equivalence class of total orderings of F, two orderings being equivalent if they differ by an even permutation of the vertices. Denote by $[v_0, \ldots, v_q]$ the oriented q-simplex consisting of the q-simplex $F = \{v_0, \ldots, v_q\}$, together with the equivalence class of orderings containing $v_0 < v_1 < \ldots < v_q$. Fix a ring A (commutative with 1). Let $C_q(\Delta)$ be the free A-module with basis consisting of the oriented q-simplices in Δ, modulo the relations $\sigma_1 + \sigma_2 = 0$ whenever σ_1 and σ_2 are different oriented q-simplices corresponding to the same q-simplex of Δ. Thus $C_q(\Delta) = O$ for $q < 0$, and for $q \geq 0$ the module $C_q(\Delta)$ is a free A-module with rank equal to the number of q-simplices of Δ. If Δ is empty, then $C_q(\Delta) = O$ for all q.

We define homomorphisms $\partial_q \colon C_q(\Delta) \to C_{q-1}(\Delta)$ for $q \geq 1$ by defining them on the basis elements by

$$\partial_q[v_0, v_1, \ldots, v_q] = \sum_{i=0}^{q} (-1)^i \, [v_0, v_1, \ldots, \hat{v}_i, \ldots, v_q],$$

where $\hat{v}_i$ denotes that v_i is missing. It is easily verified that ∂_q indeed extends to a homomorphism $C_q(\Delta) \to C_{q-1}(\Delta)$, and that $\partial_q \partial_{q+1} = 0$. The chain complex $C(\Delta) = \{C_q(\Delta), \partial_q\}$ is the oriented chain complex of Δ. Define an augmentation $\varepsilon \colon C_0(\Delta) \to A$ by $\varepsilon(x) = 1$ for every vertex $x \in V$. The augmented chain complex $(C(\Delta), \varepsilon)$, is the augmented oriented chain complex of Δ (over A). Then the qth *reduced homology group* of Δ with coefficients A, denoted $\tilde{H}_q(\Delta; A)$, is defined to be the qth homology group of the augmented oriented chain complex of Δ over A.

Furthermore, the reduced Euler characteristic $\tilde{\chi}(\Delta)$ of Δ is defined by

$$\tilde{\chi}(\Delta) = \sum_{q \geq -1} (-1)^q \operatorname{rank} \tilde{H}_q(\Delta; A).$$

It is independent of A and is also given by

$$\tilde{\chi}(\Delta) = -1 + f_0 - f_1 + \ldots,$$

where f_q is the number of q-simplices in Δ. If $\chi(\Delta)$ is the ordinary Euler characteristic then $\tilde{\chi}(\Delta) = \chi(\Delta) - 1$.

If $\Delta \neq \emptyset$, then $\tilde{H}_q(\Delta; A) = O$ for $q < 0$. If $\Delta = \emptyset$, then $\tilde{H}_q(\emptyset; A) \cong A$ for $q = -1$, and O for $q \neq -1$. In particular, $\tilde{\chi}(\emptyset) = -1$.

We now wish to define the homology groups of a space X, rather than a simplicial complex Δ. Let X be a topological space. Let Δ^q denote the standard q-dimensional ordered geometric simplex $\langle p_0, \ldots, p_q \rangle$ whose vertices p_i are the unit coordinate vectors in $\mathbf{R}^{q+1}$. A singular q-simplex in X is a continuous map

$$\sigma: \Delta^q \to X.$$

Let $C_q(X)$ be the free A-module generated by all singular q-simplices. The elements of C_q are formal finite linear combinations $\sum_\sigma c_\sigma \sigma$, where σ is a singular q-simplex and $c_\sigma \in A$. Given a vertex p_i of Δ_q, there is an obvious linear map $e_q^i: \Delta^{q-1} \to \Delta^q$ which sends Δ^{q-1} to the face of Δ^q opposite p_i. The ith face of σ, denoted by $\sigma^{(i)}$, is defined to be the singular $(q - 1)$-simplex which is the composite

$$\sigma^{(i)} = \sigma \circ e_q^i: \Delta^{q-1} \to \Delta^q \to X.$$

We now define a linear map ($=$ A-module homomorphism) $\partial_q: C_q \to C_{q-1}$ by

$$\partial_q(\sigma) = \sum_{i=0}^{q} (-1)^i \sigma^{(i)},$$

where σ is a singular q-simplex. It is easily checked that $\partial_{q-1}\partial_q = 0$, so $C(X) = \{C_q(X), \partial_q\}$ is a chain complex, the singular chain complex of X (over A). Define an augmentation $\varepsilon: C_0(X) \to A$ by $\varepsilon(\sigma) = 1$ for all singular 0-simplices σ. The augmented chain complex $\tilde{C}(X)$ is the augmented singular chain complex of X (over A). Then the qth reduced singular homology group of X with coefficients A, denoted $\tilde{H}_q(X; A)$, is the qth *homology group of the augmented singular chain complex* of X (over A).

Considering this case the *reduced Euler characteristic* $\tilde{\chi}(X)$ of X is defined by

$$\tilde{\chi}(X) = \sum_{q \geq -1} (-1)^q \operatorname{rank} \tilde{H}_q(X; A).$$

It is independent of A.

If Δ is a simplicial complex and Δ_1 and Δ_2 are subcomplexes of Δ, then there is an exact sequence (whose definition we omit)

$$\ldots \to \tilde{H}_q(\Delta_1 \cap \Delta_2) \to \tilde{H}_q(\Delta_1) \oplus \tilde{H}_q(\Delta_2) \to \tilde{H}_q(\Delta_1 \cup \Delta_2) \to \tilde{H}_{q-1}(\Delta_1 \cap \Delta_2) \to \ldots$$

(with all coefficients A), called the *reduced Mayer-Vietoris sequence* of Δ_1 and Δ_2. Similarly, if X is a topological space and X_1, X_2 are "nice" subspaces (e.g., if $X_1 \cup X_2$

$= (\mathrm{int}_{X_1 \cup X_2} X_1) \cup (\mathrm{int}_{X_1 \cup X_2} X_2)$, where $\mathrm{int}_Y Z$ denotes the relative interior of Z in the space Y), then we have a reduced Mayer-Vietoris sequence of X_1 and X_2 exactly analogous to that of Δ_1 and Δ_2.

We now come to the relationship between simplicial and singular homology: Let Δ be a finite simplicial complex and $X = |\Delta|$. Then there is a (canonical) isomorphism for all q:

$$\tilde{H}_q(\Delta; A) \cong \tilde{H}_q(X; A).$$

For example, let S^{d-1} denote a $(d-1)$-dimensional sphere. Then $H_q(\Delta; A) = A$ for $q = d - 1$ and O for $q \neq d - 1$.

A simplicial complex Δ or topological space X is *acyclic* (over A) if its reduced homology with coefficients A vanishes in all degrees q. (Thus the null set is not acyclic, since $\tilde{H}_{-1}(\emptyset; A) \cong A$.)

Let Y be a subspace of X. Then the singular chain module $C_q(Y)$ is a submodule of $C_q(X)$, so we have a quotient complex $C(X, Y) = C(X)/C(Y) = \{C_q(X)/C_q(Y), \bar{\partial}_q\}$. Define the relative homology of X modulo Y (with coefficients A) by

$$H_q(X, Y; A) = H_q\big(C(X, Y)\big).$$

We next want to define reduced cohomology of simplicial complexes and spaces. The simplest way (though not the most geometric) is to dualize the corresponding chain complexes.

Let $C'(\Delta) = C(\Delta, \varepsilon)$ be the augmented oriented chain complex of the simplicial complex Δ, over the ring A. The qth *reduced singular cohomology group* of Δ with coefficients A is defined to be

$$\tilde{H}^q(\Delta; A) = \tilde{H}^q\big(\mathrm{Hom}_A(C'(\Delta), A)\big),$$

where $\mathrm{Hom}_A\big(C'(\Delta), A\big)$ is the cochain complex obtained by applying the functor $\mathrm{Hom}_A(\ , A)$ to $C'(\Delta)$. Exactly analogously define $\tilde{H}^q(X; A)$ and $H^q(X, Y; A)$. Sometimes one identifies the free modules $C_q(\Delta)$ and $C^q(\Delta) = \mathrm{Hom}_A\big(C_q(\Delta), A\big)$ by identifying the basis of oriented q-chains σ of $C_q(\Delta)$ with its dual basis in $C^q(\Delta)$. Similarly one can identify $C_q(X)$ with $C^q(X)$.

There is a close connection between homology and cohomology of Δ or X arising from the "universal-coefficient theorem for cohomology". We merely mention the (easy) special case that when A is a field k, there are "canonical" isomorphisms

$$\tilde{H}_q(\Delta; k) \cong \mathrm{Hom}_k\big(\tilde{H}^q(\Delta; k), k\big),$$

$$\tilde{H}_q(X; k) \cong \mathrm{Hom}_k\big(\tilde{H}^q(X; k), k\big).$$

Thus in particular when $\tilde{H}_q(\Delta; k)$ is finite-dimensional (e.g., when Δ is finite), we have $\tilde{H}_q(\Delta; k) \cong \tilde{H}^q(\Delta; k)$ and similarly for X, but these isomorphisms are not canonical.

We recall that a *topological n-manifold* (without boundary) is a Hausdorff space in which each point has an open neighborhood homeomorphic to $\mathbf{R}^n$. An n-manifold with boundary is a Hausdorff space X in which each point has an open neighborhood which is homeomorphic with $\mathbf{R}^n$ or $\mathbf{R}^n_+ = \{(x_1, \ldots, x_n) \in \mathbf{R}^n \mid x_i \geq 0\}$. The boundary ∂X of X consists of those points with no open neighborhood homeomorphic to $\mathbf{R}^n$. It follows easily that ∂X is either void or an $(n - 1)$-manifold.

Suppose X is a compact connected n-manifold with boundary. Then one can show $H_n(X, \partial X; A)$ is either void or isomorphic to A.

A compact connected n-manifold X with boundary is *orientable* (over A) if we have $H_n(X, \partial X; A) = A$. (The usual definition of orientable is more technical but equivalent to the one given here; see also Definition 3.13 below.)

For example, every compact connect n-manifold with boundary is orientable over a field of characteristic two.

If a compact connected n-manifold X is orientable over A, then we have $H_q(X; A) \cong H^{n-q}(X; A)$. This is the so-called *Poincaré Duality Theorem*.

An n-dimensional *pseudomanifold without boundary* (resp., *with boundary*) is a simplicial complex Δ such that:

(a) Every simplex of Δ is the face of an n-simplex of Δ.

(b) Every $(n-1)$-simplex of Δ is the face of exactly two (resp., at most two) n-simplices of Δ.

(c) If F and F' are n-simplices of Δ, there is a finite sequence $F = F_1, \ldots, F_m = F'$ of n-simplices of Δ such that F_i and F_{i+1} have an $(n-1)$-face in common for $1 \leq i \leq m$.

The boundary $\partial\Delta$ of a pseudomanifold Δ consists of those faces F contained in some $(n-1)$-simplex of Δ which is the face of exactly one n-simplex of Δ.

Let Δ be a finite n-dimensional pseudomanifold with boundary. Then $H_n(\Delta, \partial\Delta; A) \cong A$ or O. In the former case we say that Δ is *orientable* over A; otherwise *non-orientable*.

Let I be the unit interval $[0, 1]$. The *suspension* ΣX of a topological space X is defined to be the quotient space of $X \times I$ in which $X \times 0$ is identified to one point and $X \times 1$ is identified to another point. The *n-fold suspension* $\Sigma^n X$ is defined recursively by $\Sigma^n X = \Sigma(\Sigma^{n-1} X)$.

For any X and q we have

$$\tilde{H}_q(X; A) \cong \tilde{H}_{q+1}(\Sigma X; A).$$

The purpose of this chapter also is to introduce a new kind of partially ordered set: *Buchsbaum poset*. The notion of a Cohen-Macaulay poset originated in Baclawski's thesis, see Baclawski [3]. It is now known that this concept provides some interesting connections among algebraic topology, combinatorics, commutative algebra and homological algebra.

Let P be a *finite poset*; that is, a partially ordered set. We need some auxiliary concepts. A *chain* of P is a totally ordered subset of P. We will usually write $x_1 < \ldots < x_n$ for a typical chain of P. The *rank* of a chain is the number of elements in it; thus $r(x_1 < \ldots < x_n) = n$. More generally, the rank of P, written $r(P)$, is the rank of the longest chain of P. The *length* of P, written $l(P)$, is given by $l(P) = r(P) - 1$. The length is a more topological notion whereas the rank seems to be more combinatorial. Apparently topologists start counting at zero while combinatorialists prefer to begin at 1. We will do both. A poset is said to be *ranked* if every maximal chain has rank $r(P)$.

Given a poset P, we will write $\hat{P}$ for the poset obtained by adjoining a new pair of elements to P, written $\hat{0}$, $\hat{1}$ such that $\hat{0} < x < \hat{1}$ for all $x \in P$. If we only require that $\hat{0}$ or $\hat{1}$ be adjoined, we will write $P_{\hat{0}}$ or $P^{\hat{1}}$ respectively. We use the convention that $\hat{0}$ or $\hat{1}$ is never an element of P. The context should indicate to which poset $\hat{0}$ or $\hat{1}$ is to be adjoined.

A subset $J \subseteq P$ will be called an *order-ideal* if for every $x \in J$, $y \leq x$ implies $y \in J$. The dual definition gives the concept of an *order-filter*. The order-ideal generated by a subset $S \subseteq P$ will be noted $J(S)$ or $J_P(S)$; while $V(S) = V_P(S)$ denotes the order-filter generated by S. The special case $J(x)$ for $x \in P$ can also be denoted $(\hat{0}, x]$. If P is ranked, then so is every subset $J(x)$, and we write $r(x)$ for $r(J(x))$. The function r takes values in the set $[r(P)]$ which by definition denotes $\{1, 2, \ldots, r(P)\}$. The length of an open intervall will be denoted $l(x, y)$ instead of $l((x, y))$.

We will often use the *Möbius function*. For a poset P we write $\mu(P)$ for $\mu(\hat{0}, \hat{1})$ as computed in $\hat{P}$. For $x \in P$ we will write $\mu(x)$ or $\mu_P(x)$ for $\mu(J(x))$. Finally, for $x \leq y$ in P we will think of $\mu(x, y)$ as an abbreviation for $\mu((x, y))$.

For a finite set S, let $B(S)$ denote the poset of nonempty subsets of S. A *finite simplicial complex* is an order-ideal of $B(S)$. The minimal elements are called *vertices* and elements in general are called *simplices*. Much of what we do in the sequel may be extended routinely to simplicial complexes. As we have defined it, a simplicial complex is a special kind of poset. However, given a finite poset P, we can define the *order complex* of P, denoted $\Delta(P)$, to be the subset of $B(P)$ consisting of the nonempty chains of P. By this device one may view posets as a special kind of simplicial complex.

We now review the correspondence between monomial ideals of the polynomial ring $S = K[x_0, \ldots, x_n]$ and finite simplicial complexes. Let Δ_n denote the standard n-simplex; that is, the complete simplicial complex on $(n+1)$-vertices which we label as $x_0, \ldots, x_n$. Recall that this means that Δ_n is the set of all subsets of $\{x_0, x_1, \ldots, x_n\}$. Let K be a field and I an ideal of S. Let $V(I)$ be the subset of K^{n+1} where the elements of I vanish.

If $I = (x_{i_0}, \ldots, x_{i_r})$, we refer to $V(I)$ as a coordinate hyperplane. Then we get 1-1 correspondences between:

$S_1 = \{$subcomplexes of $\Delta_n\}$,

$S_2 = \{$ideals of S generated by square-free monomials$\}$,

$S_3 = \{$unions of coordinate hyperplanes in $K^{n+1}\}$.

We describe some of these 1-1 correspondences in detail. The correspondence $S_1 \to S_2$ is defined by

$$\Sigma \rightsquigarrow I_\Sigma$$

where Σ denotes a subcomplex of Δ_n, and I_Σ denotes the ideal of S generated by the monomials $x_{i_0} \cdots x_{i_r}$, $i_0 < \ldots < i_r$, such that the simplex $\{x_{i_0}, \ldots, x_{i_r}\}$ is *not* in Σ.

The correspondence $S_1 \leftarrow S_2$ is defined by $\Sigma_I \leftsquigarrow I$ where I denotes a square-free monomial ideal of S, and Σ_I denotes the subcomplex of Δ_n consisting of all simplices $\{x_{i_0}, \ldots, x_{i_r}\}$ such that the monomial $x_{i_0} \ldots x_{i_r}$ is *not* in I.

For the correspondence

$$S_2 \to S_3, \quad I \rightsquigarrow V(I)$$

we simply associate to I its vanishing locus in K^{n+1}.

The correspondence $S_3 \to S_1$ is given by

$$H \rightsquigarrow \Sigma_H .$$

Here H denotes a union of coordinate hyperplanes in K^{n+1}, and Σ_H denotes the subcomplex of Δ_n consisting of all simplices $\{x_{i_0}, \ldots, x_{i_r}\}$ such that the element of K^{n+1} whose $x_{i_0}, \ldots, x_{i_r}$ coordinates are 1 and whose other coordinates are 0 is in H.

If the field K is the real field $\mathbf{R}$, we may view this last correspondence in terms of geometric realizations. First we recall:

Definition 1.3. Let the vertices $x_0, \ldots, x_n$ of Δ_n be identified with the canonical basis of $\mathbf{R}^{n+1}$. Then if Σ is any subcomplex of Δ_n, its geometric realization $|\Sigma|$ is defined as $\bigcup_{\sigma \in \Sigma} \text{cx}(\sigma) \subset \mathbf{R}^{n+1}$; $|\Sigma| \subset \mathbf{R}^{n+1}$ is a topological space (see McMullen-Shephard [1]). Then if $K = \mathbf{R}$ the above correspondence may be viewed as $S_3 \to S_1$ given by

$$H \rightsquigarrow |\Delta_n| \cap H = |\Sigma_H|.$$

We now relate these correspondences to primary decomposition.

Lemma 1.4.

(i) *If $P \subset S$ is a square-free monomial ideal of the form $(x_{i_1}, \ldots, x_{i_h})$, then the associated simplicial complex $\Sigma_P \subset \Delta_n$ is a simplex of codimension $h := n - \dim \Sigma_P$ in Δ_n.*

(ii) *If $I \subset S$ in any square-free monomial ideal, let $P_1 \cap \ldots \cap P_m$ be an irredundant primary decomposition of I. Then each P_i is of the type described in* (i), *and we have $\Sigma_I = \bigcup_{i=1}^{m} \Sigma_{P_i}$ as the decomposition of Σ_I into its maximal simplicies.*

(iii) *If $I \subset S$ is any square-free monomial ideal,* $\text{ht } I = \text{codim}(\Sigma_I, \Delta_n)$.

Proof: (i) is obvious from the correspondence $S_2 \to S_1$ above.

(ii) That each P_i is of type (i) follows from the splitting lemma. The fact that the intersection of $P_1 \cap \ldots \cap P_m$ corresponds to the union $\bigcup_{i=1}^{m} \Sigma_{P_i}$ follows from the correspondences $S_2 \to S_3 \to S_1$. The maximality of the simplices Σ_{P_i} in Σ_I follows from the irredundancy of the primary decomposition.

(iii) is a direct consequence of (i) and (ii).

Definition 1.5. If Σ is a finite simplicial complex, we define the *codimension of* Σ (written codim Σ) to be $\text{codim}(\Sigma, \Delta_{v-1})$ where v is the number of vertices in Σ.

Lemma 1.6. *Let $I \subset S$ be any square-free monomial ideal, and let $\Sigma_I \subset \Delta_n$ be the associated simplicial complex. Then*

$$\text{codim } \Sigma_I \leq \text{ht } I.$$

Proof: This follows immediately from Lemma 1.4(iii).

The following definitions are fundamental in the sequel.

Definition 1.7. If σ is a maximal simplex in a simplicial complex we call it a *facet* of Σ.

Definition 1.8. If $\sigma \in \Sigma$ is any simplex, we define the *star* of σ to be the subcomplex

$$\text{Star}_\Sigma \sigma = \{s \in \Sigma \mid \sigma \cup s \in \Sigma\}.$$

Definition 1.9. If $\sigma \in \Sigma$ is any simplex, we define the *link* of σ to be the subcomplex

$$\text{lk}_\Sigma \sigma = \{s \in \Sigma \mid \sigma \cap s = \emptyset \text{ and } \sigma \cup s \text{ is in } \Sigma\}.$$

Lemma 1.10. *Let Σ be a simplicial complex. Then:*

α) *Any link* $\mathrm{lk}_\Sigma \sigma$ *is an iterated link of vertices (the vertices in* σ*).*

β) *If v is a vertex of Σ,* $\mathrm{Star}_\Sigma v$ *is the simplicial cone with base* $\mathrm{lk}_\Sigma v$ *and vertex v.*

γ) *Any* $\mathrm{Star}_\Sigma \sigma$ *is an iteration of cones over* $\mathrm{lk}_\Sigma \sigma$ *(by the vertices in* σ*).*

Lemma 1.11. *Let Σ be a simplicial complex which is equi-dimensional of dimension d. Then:*

α) *If σ is any simplex of Σ,* $\mathrm{Star}_\Sigma \sigma$ *is an equi-dimensional complex of dimension d.*

β) *If σ is a k-simplex of Σ,* $\mathrm{lk}_\Sigma \sigma$ *is an equi-dimensional complex of dimension $d-k-1$.*

Definition 1.12. If Σ is a simplicial complex let $\mathrm{Cone}_P(\Sigma)$ denote the simplicial complex given by the simplices

$$\{\sigma \mid \sigma \text{ is a simplex of } \Sigma\} \cup \{(\sigma, P) \mid \sigma \text{ is a simplex of } \Sigma\} \cup \{P\}.$$

Note that $\mathrm{Cone}_P(\Sigma)$ contains Σ as a subcomplex. $\mathrm{Cone}_P(\Sigma)$ is called "the simplicial cone over Σ" or "the simplicial cone with base Σ and vertex P".

The following lemmas are of basic importance.

Lemma 1.13. *Σ is equi-dimensional* $\Leftrightarrow$ $\mathrm{Cone}_P(\Sigma)$ *is equi-dimensional (of dimension one more than Σ).*

Proof: Observe that $(\ ,P)$ induces a 1-1 correspondence

$$\{\text{facets of } \Sigma\} \xrightarrow[\text{1-1}]{(\ ,P)} \{\text{facets of } \mathrm{Cone}_P(\Sigma)\}.$$

Since $(\ ,P)$ raises the dimension of any simplex by exactly one we conclude that one set of facets is equi-dimensional if and only if this is true for the other set.

Lemma 1.14. *Let T be a simplex of $C = \mathrm{Cone}_P(\Sigma)$. Then there are three possibilities for* $\mathrm{lk}_C T$*:*

α) *If $T = P$, then* $\mathrm{lk}_C T = \Sigma$.

β) *If $T \ni P$ (i.e. $T = (\sigma, P)$ for some simplex $\sigma \in \Sigma$), then* $\mathrm{lk}_C T = \mathrm{lk}_\Sigma \sigma$.

γ) *If $T \not\ni P$ (i.e. $T \in \Sigma$), then* $\mathrm{lk}_C T = \mathrm{Cone}_P(\mathrm{lk}_\Sigma T)$.

Proof: α) This is obvious from the definition of $\mathrm{Cone}_P(\Sigma)$.

β)
$$\begin{aligned} \mathrm{lk}_C(\sigma, P) &= \mathrm{lk}_{\mathrm{lk}_C \sigma} P && \text{(recall Lemma 1.10)} \\ &= \mathrm{lk}_{\mathrm{Cone}_P(\mathrm{lk}_\Sigma \sigma)} P && (\text{by } \gamma)) \\ &= \mathrm{lk}_\Sigma \sigma && (\text{by } \alpha)). \end{aligned}$$

γ) *Step 1:* $\sigma \in \Sigma$ and $\sigma \cap T = \emptyset \Leftrightarrow (\sigma, P) \in C$ and $(\sigma, P) \cap T = \emptyset$.

Step 2: $\sigma \cup T \in \Sigma \Leftrightarrow (\sigma, P) \cup T \in C$.

We now present two examples of simplicial complexes.

Example 1.15.

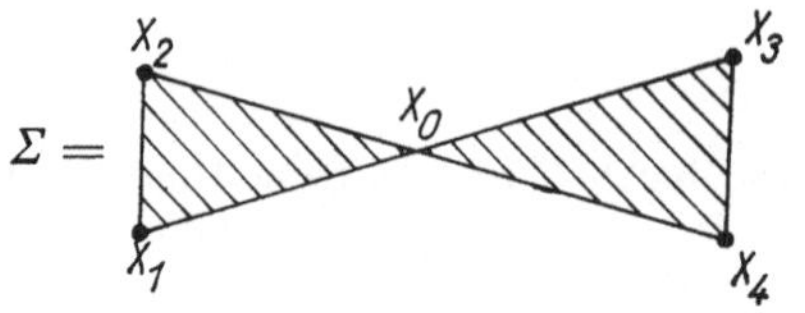

For each $X_i \neq X_0$, $\mathrm{lk}_\Sigma X_i =$  and $\mathrm{Star}_\Sigma X_i =$ 

However, for $X_i = X_0$, $\mathrm{lk}_\Sigma X_0 =$ 

and $\mathrm{Star}_\Sigma X_0 = \Sigma$.

Σ is the simplicial cone with base 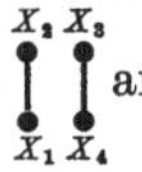

and vertex X_0.

Example 1.16.

$\Sigma =$ 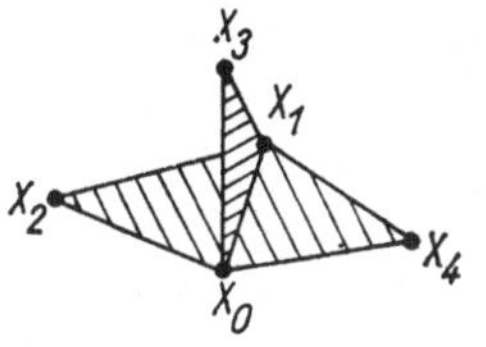

For $X_i \neq X_0, X_1$, $\quad$ lk $X_i =$ 

For $X_i = X_0$ or X_1, $\quad$ lk $X_i =$ 

For edges $e \neq \overline{X_0X_1}$, $\quad$ $\mathrm{lk}_\Sigma e =$ 

For $e = \overline{X_0X_1}$, $\quad$ $\mathrm{lk}_\Sigma e =$ 

Σ is the simplicial cone over , q.e.d.

Stars and links are important in the study of simplicial manifolds, as we shall see in the following sections.

§ 2. The homological Cohen-Macaulay criterion of Reisner

Let $\Sigma \subset \Delta_n$ be any finite simplicial complex. We denote by $K[\Sigma]$ the quotient ring S/I_Σ. We will see that the graded K-algebra $K[\Sigma]$ is closely related to the combinatorical and topological properties of Σ. Therefore we have interactions between commutative algebra and combinatorics. By combining these techniques we open the way to a deeper study of $K[\Sigma]$ and Σ. First, we begin by presenting Reisner's Cohen-Macaulay criterion.

$\tilde{H}_i(\Sigma; K)$ will always denote the ith reduced homology group of Σ with coefficients in K (see, for example, E. H. Spanier [1], p. 168).

In 1976, G. A. Reisner [1] proved the following statement:

Theorem 2.1 (the Reisner Cohen-Macaulay criterion). *Let $\Sigma \subset \Delta_n$ be any finite simplicial complex. Then the following conditions are equivalent:*

(i) *$K[\Sigma]$ is Cohen-Macaulay.*

(ii) *$K[\mathrm{lk}_\Sigma x_i]$ is Cohen-Macaulay for every vertex $x_i \in \Sigma$; and $\tilde{H}_i(\Sigma; K) = 0$, $i < \dim \Sigma$.*

(iii) *$H_i(\mathrm{lk}_\Sigma \sigma; K) = 0$, $i < \dim \mathrm{lk}_\Sigma \sigma$, for all simplices $\sigma \in \Sigma$; and $\tilde{H}_i(\Sigma, K) = 0$, $i < \dim \Sigma$.*

Proof: Reisner [1], Theorem 1 (see also Weibel [1]).

If the geometric realization $|\Sigma|$ of Σ (see again E. H. Spanier [1]) is a manifold (with or without boundary) the condition on the links holds automatically. This follows by using excision (see E. H. Spanier [1]). Hence we have:

Corollary 2.2. *If $\Sigma \subset \Delta_n$ is a simplicial complex such that $|\Sigma|$ is a topological manifold-with-∂, then the following conditions are equivalent:*

(i) *$K[\Sigma]$ is Cohen-Macaulay.*

(ii) *$\tilde{H}_i(\Sigma; K) = 0$, $i < \dim \Sigma$.*

Furthermore, by proving Theorem 2.1 Reisner has also investigated the property for Σ to have Cohen-Macaulay links of its vertices. From the point of view of commutative algebra the key is the following simple lemma.

Lemma 2.3. *Let R be a finitely generated graded K-algebra with $R_0 = K$. Suppose $R = K[x_0, \ldots, x_n]$ where $x_0, \ldots, x_n$ are forms of degree 1. Then the following conditions are equivalent:*

(i) *R_P is Cohen-Macaulay for all prime ideals P except, perhaps, the irrelevant maximal ideal $(x_0, \ldots, x_n)$.*

(ii) *$R/(x_i - 1)$ is Cohen-Macaulay for $i = 0, \ldots, n$.*

Proof: Assume (ii), and let $P \neq (x_0, \ldots, x_n)$ be a prime ideal. Choose $x_j \notin P$. It suffices to show that R_{x_i} (i.e. R localized at the powers of x_i) is Cohen-Macaulay. But this follows from the well-known isomorphisms (see Grothendieck [2], Chapter II, 2.2.5)

$$R_{x_i} \cong (R_{x_i})_0 [(1/x_i), x_i], \qquad (R_{x_i})_0 \cong R/(x_i - 1)$$

and the fact that a ring S is Cohen-Macaulay if and only if $S[t, 1/t]$ is Cohen-Macaulay with t an indeterminate.

Assume (i). Then the above isomorphisms and since $x_i - 1$ is not a zero divisor in R_{x_i} imply (ii).

We will be interested in Lemma 2.3 when $R = K[\Sigma]$ for some $\Sigma \subset \Delta_n$. The preceding will be used in proving the following Buchsbaum criterion.

Theorem 2.4 (the Reisner locally Cohen-Macaulay criterion). *Let $\Sigma \subset \Delta_n$ be any finite simplicial complex. Then the following conditions are equivalent:*

(i) *$(K[\Sigma])_P$ is a Cohen-Macaulay ring for all prime ideals P different from the irrelevant ideal $(x_0, \ldots, x_n) \cdot K[\Sigma]$.*

(ii) *For all simplices $\sigma \in \Sigma$ we have*

$$\tilde{H}_i(\mathrm{lk}_\Sigma \sigma; K) = 0 \quad \text{if} \quad i \neq \dim \mathrm{lk}_\Sigma \sigma.$$

(iii) *$K[\Sigma]$ is a (local) Buchsbaum ring.*

Proof: The equivalence of (i) and (ii) is proved by G. A. Reisner [1]. The implication (iii) $\Rightarrow$ (i) is trivially true because a Buchsbaum ring is locally a Cohen-Macaulay ring. The converse (i) $\Rightarrow$ (iii) results immediately from Proposition I.3.10 and the following Lemma 2.5(ii).

Lemma 2.5. *Let $\Sigma \subset \Delta_n$ be any finite simplicial complex. Set $R = K[\Sigma]$. Then we have for the local cohomology module $\underline{H}^i_{\mathfrak{m}}(R)$ the following properties:*

(i) $[\underline{H}^i_{\mathfrak{m}}(R)]_n = O$ *for all* $n > 0$ *and* $i \in \mathbf{Z}$.

(ii) *Assuming that* $\underline{H}^i_{\mathfrak{m}}(R)$ *is of finite length then*

$$\underline{H}^i_{\mathfrak{m}}(R) = [\underline{H}^i_{\mathfrak{m}}(R)]_0.$$

Proof: In order to prove these facts on local cohomology we make use of the techniques of Hochster and Roberts about the purity of the Frobenius homomorphism (in characteristic $p \neq 0$) and the fact that R has a presentation of relative graded F-pure type (in characteristic zero). The key is then the following assertion. (We use terminology from Hochster-Roberts [2].)

Claim. *Let K denote a perfect field of prime characteristic $p > 0$. Then $K[\Sigma]$ is F-pure. For an arbitrary field K, $K[\Sigma]$ has a presentation of relative graded F-pure type.*

Proof of claim: First let K be a perfect field of prime characteristic $p > 0$. We show that the Frobenius map $F: R \to R$ is pure. To this end we prove that $F(R)$ is a direct summand of R as $F(R)$-module. This is equivalent to showing that there is an R-retraction, say $r: R \to F(R)$. Here the point is that we can define r for monomials by

$$r(x_0^{k_0} \cdots x_n^{k_n}) = \begin{cases} x_0^{k_0} \cdots x_n^{k_n} & \text{for } k_i \equiv 0 \bmod p,\ 0 \leq i < n, \\ 0 & \text{otherwise} \end{cases}$$

and extend it K-linearly to R. Since K is a perfect field we get $r(R) = F(R)$. To complete the proof we note that r is the identity on $F(R)$ and r is an $F(R)$-module homomorphism. For an arbitrary field we get from $\mathbf{Z}[x_0, \ldots, x_n]/I_\Sigma$ a presentation of relative graded F-pure type, see M. Hochster and J. L. Roberts [2].

Having proven this claim we can now apply conclusions on local cohomology from Hochster-Roberts [2]. Therefore our Lemma 2.5 results immediately as a consequence from this important paper, q.e.d.

Analyzing the proof of Lemma 2.5 it becomes desirable to look for more precise information. Before stating results in this direction, we make some general observations and collect some known results from Hochster-Roberts [2] which will be needed.

In the following we call a homomorphism of rings $R \to S$ *pure*, if

$$M \to M \otimes_R S \quad \text{via} \quad m \mapsto m \otimes 1$$

is injective for every R-module M. A ring R of prime characteristic $p > 0$ is called *F-pure*, if the Frobenius map

$$F: R \to R, \quad r \mapsto r^p,\ r \in R,$$

is pure. F^e denotes F iterated e-times. We note that R is F-pure if and only if $F^e: R \to R$ is pure for all $e \geq 1$. If R is F-pure, it follows that it is reduced. A pure homomorphism of rings $R \to S$ induces a homomorphism of complexes

$$f^{\cdot}: X^{\cdot} \to X^{\cdot} \otimes_R S$$

for a complex of R-modules $X^{\cdot}$. In fact, it follows by M. Hochster and J. L. Roberts [1], Proposition 6.5, that the induced homomorphisms

$$H^i(f^{\cdot}): H^i(X^{\cdot}) \to H^i(X^{\cdot} \otimes_R S)$$

are injective for all $i \in \mathbf{Z}$. Assume that $R \to S$ is a pure homomorphism of graded rings such that $R \to S$ multiplies degrees by d.

Let $K^{\cdot}(\underline{x}^t; R)$ be the Koszul complex of R with respect to $\underline{x}^t = (x_1^t, \ldots, x_n^t)$, $t \geq 1$, where $(x_1, \ldots, x_n)$ is a system of forms with $\operatorname{Rad}(x_1, \ldots, x_n)R = \mathfrak{m}$, the irrelevant maximal ideal of R. Then the maps

$$[H^i(K^{\cdot}(\underline{x}^t; R))]_k \to [H^i(K^{\cdot}(\underline{x}'^t; S))]_{kd}$$

are injective for all $i \in \mathbf{Z}$ and for all $k \in \mathbf{Z}$. Here, $\underline{x}' = (x_1', \ldots, x_n')$ denotes the image of $\underline{x} = (x_1, \ldots, x_n)$ by the pure homomorphism $R \to S$. By taking the direct limit of the cohomology of the Koszul complexes it follows that the maps

$$[\underline{H}^i_{\mathfrak{m}}(R)]_k \to [\underline{H}^i_{\mathfrak{m}'S}(S)]_{kd}$$

are injective for all $k \in \mathbf{Z}$ and for all $i \in \mathbf{Z}$, where $\mathfrak{m}'S$ denotes the image of $\mathfrak{m}$ in S. In particular let R be a F-pure graded ring. Then the maps

$$[\underline{H}^i_{\mathfrak{m}}(R)]_k \to [\underline{H}^i_{\mathfrak{m}}(R)]_{kp^e}$$

are injective for all $k \in \mathbf{Z}$, $i \in \mathbf{Z}$, and $e \geq 1$, since $F^e(\mathfrak{m})$ has radical $\mathfrak{m}$. This implies the following lemma due to M. Hochster and J. L. Roberts [2].

Lemma 2.6. *If R is a graded F-pure ring, then we have $[\underline{H}^i_{\mathfrak{m}}(R)]_n = 0$ for all $n > 0$ and $i \in \mathbf{Z}$.*

Suppose that the local cohomology module $\underline{H}^i_{\mathfrak{m}}(R)$ is of finite length. Then $\underline{H}^i_{\mathfrak{m}}(R) = [\underline{H}^i_{\mathfrak{m}}(R)]_0$ also follows.

Proof: Using the fact that $\underline{H}^i_{\mathfrak{m}}(R)$ are artinian R-modules, i.e. $[\underline{H}^i_{\mathfrak{m}}(R)]_n = 0$ for $n \gg 0$, the first part of the lemma is obtained by previous considerations. If the local cohomology module is of finite length, then we have in particular $[\underline{H}^i_{\mathfrak{m}}(R)]_n = 0$ for $n \ll 0$. Thus, the second statement follows the same way, q.e.d.

In characteristic zero M. Hochster and J. L. Roberts [2], Lemma 4.7, proved a corresponding result for rings R which have presentations of certain F-pure types. For the definition and related technical results — in particular the definition of "R has a presentation of relative graded F-pure type" — we refer to the fundamental paper of M. Hochster and J. L. Roberts [2]. In fact, Lemma 2.6 and the corresponding result in characteristic zero lead to a number of Buchsbaum rings.

Theorem 2.7. *Let R be an equi-dimensional graded k-algebra such that $R_{\mathfrak{p}}$ is a Cohen-Macaulay ring for all prime ideals different from the irrelevant ideal $\mathfrak{m}$. Assume that R is F-pure resp. has a presentation of relative graded F-pure type. Then it follows that R is a Buchsbaum ring.*

Proof: If we show that all the local cohomology modules $\underline{H}^i_{\mathfrak{m}}(R)$, $0 \leq i < \dim R$, are modules of finite length, the proof will follow. From this property we get by Lemma 2.6:

$$[\underline{H}^i_{\mathfrak{m}}(R)]_n = 0 \quad \text{for all } n \neq 0 \text{ and } 0 \leq i < \dim R.$$

Then Proposition I.3.10 does imply that R is a Buchsbaum ring. Now, it is clear that the finite length of local cohomology modules $\underline{H}^i_{\mathfrak{m}}(R)$, $0 \leq i < \dim R$, is equivalent to

(see also Corollary 0.4.15)

(i) R is equi-dimensional and

(ii) $R_{\mathfrak{p}}$ is a Cohen-Macaulay ring for all prime ideals $\mathfrak{p} \neq \mathfrak{m}$.

Hence Theorem 2.7 is proved, q.e.d.

Next we investigate a geometrical interpretation of Theorem 2.7. To this end we state an interpretation of Lemma 2.6. Let $X \subset \mathbf{P}_k^n =: P$ be a projective scheme such that for a coherent sheaf $\mathcal{F}$ and an integer t:

(a) The canonical map

$$M_n \to H^0(X, \mathcal{F}(n))$$

is bijective for all $n \neq t$, where M denotes the graded module associated to $\mathcal{F}$, and

(b) $H^i(X, \mathcal{F}(n)) = 0$ for all $n \neq t$ and $0 < i < \dim \mathcal{F}$.

Then $\mathcal{F}$ is arithmetically Buchsbaum, i.e., M is a graded Buchsbaum module.

Using this notion, our Theorem 2.7 says: *Let R be an F-pure graded k-algebra resp. R has a presentation of relative graded F-pure type. If $(X, \mathcal{O}_X) = \operatorname{Proj}(R)$ is a pure dimensional (locally) Cohen-Macaulay scheme, then X is arithmetically Buchsbaum.*

Now we will examine some Buchsbaum rings which arise from the purity of the Frobenius.

Example 2.8. Let k be a field of characteristic $p > 0$ resp. of characteristic zero, and let R be a graded k-algebra which is F-pure resp. has a presentation of relative graded F-pure type. Suppose R is a domain such that $R_{\mathfrak{p}}$ is regular for every prime ideal $\mathfrak{p}$ different from the irrelevant ideal $\mathfrak{m}$. Let $S \subset R$ be a graded k-subalgebra which is pure in R. Then S is a Buchsbaum ring.

M. Hochster and J. L. Roberts [2], § 5, showed that S is F-pure resp. has a presentation of relative graded F-pure type. Also, they showed that $S_{\mathfrak{p}}$ is a Cohen-Macaulay ring for all prime ideals $\mathfrak{p}$ different from the irrelevant ideal which follows from the main result of Hochster-Roberts [1], i.e., a pure subring of a regular ring with characteristic $p > 0$ is a Cohen-Macaulay ring. From this it follows that the ring of invariants of linearly reductive affine linear algebraic groups acting on regular rings are Cohen-Macaulay rings, compare Hochster-Roberts [1].

Therefore, our Theorem 2.7 shows that rings of invariants of those groups acting on certain singular rings are Buchsbaum rings.

Example 2.9. Let k, R be as in the previous example. Let G be a linearly reductive affine linear algebraic group over k acting on R by preserving degrees. Then the ring of invariants $S = R^G$ is a Buchsbaum ring.

We note that, $S = R^G$ is in general not a Cohen-Macaulay ring, if R is singular. For this we consider an example of M. Hochster-J. L. Roberts [1]. Let

$$R = k[x_1, \ldots, x_n]/(x_1^n + \cdots + x_n^n).$$

Then R is F-pure for a perfect field k with characteristic $p \equiv 1 \bmod n$. Therefore we have a presentation of relative graded F-pure type, if k is a field of characteristic zero. Let $S = k[y_1, y_2]$. Let $G = \mathrm{Gl}(l, k) = k \setminus \{0\}$ acting on R resp. S by multiplication of a form of degree m by g^m resp. g^{-m} for $g \in G$. The tensor product

$$R \otimes_k S = k[x_1, \ldots, x_n, y_1, y_2]/(x_1^n + \ldots + x_n^n)$$

is F-pure resp. has a presentation of relative graded F-pure type. Furthermore, all the assumptions of Example 2.9 above are fulfilled. Thus $(R \otimes_k S)^G$ is a Buchsbaum ring. In fact, $(R \otimes_k S)^G$ is the Segre product of R and S. Because R is improper in the sense of W. L. Chow [1], i.e., (see Chapter IV, Corollary 4.4 and 4.5) $[\underline{H}_{\mathfrak{m}}^d(R)]_0 \neq O$, $d = \dim R$, we obtain $(R \otimes_k S)^G$ is not a Cohen-Macaulay ring, compare Chapter I, § 4.

More generally, we can show that certain Segre products are Buchsbaum rings. For this we extend the investigate of Chapter I, § 4. Let R_j, $j = 1, 2$, be two graded k-algebras over a field k with $\dim R_j = 1$. Let S denote their Segre product. Let $(X_j, \mathcal{O}_{X_j}) = \mathrm{Proj}(R_j)$, $j = 1, 2$, and $(W, \mathcal{O}_W) = \mathrm{Proj}(S)$. Then we have

$$W \cong X_1 \times_k X_2 \quad \text{and} \quad \mathcal{O}_W(n) = p_1^* \mathcal{O}_{X_1}(n) \otimes_k p_2^* \mathcal{O}_{X_2}(n),$$

where $p_i \colon W \to X_i$, $i = 1, 2$, denote the canonical projections.

Proposition 2.10. *Let $r \geq 0$ be an integer. Assume that the following conditions are fullfiled for $i = 1, 2$:*

a) *The canonical map*

$$[R_j]_n \to H^0\big(X_j, \mathcal{O}_{X_j}(n)\big), \quad n \neq r,$$

   is bijective.

b) $H^i\big(X_j, \mathcal{O}_{X_j}(n)\big) = O$ *for all $n \neq r$ and $0 < i < \dim X_j$.*

c) $H^{d_j}\big(X_j, \mathcal{O}_{X_j}(n)\big) = O$ *for all $n \geq 0$, $n \neq r$ and $d_j = \dim X_j$.*

Then W is arithmetically Buchsbaum. Additionally W satisfies conditions a), b), c).

Proof: By virtue of the above remark, it is enough to show that conditions a), b), c) are satisfied on the cohomology of W.

Using the Künneth formula

$$H^s\big(W, \mathcal{O}_W(n)\big) \cong \bigoplus_{a+b=s} H^a\big(X_1, \mathcal{O}_{X_1}(n)\big) \otimes_k H^b\big(X_2, \mathcal{O}_{X_2}(n)\big),$$

compare Proposition 0.2.10, this follows immediately, q.e.d.

Thus by analyzing the proof of Lemma 2.5 we have obtained our results 2.6−2.10 on Frobenius purity and the arithmetical Buchsbaum property. These assertions are also contained in the paper by P. Schenzel [1], 4.4.

Now we return to Reisner's locally Cohen-Macaulay criterion of Theorem 2.4. By virtue of this statement it is possible to construct many Buchsbaum rings arising from simplicial complexes.

Example 2.11 (Reisner's example from [1]). The fact that the Cohen-Macaulay property of $K[\Sigma]$ depends upon K follows now immediately from Theorem 2.4. Take, for example, a triangulated manifold M, whose only nonzero homology is pure p-torsion (for some prime p). Then $K[x_1, \ldots, x_n]/I_M$ is Cohen-Macaulay if K has characteristic other than p and is not Cohen-Macaulay if char $K = p$. Examples of such manifolds are Lens spaces (see, e.g. Hilton-Wylie [1], p. 223). For a simpler example, one can take M to be the projective plane. In particular, if we consider the minimal triangulation of the projective plane (Fig. 1) then

$$I_M = (x_1x_2x_3, x_1x_2x_4, x_1x_3x_5, x_1x_4x_6, x_1x_5x_6, x_2x_3x_6, x_2x_4x_5, x_2x_5x_6, x_3x_4x_5, x_3x_4x_6).$$

In this case $K[\Sigma]$ is not Cohen-Macaulay if char $K = 2$ and it is Cohen-Macaulay if char $K \neq 2$. It will also follow from Corollary 2.12 below that $K[\Sigma]$ is not Cohen-Macaulay but Buchsbaum for char $K = 2$. We also could have taken Δ_n to be a finite triangulation of the real projective n-space $\mathbf{P}^n_R$. We then get for the reduced simplicial

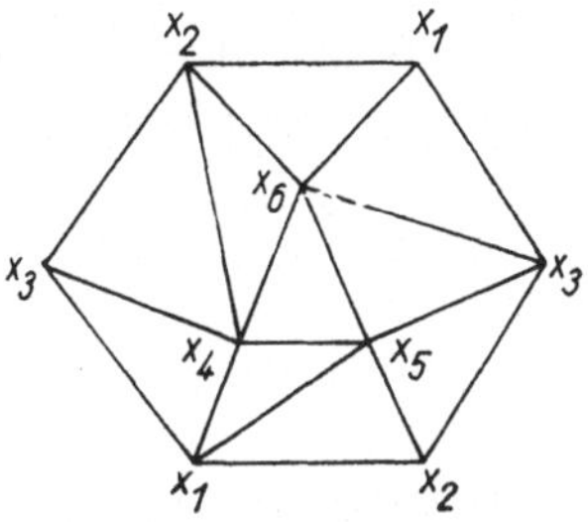

Fig. 1

homology of Δ_n with coefficients in an abelian group G (see, e.g., Hilton-Wylie [1], 3.9.4):

$$H_i(\Delta_n; G) \cong \begin{cases} G/(2)\,G & \text{if } i \text{ is odd,} \\ T_2(G) & \text{if } i \text{ is even.} \end{cases}$$

where $T_2(G) = \{g \in G \text{ such that } 2g = 0\}$ denotes the 2-torsion part of G. For a field K we get for $0 < i < n$:

$$H_i(\Delta_n; K) \cong \begin{cases} K & \text{if char } K = 2, \\ 0 & \text{otherwise.} \end{cases}$$

Therefore $K[\Delta_n]$ is a Cohen-Macaulay ring of dimension $n + 1$ if char $K \neq 2$. If char $K = 2$ then $K[\Delta_n]$ is a Buchsbaum ring by applying again Corollary 2.12. Furthermore it follows in this case that dim $K[\Delta_n] = n + 1$, depth $K[\Delta_n] = 2$, and the invariant of the Buchsbaum ring $K[\Delta_n]$, $I(K[\Delta_n]) = 2^n - (n + 1)$.

This example therefore shows that the invariant of the Buchsbaum ring $K[\Delta_n]$ depends on the ground field K. This fact was examined by P. Schenzel [1], 6.2.3, by considering Reisner's example in case $n = 2$. Moreover, S. Šolčan [1] showed that the Buchsbaum property of $K[\Sigma]$ depends on the characteristic of the ground field K by extending Reisner's Example 2.11. Other examples are constructed by L. T. Hoa [2] by using Segre products of affine semigroup rings (see also Gräbe [1]). Šolčan's example is as follows:

Take

$$I_S := (x_0, x_2, x_3) \cap (x_0, x_2, x_4) \cap (x_0, x_3, x_5) \cap (x_0, x_4, x_6) \cap (x_0, x_5, x_6) \cap I_M$$
$$\subset K[x_0, \ldots, x_6]$$

where I_M is Reisner's Example 2.11. Introduce a second copy $K[x_7, \ldots, x_{13}]$ of $K[x_0, \ldots, x_6]$ and denote by I'_S the ideal in $K[x_7, \ldots, x_{13}]$ corresponding to I_S. Let

$$A = K[x_0, \ldots, x_{13}]/I$$

where

$$I = (I_S, x_7, \dots, x_{13}) \cap (I'_S, x_0, \dots, x_6) \subset K[x_0, \dots, x_{13}].$$

Then it follows that A is Buchsbaum if char $K \neq 2$. For char $K = 2$ we get that A is not a Buchsbaum ring. This fact is obtained from the following localization and by applying Example 2.11:

If char $K \neq 2$ then $A_{(x_0,\dots,\hat{x}_i,\dots,x_{13})}$ is a Cohen-Macaulay ring for all $i = 0, \dots, 13$.

If char $K = 2$ then $A_{(x_0,\dots,\hat{x}_i,\dots,x_{13})}$ is not a Cohen-Macaulay ring for $i = 0, 1, 7, 8$.

For proving the Buchsbaum property of $K[\Delta_n]$ the key is the following corollary of Theorem 2.4. It is precisely the phenomenon exhibited here, which makes the theory of local Buchsbaum rings interesting in algebraic topology.

Corollary 2.12. *Let Σ be any finite simplicial complex.*

(i) *If the geometric realization $|\Sigma|$ is a connected (topological) manifold then $K[\Sigma]$ is Buchsbaum.*

(ii) *The following conditions are equivalent:* (a) *$|\Sigma|$ is a homology manifold.* (b) *$K[\Sigma]$ is Buchsbaum.*

Proof: (i) If $|\Sigma|$ is a topological connected manifold then the condition (ii) of Theorem 2.4 regarding links holds automatically; that is, $K[\Sigma]$ is Buchsbaum by applying Theorem 2.4. (See also Corollary 2.2.)

(ii) This follows immediately from the definition of a homology manifold and Theorem 2.4 since the linked complex for each simplex of $|\Sigma|$ has the homology groups of a sphere (see, e.g., Hilton-Wylie [1], p. 156), q.e.d.

Finally, let us collect some well-known examples and facts.

Example 2.13. (i) Take for $X = |\Sigma|$ a cylinder. Consider the triangulation in Fig. 2 which defines in $K[x_0, \dots, x_5]$:

$$I_\Sigma = (x_0x_3, x_1x_4, x_2x_5, x_0x_2x_4, x_1x_3x_5)$$

(see Eisenreich [1]). By Corollary 2.12 it follows that $K[\Sigma]$ is a non-Cohen-Macaulay Buchsbaum ring since $\tilde{H}_1(\Sigma; K) \neq 0$.

(ii) Take for $X = |\Sigma|$ the torus and a triangulation which defines in $K[x_0, \dots, x_8]$

$$I_\Sigma = (x_0x_4, x_0x_8, x_1x_5, x_1x_6, x_2x_3, x_2x_7, x_3x_7, x_4x_8, x_5x_6, x_0x_1x_2, x_3x_4x_5, x_6x_7x_8, x_0x_5x_7, x_1x_3x_8, x_2x_4x_6, x_0x_3x_6, x_1x_4x_7, x_2x_5x_8)$$

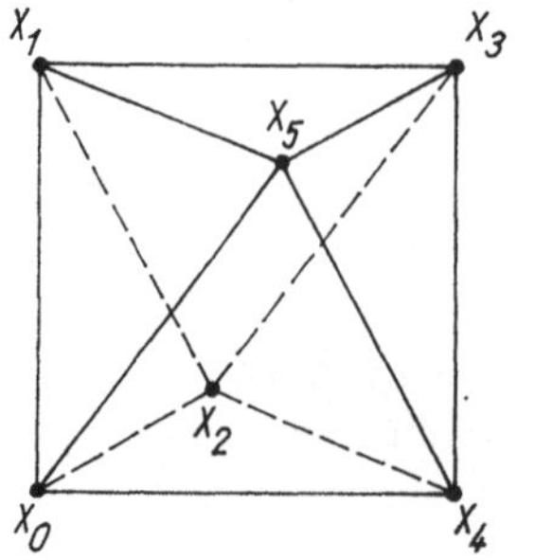

Fig. 2

(see our Introduction). $K[\Sigma]$ is a non-Cohen-Macaulay Buchsbaum ring. This assertion results immediately from Corollary 2.12 and the computation of the homology of a torus (see, e.g., Hilton-Wylie [1], p. 64). Furthermore, it follows from the homology groups that the invariant $I(K[\Sigma])$ of this Buchsbaum ring is 2.

(iii) Take for $X = |\Sigma|$ the Möbius band. Consider the triangulation in Fig. 3. Here we get in $K[x_1, \ldots, x_6]$

$$I_\Sigma = (x_1x_3, x_1x_5, x_2x_6, x_2x_3x_4, x_3x_4x_5, x_4x_5x_6).$$

$K[\Sigma]$ is a non-Cohen-Macaulay Buchsbaum ring with $I(K[\Sigma]) = 1$.

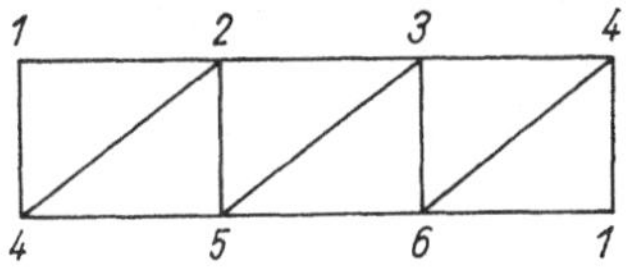

Fig. 3

Example 2.14.

(i) Spheres and discs are always Cohen-Macaulay.

(ii) If $\dim \Sigma = 0$ then $K[\Sigma]$ is Cohen-Macaulay.

(iii) If $\dim \Sigma = 1$, $K[\Sigma]$ is Cohen-Macaulay if and only if Σ is connected.

(iv) Let Σ be the complex of Example 1.16. Then Σ is not a manifold, but $K[\Sigma]$ is Cohen-Macaulay.

We also mention the following lemma proved by Hochster, see, e.g., Schwartau [1]:

Lemma 2.15. *Let $\Sigma \subset \Delta_n$ be any finite simplicial complex, and let x_i be any vertex of Σ. As usual identify x_i with a variable of $S = K[x_0, \ldots, x_n]$. Then we have:*

$$K[\mathrm{lk}_\Sigma x_i] \cong (K[\Sigma])_{(x_i)} \cong (S/I_\Sigma)_{(x_i)}.$$

Since $\mathrm{Spec}(S/I_\Sigma)_{(x_i)}$ form an open affine cover of $\mathrm{Proj}(S/I_\Sigma)$, Lemma 2.15 shows directly that the subscheme $V \subset \mathbf{P}_K^n$ defined by I_Σ will be locally Cohen-Macaulay if the links $\mathrm{lk}_\Sigma x_i$ are Cohen-Macaulay for each vertex $x_i \in \Sigma$.

Remark 2.16. We recall the polarization $I \rightsquigarrow \mathrm{Polar}\, I$ of Definition 1.1. Löfwal and Weyman have pointed out that this association allows one to extend Reisner's Cohen-Macaulay criterion to the non-square-free case. The reason for this is that I and Polar I differ only by a regular sequence (to obtain I from Polar I simply set the new generations of variables equal to the original generation again). Therefore I determines a Cohen-Macaulay quotient if and only if Polar I does; but Polar I is square-free.

Thus we may extend Reisner's Cohen-Macaulay criterion by the association

$$I \rightsquigarrow \Sigma_{\mathrm{Polar}\, I}.$$

We call $\Sigma_{\mathrm{Polar}\, I}$ the *Reisner complex* associated to I.

§ 3. The topological Cohen-Macaulay criterion of Schwartau

First we describe the combinatorical singularities of simplicial complexes.

All simplicial complexes are assumed to be finite. We interpret $\emptyset$ as a (-1)-sphere and not a (-1)-disc, thus we never allow $\emptyset$ as a simplex.

Definition 3.1. Let Σ be a simplicial complex, and σ a simplex of Σ. Then we say that $\mathrm{lk}_\Sigma\sigma$ is a *sphere* or a *disc* if $|\mathrm{lk}_\Sigma\sigma|$ is homeomorphic to a sphere or a disc. If $\mathrm{lk}_\Sigma\sigma$ is not a sphere or a disc we say σ is *singular*, or a *singularity* of Σ.

Sing Σ is the subcomplex of Σ generated by the singular simplices of Σ.

Remark 3.2. If we have a simplex $\sigma \in \mathrm{Sing}\,\Sigma$, this does *not* imply that $\mathrm{lk}_\Sigma\sigma$ fails to be a sphere or disc. Sing Σ is the complex *generated* by the singular simplices, thus all we can say is that $\sigma \sqsubset \sigma'$ where $\mathrm{lk}_\Sigma\sigma'$ fails to be a sphere or disc. However, we always know that *facets* of Sing Σ have bad links.

Definition 3.3. A simplicial complex Σ will be called a *combinatorial manifold-with-∂*, or simply a *manifold*, if:

a) Σ is equi-dimensional, and

b) $\mathrm{Sing}\,\Sigma = \emptyset$.

Definition 3.4. A *combinatorial sphere* (resp. *combinatorial disc*) will mean a combinatorial manifold-with-$\partial\Sigma$ such that $|\Sigma|$ is homeomorphic to a sphere (resp. a disc). We will write c-sphere, c-disc, for short notation.

Lemma 3.5. *Let Σ be an equi-dimensional simplicial complex. Then the following conditions are equivalent:*

(i) *Σ is a manifold.*

(ii) *For all simplices $\sigma \in \Sigma$, $\mathrm{lk}_\Sigma\sigma$ is a sphere or disc.*

(iii) *For all simplices $\sigma \in \Sigma$, $\mathrm{lk}_\Sigma\sigma$ is a c-sphere or c-disc.*

(iv) *For all simplices $\sigma \in \Sigma$, $\mathrm{Star}_\Sigma\sigma$ is a c-disc.*

Proof: (i) $\Leftrightarrow$ (ii) by definition of manifolds. (ii) $\Leftrightarrow$ (iii): The implication ($\Leftarrow$) is trivial. The implication ($\Rightarrow$) results from the fact that the link of any simplex in Σ is an iteration of links of vertices (see Lemma 1.10).

(iii) $\Leftrightarrow$ (iv): The implication ($\Leftarrow$) is easy to prove, since $\mathrm{lk}_\Sigma\sigma$ is just $\mathrm{lk}_{\mathrm{Star}_\Sigma\sigma}\sigma$, and $\mathrm{Star}_\Sigma\sigma$ is a combinatorial manifold by hypothesis. To prove the implication ($\Rightarrow$) recall that $\mathrm{Star}_\Sigma\sigma$ is just an iteration of simplicial cones over $\mathrm{lk}_\Sigma\sigma$. Thus it suffices to show that if Σ is a c-sphere or c-disc, then $\mathrm{Cone}_P(\Sigma)$ is a c-disc. But this follows from the link formulas of Lemma 1.14, q.e.d.

Remark 3.6. According to our definitions, an equi-dimensional simplicial complex is a combinatorial manifold if and only if $\mathrm{Sing}\,\Sigma = \emptyset$, i.e. the link of every simplex is homeomorphic to a sphere or disc. This is weaker than the usual definition, which would have every link PL-homeomorphic to the standard simplicial sphere or disc (it is enough to require this for links of vertices). Our definition is a hybrid of the topological and PL approaches which makes it easier to detect the singular simplices.

Proposition 3.7. *Let Σ be a simplicial complex. Then if Σ is a manifold, $|\Sigma|$ is a topological manifold-with-∂.*

Proof: Let P be any point of $|\Sigma|$. We have to show that P has an open neighborhood U in $|\Sigma|$ homeomorphic to an open subset of the closed half-space $\mathbf{H}^d \subset \mathbf{R}^d$ where d is the dimension of Σ. Now, P lies in the interior of some simplex σ of Σ. Note that we may perform a modified barycentric subdivision of Σ (with P as the barycenter of $|\sigma|$) to make P into a vertex in a new triangulation Σ' of $|\Sigma|$. Under such a barycentric subdivision we have $|\mathrm{Star}_\Sigma \sigma| \approx |\mathrm{Star}_{\Sigma'} P|$. Since Σ is by hypothesis a manifold, it follows from Lemma 3.5(iv) that $|\mathrm{Star}_{\Sigma'} P|$ is homeomorphic to a disc (of dimension d). Thus the required neighborhood U above is provided by $|\mathrm{Star}_{\Sigma'} P|$, q.e.d.

Now we want to analyze the singularities of cones and of stars.

Proposition 3.8. *Let Σ be an equi-dimensional simplicial complex. Then we have:*

(i) *(singularities of cones). There are three possibilities for the simplicial cone* $\mathrm{Cone}_P(\Sigma)$:

1) a) *Σ is a c-sphere or c-disc and* $\mathrm{Cone}_P(\Sigma)$ *is a c-disc.*

b) *Σ is a manifold other than a c-sphere or c-disc and* $\mathrm{Sing\ Cone}_P(\Sigma) = \{P\}$.

2) $\mathrm{Sing}\ \Sigma \neq \emptyset$ *and* $\mathrm{Sing\ Cone}_P(\Sigma) = \mathrm{Cone}_P(\mathrm{Sing}\ \Sigma) \neq \emptyset$.

(ii) *(singularities of stars). Let σ be any simplex of Σ. Then for any vertex $v \in \sigma$, $\sigma \in$* $\mathrm{Sing}\ \Sigma$ *if and only if $\sigma \in$* $\mathrm{Sing\ Star}_\Sigma v$.

Proof: (i) This proof depends on the link formulas α, β, γ of Lemma 1.14.

(ii) Note that $v \in \sigma \Rightarrow \sigma \in \mathrm{Star}_\Sigma v$, thus the statement is meaningful. The proof follows by showing that

$$(*) \qquad \mathrm{lk}_\Sigma\, \sigma = \mathrm{lk}_{\mathrm{Star}_\Sigma v}\, \sigma.$$

To show this, note that in any complex, the link of a simplex σ is obtained by finding all simplices of the complex containing σ and "pulling off" σ. But since $v \in \sigma$, {simplices of Σ containing σ} = {simplices of $\mathrm{Star}_\Sigma v$ containing σ} and $(*)$ is proven. We may now apply $(*)$ to show that $\mathrm{Sing}\ \Sigma$, $\mathrm{Sing\ Star}_\Sigma v$ have the same facets; hence we have an equality of complexes $\mathrm{Sing}\ \Sigma = \mathrm{Sing\ Star}_\Sigma v$, q.e.d.

Next we need to make some observations and collect known results on quasimanifolds.

Definition 3.9. A simplicial complex Σ is a *quasi-manifold* if

(i) Σ is equi-dimensional, and

(ii) $\mathrm{Sing}\ \Sigma$ contains no codimension 1 simplex of Σ.

Remark 3.10. This definition means that the link of every codimension 1 simplex is a 0-sphere or 0-disc; i.e. every codimension 1 simplex is contained in at most two facets. This constitutes only part of the usual definition of "pseudo-manifold" found in the literature.

Definition 3.11. A simplicial complex Σ is called a *regular non-quasi-manifold* (RNQM) if

(i) Σ is equi-dimensional, and

(ii) $\mathrm{Sing}\ \Sigma \neq \emptyset$ and is a union of codimension 1 simplices of Σ.

Examples 3.12.

(i) Any manifold is a quasi-manifold.

(ii) The complex Σ of Example 1.15 is not a manifold, but is a quasi-manifold.

(iii) The complex Σ of Example 1.16 is not a quasi-manifold, but is a regular non-quasi-manifold.

(iv) Let Σ be an equi-dimensional simplicial complex. If $|\Sigma|$ is a topological manifold-with-∂, then Σ is a quasi-manifold.

Proof: Suppose that Σ is not a quasi-manifold. But since Σ is equi-dimensional, so there must exist a simplex $\sigma \in \operatorname{Sing} \Sigma$ of codimension 1 in Σ. Let P be any point of $|\sigma|$. Then any meighborhood of P in $|\mathrm{Star}_\Sigma \sigma|$ is a neighborhood of P in $|\Sigma|$. Therefore if we show that $|\mathrm{Star}_\Sigma \sigma|$ is not a topological manifold at P then $|\Sigma|$ is not a topological manifold at P, and the proof is finished. But σ is contained in ≥ 3 facets of Σ, therefore $|\mathrm{Star}_\Sigma \sigma|$ is a "fan". Separations theorems from topology guarantee that a fan cannot be a topological manifold. This does prove (iv), q.e.d.

We still need more facts on orientable simplicial complexes.

Definition 3.13. Let σ be a simplex in a simplicial complex Σ. Then σ becomes an *oriented* simplex if we choose an arbitrary fixed ordering of the vertices. The equivalence class of even permutations of this fixed ordering is called the *positively oriented simplex* $+\sigma$; the equivalence class of odd permutations is the *negatively oriented simplex* $-\sigma$. Notice that an orientation of σ induces through the simplicial boundary operator an orientation on each boundary-face of σ.

Definition 3.14. Let Σ be a quasi-manifold. Then we say Σ is *coherently oriented* if the facets of Σ are oriented in such a way that any two facets meeting in a codimension 1 simplex induce opposite boundary orientations on that simplex.

Definition 3.15. Let Σ be a quasi-manifold. Then we say that Σ is *orientable* if it is possible to orient Σ coherently; otherwise we say Σ is *non-orientable.*

Remark 3.16. Note that if Σ is an orientable quasi-manifold, and if $\Sigma' \subset \Sigma$ is a quasi-manifold of the same dimension, then Σ' must be orientable.

We now extend the definition of orientability to arbitrary simplicial complexes.

Definition 3.17. Let Σ be a simplicial complex. Then we say that Σ is *orientable* if:

a) Σ is equi-dimensional of dimension d.

b) Every d-dimensional quasi-manifold $\Sigma' \subset \Sigma$ is orientable.

Examples 3.18.

(i) Any triangulation of a cylinder is an orientable quasi-manifold of dimension 2.

(ii) Any triangulation of a Möbius band is a non-orientable quasi-manifold of dimension 2.

(iii) The quasi-manifold of Example 3.12(ii) is orientable.

(iv) The RNQM of Example 3.12(iii) is orientable.

Lemma 3.19. *Let $\Sigma' \subset \Sigma$ be equi-dimensional simplicial complexes of the same dimension. Then:*

$$\Sigma \text{ is orientable} \Rightarrow \Sigma' \text{ is orientable.}$$

Proof: Immediate from the definition of orientability.

Corollary 3.20. *Let Σ be a simplicial complex, and v a vertex of Σ. Then:*

$$\Sigma \text{ is orientable} \Rightarrow \mathrm{Star}_\Sigma v \text{ is orientable.}$$

Lemma 3.21. *Let Σ be a simplicial complex. Then:*

$$\Sigma \text{ is orientable} \Leftrightarrow \mathrm{Cone}_P(\Sigma) \text{ is orientable.}$$

Proof: As in the proof of Lemma 1.13, we need the 1-1 correspondence:

$$\{\text{facets of } \Sigma\} \xleftrightarrow[\text{1-1}]{(\ ,P)} \{\text{facets of } \mathrm{Cone}_P(\Sigma)\}.$$

This correspondence preserves equi-dimensionality, and preserves coherent orientations in a natural way; thus the Lemma 3.21 is proven.

Corollary 3.22. *Let Σ be a simplicial complex, and v a vertex of Σ. Then:*

$$\Sigma \text{ is orientable} \Rightarrow \mathrm{lk}_\Sigma v \text{ is orientable.}$$

Proof: $\mathrm{lk}_\Sigma v$ is the base of the simplicial cone $\mathrm{Star}_\Sigma v$. We now apply Corollary 3.20 and Lemma 3.21.

Corollary 3.23. *Let Σ be a simplicial complex and σ any simplex of Σ. If Σ is orientable then $\mathrm{lk}_\Sigma \sigma$ is orientable.*

Proof: As in Lemma 1.10, any link $\mathrm{lk}_\Sigma \sigma$ is just an iterated link of vertices. Now apply Corollary 3.22, q.e.d.

It is not too difficult to show that equi-dimensional simplicial complexes of codimension 0 or 1 are all c-spheres or c-discs. Therefore it was Schwartau's idea to investigate simplicial complexes of codimension 2. All his results depend on the following decomposition theorem:

Theorem 3.24 (Schwartau [1], Theorem 165). *Let $d > 1$, Σ an equi-dimensional d-dimensional simplicial complex of codimension 2. Then there exist subcomplexes S, St, $G = S \cap St$ of Σ such that:*

(i) *$S \neq \emptyset$, $St \neq \emptyset$, and $\Sigma = S \cup St$.*

(ii) *S is a c-sphere or c-disc of dimension d.*

(iii) *St is an equi-dimensional simplicial cone of dimension d and of codimension ≤ 2.*

(iv) *G is contained in the base of the cone St.*

(v) *G is a nonvoid union of $(d-1)$- and $(d-2)$-simplices.*

By considering orientable simplicial complexes of codimension 2 Schwartau obtained the following result (see Schwartau [1], Theorem 167):

Theorem 3.25. *Let Σ be an orientable simplicial complex of codimension 2. Then if Σ is a quasi-manifold, either:*

a) *Σ is a manifold, or*

b) Sing *Σ has a facet of codimension 2 in Σ.*

The key in analyzing the singularities of codimension 2 complexes is the following interesting lemma proved by Schwartau [1], Lemma 166:

Lemma 3.26. *Let X, Y be two-dimensional simplicial ocmplexes, each of which is a manifold or an RNQM, joined along two disjoint line segments. Then if codim $(X \cup Y) \leq 2$, $X \cup Y$ must be a Möbius band.*

We can now sharpen Theorem 3.25 and we get the topological Cohen-Macaulay criterion of Schwartau [1], Theorems 172, 173, 174 and 175.

Theorem 3.27 (Schwartau's topological Cohen-Macaulay criterion). *Let Σ be an equi-dimensional orientable simplicial complex of codimension 2 and dimension $d \geq 2$. Then the following conditions are equivalent:*

(i) *Σ is Cohen-Macaulay.*

(ii) Sing *Σ is equi-dimensional of dimension $d-1$ or is $\emptyset$.*

(iii) *Σ is a manifold.*

(iv) *Σ is a regular non-quasi-manifold.*

This is Schwartau's beautiful Cohen-Macaulay criterion.

Example 3.28. The complex of Example 1.15 is not Cohen-Macaulay, but the complex of Example 1.16 is Cohen-Macaulay.

This is now a direct consequence of Theorem 3.27.

In the following observations we show how to put Schwartau's topological Cohen-Macaulay criterion into practice. Let Σ be an equi-dimensional simplicial complex of codimension ≤ 2 and dimension d. Then by definition Σ is a subcomplex of Δ_{d+2}, the standard $(d+2)$-simplex. It follows that each facet of Σ contains all the vertices of Δ_{d+2} save two. Thus we may denote each facet of Σ by two parameters, namely by the vertices it does not contain. If we now draw vertices to indicate the vertices of Δ_{d+2}, each facet of Σ may be represented by an edge connecting two vertices. This is the so-called associated graph of Σ.

Example 3.29. Consider the ideal $I = (x_0, x_1) \cap (x_2, x_3)$ in $S = K[x_0, \ldots, x_3]$. Then I is a square-free monomial ideal of height 2. Therefore the Reisner complex Σ_I (see § 1) has codimension ≤ 2. We have a decomposition of Σ_I into its facets given by

$$\Sigma_I = \Sigma_{(x_0, x_1)} \cup \Sigma_{(x_2, x_3)}.$$

By the definition of Reisner complexes, $\Sigma_{(x_0,x_1)}$ is the simplex given by all the variables of S except x_0, x_1. Therefore in the graph for Σ_I this simplex is represented as ;
similary, the simplex $\Sigma_{(x_2,x_3)}$ is represented as . The total graph for Σ_I is thus:

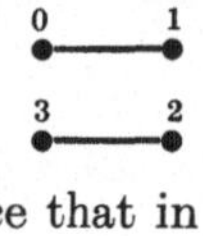

Notice that in $\mathbf{P}^3_K$ the ideal I defines two skew lines; by coincidence the graph is a simplicial model of this fact (also note that the graph of Σ_I in this case happens to coincide with Σ_I itself).

Remark 3.30. In case that a monomial ideal I of S is not a square-free monomial ideal, we must take the polarization of I before obtaining the Reisner complex (see Definition 1.1).

Now suppose $\dim \Sigma \geq 2$ and T is a simplex of codimension 2 in Σ. Since $\operatorname{lk}_\Sigma T$ is equi-dimensional of dimension 1 and since codim $\Sigma \leq 2$ it follows that codim $\operatorname{lk}_\Sigma T \leq 2$ (linking reduces vertices by at least as much as dimension). Thus $\operatorname{lk}_\Sigma T$ must be a 1-dimensional subcomplex of Δ_3, the standard 3-simplex. It follows that $\operatorname{lk}_\Sigma T$ is either a 1-sphere, or a 1-disc, or  , or contains a subcomplex of the form .

In the last case (T, v) is a singular simplex of Σ containing T. Thus if T is a facet of Sing Σ this cannot occur. Alternately $\mathrm{lk}_\Sigma T$ could also not be a sphere or dics. Thus we must have

$\mathrm{lk}_\Sigma T =$ 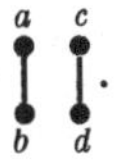 .

Conversely, if T has such a link, T must be a facet of Sing Σ. For T is clearly in Sing Σ, so we only need to check that any codimension 1 simplex of Σ containing T is non-singular. But the link in Σ of any such simplex is simply the link of a vertex in $\mathrm{lk}_\Sigma T$. This is always a 0-disc and we are done. Thus we have shown:

T is a facet of Sing Σ of codimension 2 in $\Sigma \Leftrightarrow \mathrm{lk}_\Sigma T =$ 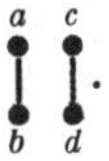 .

Notice that T has such a link if and only if (T, a, b), (T, c, d) are facets of Σ and (T, a, c), (T, a, d), (T, b, c), (T, b, d) are not. Thus T is an isolated codimension 2 singularity of Σ if and only if the edges ab, cd appear in the associated graph of Σ and the edges ac, ad, bc, bd do not. Notice that the four missing edges are the only possible edges which could connect the disjoint edges ab, cd to each other.

Conversely, suppose there exist disjoint edges ab, cd in the graph which are not connected to each other by any other edge of the graph. The facets of Σ represented by ab, cd must then intersect along a codimension 2 simplex T of Σ (given by $\Delta_{d+2} - \{a, b, c, d\}$), and $\mathrm{lk}\, T$ must be precisely 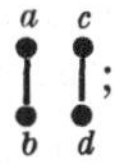; hence T is a facet of Sing Σ of codimension 2 in Σ. Therefore we have shown that there is a 1-1 correspondence:

$$\{\text{Facets of Sing of codimension 2 in } \Sigma\} \overset{1\text{-}1}{\longleftrightarrow} \left\{\begin{matrix}\text{pairs of disjoint edges in the} \\ \text{graph of } \Sigma \text{ not connected to} \\ \text{each other by any other edge}\end{matrix}\right\}.$$

Therefore Sing Σ has no facets of codimension 2 in Σ if and only if given any two edges of the associated graph, either they are connected to each other or else there exists a third edge connected to both of them.

Let Σ be an equi-dimensional simplicial complex of codimension ≤ 2, and write G for the associated graph of Σ. Consider the dual graph G^*: each edge of G becomes a vertex in G^*, and two vertices of G^* are connected with an edge if and only if the two corresponding edges of G meet at a vertex.

Recall that the *diameter* of a graph is the largest number of edges needed to connect any two given vertices of the graph. In this context, we have proven:

Theorem 3.31. *Let Σ be an equi-dimensional simplicial complex of codimension ≤ 2, G the associated graph, and G^* the dual graph of G. Then:*

1. Sing Σ *contains no facets of codimension* 1 *in* $\Sigma \overset{\left(\substack{\text{assume} \\ \dim \Sigma \geq 1}\right)}{\Longleftrightarrow}$ G *contains no triangles,*

2. Sing Σ *contains no facets of codimension* 2 *in* $\Sigma \overset{\left(\substack{\text{assume} \\ \dim \Sigma \geq 2}\right)}{\Longleftrightarrow}$ G^* *has diameter* ≤ 2.

Corollary 3.32. *Let Σ be an equi-dimensional orientable simplicial complex of codimension ≤ 2; G the associated graph, and G^* the dual graph. Then:*

A) *Σ is Cohen-Macaulay $\Leftrightarrow$ G^* has diameter ≤ 2.*

B) *Σ is a c-sphere or a c-disc $\Leftrightarrow$ G^* has diameter ≤ 2 and G contains no triangles.*

Proof: If $\dim \Sigma = 0$ then Σ is always Cohen-Macaulay (see Example 2.14(ii)), and the result follows.

If $\dim \Sigma = 1$, Σ must be a subcomplex of Δ_3. Σ is Cohen-Macaulay if and only if Σ is connected (see Example 2.14(iii)), and again the result follows directly.

Thus we may assume $\dim \Sigma \geq 2$, whence both parts of Theorem 3.31 apply. If $\operatorname{codim} \Sigma < 2$, the corollary is a trivial consequence of Theorem 3.31 by use of our observations after Corollary 3.23. Thus we may assume $\operatorname{codim} \Sigma = 2$. But then we finish the proof as follows:

A) $\Rightarrow$: Theorem 3.31 and Theorem 3.27.

A) $\Leftarrow$: Theorem 3.31, Theorem 3.27 and an application of the following fact which improves Theorem 3.25:

Let Σ be an equi-dimensional orientable simplicial complex of codimension 2. Then either Σ is a manifold or else $\operatorname{Sing} \Sigma$ is a non-empty union of codimension 1 and codimension 2 simplices of Σ.

B) $\Rightarrow$: Theorem 3.31.

B) $\Leftarrow$: Theorem 3.31 and the just mentioned fact. We also note that if Σ is a manifold then Σ is a c-sphere or c-disc, q.e.d.

Now let $I \subset S = K[x_0, \ldots, x_n]$ be a monomial ideal of height 2; we relate the above theory to the Cohen-Macaulay property for S/I. First note that S/I cannot be Cohen-Macaulay unless I is equi-dimensional and unmixed; that is, every associated prime of I has height 2. It follows that the Reisner complex Σ associated to I is equi-dimensional and of codimension ≤ 2. In fact by these results, each facet of Σ is given by a Reisner complex $\Sigma_{(x_a, x_b)} \subset \Delta_N$ for (x_a, x_b) an associated prime of Polar $I \subset S^{(N)} = k[x_0, \ldots, x_N]$. By the definition of Reisner complexes, x_a, x_b are the two vertices of Δ_N *not* contained in the facet $\Sigma_{(x_a, x_b)}$; hence this facet correspond to the edge $\underset{a}{\bullet}\!\!-\!\!\!-\!\!\!-\!\!\underset{b}{\bullet}$ in the graph G which we associate to Σ. In short, G (or G^*) may be obtained directly from the primary decomposition of Polar I. Thus we now refer to G^* as "the graph or the primary decomposition of Polar I". In addition, we now call I *orientable* or a *quasi-manifold*, etc., if the Reisner complex Σ of I has these properties.

It follows immediately from Corollary 3.32 A):

Corollary 3.33. *Let $I \subset S$ be an orientable monomial ideal of height 2, and let G^* be the graph of the primary decomposition of the polarization of I. Then:*

S/I is Cohen-Macaulay if and only if

(i) *I is equi-dimensional and unmixed, and*

(ii) *G^* has diameter ≤ 2.*

Example 3.34. Let $I \subset S$ be the square-free monomial ideal $(x_0, x_1) \cap (x_1, x_2) \cap \ldots \cap (x_{k-1}, x_k)$. It is not difficult to see that I is an orientable quasi-manifold. Thus Corollary 3.33 applies. The conclusion is:

S/I is Cohen-Macaulay if and only if $k \leq 3$.

We now apply Schwartau's theory to answer the following question:
Consider the ideal

$$I := (x_0, x_1)^a \cap (x_1, x_2)^b \cap (x_2, x_3)^c \cap (x_3, x_0)^d \subset S := K[x_0, \ldots, x_3].$$

Which exponents a, b, c, d (≥ 0) define arithmetically Cohen-Macaulay curves in $\mathbf{P}^3_K$. Examining the dual graph G^* of the Reisner complex of I we get the following general result:

Example 3.35. Let G^* be the graph associated to Polar I. Then G^* has diameter ≤ 2 if and only if:

Case 1. $a, b, c, d > 0$: $a + c = b + d + \varepsilon$, for $\varepsilon = -1, 0, 1$.

Case 2. $a, b, c > 0$, $d = 0$: $a + c \leq b + 1$.

Case 3a. $a, b > 0$, $c = d = 0$: always.

Case 3b. $a, c > 0$, $b = d = 0$: never.

Case 4. $a > 0$, $b = c = d = 0$: always.

Proof: Left to the reader (see also Schwartau [1], Theorem 186).

Now we recall the original motivation for this section: Consider the following ideals in S:

$$I = (x_0, x_1)^n \cap (x_1, x_2)^n \cap (x_2, x_3)^n \cap (x_3, x_0)^{n-1},$$

and

$$I' = (x_0, x_1)^n \cap (x_1, x_2)^{n-1} \cap (x_2, x_3)^n \cap (x_3, x_0)^{n-1}.$$

We will see from liaison addition in Chapter III that S/I is always Cohen-Macaulay, and S/I' is always Buchsbaum.

Example 3.35 provides also an explanation for this phenomenon independent of liaison addition. By Example 3.35 the rings S/I' are not Cohen-Macaulay since the associated Reisner complex Σ has isolated codimension 2 singularities (i.e. Sing Σ has facets of codimension 2 in Σ). In addition it follows that the Reisner complex for the ideal I' has precisely n isolated codimension 2 singularities. In order to show that S/I' is a Buchsbaum ring with invariant n we have to use another method. Applying Schwartau's liaison addition from Chapter III we will prove that the curves in $\mathbf{P}^3_K$ defined by I' have liaison invariant $\underline{K}^n$ (up to shift); that is, S/I' is always Buchsbaum.

Furthermore, by constructing the minimal graded free resolution of the ideal from Example 3.35 we can sharpen the result of Example 3.35 (see Schwartau [1], Chap. 3):

Proposition 3.36. *The ideal* $(x_0, x_1)^a \cap (x_1, x_2)^b \cap (x_2, x_3)^c \cap (x_3, x_0)^d$ *defines an arithmetically Cohen-Macaulay curve in* $\mathbf{P}^3_K$ *if and only if:*

Case 1. $a, b, c, d > 0$: $a + c = b + d + \varepsilon$, *for* $\varepsilon = -1, 0, 1$.

Case 2. $a, b, c > 0$, $d = 0$: $a + c \leq b + 1$.

Case 3a. $a, b > 0$, $c = d = 0$: *always.*

Case 3b. $a, c > 0$, $b = d = 0$: *never.*

Case 4. $a > 0$, $b = c = d = 0$: *always.*

§ 4. Further applications to algebraic topology and combinatorics

In this paragraph we will investigate properties of Buchsbaum complexes of simplicial complexes. For example, following P. Schenzel [1], we will examine a term which can be interpreted as measuring the error when we no longer have the Cohen-Macaulay case of simplicial complexes. We note that we actually do not need Schenzel's approach for these applications to combinatorics (see, for example, Corollary 4.7′ and Lemma 4.14′).

The first topic of this paragraph will be the characterization of Buchsbaum modules using dualizing complexes. For this we will follow an idea initiated by R. Kiehl [1]. P. Schenzel [1] obtained some generalizations of Kiehl's approach by improving the locally Cohen-Macaulay criterion of Reisner of Theorem 2.4. We will first of all review the relevant commutative algebra.

Let $M^\cdot: \ldots \to M^k \to M^{k+1} \to \ldots$ be a complex of A-modules. By $\tau^s M^\cdot$, resp. $\tau_r M^\cdot$, we denote the truncated complex

$$\ldots \to M^k \to \ldots \to M^{s-1} \to \operatorname{Im}(M^{s-1} \to M^s) \to 0$$

resp.

$$0 \to \operatorname{Im}(M^{r+1} \to M^{r+2}) \to M^{r+1} \to \ldots \to M^k \to \ldots$$

If $r < s$ we have

$$H^k(\tau_r^s M^\cdot) \cong \begin{cases} 0 & \text{for } k \leq r \text{ or } k \geq s, \\ H^k(M^\cdot) & \text{for } r < k < s, \end{cases}$$

where $\tau_r^s(M^\cdot) = \tau_r(\tau^s M^\cdot)$. Now $M^\cdot[t]$ will denote the complex $M^\cdot$ shifted t places to the left with changed sign of the boundary map, that is, $(M^\cdot[t]) = M^{n+t}$ and $d_{M^\cdot[t]} = (-1)^t d_{M^\cdot}$, $n \in \mathbf{Z}$. To state the first main result we will use some notation from Chapter 0, § 3. The following theorem was first published by P. Schenzel [1].

Theorem 4.1. *Let M denote a Noetherian A-module with $d := \dim_A M > 0$. Then the following conditions are equivalent:*

(i) *M is a Buchsbaum module,*

(ii) *$\tau^d \underline{R}\Gamma_{\mathfrak{m}}(M)$ is quasi-isomorphic to a complex of k-vector spaces,*

(iii) *$\tau^d \underline{R}\Gamma_{\mathfrak{m}}(M) \xrightarrow{\sim} C^\cdot(M)$, where $C^\cdot(M)$ is a complex of k-vector spaces with*

$$C^i(M) = \begin{cases} H^i_{\mathfrak{m}}(M) & \text{for } 0 \leq i < d, \\ 0 & \text{otherwise} \end{cases}$$

and trivial boundary homomorphisms.

If A possesses a dualizing complex $I^\cdot_A$, then the above conditions are equivalent to:

(iv) *$\tau_{-d} \operatorname{Hom}_A(M, I^\cdot_A)$ is quasi-isomorphic to a complex of k-vector spaces.*

Before we will embark on the proof, we will show a lemma which is of interest in its own right.

Lemma 4.2. *Let A be a local ring possessing a dualizing complex $I^\cdot_A$. Let $F^\cdot$ be a bounded complex of finitely generated free A-modules such that $F^i = O$ for $i < 0$. Assume that the cohomology modules $H^i(F^\cdot \otimes M)$, $i \in \mathbf{Z}$, are A-modules of finite length for a Noetherian A-module M. If $\tau_{-d} \operatorname{Hom}_A(M, I_A)$ is quasi-isomorphic to a complex of k-vector spaces, then $\tau^d(F^\cdot \otimes M)$ is quasi-isomorphic to a complex of k-vector spaces.*

Proof: By the definition of the dualizing complex we get

$$\mathrm{Hom}_A(M, I^i_A) = 0 \quad \text{for } i < -\dim M.$$

Therefore $K_M := H^{-d}(\mathrm{Hom}_A(M, I^{\cdot}_A))$ is the first non-vanishing cohomology module by virtue of the Local Duality Theorem. Hence we have the following short exact sequence of complexes

$$0 \to K_M[d] \to \mathrm{Hom}_A(M, I^{\cdot}_A) \to \tau_{-d}\mathrm{Hom}_A(M, I^{\cdot}_A) \to 0.$$

Applying the derived functor $\underline{R}\,\mathrm{Hom}(F^{\cdot}, \ )$ and $\mathrm{Hom}_A(\ , E)$ we get

$$0 \to \mathrm{Hom}_A(\underline{R}\,\mathrm{Hom}(F^{\cdot}, \tau_{-d}\mathrm{Hom}_A(M, I^{\cdot}_A), E)) \to \underline{R}\Gamma_{\mathfrak{m}}(F^{\cdot} \otimes M)$$
$$\to \mathrm{Hom}_A(\underline{R}\,\mathrm{Hom}(F^{\cdot}, K_M[d]), E) \to 0$$

by virtue of the Local Duality Theorem. Here E denotes the injective hull of the residue field k. Now we remark

$$\mathrm{Hom}_A(F^{\cdot}, K_M[d]) \simeq \underline{R}\,\mathrm{Hom}_A(F^{\cdot}, K_M[d])$$

and

$$\mathrm{Hom}_A(\underline{R}\,\mathrm{Hom}(F^{\cdot}, K_M[d]), E) \simeq \mathrm{Hom}_A(\mathrm{Hom}_A(F^{\cdot}, K_M[d]), E).$$

From this it follows that

$$(\mathrm{Hom}_A(\underline{R}\,\mathrm{Hom}(F^{\cdot}, K_M[d]), E))^i = 0 \quad \text{for all } i < d.$$

Therefore, the short exact sequence induces a quasi-isomorphism

$$\tau^d(F^{\cdot} \otimes \mathrm{Hom}_A(\tau_{-d}\mathrm{Hom}_A(M, I^{\cdot}_A), E)) \simeq \tau^d\underline{R}\Gamma_{\mathfrak{m}}(F^{\cdot} \otimes M). \tag{$*$}$$

Because all the cohomology modules $H^i(F^{\cdot} \otimes M)$ are modules of finite length, it follows that $\underline{R}\Gamma_{\mathfrak{m}}(F^{\cdot} \otimes M) = F^{\cdot} \otimes M$. If $\tau_{-d}\mathrm{Hom}_A(M, I^{\cdot}_A)$ is quasi-isomorphic to a complex of k-vector spaces, the same is true for

$$F^{\cdot} \otimes \mathrm{Hom}_A(\tau_{-d}\mathrm{Hom}_A(M, I^{\cdot}_A), E).$$

Thus, our statement follows from the quasi-isomorphism given in $(*)$.

Proof of Theorem 4.1. First of all we remark that a complex $X^{\cdot}$ is quasi-isomorphic to a complex of k-vector spaces if and only if $X^{\cdot} \otimes \hat{A}$ is quasi-isomorphic to a complex of k-vector spaces. Furthermore, a Noetherian A-module M is a Buchsbaum module if and only if $M \otimes_A \hat{A}$ is a Buchsbaum module over $\hat{A}$, see Lemma I.1.13. That is, without loss of generality we can assume $A = \hat{A}$ and $M = \hat{M}$, where $\hat{\ }$ denotes the $\mathfrak{m}$-adic completion. That means, we can assume A to be a quotient of a regular local ring R by the Cohen Structure Theorem, compare for example Matsumura [1]. Let $I^{\cdot}_A$ be the dualizing complex. Then we have

$$\underline{R}\Gamma_{\mathfrak{m}}(M) \simeq \mathrm{Hom}_A(\underline{R}\,\mathrm{Hom}_A(M, I^{\cdot}_A), E).$$

Therefore, the conditions (ii) and (iv) are equivalent. Also, a complex of k-vector spaces s quasi-isomorphic to its cohomology complex. Hence, (ii) $\Leftrightarrow$ (iii). Now, we will show (i) $\Rightarrow$ (iii). To this end we make use of the following:

Proposition 4.3. *Let $f^{\cdot}: K^{\cdot} \to L^{\cdot}$ be a homomorphism of complexes of A-modules such that*
(a) *$K^{\cdot}$ is a complex of k-vector spaces, and*
(b) *$H^i(f^{\cdot}): H^i(K^{\cdot}) \to H^i(L^{\cdot})$, $i \in \mathbf{Z}$, is a surjective homomorphism.*

Then the complex $L^{\cdot}$ is quasi-isomorphic to a complex of k-vector spaces.

Proof: We denote by $B^i_{K^{\cdot}}$, $B^i_{L^{\cdot}}$ the image of the homomorphism $K^{i-1} \to K^i$ resp. $L^{i-1} \to L^i$, and we denote by $Z^i_{K^{\cdot}}$, $Z^i_{L^{\cdot}}$ the kernel of the homomorphism $K^i \to K^{i+1}$ resp. $L^i \to L^{i+1}$. Then we have the following commutative diagram with exact rows

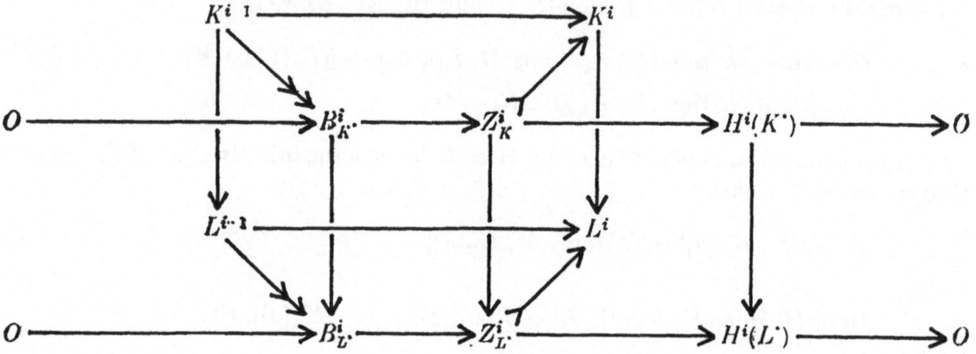

where the homomorphisms $B^i_{K^{\cdot}} \to B^i_{L^{\cdot}}$, $Z^i_{K^{\cdot}} \to Z^i_{L^{\cdot}}$, and $H^i(K^{\cdot}) \to H^i(L^{\cdot})$ are induced by $f^{\cdot}$. Because K is a complex of k-vector spaces, the canonical homomorphism $Z^i_K \to H^i(K^{\cdot})$ splits. Furthermore, $H^i(f^{\cdot}): H^i(K^{\cdot}) \to H^i(L^{\cdot})$ is a surjective homomorphism of k-vector spaces, i.e., it splits. That is, we get a homomorphism $H^i(L^{\cdot}) \to L^i$. In fact, it is a homomorphism of the cohomology complex $H^{\cdot}(L^{\cdot})$ of $L^{\cdot}$ to $L^{\cdot}$, which induces isomorphisms on the cohomology modules. This proves Proposition 4.3.

Next we prove (i) $\Rightarrow$ (iii) of Theorem 4.1. By Theorem I.2.15, for a Buchsbaum module M, the canonical homomorphism of complexes

$$\tau^d K^{\cdot}(\mathfrak{m}; M) \to \tau^d K^{\infty}(\mathfrak{m}; M)$$

induces surjective homomorphisms on the cohomology modules. Because $K^{\cdot}(\mathfrak{m}; M)$ is a complex of k-vector spaces it follows that $\tau^d K^{\infty}(\mathfrak{m}; M)$ is quasi-isomorphic to a complex of k-vector spaces, see Proposition 4.3. Because $K^{\infty}(\mathfrak{m}; M) = \underline{R}\Gamma_{\mathfrak{m}}(M)$ in the derived category, we have that $\tau_{-d}\underline{R}\operatorname{Hom}_A(M, I^{\cdot}_A)$ is quasi-isomorphic to a complex of k-vector spaces by virtue of the Local Duality Theorem, Theorem 0.3.4. For the proof of Theorem 4.1 it remains to show (iv) $\Rightarrow$ (i). To this end we consider the Koszul complex $K^{\cdot}(\mathfrak{q}; M)$ of M with respect to an arbitrary parameter ideal $\mathfrak{q} = (x_1, \ldots, x_d)$ of M. By the definition we have

$$K^{\cdot}(\mathfrak{q}; M) = K^{\cdot}(\mathfrak{q}; A) \otimes_A M,$$

where $K^{\cdot}(\mathfrak{q}; A)$ is a complex of finitely generated free A-modules such that the cohomology modules $H^i(\mathfrak{q}; M)$ of $K^{\cdot}(\mathfrak{q}; M)$ are A-modules of finite length, compare Chapter 0, § 1 for properties of Koszul complexes. By virtue of our Lemma 4.2 it follows from (iv) that $\tau^d K^{\cdot}(\mathfrak{q}; M)$ is quasi-isomorphic to a complex of k-vector spaces. In particular we have

$$\mathfrak{m} H^{d-1}(\mathfrak{q}; M) = 0$$

for an arbitrary parameter ideal $\mathfrak{q}$ of M. We set $\mathfrak{q}' = (x_1, \ldots, x_{d-1})$. Then we have the following short exact sequence

$$0 \to H^{d-2}(\mathfrak{q}'; M) \otimes_A A/x_d A \to H^{d-1}(\mathfrak{q}; M) \to \mathfrak{q}'M : x_d/\mathfrak{q}'M \to 0,$$

compare Chapter 0, Lemma 1.6. Therefore we get $\mathfrak{m} \cdot (\mathfrak{q}'M : x_d) \subseteq \mathfrak{q}'M$, i.e., M is a Buchsbaum module by virtue of Proposition I.1.10.

The following corollaries indicate, in our opinion, how substantial Theorem 4.1 is. First, we recall that an ideal $\mathfrak{a}$ of a regular local ring R is called *perfect*, if $R/\mathfrak{a}$ is a Cohen-Macaulay ring. That is equivalent to the vanishing of

$$H^i(\underline{R} \operatorname{Hom}(R/\mathfrak{a}, R)) \cong \operatorname{Ext}^i_R(R/\mathfrak{a}, R)$$

for all $i = \dim R - \dim R/\mathfrak{a} =: n$. For Buchsbaum rings a corresponding result is valid.

Corollary 4.4. *The local ring $R/\mathfrak{a}$ is a Buchsbaum ring if and only if $\tau_n \underline{R} \operatorname{Hom}(R/\mathfrak{a}, R)$ is quasi-isomorphic to a complex of k-vector spaces, where k denotes the residue field of R.*

The *proof* follows from Theorem 4.1 because $\underline{R} \operatorname{Hom}(R/\mathfrak{a}, R)$ is up to a shift isomorphic to the (normalized) dualizing complex of $R/\mathfrak{a}$. If M is a Buchsbaum module, then it follows that

$$\tau_{-d} \operatorname{Hom}_A(M, I^{\cdot}_A) \cong \operatorname{Hom}_k(C^{\cdot}(M), k))$$

by the Local Duality Theorem and Matlis duality.

Corollary 4.5. *Let M be a Buchsbaum module. Let $F^{\cdot}$ be a bounded complex of finitely generated free A-modules such that $F^i = 0$ for $i < 0$ and $F^{\cdot} \otimes_A k$ has trivial boundary maps. Assume that $H^i(F^{\cdot} \otimes M)$, $i \in \mathbf{Z}$, are modules of finite length. Then we get*

$$\mathfrak{m} \cdot H^i(F^{\cdot} \otimes M) = 0$$

and

$$\dim_k H^i(F^{\cdot} \otimes M) = \sum_{v=0}^{i} \operatorname{rank} F^{i-v} \dim_k H^v_{\mathfrak{m}}(M) \quad \textit{for } 0 \leq i < \dim M.$$

Proof: First we notice that $\mathfrak{m} \cdot H^i(F^{\cdot} \otimes M) = 0$, $i < \dim M$, follows from Lemma 4.2. Then the quasi-isomorphism $(*)$ given in the proof of Lemma 4.2 implies

$$H^i(F^{\cdot} \otimes M) = H^i(F^{\cdot} \otimes C^{\cdot}(M)), \quad i < \dim M.$$

Because $F^{\cdot} \otimes k$ has trivial boundary homomorphisms it follows that

$$\dim_k H^i(F \otimes C^{\cdot}(M)) = \sum_{v=0}^{i} \operatorname{rank} F^{i-v} \dim_k H^v_{\mathfrak{m}}(M), \quad i < \dim M.$$

Next we will apply Corollary 4.5 to the Koszul complex $K^{\cdot}(\mathfrak{q}; M)$ of M with respect to a parameter ideal $\mathfrak{q} = (x_1, \ldots, x_d)$, $d = \dim M$, of M.

Corollary 4.6. *Let M be a Buchsbaum module, and let $\mathfrak{q} = (x_1, \ldots, x_d)$, $d = \dim M$, be an arbitrary parameter ideal of M. Then we get*

$$\mathfrak{m} H^i(\mathfrak{q}; M) = 0$$

and

$$\dim_k H^i(\mathfrak{q}; M) = \sum_{v=0}^{i} \binom{d}{i-v} \dim_k H^v_{\mathfrak{m}}(M) \quad \textit{for } 0 \leq i < \dim M$$

By results of Auslander-Buchsbaum [1] and Serre [2] we know that

$$l_A(M/\mathfrak{q}M) - e_0(\mathfrak{q}; M) = \sum_{i=1}^{d} (-1)^{i-1} l_A(H^i(\mathfrak{q}; M))$$

for an arbitrary parameter ideal $\mathfrak{q} = (x_1, \ldots, x_d)$, $d = \dim M$, of a Noetherian A-module M. That is, in the case of a Buchsbaum module M we have

$$l_A(M/\mathfrak{q}M) - e_0(\mathfrak{q}; M) = \sum_{v=0}^{d-1} \binom{d-1}{v} \dim_k H^v_{\mathfrak{m}}(M),$$

which is another proof of Proposition I.2.6. Also Theorem 4.1 provides a Koszul complex characterization of Buchsbaum modules.

Corollary 4.7. *Let M denote a Noetherian A-module, $d = \dim M$. Then the following conditions are equivalent:*

(i) *M is a Buchsbaum module,*

(ii) *$\mathfrak{m}H^{d-1}(\mathfrak{q}; M) = 0$ for every parameter ideal $\mathfrak{q} = (x_1, \ldots, x_d)$ of M,*

(iii) *$\mathfrak{m}H^r(\mathfrak{q}; M) = 0$ for every parameter ideal $\mathfrak{q}$ of M and all $0 \leq r < \dim M$, and*

(iv) *for every parameter ideal $\mathfrak{q}$ of M, $\tau^d K^{\cdot}(\mathfrak{q}; M)$ is quasi-isomorphic to a complex of k-vector spaces.*

Proof: First we note that we can assume $A = \hat{A}$ and $M = \hat{M}$ without loss of generality. That is, we can assume that A possesses a dualizing complex. Then, (i) $\Rightarrow$ (iv) follows by Lemma 4.2. The implications (iv) $\Rightarrow$ (iii) $\Rightarrow$ (ii) are trivial. Finally, (ii) $\Rightarrow$ (i) is proven by the same way as the statement (ii) $\Rightarrow$ (i) of Theorem 4.1.

The analogy of the previous corollary to the Koszul complex characterization of Cohen-Macaulay modules is obvious. Furthermore, we want to mention here that Suzuki [3] also gave a characterization of a Buchsbaum module by using the Koszul complex generated by a system of parameters of the module and the standard properties of systems of parameters of Buchsbaum modules. To this end he obtained a new statement concerning the cycles and the boundaries of the Koszul complex.

In view of our remark given in the preface that the theory of derived categories is not needed for our applications to algebraic topology and combinatorics we want to give a new and simple proof of Corollary 4.7. This approach will enable us to give an elementary proof of Lemma 4.14 below. It is precisely this Lemma 4.14 which does supply the key for applications to combinatorics (see, for example, the proof of Theorem 4.19 below).

Corollary 4.7'. *Let M be a Noetherian A-module of dimension $d \geq 1$. Then the following conditions are equivalent:*

(i) *M is a Buchsbaum module.*

(ii) *$\mathfrak{m} \cdot H^{d-1}(x_1, \ldots, x_d; M) = 0$ for all systems of parameters $x_1, \ldots, x_d$ of M.*

(iii) *$\mathfrak{m} \cdot H^i(x_1, \ldots, x_d; M) = 0$ for all systems of parameters $x_1, \ldots, x_d$ of M and all $i < d$.*

(iv) *$\mathfrak{m} \cdot H^i(x_1, \ldots, x_r; M) = 0$ for all parts of systems of parameters $x_1, \ldots, x_r$, $r \leq d$, of M and all $i < r$.*

Proof: The implications (iv) $\Rightarrow$ (iii) $\Rightarrow$ (ii) are trivial. For the proof of (ii) $\Rightarrow$ (i) we have that the top row of the commutative diagram of Lemma 0.1.6 does yield an epimorphism:

$$H^{d-1}(x_1, \ldots, x_d; M) \to H^0\big(x_d; H^{d-1}(x_1, \ldots, x_{d-1}; M)\big)$$
$$\cong (x_1, \ldots, x_{d-1}) \cdot M : x_d/(x_1, \ldots, x_{d-1}) \cdot M.$$

Hence our assumption now provides that

$$\mathfrak{m} \cdot \big((x_1, \ldots, x_{d-1}) \cdot M : x_d/(x_1, \ldots, x_{d-1}) \cdot M\big) = 0.$$

By applying Proposition I.1.10 we therefore get the result (i).

It remains to prove (i) $\Rightarrow$ (iv). We will prove the following result: Let $M, x_1, \ldots, x_r$ be as above. We denote by $d^{\cdot}$ the differentiation of the Koszul complex $K^{\cdot}(x_1, \ldots, x_r; M)$. If $0 \leq i < r$ and $y_1, \ldots, y_s$, $s \leq d - i$, are elements of $\mathfrak{m}$ such that $y_1, \ldots, y_s, x_{j_1}, \ldots, x_{j_i}$ is a part of a system of parameters of M for all $1 \leq j_1 < \ldots < j_i \leq r$, then we show the following claim to be true:

Claim. $\operatorname{Ker} d^i \cap (y_1, \ldots, y_s)\, K^i(x_1, \ldots, x_r; M) \subseteq \operatorname{Im} d^{i-1}$.

Having this one obtains the implication (i) $\Rightarrow$ (iv) as follows: First, choose a $\left(\bigoplus_{1 \leq j_1 < \ldots < j_i \leq r} M/(x_{j_1}, \ldots, x_{j_r}) \cdot M\right)$-basis $y_1, \ldots, y_t$ of $\mathfrak{m}$ (see our Definition I.1.7 and Proposition I.1.9). Let $1 \leq l \leq t$. Then the claim implies

$$y_l \cdot \operatorname{Ker} d^i \subseteq \operatorname{Ker} d^i \cap (y_l) \cdot K^i(x_1, \ldots, x_r; M) \subseteq \operatorname{Im} d^{i-1};$$

that is, $y_l \cdot H^i(x_1, \ldots, x_r; M) = 0$ for all $l = 1, \ldots, t$. Hence we get condition (iv).

Now we prove the claim: We will use induction on r. If $r = 1$ we have only to consider the case $i = 0$:

$$\begin{aligned}\operatorname{Ker} d^0 \cap (y_1, \ldots, y_s)\, K^0(x_1, \ldots, x_r; M) &= \big(0 :_M (x_1, \ldots, x_r)\big) \cap (y_1, \ldots, y_s) \cdot M \\ &= (0 :_M \mathfrak{m}) \cap (y_1, \ldots, y_s) \cdot M \\ &= 0 = \operatorname{Im} d^{-1}\end{aligned}$$

(see Lemma I.1.14).

If $1 < r \leq d$ we put

$$K^{\cdot} := K^{\cdot}(x_1, \ldots, x_r; M) \quad \text{and} \quad L^{\cdot} := K^{\cdot}(x_1, \ldots, x_{r-1}; M)$$

with differentiation $\tilde{d}^{\cdot}$. Then $K^i = L^i \bigoplus L^{i-1}$, and for the element $(a, b) \in K^i$ with $a \in L^i$ and $b \in L^{i-1}$ we have:

$$d^i(a, b) = \big(\tilde{d}^i(a), (-1)^i x_r a + \tilde{d}^{i-1}(b)\big).$$

If $(a, b) \in \operatorname{Ker} d^i \cap (y_1, \ldots, y_s) \cdot K^i$ then we have:

(i) $a \in \operatorname{Ker} \tilde{d}^i \cap (y_1, \ldots, y_s) \cdot L^i$

and

(ii) $b \in (y_1, \ldots, y_s) \cdot L^{i-1}, \quad (-1)^i x_r a + \tilde{d}^{i-1}(b) = 0.$

If $i < r - 1$ then (i) and the induction hypothesis imply that

$$a = \tilde{d}^{i-1}(a') \text{ with } a' \in L^{i-1}.$$

If $i = r - 1$ we have $L^{r-1} = M$ and $\operatorname{Im} \tilde{d}^{r-2} = (x_1, \ldots, x_{r-1}) \cdot M$. Therefore with the equation in (ii) one obtains:

$$\begin{aligned} a &\in (\operatorname{Im} \tilde{d}^{r-2} :_M x_r) \cap (y_1, \ldots, y_s) \cdot M \\ &= \big((x_1, \ldots, x_{r-1})\, M : x_r\big) \cap (y_1, \ldots, y_s) \cdot M \\ &= \big((x_1, \ldots, x_{r-1}) \cdot M : \mathfrak{m}\big) \cap (y_1, \ldots, y_s) \cdot M \\ &\subseteq (x_1, \ldots, x_{r-1}) \cdot M \end{aligned}$$

which by applying Lemma I.1.14, equals $\operatorname{Im} \tilde{d}^{r-2}$; that is, $a = \tilde{d}^{r-2}(a')$.

Hence we obtain for all $i < r$ that $a = \tilde{d}^{i-1}(a')$. Then

$$\tilde{d}^{i-1}\big((-1)^i\, x_r a' + b\big) = (-1)^i\, x_r a + \tilde{d}^{i-1}(b) = 0;$$

that is,

$$(-1)^i\, x_r a' + b \in \operatorname{Ker} \tilde{d}^{i-1} \cap (y_1, \ldots, y_s, x_r) \cdot L^{i-1} \subseteq \operatorname{Im} \tilde{d}^{i-2}.$$

Thus we see that $b = (-1)^{i-1}\, x_r a' + \tilde{d}^{i-2}(b')$ with $b' \in L^{i-2}$; that is, $(a, b) = d^{i-1}(a', b') \in \operatorname{Im} d^{i-1}$. This proves the claim and therefore also Corollary 4.7′.

Next we are interested in the Buchsbaum property of the canonical module of a Buchsbaum ring. We recall the definition of the canonical module.

Definition 4.8. A Noetherian A-module K_A is called a *canonical module* of A if

$$K_A \otimes_A \hat{A} \cong \operatorname{Hom}_A\big(H^d_{\mathfrak{m}}(A), E\big), \qquad d = \dim A,$$

as $\hat{A}$-modules, where $E = E(k)$ denotes the injective hull of the residue field k of A.

The notion of the canonical (or dualizing) module was introduced by Herzog and Kunz in [1]. It is uniquely determinated if it exists. In case A possesses a dualizing complex $I^{\cdot}_A$, A has a canonical module. More precisely,

$$K_A \cong H^{-d}(I^{\cdot}_A), \qquad d = \dim A.$$

This follows from the Local Duality Theorem, compare § 3, Chapter 0.

Theorem 4.9. *Let A denote a d-dimensional Buchsbaum ring which has a canonical module K_A. Then K_A is a Buchsbaum module with*

$$H^i_{\mathfrak{m}}(K_A) \cong \operatorname{Hom}_A\big(H^{d-i+1}_{\mathfrak{m}}(A), E\big) \quad \textit{for } 2 \le i < d.$$

Proof: Since all statements are preserved by passing to the completion $\hat{A}$ of A we can assume without loss of generality A complete. Now we apply the derived functor to the short exact sequence of complexes

$$0 \to K_A[d] \to I^{\cdot}_A \to \tau_{-d} I^{\cdot}_A \to 0.$$

By the Local Duality Theorem we get

$$0 \to \underline{R}\Gamma_{\mathfrak{m}}(K_A[d]) \to E \to \underline{R}\Gamma_{\mathfrak{m}}(\tau_{-d} I^{\cdot}_A) \to 0.$$

That means, the complex $\underline{R}\Gamma_{\mathfrak{m}}(\tau_{-d} I^{\cdot}_A)$ is isomorphic to the mapping cone of $\underline{R}\Gamma_{\mathfrak{m}}(K_A[d]) \to E$. Because E is concentrated in degree zero we obtain

$$\tau^{-1}\underline{R}\Gamma_{\mathfrak{m}}(\tau_{-d} I^{\cdot}_A) \cong \big(\tau^d \underline{R}\Gamma_{\mathfrak{m}}(K_A)\big)\,[d+1] \qquad (**)$$

by a simple calculation. Now, if A is a Buchsbaum ring, $\tau_{-d}I^{\cdot}_A$, and therefore also the left complex in (**), is isomorphic to a complex of k-vector spaces. From (**) it follows that $\tau^d \underline{R}\Gamma_{\mathfrak{m}}(K_A)$ is isomorphic to a complex of k-vector spaces. By our Theorem 4.1 this proves the first statement. Because

$$\underline{R}\Gamma_{\mathfrak{m}}(\tau_{-d}I^{\cdot}_A) \cong \tau_{-d}I^{\cdot}_A,$$

the isomorphisms for the local cohomology modules follow from (**) by the Local Duality Theorem, q.e.d.

Theorem 4.9 answers affirmatively a question posed by Goto and Shimoda in their paper [1], where the result is proven for the case $\dim A = 3$. Naoyoshi Suzuki [1] was the first who asked this question in 1978 at the first symposium on commutative algebra in Japan. He also proved in his report at the symposium Theorem 4.9 in case $\dim A = 3$. In his talk at the 4th symposium, which took place during the period 3–6 November 1982, N. Suzuki [6] gave an elementary proof of Theorem 4.9 by using extensively a lemma on Buchsbaum rings proven by S. Goto. We will give a brief sketch of Suzuki's proof.

First we mention the following interesting lemma by Shiro Goto.

Goto's Lemma. *Let M be a Buchsbaum module for dimension $d \geq 2$ and let $y_1, \ldots, y_r$, $r \leq d$ be a part of a system of parameters of M and $s_1, \ldots, s_r$ positive integers. Then*

$$(y_1^{s_1}, \ldots, y_r^{s_r})\, M :_M \mathfrak{m} = \sum_{k=0}^{r} \sum_{1 \leq i_1 < \ldots < i_k \leq r} y_{i_1}^{s_{i_1}-1} \cdot \ldots \cdot y_{i_k}^{s_{i_k}-1}\big((y_{i_1}, \ldots, y_{i_k})\, M :_M \mathfrak{m}\big).$$

Proof: Using an easy induction argument it is sufficient to prove the following statement:

Let N be a Noetherian A-module and $\mathfrak{a} \subset A$ be an ideal so that $N/\mathfrak{a}N$ is a Buchsbaum module of dimension ≥ 2. Then for every parameter element a with respect to $N/\mathfrak{a}N$ and all $n > 0$ we have

$$(\mathfrak{a}N + a^n N) :_N \mathfrak{m} = \mathfrak{a}N :_N \mathfrak{m} + a^{n-1}\big((\mathfrak{a}N + aN) :_N \mathfrak{m}\big).$$

Passing to $M := N/\mathfrak{a}N$ we may assume without loss of generality that $\mathfrak{a} = 0$. Then

$$a^n M : \mathfrak{m} \subseteq a^n M : a = a^{n-1}M + 0 : a = a^{n-1}M + 0 : \mathfrak{m},$$

hence

$$\begin{aligned} a^n M : \mathfrak{m} &= (a^n M : \mathfrak{m}) \cap (a^{n-1}M + 0 : \mathfrak{m}) = 0 : \mathfrak{m} + \big(a^{n-1}M \cap (a^n M : \mathfrak{m})\big) \\ &= 0 : \mathfrak{m} + a^{n-1}\big((a^n M : \mathfrak{m}) : a^{n-1}\big) = 0 : \mathfrak{m} + a^{n-1}\big((a^n M : a^{n-1}) : \mathfrak{m}\big) \\ &= 0 : \mathfrak{m} + a^{n-1}\big((aM + 0 : \mathfrak{m}) : \mathfrak{m}\big) = 0 : \mathfrak{m} + a^{n-1}(aM : \mathfrak{m}), \end{aligned}$$

q.e.d.

Now we sketch Suzuki's *proof* of Theorem 4.9 for modules: The canonical module K_M of a Buchsbaum module M is also a Buchsbaum module.

We set $D^i(\cdot) = \mathrm{Hom}_A\big(H^i_{\mathfrak{m}}(\cdot), E(k)\big)$.

We may assume that $A = \hat{A}$ and we procede by induction on $d = \dim M$. Let $d \geq 3$ and additionally that depth $M > 0$ since $H^d_{\mathfrak{m}}(M) \cong H^d_{\mathfrak{m}}\big(M/H^0_{\mathfrak{m}}(M)\big)$. Let $a_1, \ldots, a_d$ be any system of parameters for K_M. We have an exact sequence

$$0 \to K_M/\mathfrak{a}K_M \to K_{(M/\mathfrak{a}M)} \xrightarrow{\pi} D^{d-1}(M) \to 0, \quad \text{where } \mathfrak{a} = (a_1, \ldots, a_d).$$

Consider the long exact sequence of Koszul homology modules with respect to $\mathfrak{a}' = \{a_2, \ldots, a_d\}$,

$$H_1(\mathfrak{a}'; K_{M'}) \to H_1(\mathfrak{a}'; V) \to K_M/(\mathfrak{a}) K_M \to K_{M'}/(\mathfrak{a}') K_{M'} \to V \to 0$$

where $M' = M/\mathfrak{a}M$ and $V = D^{d-1}(M)$. If we have that the mapping

$$H_1(\mathfrak{a}'; \pi) : H_1(\mathfrak{a}'; K_{M'}) \to H_1(\mathfrak{a}'; V)$$

is a zero map, then the equality

$$l_A(K_M/(\mathfrak{a}) K_M) = l(K_{M'}/(\mathfrak{a}') K_{M'}) + (d-1)(\dim_K V)$$

holds. On the other hand, we have

$$e_0(\mathfrak{a}; K_M) = e_0(\mathfrak{a}'; K_M/\mathfrak{a}K_M) = e_0(\mathfrak{a}'; K_{M'})$$

and therefore, by the induction assumption, we can conclude that the difference

$$l_A(K_M/(\mathfrak{a}) K_M) - e_0(\mathfrak{a}; K_M)$$

does not depend on the choice of the system of parameters $a_1, \ldots, a_d$ for M; which is the definition of the Buchsbaumness of K_M.

Consider the direct system $\{H^i(a_1^n, \ldots, a_d^n; M), \Phi^i_{n,n+1}\}$ with the limit $H^i_{\mathfrak{m}}(M)$ (see Chapter 0, § 1), where

$$H_{d-i}(\mathfrak{a}; M) = H^i(\operatorname{Hom}_A(K.(\mathfrak{a}; A), M)) \cong H^i(\mathfrak{a}; M).$$

Let $\mathfrak{a}^n := \{a_1^n, \ldots, a_d^n\}$ and $\mathfrak{a}'^n := \{a_2^n, \ldots, a_d^n\}$. A commutative diagram is induced by the exact sequence (see Lemma 0.1.5) $0 \to \mathfrak{a}M \to M \xrightarrow{\pi} M/\mathfrak{a}M \to 0$.

$$\begin{array}{ccc}
H^{d-1}(\mathfrak{a}^n, M) & \longrightarrow & H^{d-1}(\mathfrak{a}^n; M') \\
\wr\| & & \wr\| \\
H_1(\mathfrak{a}^n; M) & \xrightarrow{H_1(\mathfrak{a}^n;\pi)} & H_0(\mathfrak{a}'^n; M') \oplus H_1(\mathfrak{a}'^n; M') \\
\downarrow \Phi^n_M & & \downarrow \Phi^n_{M'} \\
H^{d-1}_{\mathfrak{m}}(M) & \xrightarrow{\lambda} & H^{d-1}_{\mathfrak{m}}(M')
\end{array}$$

Let $(f_2, \ldots, f_d) \in Z_1(a_2, \ldots, a_d; K_{M'})$, namely, for $j = 2, \ldots, d$,

$$f_j \in K_{M'} = \operatorname{Hom}_A(H^{d-1}(M/\mathfrak{a}M), E_A(k)) \quad \text{and} \quad \sum_{j=2}^{d} a_j f_j = 0. \tag{\#}$$

It suffices to show that for any $j = 2, \ldots, d$, $f_j \cdot \lambda = 0$. Let $z \in H^{d-1}_{\mathfrak{m}}(M)$. Then there exists $(u, v) \in Z_1(\mathfrak{a}^n; M) \subset K_0(\mathfrak{a}'^n, M) \oplus K_1(\mathfrak{a}'^n; M)$ such that $\Phi^n_M(|u, v|) = z$. Note first that the cycle condition implies that $a_n^1 u \in (a_2^n, \ldots, a_d^n) M$; hence

$$u \in U_M(a_2^n, \ldots, a_d^n). \tag{\#\#}$$

We claim that $f_j \cdot \Phi^n_{M'} \cdot H_1(\mathfrak{a}^n; \pi)(|u, v|) = 0$. Here we use the notation $|c|$ for the homology class of a cycle c. Let

$$|\bar{u}, \bar{v}| = (|\bar{u}|, |\bar{v}|) := H_1(\mathfrak{a}^n; \pi)(|u, v|) \in H_0(\mathfrak{a}'^n; M') \oplus H_1(\mathfrak{a}'^n; M').$$

It is not too difficult to see that

$$\Phi^n_{M'}(|\bar{u}, \bar{v}|) = (|(a_2, \ldots, a_d)\,\tilde{u}|, |\tilde{0}|) \in H_0(\mathfrak{a}'^{n+1}; M') \oplus H_1(\mathfrak{a}'^{n+1}; M').$$

Therefore it suffices to show that

$$f_j \cdot \Phi^n_{M'}(|\bar{u}|, |\bar{v}|) = f_j \cdot \Phi^{n+1}_{M'}(|a_2, \ldots, a_d\tilde{u}|, |\tilde{0}|) = 0,$$

with $|\tilde{u}| \in H_0(a_2^{n+1}, \ldots, a_d^{n+1}; M/\mathfrak{a}M) = M/(\mathfrak{a}, a_2^{n+1}, \ldots, a_d^{n+1})\,M$.

By Goto's Lemma, from (# #) we obtain the following expression:

$$u = \sum_{I \subseteq \{2, \ldots, d\}} a_I^{n-1} u_I$$

with $u_I \in U_M(a_i; i \in I)$ and $a_I = \prod_{i \in I} a_i$ and $a_\emptyset = 1$. We must show that for any subset I of $\{a_2, \ldots, a_d\}$

$$f_j \cdot \Phi^{n+1}_{M'}(|(a_2, \ldots, a_d)\,a_I^{n-1}\tilde{u}_I|) = 0.$$

If $I \neq \{2, \ldots, d\}$, there exists $j \notin I$, hence

$$(a_2, \ldots, a_d)\,a_I^{n-1}u_I = \Big(\prod_{\substack{l \neq j\\ l \notin I}} a_l\Big)\, a_I^n a_j u_I \in (a_i^{n+1}; i \in I)\,M \subset (a_2^{n+1}, \ldots, a_d^{n+1})\,M.$$

This means that

$$|(a_2, \ldots, a_d)\,a_I^{n-1}\tilde{u}_I| = 0 \quad \text{in } M/(\mathfrak{a}, a_2^{n+1}, \ldots, a_d^{n+1})\,M.$$

For $I = \{2, \ldots, d\}$,

$$f_j \cdot \Phi^{n+1}_{M'}\big(|(a_2, \ldots, a_d)^{n-1}\,a_I\tilde{u}_I|\big) = f_j \cdot \Phi^{n+1}_{M'}\big(|(a_2, \ldots, a_d)^n\,\tilde{u}_I|\big)$$

$$= a_j f_j \cdot \Phi^{n+1}_{M'}\big(|(a_2, \ldots, \hat{a}_j, \ldots, a_d)^n\,a_j^{n-1}\tilde{u}_I|\big),$$

by the cycle condition (#),

$$= -\sum_{i \neq j} a_i f_i \cdot \Phi^{n+1}_{M'}\big(|(a_2, \ldots, \hat{a}_j, \ldots, a_d)^n\,a_j^{n-1}\tilde{u}_I|\big)$$

$$= -\sum_{i \neq j} f_i \cdot \Phi^{n+1}_{M'}\big(|(a_2, \ldots, \hat{a}_i\hat{a}_j, \ldots, a_d)^n\,a_j^{n-1}a_i^{n+1}\tilde{u}_I|\big).$$

Since

$$|a_i^{n+1}\tilde{u}_I| = 0 \quad \text{in } M/(a_2^{n+1}, \ldots, a_d^{n+1})\,M',$$

we conclude in this case also that

$$f_j \cdot \Phi^{n+1}_{M'}\big(|(a_2, \ldots, a_d)^{n-1}\,a_I\tilde{u}_I|\big) = 0$$

as required, q.e.d.

Our Theorem 4.9, proved by P. Schenzel in [1], is an extension of the main result of Kiehl's paper [1] to Buchsbaum rings. Using the criterion of Theorem I.2.10, Kiehl showed that K_A is a Buchsbaum module, if $\tau_{-d}I^{\cdot}_A$ is isomorphic to a complex of k-vector spaces. If A is a two-dimensional local ring admitting a canonical module K_A, then K_A is a Cohen-Macaulay module. Therefore, the converse of Theorem 4.9 is not true in general, even if we assume that A has small dimension or that K_A is a Cohen-Macaulay module.

Theorem 4.10. *Let A denote a local ring having the canonical module K_A. Assume that $\hat{A}$ satisfies S_2, i.e.,*

$$\operatorname{depth} \hat{A}_{\mathfrak{p}} \geq \min(2, \dim \hat{A}_{\mathfrak{p}})$$

for all $\mathfrak{p} \in \operatorname{Spec} \hat{A}$. If K_A is a Buchsbaum module, then A is a Buchsbaum ring.

Proof: First without loss of generality we can assume $A = \hat{A}$, i.e., A possesses the dualizing complex $I^{\cdot}_A$ by virtue of the Cohen Structure Theorem. For $\dim A \leq 2$ there is nothing left to prove. Therefore, we can assume $\dim A > 2$. By Proposition 0.3.6, we have

$$\operatorname{Supp} K_A = \operatorname{Spec} A .$$

It follows from Theorem 4.9 that

$$H^i_{\mathfrak{m}}(K_A) \cong \operatorname{Hom}_A\big(H^{d-i+1}_{\mathfrak{m}}(A), E\big) \quad \text{for } 2 \leq i < d .$$

Since K_A is a Buchsbaum module $H^i_{\mathfrak{m}}(K_A)$ is isomorphic to a k-vector space. Hence this is also true for $H^{d-i+1}_{\mathfrak{m}}(A)$ for $2 \leq i < d$. By this property and the assumption of 4.10 we get that $H^i_{\mathfrak{m}}(A)$ has finite length over $A/\mathfrak{m}$. Therefore we get that $A_{\mathfrak{p}}$ is a Cohen-Macaulay ring for all prime ideals $\mathfrak{p} \neq \mathfrak{m}$. Since

$$\dim A = \dim A/\mathfrak{p} + \dim A_{\mathfrak{p}}$$

for all prime ideals $\mathfrak{p}$, it follows by Proposition 16 of the Appendix, that $\tau_{-d} I^{\cdot}_A$ is a complex whose cohomology modules are modules of finite length. Therefore

$$\tau_{-d} I^{\cdot}_A \cong \underline{R}\Gamma_{\mathfrak{m}}(\tau_{-d} I^{\cdot}_A) .$$

By the isomorphism (**) given in the proof of Theorem 4.10 we get

$$\big(\tau^d \underline{R}\Gamma_{\mathfrak{m}}(K_A)\big)\,[d+1] \cong \tau^{-1}_{-d} I^{\cdot}_A \cong \tau_{-d} I^{\cdot}_A ,$$

where we have used the fact depth $A \geq 2$, i.e.,

$$H^i(\tau_{-d} I^{\cdot}_A) = 0, \quad i = 0, -1 .$$

Now, if K_A is a Buchsbaum module, then the left complex in the above chain of isomorphisms is quasi-isomorphic to a complex of k-vector spaces. Therefore this is also true for $\tau_{-d} I^{\cdot}_A$ which proves our statement by Theorem 4.1, q.e.d.

Proposition 4.11. *If in addition K_A in Theorem 4.10 is a Cohen-Macaulay module, then A itself is a Cohen-Macaulay ring.*

Proof: If K_A is a Cohen-Macaulay module, then it follows from the proof of Theorem 4.10 that $H^i(\tau_{-d} I^{\cdot}_A) = 0$ for all $i \in \mathbf{Z}$. By the Local Duality Theorem we have $H^i_{\mathfrak{m}}(A) = 0$ for all $i \neq \dim A$. Thus, A is a Cohen-Macaulay ring, q.e.d.

Having Theorem 4.1 we can now sharpen Reisner's local Cohen-Macaulay criterion of Theorem 2.4.

Theorem 4.12. *Let Σ be a finite connected simplicial complex with* $\dim \Sigma = d - 1$. *Then the following conditions are equivalent:*

(i) *$(K[\Sigma])_P$ is a Cohen-Macaulay ring for all prime ideals P different from the irrelevant ideal $\mathfrak{m}$.*

(ii) *For all* $\emptyset \neq \sigma \in \Sigma$ *we have* $\tilde{H}_i(\mathrm{lk}_\Sigma\sigma; K) = 0$ *if* $i \neq \dim \mathrm{lk}_\Sigma\sigma$.

(iii) *If* $X = |\Sigma|$ *is the geometric realization of* Σ, *then* $H_i(X, X - p; K) = 0$ *for all points* $p \in X$ *and all* $i \neq \dim X$.

(iv) *There is a canonical isomorphisms*

$$\tau_{-d} D_\Sigma^{\cdot} \cong \tau_{-d}(C.(\Sigma; K)^* [1]),$$

where $D_\Sigma^{\cdot}$ *denotes the dualizing complex of* $K[\Sigma]$ *and* $C.(\Sigma; K)$ *the reduced simplicial chain complex with coefficients in* K, *i.e.* $C_i(\Sigma; K)$ *is the* K*-vector space of* i*-faces of* Σ.

(v) Σ *is a Buchsbaum complex, i.e.,* $K[\Sigma]$ *is a Buchsbaum ring.*

Proof: It follows from Theorem 2.4 that (i) $\Leftrightarrow$ (ii) $\Leftrightarrow$ (v). The equivalence of (ii) and (iii) is even an exercise in topology. For this we note that if σ is non-empty and p is a point interior to σ, we have the following isomorphisms by excision:

$$H_j(X, X - p; K) \cong H_j\left(\overline{\mathrm{star}_\Sigma}\,\sigma, (\overline{\mathrm{star}_\Sigma}\,\sigma) - p; K\right),$$

where $\overline{\mathrm{star}_\Sigma}\,\sigma$ denotes the closed star of σ in Σ (see our Definition 1.8).

$$\cong H_j\left(\overline{\mathrm{star}_\Sigma}\,\sigma, (\mathrm{bd}\,\sigma) * \left(\mathrm{lk}(\sigma)\right), K\right)$$

by deformation retraction, where $\mathrm{lk}\,\sigma$ denotes $\mathrm{lk}_\Sigma\,\sigma$, $*$ is the join operation and $\mathrm{bd}\,\sigma$ denotes the boundary of σ,

$$\cong H_{j-1}\left((\mathrm{bd}\,\sigma) * (\mathrm{lk}\,\sigma); K\right)$$

by the long exact cohomology sequence,

$$\cong H_{j-\dim\sigma-1}(\mathrm{lk}\,\sigma; K)$$

by the suspension isomorphism.

This proves the equivalence of (ii) and (iii).

(iv) $\Rightarrow$ (v): $\tau_{-d} D_\Sigma^{\cdot}$ is isomorphic to a complex of K-vector spaces in the derived category. Therefore the assertion is immediate from Theorem 4.1 (see also the proof of Corollary 4.4). Hence it is enough to show (i) $\Rightarrow$ (iv) for proving the theorem.

(i) $\Rightarrow$ (iv): We apply again some ideas of G. A. Reisner. Under the assumption (i) Reisner [1], Theorem 2 proved

$$\mathrm{Hom}_K\left(K_0^{\cdot}(\underline{x}; K[\Sigma]); K\right) \cong C.(\Delta; K)\,[1],$$

where $K_0^{\cdot}(\underline{x}; K[\Sigma])$ denotes the 0th graded piece of the Koszul complex $K^{\cdot}(\underline{x}; K[\Sigma])$ with respect to the elements $\underline{x} = \{x_1, \ldots, x_n\}$ which generate the irrelevant ideal of $K[\Sigma]$. From this we get

$$K_0^{\cdot}(\underline{x}; K[\Sigma]) \cong C^{\cdot}(\Sigma, K)\,[-1].$$

Since $K[\Sigma]$ is F-pure resp. has a presentation of relative graded F-pure type it follows from Hochster-Roberts [2], Theorem 1.1 and Theorem 4.8 that

$$\tau^d K_0^{\cdot}(\underline{x}; K[\Sigma]) \cong \tau^d K_0^{\cdot},$$

where $K_0^{\cdot}$ denotes the 0th graded piece of the complex

$$K^{\cdot} = \varinjlim_t K^{\cdot}(\underline{x}^t; K[\Sigma]).$$

Also we know that

$$\tau^d K_0^{\cdot} \cong \tau^d K^{\cdot} \text{ in the derived category } D(K[\Sigma])$$

since $H^i(K^{\cdot}) \cong \underline{H}_{\mathfrak{m}}^i(K[\Sigma]) = [H_{\mathfrak{m}}^i(K[\Sigma])]_0$ for $0 \leq i < d$ by using Lemma 2.5 (ii). We now use our assumption (i) which implies the finiteness of the local cohomology modules $\underline{H}_{\mathfrak{m}}^i(K[\Sigma])$, $i \neq \dim K[\Sigma]$ (see Proposition I.3.4). Therefore we obtain

$$\tau^d K^{\cdot} \cong \tau^d\big(C^{\cdot}(\Sigma, K)\,[-1]\big).$$

The assertion (iv) now results by virtue of the Local Duality Theorem, q.e.d.

By the Local Duality Theorem and Theorem 4.12 we get a result which relates the local cohomology modules of $K[\Sigma]$ to the reduced simplicial homology of Σ with coefficients in K.

Corollary 4.13. *Let Σ be a Buchsbaum complex. Then there are isomorphisms*

$$\underline{H}_{\mathfrak{m}}^i(K[\Sigma]) \cong \tilde{H}_{i-1}(\Sigma; K), \quad 1 \leq i \leq \dim \Sigma, \quad \textit{and} \quad \underline{H}_{\mathfrak{m}}^0(K[\Sigma]) = 0.$$

Following P. Schenzel [1], 6.3, we next will present some further applications of the purity of the Frobenius in the case of the graded k-algebra $k[\Sigma]$ if Σ is a Buchsbaum complex. This is of some interest for our investigations regarding the Upper Bound result for connected manifolds.

Let N be a graded module and $p \in \mathbf{Z}$ an integer. We denote again by $N(p)$ the graded R-module whose underlying module is the same as that of N and whose grading is given by $[N(p)]_i = [N]_{p+i}$, $i \in \mathbf{Z}$. The graded ring R is assumed to be a graded k-algebra which is equi-dimensional and such that $R_{\mathfrak{p}}$ is a Cohen-Macaulay ring for all prime ideals $\mathfrak{p}$ different from the irrelevant ideal $\mathfrak{m}$. Furthermore, we assume that R is F-pure resp. has a presentation of relative graded F-pure type. By Theorem 4.1 we have that $\tau_{-d} D_R^{\cdot}$, $d = \dim R$, is isomorphic to a complex of k-vector spaces. Next we will prove the graded counterpart to Corollary 4.7.

Lemma 4.14. *Let R be a as defined. We denote by $\underline{x} = (x_1, \ldots, x_d) \cdot R$ a parameter ideal of R consisting of forms of degree t. We set $\underline{x}_s = \{x_1, \ldots, x_s\}$, $s = 0, 1, \ldots, d$. Then it follows for the Koszul complex $K^{\cdot}(\underline{x}_s; R)$ that $\tau^s K^{\cdot}(\underline{x}_s; R)$ is isomorphic to a complex of k-vector spaces. In particular, for the cohomology modules we have*

$$H^r(\underline{x}_s; R) \cong \bigoplus_{i=0}^{r} k^{e_i}\big((r-i)\,t\big), \quad 0 \leq r < s,$$

where

$$e_i = \binom{s}{r-i} \dim_k[\underline{H}_{\mathfrak{m}}^i(R)]_0.$$

Proof: First of all we show the assertion for $s = d$. In this case the cohomology modules $H^r(\underline{x}; R)$, $0 \leq r < d$, are modules of finite length. We have a short exact sequence of complexes

$$0 \to K_R[d] \to D_R^{\cdot} \to \tau_{-d} D_R^{\cdot} \to 0,$$

where $K_R[d]$ denotes the canonical module K_R shifted d places to the left. Applying the functors $\underline{R}\operatorname{Hom}(K^{\cdot}(\underline{x}; R),\ )$ and $\operatorname{Hom}(\ , E)$ we obtain

$$0 \to \operatorname{Hom}(\underline{R}\operatorname{Hom}(K^{\cdot}(\underline{x}; R), \tau_{-d}D_R^{\cdot}), E) \to \underline{R}\Gamma_{\mathfrak{m}}(K^{\cdot}(\underline{x}; R))$$
$$\to \operatorname{Hom}(\underline{R}\operatorname{Hom}(K^{\cdot}(\underline{x}; R), K_R[d]), E) \to 0$$

by virtue of the Local Duality Theorem. Since

$$\big(\operatorname{Hom}(\underline{R}\operatorname{Hom}(K^{\cdot}(\underline{x}; R), K_R[d]), E)\big)^i = 0$$

for all $i < d$, the exact sequence induces an isomorphism

$$\tau^d \operatorname{Hom}(\underline{R}\operatorname{Hom}(K^{\cdot}(\underline{x}; R), \tau_{-d}D_R^{\cdot}), E) \cong \tau^d\underline{R}\Gamma_{\mathfrak{m}}(K^{\cdot}(\underline{x}; R))$$

in the derived category of R. For the left complex we have

$$\tau^d K^{\cdot}(\underline{x}; R) \otimes_R \operatorname{Hom}(\tau_{-d}D_R, E).$$

By considering the structure of the dualizing complex it follows

$$\tau^d K^{\cdot}(\underline{x}; R) \otimes_R C^{\cdot}(R) \cong \tau^d K^{\cdot}(\underline{x}; R)$$

where we have used $\tau^d\underline{R}\Gamma_{\mathfrak{m}}(K^{\cdot}(\underline{x}; R)) \cong \tau^d K^{\cdot}(\underline{x}; R)$ since $\tau^d K^{\cdot}(\underline{x}; R)$ is a complex whose cohomology modules $H^r(\underline{x}; R)$, $0 \leq r < d$, are modules of finite length. By our assumption $C^{\cdot}(R)$ is a complex of k-vector spaces, therefore also $K^{\cdot}(\underline{x}; R) \otimes_R C^{\cdot}(R)$ is isomorphic to a complex of k-vector spaces. This proves our first statement in the case $s = d$. Taking into account that

$$K^i(\underline{x}; R) = R^{\binom{d}{i}}(it) \quad \text{and} \quad K^{\cdot}(\underline{x}; R) \otimes_R C^{\cdot}(R)$$

has trivial boundary homomorphisms, the formula for the cohomology modules follows immediately. For an arbitrary $0 \leq s \leq d$ and an integer $i \geq 1$ we have an exact sequence for the cohomology modules of the Koszul complexes

$$0 \to H^{r-1}(\underline{x}_{s-1}; R) \otimes_R R/x_s^i R \to H^r(\underline{x}_{s-1}, x_s^i; R).$$

If $s = d$ and $r < d$, we saw that $\mathfrak{m}$ *annihilates* the left modules for all $i \geq 1$. By Nakayama's Lemma it follows that $\mathfrak{m} \cdot H^{r-1}(\underline{x}_{s-1}; R) = 0$, i.e., for $s - 1 = d - 1$ and $r < d$ the cohomology modules are of finite length. By an induction argument it follows that $H^r(\underline{x}_s; R)$, $0 \leq r < s$, are modules of finite length, in fact k-vector spaces. Repeating the above arguments we have, as in the case $s = d$, that $\tau^s K^{\cdot}(\underline{x}_s; R)$ is isomorphic to a complex of k-vector spaces. Thus, the formula for the cohomology modules can be derived here also in a similar manner, q.e.d.

Lemma 4.14 provides a key to applications in combinatorics by use of homological algebra in derived categories. We want to describe an elementary approach to these applications by giving a new and simple proof of Lemma 4.14. Also our Lemma 4.14′ improves Lemma 4.14.

Lemma 4.14′. a) *Let M be a Buchsbaum module of dimension d and let $\{x_1, \ldots, x_d\}$ be a system of parameters of M. Then we have*

(i) $H^i_{\mathfrak{m}}(M/(x_1, \ldots, x_j)M) \cong \bigoplus_{l=0}^{j} \Big(\bigoplus_{\binom{j}{l}} H^{l+i}_{\mathfrak{m}}(M)\Big)$ *for all i and j with $i + j < d$.*

(ii) $H^i(x_1, \ldots, x_j; M) \cong \bigoplus_{l=0}^{i} \Big(\bigoplus_{\binom{j}{l}} H_{\mathfrak{m}}^{i-l}(M)\Big)$ *for all i and j with* $i < j$.

b) *Let* $R = \bigoplus_{i \geq 0} R_i$ *be a Noetherian graded ring with* $R_0 = K$ *a field,* $\mathfrak{m} := \bigoplus_{i \geq 1} R_i$. *Let* M *be a Noetherian graded R-module of dimension d. Let* $\{x_1, \ldots, x_d\}$ *be a system of homogeneous parameters of M with* $t_i :=$ *degree of* x_i *for* $i = 1, \ldots, d$. *Then we have*

(i') $\underline{H}_{\mathfrak{m}}^i\big(M/(x_1, \ldots, x_j) \cdot M\big) \cong \bigoplus_{l=0}^{j} \Big(\bigoplus_{1 \leq p_1 < \ldots < p_l \leq j} \underline{H}_{\mathfrak{m}}^{l+i}(M)\,(-t_{p_1} - \ldots - t_{p_l})\Big)$
for all i and j with $i + j < d$.

(ii') $\underline{H}^i(x_1, \ldots, x_j; M) \cong \bigoplus_{l=0}^{i} \Big(\bigoplus_{1 \leq p_1 < \ldots < p_l \leq j} \underline{H}_{\mathfrak{m}}^{i-l}(M)\,(t_{p_1} + \ldots + t_{p_l})\Big)$
for all i and j with $i < j$.

Proof: a) (i) is clear for $j = 0$. If $j > 0$ we have the following exact sequence:

$$0 \to M/(x_1, \ldots, x_{j-1}) \cdot M : \mathfrak{m} \xrightarrow{x_j} M/(x_1, \ldots, x_{j-1}) \cdot M \to M/(x_1, \ldots, x_j) \cdot M \to 0. \tag{E}$$

Applying the local cohomology functors $H_{\mathfrak{m}}^i(\;\;)$ we get an exact sequence:

$$0 \to H_{\mathfrak{m}}^i\big(M/(x_1, \ldots, x_{j-1}) \cdot M\big) \to H_{\mathfrak{m}}^i\big(M/(x_1, \ldots, x_j) \cdot M\big) \to H_{\mathfrak{m}}^{i+1}\big(M/(x_1, \ldots, x_{j-1}) \cdot M : \mathfrak{m}\big) \to 0.$$

Note that

$$H_{\mathfrak{m}}^{i+1}\big(M/(x_1, \ldots, x_{j-1}) \cdot M : \mathfrak{m}\big) \cong H_{\mathfrak{m}}^{i+1}\big(M/(x_1, \ldots, x_j) \cdot M\big)$$

and that all modules occuring in this sequence are annihilated by $\mathfrak{m}$. Hence we get

$$H_{\mathfrak{m}}^i\big(M/(x_1, \ldots, x_j) \cdot M\big) \cong H_{\mathfrak{m}}^i\big(M/(x_1, \ldots, x_{j-1}) \cdot M\big) \oplus H_{\mathfrak{m}}^{i+1}\big(M/(x_1, \ldots, x_{j-2}) \cdot M\big)$$

and an easy induction on j proves the result (i).

(ii) This result is clear for $j = 1$ since

$$H^0(x_1; M) = 0 :_M x_1 \cong H_{\mathfrak{m}}^0(M).$$

If $j > 1$ we consider the following exact sequence (E') (see Lemma 0.1.6)

$$0 \to H^1\big(x_j; H^{i-1}(x_1, \ldots, x_{j-1}; M)\big) \to H^i(x_1, \ldots, x_j; M) \to H^0\big(x_j; H^i(x_1, \ldots, x_{j-1}; M)\big) \to 0. \tag{E'}$$

If $i < j$ we have

$$H^1\big(x_j; H^{i-1}(x_1, \ldots, x_{j-1}; M)\big) \cong H^{i-1}(x_1, \ldots, x_{j-1}; M).$$

If $i < j - 1$ we obtain

$$H^0\big(x_j; H^i(x_1, \ldots, x_{j-1}; M\big) \cong H^i(x_1, \ldots, x_{j-1}; M)$$

by Corollary 4.7'. Hence we have from (E') for $i < j - 1$

$$H^i(x_1, \ldots, x_j; M) \cong H^{i-1}(x_1, \ldots, x_{j-1}; M) \oplus H^i(x_1, \ldots, x_{j-1}; M)$$

since these modules are annihilated by $\mathfrak{m}$ (see Corollary 4.7'). If $i = j - 1$ we get

$$\begin{aligned} & H^0(x_j; H^{j-1}(x_1, \ldots, x_{j-1}; M)) \\ & \cong (x_1, \ldots, x_{j-1}) \cdot M : \mathfrak{m}/(x_1, \ldots, x_{j-1}) \cdot M \\ & \cong H^0_{\mathfrak{m}}(M/(x_1, \ldots, x_{j-1}) \cdot M) \end{aligned}$$

and we therefore have:

$$H^{j-1}(x_1, \ldots, x_j; M) \cong H^{j-2}(x_1, \ldots, x_{j-1}; M) \oplus H^0_{\mathfrak{m}}(M/(x_1, \ldots, x_{j-1}) \cdot M).$$

(i) and an easy induction on j prove (ii).

b) The exact sequence (E) of part a) has the following form:

$$\begin{aligned} & 0 \to M/(x_1, \ldots, x_{j-1}) \cdot M : \mathfrak{m}(-t_j) \xrightarrow{x_j} M/(x_1, \ldots, x_{j-1}) \cdot M \\ & \quad \to M/(x_1, \ldots, x_j) \cdot M \to 0. \end{aligned}$$

Therefore again we obtain (i') by induction on j. For (ii') we point out that the exact sequence (E') has the following form:

$$\begin{aligned} & 0 \to \underline{H}^1(x_j; \underline{H}^{i-1}(x_1, \ldots, x_{j-1}; M))\,(t_j) \to \underline{H}^i(x_1, \ldots, x_j; M) \\ & \quad \to \underline{H}^0(x_j; \underline{H}^i(x_1, \ldots, x_{j-1}; M)) \to 0. \end{aligned}$$

Furthermore we have:

$$\begin{aligned} & \underline{H}^0(x_j; \underline{H}^{j-1}(x_1, \ldots, x_{j-1}; M)) \\ & \cong ((x_1, \ldots, x_{j-1}) \cdot M : \mathfrak{m}/(x_1, \ldots, x_{j-1}) \cdot M)\,(t_1 + \ldots + t_{j-1}) \\ & \cong \underline{H}^0_{\mathfrak{m}}(M/(x_1, \ldots, x_{j-1}) \cdot M)\,(t_1 + \ldots + t_{j-1}). \end{aligned}$$

Hence our result (ii') follows by the same arguments as above, q.e.d.

Remark 4.15. It follows from Lemma 2.6 and Lemma 4.14 that

$$\underline{H}^i_{\mathfrak{m}}(R) \cong [H^i(\underline{x}; R)]_0, \quad 0 \le i < d.$$

For this it is not necessary to assume that $\underline{x} = \{x_1, \ldots, x_d\}$ is a system of parameters consisting of forms of the same degree. Following the reasoning of the proof it is seen immediately that it is enough to assume for $\underline{x} = \{x_1, \ldots, x_n\}$ to be any set of forms contained in $\mathfrak{m}$ and such that $\operatorname{Rad} \underline{x}R = \mathfrak{m}$. Or put differently, it is not necessary to take a direct limit in order to compute the local cohomology modules via Koszul complexes for those rings. This was proved by M. Hochster and J. L. Roberts in [2], Theorem 1.1(b), in case of an F-pure graded k-algebra over a perfect field k of prime characteristic p and in Hochster-Roberts [2], Theorem 4.8(b), in case R has a presentation of perfect graded F-pure type. Hence, in the case of the ground field of characteristic zero our result leads to a slight improvement of Hochster-Roberts [2], Theorem 4.8(b).

For our considerations in the following topics Lemma 4.14' has an important application with respect to quotients of certain ideals. It allows us to give an explicit description of certain Hilbert functions.

Corollary 4.16. *Let R denote as before a graded k-algebra. Let $\underline{x} = \{x_1, \ldots, x_d\}$ be a system of parameters consisting of forms of degree t. Then, for $1 \leq s \leq d$ there are isomorphisms*

$$\big((x_1, \ldots, x_{s-1}) R : x_s\big)/(x_1, \ldots, x_{s-1}) R \simeq \bigoplus_{i=0}^{s-1} \underline{k}^{r_i}(-it),$$

where $r_i = \binom{s-1}{i} \dim_k[\underline{H}^i_{\mathfrak{m}}(R)]_0$.

Proof: For the cohomology modules of Koszul complexes there are the following short exact sequences

$$0 \to H^{s-2}(\underline{x}_{s-1}; R) \to H^{s-1}(\underline{x}_s; R) \to (\underline{x}_{s-1}R : x_s/\underline{x}_{s-1}R)\big((s-1)\,t\big) \to 0,$$

where it was used that $H^{s-2}(\underline{x}_{s-1}; R)$ is annihilated by $\mathfrak{m}$. Since this short exact sequence is a sequence of k-vector spaces it splits. By virtue of Lemma 4.14′ our statement now follows, q.e.d.

In point of fact, the graded rings considered in Lemma 4.14′ are Buchsbaum rings by virtue of Theorem 2.7. Corresponding results hold for an arbitrary graded Buchsbaum ring without assuming the purity condition. However, for two reasons we restrict ourselves to this special case. Purity implies that the non-vanishing cohomology of $\tau^d \underline{R}\Gamma_{\mathfrak{m}}(R)$ is concentrated in at most the zero'th graded piece. This is no longer true for an arbitrary graded Buchsbaum ring. For our purposes here it is enough to consider pure rings for which the statements can be formulated in an easier manner.

Let Δ denote a finite simplicial complex with the vertex set $\{x_1, \ldots, x_n\}$. Let f_i be the number of i-dimensional faces of Δ. Thus $f_0 = n$ and $f_{-1} = 1$, since Δ has the unique (-1)-face $\emptyset$. The vector $f = (f_{-1}, f_0, \ldots, f_{d-1})$, $d = \dim \Delta + 1$, is called the f-vector of Δ. Since $k[\Delta]$ is a graded algebra, we can associate to it its *Hilbert function* $H(m, k[\Delta])$, that is

$$H(m, k[\Delta]) = \dim_k\big[k[\Delta]\big]_m, \quad m \in \mathbf{Z},$$

where $\big[k[\Delta]\big]_m$ denotes the k-vector space of all homogeneous forms of degree m in $k[\Delta]$. For m large, $H(m, k[\Delta])$ coincides with a polynomial, the Hilbert polynomial. R. P. Stanley [3], Proposition 3.1 showed that the f-vector describes the Hilbert function exactly.

Proposition 4.17. *For the Hilbert function $H(m, k[\Delta])$ we have*

$$H(m, k[\Delta]) = \sum_{i=-1}^{d-1} f_i \binom{m-1}{i}, \quad m \geq 0,$$

where $\binom{-1}{i} = 0$ *for* $i \neq -1$ *and* $\binom{-1}{-1} = 1$.

Proof: Let $\bar{x}_i$ denote the image of x_i by the canonical projection

$$k[x_1, \ldots, x_n] \to k[\Delta].$$

A k-basis of $\big[k[\Delta]\big]_m$ consists of all monomials $x = \bar{x}_1^{a_1} \cdot \ldots \cdot \bar{x}_n^{a_n}$ such that $\deg x = a_1 + \ldots + a_n = m$ and $\operatorname{Supp}(x) \in \Delta$, where the support $\operatorname{Supp}(x)$ is defined by

$$\operatorname{Supp}(x) = \{x_i \mid a_i > 0\}.$$

If $\sigma \in \Delta$ has exactly $i+1$ elements, then the number of monomials of degree $m \geq 0$ whose support is contained in σ coincides with $\binom{m-1}{i}$. Hence, the proposition is proved, q.e.d.

In particular by Proposition 4.17

$$\dim k[\Delta] = \dim \Delta + 1.$$

Next we will prove an estimate of the number of i-faces of certain simplicial complexes. To this end we define integers h_i by

$$(1-T)^d \sum_{m\geq 0} H(m, k[\Delta])\, T^m = h_0 + h_1 T + \ldots + h_d T^d.$$

It is easily seen that the degree of the polynomial on the left does not exceed d. The vector $h = (h_0, \ldots, h_d)$ is called the *h-vector* of Δ. In fact, the formal power series $\sum_{m\geq 0} H(m, k[\Delta])\, T^m$ is the Poincaré series of the graded k-algebra $k[\Delta]$. By the Theorem of Hilbert-Serre there is a polynomial $f(T, R)$ such that

$$F(T, R) = f(T, R)/(1-T)^d$$

for the Poincaré series $F(T, R)$ of a d-dimensional graded k-algebra R. We recall that $H(m, \ )$ and $F(T, \ )$ are additive functions on short exact sequences.

For our particular situation we remark: knowing the f-vector of Δ is equivalent to knowing its h-vector.

Proposition 4.18. *For the h-vector and f-vector of a finite simplicial complex the following relations hold:*

$$h_v = \sum_{i=0}^{v} (-1)^{v-i} \binom{d-i}{v-i} f_{i-1} \quad \textit{and} \quad f_{v-1} = \sum_{i=0}^{v} \binom{d-i}{v-i} h_i$$

for $v = 0, 1, \ldots, d$, $d = \dim \Delta + 1$.

The *proof* follows by an easy calculation. We omit it.

Now we prove one of the main results of our applications in algebraic topology and combinatorics.

Theorem 4.19. *Let Δ denote a simplicial Buchsbaum complex with n vertices and we denote* $d = \dim \Delta + 1$. *For the f-vector and the h-vector of Δ there are the following bounds*

$$f_{v-1} \leq \binom{n}{v} - \binom{d}{v} \sum_{i=-1}^{v-2} \binom{v-1}{i+1} \dim_k \tilde{H}_i(\Delta; k)$$

and

$$h_v \leq \binom{n-d+v-1}{v} - (-1)^v \binom{d}{v} \sum_{i=-1}^{v-2} (-1)^i \dim_k \tilde{H}_i(\Delta; k),$$

where $0 \leq v \leq d$ *and k denotes an arbitrary field.*

Proof: By the previous Proposition 4.18 we have an exact expression of the Hilbert function or even of the Poincaré series of $k[\Delta]$.

What now follows is another way to calculate the Poincaré series by techniques from commutative algebra. To this end we consider the associated graded k-algebra $R := k[\Delta]$, where k denotes the fixed field. Without loss of generality we can assume k an infinite field. Otherwise we extend k to the field of rational functions $k(t)$ in one variable. Then there exists a homogeneous system of parameters $\underline{x} = \{x_1, \ldots, x_d\}$ of R consisting of forms of degree 1. Let $x \in R$ be homogeneous of degree one. Then there is the following exact sequence of graded R-modules

$$0 \to O_R : x \to R \xrightarrow{x} R(1) \to (R/xR)(1) \to 0.$$

Because $F(T, \ )$ is additive, it follows that

$$(1 - T) F(T, R) = F(T, R/xR) - TF(T, O_R : x).$$

Iterating this argument d-times we obtain

$$(1 - T)^d F(T, R) = F(T, R/\underline{x}R) - \sum_{i=0}^{d-1} T(1 - T)^i F(T, Q_i),$$

where Q_i, $0 \leq i \leq d - 1$, denotes the quotients

$$\big((x_1, \ldots, x_{d-1-i})R : x_{d-i}/(x_1, \ldots, x_{d-1-i})\big) R.$$

By virtue of Corollary 4.16 we get

$$F(T, Q_i) = \sum_{j=0}^{d-i-1} \binom{d-i-1}{j} \dim_k[\underline{H}^j_{\mathfrak{m}}(R)]_0 T^j.$$

It now follows from simple calculations that

$$(1 - T)^d F(T, R) = F(T, R/\underline{x}R) - \sum_{v=1}^{d} \binom{d}{v} \left(\sum_{i=0}^{v-1} (-1)^{v-i-1} \dim_k[\underline{H}^i_{\mathfrak{m}}(R)]_0\right) T^v.$$

In particular, the last equality shows that

$$F(T, R/\underline{x}R) = g_0 + g_1 T + \ldots + g_d T^d$$

does not depend on the system of parameters $\underline{x}$ (consisting of forms of degree 1). We define

$$g = (g_0, g_1, \ldots, g_d)$$

the *g-vector* of the simplicial Buchsbaum complex Δ. By the definition of the h-vector it follows that

$$h_v = g_v - \binom{d}{v} \sum_{i=0}^{v-1} (-1)^{v-i-1} \dim_k[\underline{H}^i_{\mathfrak{m}}(R)]_0,$$

where $0 \leq v \leq d$. Using the second formula of Proposition 4.18 we get

$$f_{v-1} = \sum_{i=0}^{v} \binom{d-i}{v-i} g_v - \binom{d}{v} \sum_{i=0}^{v-1} \binom{v-1}{i} \dim_k[\underline{H}^i_{\mathfrak{m}}(R)]_0,$$

$v = 0, 1, \ldots, d$, by some simple calculations. In point of fact, the g-vector of Δ is the Hilbert function of the 0-dimensional graded k-algebra $R/\underline{x}R$, i.e.

$$g_v = \dim_k[R/\underline{x}R]_v, \quad 0 \leq v \leq d.$$

Therefore, g_v is not larger than the number of all linearly independent forms of degree v in $n - d$ variables. That is,

$$g_v \leq \binom{n-d+v-1}{v}, \qquad 0 \leq v \leq d.$$

Since

$$\sum_{i=0}^{v} \binom{d-i}{v-i} g_v \leq \sum_{i=0}^{v} \binom{d-i}{v-i} \binom{n-d+v-1}{v} = \binom{n}{v},$$

the two inequalities follow by replacing the local cohomology modules by the reduced simplicial homology, see Corollary 4.13, q.e.d.

We know (see Corollary 2.12(i)) that the graded k-algebra $k[\Delta]$ is a Buchsbaum ring, if $X = |\Delta|$ is a connected manifold. This implies:

Corollary 4.20. *If the geometric realization $X = |\Delta|$ of a finite simplicial complex Δ is a manifold, then we have*

$$f_v \leq \binom{n}{v+1} - \binom{d}{v+1} \sum_{i=-1}^{v-1} \binom{v}{i+1} \dim_k \tilde{H}_i(\Delta; k),$$

for $v = 0, 1, \ldots, \dim \Delta$, resp.

$$h_v \leq \binom{n-d+v-1}{v} - (-1)^v \binom{d}{v} \sum_{i=-1}^{v-2} (-1)^i \dim_k \tilde{H}_i(\Delta; k),$$

for $v = 0, 1, \ldots, \dim \Delta + 1$, where $\tilde{H}_i(\Delta; k)$ denotes the reduced simplicial cohomology of Δ with coefficients in an arbitrary fixed field k.

R. P. Stanley [3] characterized those numerical functions H which are Hilbert functions of a certain graded k-algebra. (This result goes back to F. S. Macaulay.) This leads to a slight sharpening of Theorem 4.19.

In the expression for the upper bound of the h-vector the sum

$$\sum_{i=-1}^{v-2} (-1)^i \dim_k \tilde{H}_i(\Delta; k)$$

is a partially reduced Euler-Poincaré characteristic of Δ. It is not surprising that the homology of Δ appears in the expression for the upper bound of f_v. If the homology does not vanish, there are cycles which are not boundaries. That is, some faces are deleted in Δ. The sum

$$\sum_{i=-1}^{v-2} \binom{v-1}{i+1} \dim_k \tilde{H}_i(\Delta; k)$$

in the bound of the f_v's coincides with $\dim_k Q_{d-v}$, i.e.,

$$f_{v-1} \leq \binom{n}{v} - \binom{d}{v} \dim_k\big((x_1, \ldots, x_{v-1})R : x_v/(x_1, \ldots, x_{v-1}) R\big).$$

If Δ is a Cohen-Macaulay complex, then the reduced simplicial homology modules $\tilde{H}_i(\Delta; k)$, $0 \leq i \leq d-2$, vanish. In this case we obtain the formulas proved by R. P. Stanley [1] and [8].

The purpose of the sequel is to introduce some of the foregoing ideas to partially ordered sets, posets for short. Let P be a finite poset with rank function ϱ. We say that P is a *Cohen-Macaulay poset*, if the associated simplicial complex $\Delta(P)$ is a Cohen-Macaulay complex. This concept originated with K. Baclawski [3]. In particular, he showed that the simplicial cohomology of $\Delta(P)$ with coefficients in a field k coincides with a cohomology theory on the poset with coefficients in certain diagrams, compare K. Baclawski [3, 4, 5]. Related to the foregoing we say that P is a *bouquet*, if it has non-zero reduced simplicial cohomology at most in the highest possible dimension. For our purposes we need a more general notion than that of a Cohen-Macaulay poset.

Definition and Proposition 4.21. *Let P denote a finite poset. Then the following conditions on P are equivalent:*

(i) $H_i(X, X - p; k) = 0$ *for every point* $p \in X$ *and every* $i \neq \dim X$, *where* X *denotes the geometric realization of* $\Delta(P)$,

(ii) *Every open intervall* (x, y) *of* $\hat{P}$ *is a bouquet, except possible for* $(x, y) = (\hat{0}, \hat{1})$.

(iii) $k[\Delta(P)]$ *is a Buchsbaum ring.*

If P satisfies one of these equivalent conditions, we say P is a *Buchsbaum poset.*

Proof: The equivalence of (i) and (ii) follows from results of J. Munkres [1], Lemma 2.2, and K. Baclawski [3], Proposition 3.3. That (i) and (iii) are equivalent was proved in our Theorem 4.12, q.e.d.

We remark that K. Baclawski [3, 5] considered this kind of posets under the name "almost Cohen-Macaulay". In his papers it is a useful concept for induction proofs. We prefer the name Buchsbaum poset, because $k[\Delta(P)]$ is a Buchsbaum ring which has additional structure by virtue of our results. Examples of Buchsbaum posets are simplicial complexes Δ such that $|\Delta|$ is a (connected) manifold, see Corollary 2.12. Next we will discuss some Buchsbaum preserving operations on posets. Let P and Q be finite posets. The order-dual of P, denoted by P^*, is the poset obtained by reversing the order of P. The product of P and Q, denoted by $P \times Q$, is the cartesian product of P and Q with its order defined by $(x, y) \leq (x', y')$ if and only if $x \leq x'$ and $y \leq y'$. The *interval poset* of P, denoted by $\mathrm{Int}(P)$, is the poset of closed intervals of P, ordered by inclusion, i.e.,

$$[x, y] \leq [x', y']$$

if and only if $x' \leq x$ and $y \leq y'$. This means, $\mathrm{Int}(P)$ is given by the induced order as a subset of $P^* \times P$. Next we recall some results of K. Baclawski [3], § 7.

Theorem 4.22. *Let P, Q denote Buchsbaum posets. Then P^*, $P \times Q$, and $\mathrm{Int}(P)$ are again Buchsbaum posets.*

Proof: If P is a Buchsbaum poset, then P^* is a Buchsbaum poset too. Because $\Delta(P) = \Delta(P^*)$ and (x, y) is an open interval in P^*. For the rest we refer to K. Baclawski [3], § 7, q.e.d.

Next we discuss some more terminology for posets.

Let μ denote the Möbius function of P defined by

$$\mu\colon \{(x, y) \in \hat{P} \times \hat{P} \mid x \leq y\} \to \mathbf{Z}$$

such that

(a) $\mu(x, x) = 1$ for $x \in \hat{P}$,

(b) $\sum_{x \leq y \leq z} \mu(x, y) = 0$ for all fixed pairs $x < z$ in $\hat{P}$.

Let $S \subset \{1, \ldots, d\} =: [d]$ be a set consisting of ranks of P. The *rank selected subposet* with respect to S is defined by

$$P_S = \{x \in P \mid \varrho(x) \in S\}.$$

In addition, we denote by $\alpha(P, S)$ the number of chains $x_1 < \ldots < x_s$ in P such thaf $\{\varrho(x_1), \ldots, \varrho(x_s)\} = S$. It follows from the definition, that $\alpha(P, S)$ denotes the number ot maximal chains in P_S. Now we define

$$\beta(P, S) = \sum_{T \subset S} (-1)^{\operatorname{card} S \setminus T} \alpha(P, T),$$

where the sum is taken over all subsets T of S. Conversely we have

$$\alpha(P, S) = \sum_{T \subset S} \beta(P, T).$$

The numbers $\alpha(P, S)$ and $\beta(P, S)$ were introduced and examined by R. P. Stanley [4, 5] for various posets. Another related invariant of P is its zeta polynomial, see Stanley [5]. If m is a non-negative integer, define $Z(P, m)$ to be the number of chains

$$\hat{0} \leq x_0 \leq x_1 \leq \ldots \leq x_m \leq \hat{1} \quad \text{in } \hat{P}.$$

$Z(P, m)$ is a polynomial function of degree $d + 1$ of m. It follows the existence of constants $e_0, e_1, \ldots, e_d$ and $h_0, h_1, \ldots, h_d$ such that

$$Z(P, m) = \sum_{i=1}^{d+1} e_{i-1} \binom{m}{i}$$

and

$$(1 - T)^{d+2} \sum_{m \geq 0} Z(P, m) T^m = h_0 T + h_1 T^2 + \ldots + h_d T^{d+1},$$

see Stanley [3]. In this reference it is also notet that

$$e_i = \sum_{\substack{S \subset [d] \\ \operatorname{card} S = i}} \alpha(P, S) \quad \text{and} \quad h_i = \sum_{\substack{S \subset [d] \\ \operatorname{card} S = i}} \beta(P, S).$$

Let $f = (f_{-1}, f_0, \ldots, f_{d-1})$ denote the f-vector of $\Delta(P)$, i.e., f_i is equal to the number of chains $x_0 < x_1 < \ldots < x_i$ in P. It is clear that $f_{i-1} = e_i$.

Theorem 4.23. *Let P denote a Buchsbaum poset with* $\dim P + 1 = d$. *Then we have*

$$h_v = \sum_{\substack{S \subset [d] \\ \operatorname{card} S = v}} \dim_k \tilde{H}_{v-1}(P_S; k) - (-1)^v \binom{d}{v} \sum_{i=1}^{v-2} (-1)^i \dim_k \tilde{H}_i(P; k)$$

for $v = 0, 1, \ldots, d$ and an arbitrary fixed field k.

Proof: By Philip Hall's Theorem, see Rota [1], Proposition 6, it follows that

$$\beta(P, S) = (-1)^{1 + \operatorname{card} S} \mu(P_S) = (-1)^{1 + \operatorname{card} S} \tilde{\chi}(P_S),$$

where $\tilde{\chi}(P_S)$ denotes the reduced Euler characteristic of P_S, i.e.,

$$\tilde{\chi}(P_S) = \sum_{i \geq 0} (-1)^i \dim_k \tilde{H}_i(P_S; k).$$

Now, by the results of K. Baclawski [3] on rank selected subposets, Theorem 6.5, we get

$$\tilde{H}_i(P_S; k) \cong \tilde{H}_i(P; k) \quad \text{for } i < \operatorname{card} S - 1.$$

A corresponding result was proven implicitely by J. Munkres [1], Theorem 6.4. By virtue of the above formula for the h_i we obtain

$$h_v = \sum_{\substack{S \subset [d] \\ \operatorname{card} S = v}} (-1)^{1+v} \left(\sum_{i=-1}^{v-2} (-1)^i \dim_k \tilde{H}_i(P; k) + (-1)^{v-1} \dim_k \tilde{H}_{v-1}(P_S; k) \right).$$

Thus, our statements follows, q.e.d.

If we apply Theorem 4.23 to the case of a finite simplicial Buchsbaum complex Δ, we obtain an exact description of the g-vector, introduced in the proof of Theorem 4.19, in terms of the reduced simplicial homology.

Corollary 4.24. *Let $\Delta = \Delta(P)$ be the order complex of a Buchsbaum poset P. Then we have for the g-vector of Δ:*

$$g_v = \sum_{\substack{S \subset [d] \\ \operatorname{card} S = v}} \dim_k \tilde{H}_{v-1}(P_S; k),$$

$0 \leq v \leq d$, *where* $d = \dim \Delta + 1$.

Chapter III

On liaison among curves in projective three space

The underlying philosophy of liaison is to examine the closed subschemes of $\mathbf{P}^n_K$ by introducing equivalence relations on them, see the beautiful paper of Peskine and Szpiro [3]. This allows one to study the closed subschemes indirectly by investigating the induced equivalence classes. The liaison equivalence relation orginates from the tenet that the simplest closed subschemes of $\mathbf{P}^n_K$ are the complete intersections. Two arbitrary closed subschemes are then equated if they bear an appropriate relationship to some complete intersection.

The main subject of this chapter, but not exclusively, is liaison of curves in $\mathbf{P}^3_K$. Roughly speaking, liaison considers two curves in $\mathbf{P}^3_K$ to be equivalent if their union is a complete intersection. That this idea for curves in $\mathbf{P}^3$ has interested algebraic geometers for over a century is attested by, for instance, Rohn and Berzolari [1].

In a fundamental paper, P. Rao [1] has shown that curves in $\mathbf{P}^3_K$ which are generic complete intersections and without isolated or imbedded points are classified up to liaison by a certain graded module of finite length over $K[x_0, x_1, x_2, x_3]$. This "liaison invariant" vanishes if and only if the curve is arithmetically Cohen-Macaulay. For instance, the liaison invariant of a line in $\mathbf{P}^3_K$ is O, whereas the liaison invariant of two skew lines in $\mathbf{P}^3_K$ is K. In § 1 we show how famous classical results (see for example Apéry [1, 2], Gaeta [1], Peskine and Szpiro [3] or Artin and Nagata [1]) may be generalized for arithmetical Buchsbaum curves in $\mathbf{P}^3_K$. We also point out a different formulation of the liaison invariant by using a certain dualizing complex. This allows us to study the liaison invariant in any codimension via the theory of derived functors and categories.

Rao's work raises the following question: Does there exist a geometric addition of curves in $\mathbf{P}^3_K$ corresponding to the direct sum of their liaison invariants? This is the liaison addition problem. Merely defining the sum of two curves to be their union will not suffice, as evidenced by the preceding examples of lines. In § 2 we exhibit a solution to the liaison addition problem by using results of Phil Schwartau's [1] magnificent thesis. We therefore find that not only there is a way to add curves in $\mathbf{P}^3_K$, but that an explicit procedure is possible: that is, equations for the added curve may be written down from the equations of the curves being added. The addition procedure turns out to be quite simple, and in fact it admits a purely intrinsic formulation reminiscent of liaison itself.

Our aim in § 3 is to study curves linked to lines in $\mathbf{P}^3_K$. Our results deal mainly, but not exclusively, with the case of a line lying on a nonsingular quadric surface. By using a very explicit description we may construct several examples of non-Cohen-Macaulay and non-Buchsbaum curves in $\mathbf{P}^3_K$. Coupling these methods with the notion of liaison we are able to produce an infinite family of counter-examples to a question of

R. Hartshorne on set-theoretic complete intersections in $\mathbf{P}^3_K$ and also comment further on a counter-example to Hartshorne's question given by Peskine and Szpiro [3].

In § 4 we apply the results of the third paragraph to the notion of "self-linked" curves, recently studied by Rao [2], and we are able to give new examples of non-self-linked curves in $\mathbf{P}^3_K$. It is still an open question to determine those liaison classes in $\mathbf{P}^3_K$ which contain a self-linked curve. Liaison addition of § 2 provides one step toward the solution of this problem by use of Phil Schwartau's [1] thesis.

§ 1. On liaison among arithmetical Buchsbaum curves in $\mathbf{P}^3$

It seems that A. Cayley [1], p. 152, in 1847 was the first who posed the problem to describe the liaison class of a complete intersection. Nowadays it is well-known that a curve in $\mathbf{P}^3$ is in the liaison class of a complete intersection if and only if the curve is arithmetically Cohen-Macaulay.

Furthermore, L. Gruson and C. Peskine [1] have given a complete classification of arithmetical Cohen-Macaulay curves in $\mathbf{P}^3$. Discussing some results from Bresinsky, Schenzel and Vogel [1] we will begin in this paragraph to investigate the next simple case, that is the liaison classes which are characterized by finite-dimensional vector spaces of dimension ≥ 1. By using the theory of Buchsbaum rings this means we will study liaison of arithmetical Buchsbaum curves in $\mathbf{P}^3$, see Peskine and Szpiro [3] and Rao [1].

Definition 1.1. Let R be a Cohen-Macaulay local ring and let $\mathfrak{a}$ and $\mathfrak{b}$ be two ideals of R. The ideals $\mathfrak{a}$ and $\mathfrak{b}$ are said to be (*algebraically*) *linked* (written $\mathfrak{a} \sim \mathfrak{b}$) if there is an R-sequence $x_1, \ldots, x_n$ in $\mathfrak{a} \cap \mathfrak{b}$ such that $\mathfrak{a} = \big((x_1, \ldots, x_n)\big) : \mathfrak{b}$, and $\mathfrak{b} = \big((x_1, \ldots, x_n)\big) : \mathfrak{a}$.

We will make use of the following equivalent definition:

Let $\mathfrak{a}, \mathfrak{b}$ be two ideals of the local Gorenstein ring R. The ideals $\mathfrak{a}$ and $\mathfrak{b}$ are algebraically linked by a complete intersection $\mathfrak{x} := (x_1, \ldots, x_g) \subset \mathfrak{a} \cap \mathfrak{b}$ if

(i) $\mathfrak{a}$ and $\mathfrak{b}$ are ideals of pure height g, and

(ii) $\mathfrak{a}/\mathfrak{x} \cdot R \cong \operatorname{Hom}_R(R/\mathfrak{b}, R/\mathfrak{x} \cdot R)$

and

$\mathfrak{b}/\mathfrak{x} \cdot R \cong \operatorname{Hom}_R(R/\mathfrak{a}, R/\mathfrak{x} \cdot R)$.

We now consider the equivalence class of ideals generated by linkage, that is we consider the transitive hull of this relation.

We say an ideal $\mathfrak{b}$ is in the *linkage class* of $\mathfrak{a}$ if there are ideals $\mathfrak{c}_1, \ldots, \mathfrak{c}_n$ such that

$$\mathfrak{a} \sim \mathfrak{c}_1 \sim \mathfrak{c}_2 \sim \ldots \sim \mathfrak{c}_n \sim \mathfrak{b}.$$

We remark that the linkage classes of complete intersections are reasonably well-understood in codimension ≤ 3. Let $\mathfrak{a}$ be an ideal of the local ring R. If R is normal, then if $\mathfrak{a}$ is an unmixed ideal of codimension 1 is in the linkage class of a complete intersection if and only if the associated element $\mathfrak{a}$ in the class group of R is trivial. If R is regular, and $\mathfrak{a}$ is an unmixed ideal of codimension 2, $\mathfrak{a}$ is in the linkage class of a complete intersection if and only if $R/\mathfrak{a}$ is Cohen-Macaulay, see Peskine-Szpiro [3]. If R is regular and $\mathfrak{a}$ is an ideal of codimension 3 such that $R/\mathfrak{a}$ is Gorenstein, then $\mathfrak{a}$ is in the linkage class of a complete intersection (see Watanabe [1]).

Moreover, $\mathfrak{a}$ and $\mathfrak{b}$ are linked (geometrically) by a complete intersection $\mathfrak{x} := (x_1, \ldots, x_g)$ if $\mathfrak{a}$ and $\mathfrak{b}$ have no components in common and $\mathfrak{a} \cap \mathfrak{b} = \mathfrak{x} \cdot R$.

Two subschemes of $\mathbf{P}^n$ of codimension 2 are said to be *linked geometrically* if their scheme theoretic union is the complete intersection of two hypersurfaces. For instance, the twisted quartic curve C in $\mathbf{P}^3$ given parametrically by (t^4, t^3u, tu^3, u^4) and the union X of two skew lines in $\mathbf{P}^3$ are linked since we have for the defining ideals $I(C)$ and $I(X) = (x_0, x_1) \cap (x_2, x_3)$ of C and X, respectively:

$$I(C) \cap I(X) = (x_0x_3 - x_1x_2, x_0x_2^2 - x_1^2x_3).$$

This liaison of C and X was discovered by G. Salmon [1], p. 40, already in 1848 and a little later in 1857 again by J. Steiner [1], p. 138.

Nowadays we know that algebraic and geometric linkage generate the same equivalence relation, see Rao [1], Theorem 1.7. This equivalence relation generated by linkage is called *liaison*.

Now let $C \subset \mathbf{P}^3_K$ be a curve, that is, C is an one-dimensional subscheme of $\mathbf{P}^3_K$ over an algebraically closed field K, equidimensional, locally Cohen-Macaulay and a generic complete intersection. A. P. Rao [1] studied the following invariant, due to R. Hartshorne:

$$M(C) := \bigoplus_v H^1(\mathbf{P}^3, \mathcal{J}_C(v))$$

where $\mathcal{J}_C(v)$ is the twisted ideal sheaf of C. Then $M(C)$ is a graded $S := K[x_0, x_1, x_2, x_3]$-module of finite length. Furthermore, $M(C)$ is invariant up to duals and shifts in gradings, under liaison. It follows from Rao [1] that for each graded S-module of finite length, there is a liaison equivalence class, and that the module determines that equivalence class.

We recall that a curve $C \subset \mathbf{P}^3$ with defining ideal $I(C)$ is said to be arithmetically Cohen-Macaulay or arithmetical Buchsbaum if the local ring $S_{(x_0,\ldots,x_3)}/I(C) =: A$ of the vertex of the affine cone over C is a Cohen-Macaulay or Buchsbaum local ring, resp. Since $M(C) = \underline{H}^1_{(x_0,\ldots,x_3)}(S/I(C)) = H^1_{(x_0,\ldots,x_3)\cdot A}(A)$ we see that C is arithmetically Cohen-Macaulay if $M(C) = O$. Moreover we get from Proposition I.2.12 that C is arithmetically Buchsbaum if and only if $\mathfrak{m} \cdot H^1_\mathfrak{m}(A) = O$ where $\mathfrak{m} = (x_0, x_1, x_2, x_3) \cdot A$; that is, $M(C)$ is a finite-dimensional vector space over $A/\mathfrak{m}$. Also it is known (see, e.g. Peskine-Szpiro [3]) that liaison respects the Cohen-Macaulay property. From local algebra we get the following more general result:

Theorem 1.2. *Let R be a local Gorenstein ring of dimension $d \geq 1$ and with maximal ideal $\mathfrak{m}$. Suppose that the ideals $\mathfrak{a}, \mathfrak{b} \subset R$ are linked. Then we have:*

(a) *$R/\mathfrak{a}$ is a Buchsbaum ring if and only if $R/\mathfrak{b}$ is a Buchsbaum ring.*

(b) *In addition, assume that the local cohomology modules $H^i_\mathfrak{m}(R/\mathfrak{a})$ have finite length over R for all integers $i = 0, 1, \ldots, t-1$ where $t :=$ dim $R/\mathfrak{a} =$ dim $R/\mathfrak{b}$. Then we get*

$$H^{t-i}_\mathfrak{m}(R/\mathfrak{b}) = \mathrm{Hom}_R(H^i_\mathfrak{m}(R/\mathfrak{a}), E) \quad \text{for all } i = 1, \ldots, t-1,$$

where E is the injective hull of the residue field of R.

Remark. The focus of this paragraph on liaison among arithmetical Buchsbaum curves is on the statement of this theorem. In proving it the key is Theorem II.4.1 by applying homological algebra in derived categories. In the preface we did mention that we will describe a different and up to now unpublished proof of Theorem 1.2. Our proof is elementary by virtue of the fact that it uses only some basic concepts from Chapter I. Proposition 1.28 below immediately opens the way to our approach (see the proof of Corollary 1.30).

Proof of Theorem 1.2: First we prove the statement (a) of the theorem. For this we need a new invariant under liaison extending $M(C)$ of a curve C. Looking at it by way of commutative algebra this invariant is defined for an arbitrary ideal $\mathfrak{a}$ of a local Gorenstein ring R. We will use the theory of derived functors and categories, see Chapter 0, § 3. If $\mathfrak{a}$ and $\mathfrak{b}$ are linked by a complete intersection $\mathfrak{x} = (x_1, \ldots, x_g)$ then we have for the canonical or dualizing module $K_\mathfrak{a}$ of $R/\mathfrak{a}$:

$$K_\mathfrak{a} = \mathrm{Ext}^g_R(R/\mathfrak{a}, R) \cong \mathrm{Hom}_R(R/\mathfrak{a}, R/(\mathfrak{x})) = \mathfrak{b}/\mathfrak{x}\cdot R,$$

and

$$K_\mathfrak{b} = \mathrm{Ext}^g_R(R/\mathfrak{b}, R) \cong \mathrm{Hom}_R(R/\mathfrak{b}, R/\mathfrak{x}) = a/\mathfrak{x}\cdot R.$$

Hence we have the following exact sequence:

$$0 \to K_\mathfrak{a} \to R/\mathfrak{x}\cdot R \to R/\mathfrak{b} \to 0.$$

From this we get the sequence:

$$0 \to K_\mathfrak{a}[-g] \to \underline{R}\,\mathrm{Hom}_R(R/\mathfrak{x}\cdot R, R) \to R/\mathfrak{b}[-g] \to 0$$

by use of the isomorphism

$$R/\mathfrak{x}\cdot R \cong \underline{R}\,\mathrm{Hom}_R(R/\mathfrak{x}\cdot R, R)\,[g].$$

Let $I^\cdot_\mathfrak{a}$ be the dualizing complex of $R/\mathfrak{a}$, that is,

$$I^\cdot_\mathfrak{a} = \mathrm{Hom}_R(R/\mathfrak{a}, E^\cdot_R)$$

where $E^\cdot_R$ is a minimal injective resolution of R over itself (see Chapter 0, § 3). Note that in the derived category the complex $I^\cdot_\mathfrak{a}$ is isomorphic to $\underline{R}\,\mathrm{Hom}_R(R/\mathfrak{a}, R)$. Factoring out the first non-vanishing cohomology module $K_\mathfrak{a}$ of $I^\cdot_\mathfrak{a}$ we get a short exact sequence:

$$0 \to K_\mathfrak{a}[-g] \to I^\cdot_\mathfrak{a} \to J^\cdot_\mathfrak{a} \to 0$$

where $J^\cdot_\mathfrak{a}$ is up to a shift in grading the truncated dualizing complex of $R/\mathfrak{a}$. Using the canonical map

$$I^\cdot_\mathfrak{a} \cong \underline{R}\,\mathrm{Hom}_R(R/\mathfrak{a}, R) \to \underline{R}\,\mathrm{Hom}_R(R/\mathfrak{x}\cdot R, R)$$

induced by the canonical epimorphism

$$R/\mathfrak{x}\cdot R \to R/\mathfrak{a} \to 0$$

we obtain the following commutative diagram of complexes with exact rows:

$$\begin{array}{ccccccccc}
O & \to & K_\mathfrak{a}[-g] & \to & \underline{R}\,\mathrm{Hom}_R(R/\mathfrak{x}\cdot R, R) & \to & R/\mathfrak{b}[-g] & \to & 0 \\
 & & \| & & \uparrow & & \uparrow\varphi & & \\
0 & \to & K_\mathfrak{a}[-g] & \to & I^\cdot_\mathfrak{a} & \to & J^\cdot_\mathfrak{a} & \to & O
\end{array}$$

where φ is defined in the obvious manner.

By applying the derived functor $\underline{R}$ $\mathrm{Hom}_R(\ , R)$ and using the Local Duality Theorem we get the following commutative diagram:

$$\begin{array}{ccccccccc} O \to \underline{R}\,\mathrm{Hom}_R(R/\mathfrak{b}, R)\,[g] & \to & R/\mathfrak{x}\cdot R & \to & \underline{R}\,\mathrm{Hom}_R(K_{\mathfrak{a}}, R)\,[g] \to 0 \\ \downarrow{\scriptstyle\varphi} & & \downarrow & & \| \\ 0 \to \underline{R}\,\mathrm{Hom}_R(J^{\cdot}_{\mathfrak{a}}, R) & \to & R/\mathfrak{a} & \to & \underline{R}\,\mathrm{Hom}_R(K_{\mathfrak{a}}, R)\,[g] \to O \end{array}$$

Now we will show that φ induces a quasi-isomorphism between $J^{\cdot}_{\mathfrak{b}}[g]$ and $\underline{R}$ $\mathrm{Hom}_R[J^{\cdot}_{\mathfrak{a}}, R)$. Therefore we consider the induced homomorphism on the homology modules. Hence we get the commutative diagram with exact rows:

$$\begin{array}{ccccccccc} O \to \mathrm{Ext}^g_R(R/\mathfrak{b}, R) & \to & R/\mathfrak{x}\cdot R & \to & \mathrm{Ext}^g_R(K_{\mathfrak{a}}, R) & \to & \mathrm{Ext}^{g+1}_R(R/\mathfrak{b}, R) & \to 0 \\ \downarrow & & \downarrow & & \| & & \downarrow & \\ 0 \to \mathrm{Hom}_R(J^{\cdot}_{\mathfrak{a}}, R) & \to & R/\mathfrak{a} & \to & \mathrm{Ext}^g_R(K_{\mathfrak{a}}, R) & \to & \mathrm{Ext}^1_R(J^{\cdot}_{\mathfrak{a}}, R) & \to O \end{array}$$

and isomorphisms for $i \geq 1$

$$\begin{array}{ccc} \mathrm{Ext}^{g+i}_R(K_{\mathfrak{a}}, R) & \cong & \mathrm{Ext}^{g+i+1}_R(R/\mathfrak{b}, R) \\ \| & & \\ \mathrm{Ext}^{g+i}_R(K_{\mathfrak{a}}, R) & \cong & \mathrm{Ext}^{i+1}_R(J^{\cdot}_{\mathfrak{a}}, R) \end{array}$$

By using $\mathrm{Ext}^g_R(R/\mathfrak{b}, R) \cong \mathfrak{a}/\mathfrak{x}\cdot R$ and the surjectivity of the mapping $R/\mathfrak{x}\cdot R \to R/\mathfrak{a}$ we get from the diagram:

$$\mathrm{Hom}_R(J^{\cdot}_{\mathfrak{a}}, R) = 0 \quad \text{and therefore} \quad \mathrm{Ext}^{g+1}_R(R/\mathfrak{b}, R) \cong \mathrm{Ext}^1_R(J^{\cdot}_{\mathfrak{a}}, R).$$

Hence φ induces a quasi-isomorphism:

$$J^{\cdot}_{\mathfrak{b}}[g] \cong \underline{R}\,\mathrm{Hom}_R(J^{\cdot}_{\mathfrak{a}}, R).$$

Note that $J^{\cdot}_{\mathfrak{a}}$ is therefore a new invariant (up to duality and shifts in gradings) under liaison.

Assume now that $R/\mathfrak{a}$ is a Buchsbaum ring then $J^{\cdot}_{\mathfrak{a}}$ and also $\underline{R}$ $\mathrm{Hom}_R(J^{\cdot}_{\mathfrak{a}}, R)$ are quasi-isomorphic to complexes of k-vector spaces (see Theorem II.4.1). We apply again this statement and get that $R/\mathfrak{b}$ is a Buchsbaum ring since $J^{\cdot}_{\mathfrak{a}}$ is invariant under liaison. Replacing $\mathfrak{a}$ by $\mathfrak{b}$ we obtain the converse.

For the proof of Theorem 1.2 it remains to prove the statement about the local cohomology of $R/\mathfrak{a}$ and $R/\mathfrak{b}$. In order to do this we use an observation which is of some interest, in its own right. Before proving the statement (b) of Theorem 1.2 we therefore collect some facts that follow from this observation. We also obtain the interesting Corollary 1.4.

Lemma 1.3. *Let R be a local Gorenstein ring of dimension $d \geq 1$ and with maximal ideal $\mathfrak{m}$. Suppose that the ideals $\mathfrak{a}, \mathfrak{b} \subset R$ are linked. Then we have the exact sequence:*

$$0 \to R/\mathfrak{a} \to \mathrm{Ext}^g_R(K_{\mathfrak{a}}, R) \to \mathrm{Ext}^{g+1}_R(R/\mathfrak{b}, R) \to 0$$

and canonical isomorphisms for $i > g = d - \dim R/\mathfrak{a}$

$$\mathrm{Ext}^i_R(K_{\mathfrak{a}}, R) \cong \mathrm{Ext}^{i+1}_R(R/\mathfrak{b}, R),$$

that is, for the local cohomology modules we get:

$$0 \to H^{t-1}_{\mathfrak{m}}(R/\mathfrak{b}) \to H^t_{\mathfrak{m}}(K_{\mathfrak{a}}) \to \operatorname{Hom}_R(R/\mathfrak{a}, E) \to 0,$$

where $t = \dim R/\mathfrak{a}$ *and* E *is the injective hull of the residue field of* R, *and for* $i < t$

$$H^i_{\mathfrak{m}}(K_{\mathfrak{a}}) \cong H^{i-1}_{\mathfrak{m}}(R/\mathfrak{b}).$$

Proof: Using local duality it suffices to prove this lemma for the "Ext" cohomology. We have again the exact sequence (see the proof of statement (a) of Theorem 1.2)

$$0 \to K_{\mathfrak{a}} \to R/\mathfrak{x}R \to R/\mathfrak{b} \to 0.$$

Since $\operatorname{Ext}^i_R(R/\mathfrak{x} \cdot R, R) = 0$ for $i \neq g$ and $\operatorname{Ext}^g_R(R/\mathfrak{x} \cdot R, R) \cong R/\mathfrak{x} \cdot R$ we get the following sequence by applying the functor $\underline{R} \operatorname{Hom}_R(\ , R)$:

$$0 \to \operatorname{Ext}^g_R(R/\mathfrak{b}, R) \to R/\mathfrak{x} \cdot R \to \operatorname{Ext}^g_R(K_{\mathfrak{a}}, R) \to \operatorname{Ext}^{g+1}_R(R/\mathfrak{b}, R) \to 0,$$

and isomorphisms for $i > g$

$$\operatorname{Ext}^i_R(K_{\mathfrak{a}}, R) \cong \operatorname{Ext}^{i+1}_R(R/\mathfrak{b}, R).$$

By use of $\operatorname{Ext}^g_R(R/\mathfrak{b}, R) = \mathfrak{a}/\mathfrak{x} \cdot R$ we obtain our assertion, q.e.d.

Corollary 1.4. *Let* $\mathfrak{a}$, $\mathfrak{b}$ *and* R *be as in Lemma* 1.3. *For an integer* $r \geq 2$ *the following conditions are equivalent:*

(i) $R/\mathfrak{a}$ *satisfies the Serre condition* (S_r); *that is* $\operatorname{depth} R_{\mathfrak{p}} \geq \min(r, \dim R_{\mathfrak{p}})$ *for all* $\mathfrak{p} \in \operatorname{Spec} R$.

(ii) $H^i_{\mathfrak{m}}(R/\mathfrak{b}) = 0$ *for all integers* $t - r < i < t$ *where* $t := \dim R/\mathfrak{b}$.

Proof: The exact sequence

$$0 \to K_{\mathfrak{a}}[-g] \to I^{\cdot}_{\mathfrak{a}} \to J^{\cdot}_{\mathfrak{a}} \to 0$$

results in the following canonical isomorphisms

$$H^{t+1-i}_{\mathfrak{m}}(K_{\mathfrak{a}}) \cong H^{-i}_{\mathfrak{m}}(J^{\cdot}_{\mathfrak{a}})$$

for all integers $i > 1$ by using local duality and applying the functor $\underline{R}H^0_{\mathfrak{m}}(\ )$. Lemma 1.3 therefore provides the isomorphisms

$$H^{t-i}_{\mathfrak{m}}(R/\mathfrak{b}) \cong H^{-i}_{\mathfrak{m}}(J^{\cdot}_{\mathfrak{a}})$$

for all integers $1 \leq i < t$.

It follows now from Schenzel [1], Theorem 3.2.2(iii) that $H^{-i}_{\mathfrak{m}}(J^{\cdot}_{\mathfrak{a}}) = 0$ for all integers $0 \leq i < r$ if and only if (S_r) holds for $R/\mathfrak{a}$.

Proof of statement (b) *of Theorem* 1.2: First, we note that we again have for all $i > 1$:

$$H^{t+1-i}_{\mathfrak{m}}(K_{\mathfrak{a}}) \cong H^{-i}_{\mathfrak{m}}(J^{\cdot}_{\mathfrak{a}}).$$

From our assumption and the Local Duality Theorem (see 0.3.4) we get

$$H^{-i}_{\mathfrak{m}}(J^{\cdot}_{\mathfrak{a}}) = H^{-i}(J^{\cdot}_{\mathfrak{a}}) \quad \text{for all } i \leq t - 1.$$

Hence we have for $2 \leq i \leq t-1$

$$H_{\mathfrak{m}}^{t+1-i}(K_{\mathfrak{a}}) \cong \operatorname{Hom}_R\big(H_{\mathfrak{m}}^i(R/\mathfrak{a}), E\big)$$

since $H^{-i}(J_{\mathfrak{a}}^{\cdot}) \cong \operatorname{Hom}_R\big(H_{\mathfrak{m}}^i(R/\mathfrak{a}), E\big)$ for $i \neq t$. Lemma 1.3 then yields our assertion, q.e.d.

Remark 1.5. (i) Using the fact that $R/\mathfrak{a}$ is a Cohen-Macaulay ring if and only if $J_{\mathfrak{a}}^{\cdot} = 0$ we obtain another proof that liaison respects the Cohen-Macaulay property.

(ii) Let $C \subset \mathbf{P}^3$ be a curve with defining ideal $I(C)$. Applying local duality we get:

$$M(C) \cong \underline{H}_{\mathfrak{m}}^1\big(S/I(C)\big) \cong \underline{R}\operatorname{Hom}_S(J_{I(C)}^{\cdot}, S)[-2].$$

Since $J_{I(C)}^{\cdot}$ is invariant under liaison we recover the invariance of $M(C)$.

Examples 1.6. The simplest curve in $\mathbf{P}^3$ which belongs to the liaison class corresponding to a vector space of dimension 1 is the union of two skew lines in $\mathbf{P}^3$. Having this curve the specific data in the papers from the late 19th century render examples of arithmetical Buchsbaum curves with invariant $i(C) = 1$. For instance we get $C_4^0, C_6^3, C_7^4, C_8^8$ from the paper of M. Noether [1] or $C_{10}^{11}, C_{13}^{21}, C_{16}^{34}, C_{19}^{50}$ from the paper of K. Rohn [1], where C_d^g is a irreducible curve of degree d and of genus g. We have studied the curves C_4^0 and C_6^3 in our Introduction. We recall that we have found the resolutions of the curves C_6^3. We made the claim that the curves C_6^3 have the following resolution if C_6^3 is arithmetically Buchsbaum:

$$0 \to S(-6) \to S^4(-5) \to S(-2) \oplus S^3(-4) \to S \to S/I(C_6^3) \to 0.$$

In order to construct this resolution we prove the following more general result.

Lemma 1.7. *Let C be a curve in $\mathbf{P}^3$ which is linked to two skew lines in $\mathbf{P}^3$ by two hypersurfaces of degree f and g, resp. Then we get the following free resolution of $S/I(C)$ where $S := K[x_0, \ldots, x_3]$:*

$$0 \to S(-f-g) \to S^4(-f-g+1) \to S(-f) \oplus S(-g) \oplus S^2(-f-g+2)$$
$$\to S \to S/I(C) \to 0.$$

In proving this lemma we must use for a curve C the property of being ideally the intersection of d hypersurfaces. Therefore we give the following definition.

Definition 1.8. A projective variety $V \subset \mathbf{P}_K^n$ is said to be *ideally the intersection of d hypersurfaces* if there exists a surjection

$$\bigoplus_{i=1}^{d} \mathcal{O}_{\mathbf{P}^n}(-a_i) \to \mathcal{J}_V \to 0$$

for some integers a_i, $i = 1, \ldots, d$. (Note that $\mathcal{O}_{\mathbf{P}^n}$ is the structure sheaf of $\mathbf{P}^n$ and $\mathcal{J}_V$ is the ideal sheaf of V.) This definition is equivalent to saying that there are homogeneous elements $f_1, \ldots, f_d$ in the defining ideal $I(C)$ of V such that $I(V)/(f_1, \ldots, f_d)$ is a $K[x_0, \ldots, x_n]$-module of finite length, that is,

$$V = \operatorname{Proj}\big(K[x_0, \ldots, x_n]/(f_1, \ldots, f_d)\big),$$

or

$$I(V)|_{x_i=1} = (f_1, \ldots, f_d)|_{x_i=1} \quad \text{for } i = 0, 1, \ldots, n.$$

A reason for the importance of this concept is the fact that for curves in $\mathbf{P}^3$ the property of being the zero scheme of a section of a rank-two bundle on $\mathbf{P}^3$ is connected to that of being ideally the intersection of three surfaces. Furthermore, the same argument as used by A. P. Rao in [1], suitably refined, can be used to prove the following

Proposition 1.9. *If a curve C in $\mathbf{P}_K^3$ is ideally the intersection of the three surfaces with defining equations $f_1 = f_2 = f_3 = 0$ of degrees f, g, h respectively, then we have:*

$$M(C) \cong \operatorname{Hom}_K\big(I(C)/(f_1, f_2, f_3)\,(f + g + h - 4), K\big),$$

that is, we have up to duals and shifts in gradings

$$I(C)/(f_1, f_2, f_3) \cong \underline{H}^1_{(x_0,x_1,x_2,x_3)}\big(K[x_0, x_1, x_2, x_3]/I(C)\big).$$

Proof of Lemma 1.7: Let V denote the union of two skew lines in $\mathbf{P}^3$. Let C be any curve in $\mathbf{P}^3$ which is linked to V. Knowing the resolution of V it follows from A. P. Rao [1] (see Remark 1.10) that C is ideally the intersection of three hypersurfaces. Our Theorem 1.2(a) implies that C is arithmetically Buchsbaum with an invariant $i(C) = 1$. Therefore we get from Proposition 1.9 that the defining ideal $I(C)$ is generated by precisely four elements. From these facts we get the resolution of C following the argument from Rao [1], Theorem 2.5 (see our Lemma 1.14 below), q.e.d.

Remark 1.10. Another consequence of the property of being ideally the intersection of three surfaces is that the homogeneous ideal $I(C)$ of a curve C is generated by precisely three elements if and only if C is ideally the intersection of three surfaces and C is arithmetically Cohen-Macaulay (non-complete intersection). Continuing in this vein, on discovers that curves in $\mathbf{P}^3$ are much more complicated if their homogeneous ideals are generated by precisely four elements. Related to this we want to prove the following theorem.

Theorem 1.11. *Let $C \subset \mathbf{P}^3$ be any curve. The following conditions are equivalent:*

(i) *C is arithmetically Buchsbaum (non-Cohen-Macaulay) and C is ideally the intersection of three hypersurfaces, say $f_1 = f_2 = f_3 = 0$.*

(ii) *There are homogeneous elements f_1, f_2, f_3, f_4 which provide a minimal base for $I(C)$, and $x_i \cdot f_4 \in (f_1, f_2, f_3)$ for $i = 0, 1, 2, 3$.*

Before proving this theorem, we need two lemmas.

$\mu(M)$ will denote the number of elements in a minimal basis of the module M over a local ring.

Lemma 1.12. *Let $C \subset \mathbf{P}_K^3$ be an arithmetically Buchsbaum curve with invariant $i(C) \geq 1$. Then we get for the defining ideal $I(C)$ of C:*

$$\mu\big(I(C)\big) \geq 3 \cdot i(C) + 1.$$

Lemma 1.13. *In addition to the hypothesis of Lemma 1.12, suppose that C is ideally the intersection of three surfaces. Then we have*

$$\mu\big(I(C)\big) = 4 \quad \text{and} \quad i(C) = 1.$$

In order to prove these two lemmas we apply the following striking result of A. P. Rao [1], Theorem (2.5).

Lemma 1.14. *Let C be a curve in $\mathbf{P}^3$. Let $M(C)$ have a minimal free resolution*

$$0 \to L_4 \xrightarrow{\sigma_4} L_3 \to L_2 \to L_1 \to L_0 \to M(C) \to 0.$$

Then $S(C) := K[x_0, x_1, x_2, x_3]/I(C)$ has a minimal resolution of the form

$$0 \to L_4 \xrightarrow{(\sigma_4, 0)} L_3 \oplus \bigoplus_1^r S(-l_i) \to \bigoplus_1^m S(-e_i) \to S \to S/I(C) \to 0,$$

where $S = K[x_0, x_1, x_2, x_3]$.

Proof of Lemma 1.12: We set $i = i(C)$, then we have $M(C) = \bigoplus_i \underline{K}(p_j)$. Tensoring the Koszul resolution of K we get the following minimal free resolution of the invariant $M(C)$, say

$$0 \to L_4 \to L_3 \to L_2 \to L_1 \to L_0 \to M(C) \to 0$$

where L_j are free S-modules of rank $i \cdot \binom{4}{j}$, $0 \leq j \leq 4$. Applying Rao's Lemma 1.14 we conclude that a minimal resolution of $S/I(C)$ is:

$$0 \to L_4 \to L_3 \oplus \bigoplus_1^r S(-l_i) \to \bigoplus_1^m S(-e_i) \to S \to S/I(C) \to 0,$$

where $m = \mu(I(C))$. Therefore the Euler-Poincaré characteristic of the vector spaces of this resolution implies

$$\mu(I(C)) = 3 \cdot i + 1 + r \geq 3 \cdot i + 1,$$

q.e.d.

Proof of Lemma 1.13: It follows from Proposition 1.9 that $\mu(I(C)) \leq 3 + i(C)$. This statement and Lemma 1.12 provide the assertion, q.e.d.

Proof of Theorem 1.11: First we prove the implication (i) $\Rightarrow$ (ii): Our Lemma 1.13 shows that $i(C) = 1$ and $\mu(I(C)) = 4$. Therefore we get from Proposition 1.9 that the S-module $I(C)/(f_1, f_2, f_3)$ has precisely one generator which provides one basis element, say f_4, of $I(C)$. Thus $I(C) = (f_1, f_2, f_3, f_4)$. Furthermore, we have

$$(x_0, x_1, x_2, x_3) \cdot I(C)/(f_1, f_2, f_3) = O$$

since C is arithmetically Buchsbaum, that is, $x_i \cdot f_4 \in (f_1, f_2, f_3)$ for $i = 0, 1, 2, 3$.

(ii) $\Rightarrow$ (i) of Theorem 1.11: We have $(x_0, x_1, x_2, x_3) = (f_1, f_2, f_3) : f_4$. It follows that (f_1, f_2, f_3) equals (f_1, f_2, f_3, f_4) up to a primary component belonging to (x_0, x_1, x_2, x_3), that is, C is ideally the intersection of the three hypersurfaces $f_1 = f_2 = f_3 = 0$. By assumption we have $(x_0, x_1, x_2, x_3) \cdot I(C)/(f_1, f_2, f_3) = O$. Proposition 1.9 therefore shows that C is arithmetically Buchsbaum (non-Cohen-Macaulay). This concludes the proof of Theorem 1.11, q.e.d.

The proof implies the following remark.

Remark 1.15. Let $C \subset \mathbf{P}^3$ be a curve. Assume that C is ideally the intersection of three hypersurfaces, say $f_1 = f_2 = f_3 = 0$, and $\mu(I(C)) = 4$. If C is arithmetically Buchsbaum, then there are homogeneous elements f_1, f_2, f_3, f_4 such that $I(C) = (f_1, f_2, f_3, f_4)$ and the first module of syzygies of (f_1, f_2, f_3, f_4) has a linear second syzygy.

Moreover, it is not too difficult to show the following fact by using the method of the proof of Lemma 1.12 and Local Duality:

Let $C \subset \mathbf{P}^3$ be an arbitrary curve and let

$$0 \to \bigoplus_{j=1}^{\lambda} S(-d_j) \xrightarrow{\Phi} F_2 \to F_1 \to S \to S/I(C) \to 0$$

be a minimal free resolution of $S/I(C)$, where $\lambda, d_1, \ldots, d_\lambda$ are suitable integers. Then C is an arithmetically Buchsbaum curve if and only if Φ is obtained up to an isomorphism of F_2 by multiplication with the matrix

$$\mathfrak{A} = \begin{pmatrix} x_0 & x_1 & x_2 & x_3 & 0 & 0 & 0 & 0 & \ldots\ldots\ldots\ldots\ldots\ldots & 0 \ldots 0 \\ 0 & 0 & 0 & 0 & x_0 & x_1 & x_2 & x_3 & 0 \ldots\ldots\ldots\ldots\ldots\ldots & 0 \ldots 0 \\ \vdots & \vdots & \vdots & \vdots & \vdots & & & & - - - - - - & \vdots \quad \vdots \\ 0 & 0 & 0 & 0 & 0 & \ldots\ldots\ldots\ldots\ldots & & & 0\, x_0\, x_1\, x_2\, x_3 & 0 \ldots 0 \end{pmatrix}.$$

Hence $\mu(F_2) \geq 4\lambda$ and therefore we see again that $\mu(I(C)) \geq 3i + 1$ if C is an arithmetically Buchsbaum curve with invariant $i := i(C)\ (= \lambda)$. If C is an arithmetically Buchsbaum curve with $\mu(I(C)) = 4$ then $\lambda = 1$ and $\mathfrak{A} = (x_0, x_1, x_2, x_3)$, that is, the second module of syzygies is generated by exactly one linear syzygy (see also Amasaki [1] and [2]).

The following two examples shed some light on the case when the homogeneous ideal $I(C)$ of a curve C in $\mathbf{P}^3$ is generated by precisely four elements.

Examples 1.16.

1. Take the curve C given parametrically by (s^7, s^5t^2, st^6, t^7). It follows from Renschuch [2], p. 324, that $I(C) = (f_1, f_2, f_3, f_4)$ where

$$f_1 = x_0^2x_2 - x_1^3, \qquad f_2 = x_0x_3^2 - x_1x_2^2,$$
$$f_3 = x_0x_2^3 - x_1^2x_3^2, \qquad f_4 = x_1x_3^4 - x_2^5.$$

We get from our characterization of the monomial arithmetical Buchsbaum curves in $\mathbf{P}^3$ (see Theorem 1.20 below) that C is not arithmetically Buchsbaum. It is easy to see that C is ideally the intersection of the three surfaces $f_1 = f_2 = f_4 = 0$. Therefore this example shows that curves C in $\mathbf{P}^3$ with $\mu(I(C)) = 4$ and the property of being ideally the intersection of three surfaces are not, in general, arithmetically Buchsbaum.

2. Here we give an example of a curve in $\mathbf{P}^3$ such that $\mu(I(C)) = 4$ and C is arithmetically non-Cohen-Macaulay-Buchsbaum but C is not ideally the intersection of three surfaces. The explicit description of such a curve, we believe, is not clear. The possibility of such a construction was suggested to us by David Eisenbud. It makes use of the theory of finite free resolutions. Let C be the curve in $\mathbf{P}^3$ with defining ideal

$$I(C) = \left(x_1x_2x_3(x_0x_2 - x_1x_3),\ x_1^2x_3(x_3^2 - x_0^2),\ x_1^2x_2(x_0x_1 - x_2x_3),\ x_3(x_0^2 - x_1x_3)\,(x_0x_2 - x_3^2)\right)$$
$$=: (f_1, f_2, f_3, f_4).$$

An easy computation now shows that a minimal free resolution of $S/I(C)$ has the form:

$$0 \to S(-8) \xrightarrow[\mathfrak{A}]{} S^4(-7) \xrightarrow[\mathfrak{B}]{} S^4(-5) \xrightarrow[\mathfrak{C}]{} S \to S/I(C) \to 0$$

with

$$\mathfrak{A} = (x_0, x_1, x_2, x_3), \qquad \mathfrak{C} = (f_1, f_2, f_3, f_4),$$

$$\mathfrak{B} = \begin{pmatrix} x_0x_1 & x_2^2 & x_3^2 & 0 \\ x_3^2 - x_0^2 & 0 & 0 & x_1x_2 \\ 0 & x_1x_3 - x_0x_2 & 0 & -x_1^2 \\ -x_1x_3 & -x_1x_2 & -x_0x_3 & 0 \end{pmatrix}.$$

By applying Lemma 1.14 and by our Remark 1.15 we get from this explicit resolution that C is arithmetically Buchsbaum and C is not ideally the intersection of three surfaces.

Remark 1.17. The preceding investigations show that the property of being ideally the intersection of three surfaces is useful in studying curves in $\mathbf{P}^3$. The result of Abhyankar [1] in $\mathbf{A}^3$, namely that every non-singular curve in $\mathbf{A}^3$ is ideally the intersection of three hypersurfaces, is not true in general for curves in $\mathbf{P}^3$. It was first pointed out by C. Peskine and L. Szpiro [3] that there exists a non-singular curve in $\mathbf{P}^3$ which is not ideally the intersection of three hypersurfaces. A. P. Rao [1] proved that every liaison class contains curves that are not ideally the intersection of three surfaces, and that there is a liaison class that does not contain any curve that is ideally the intersection of three surfaces. With regard to this we get the following corollary.

Corollary 1.18. *Every liaison equivalence class corresponding to a finite-dimensional vector space of dimension > 1 does not contain any curve that is ideally the intersection of three surfaces, hence contains no curves coming from sections of rank two bundles.*

Interestingly, we can also read this result as follows:

Let $C \subset \mathbf{P}^3$ be a curve that is ideally the intersection of three surfaces. Assume that (x_0, x_1, x_2, x_3) annihilates the first local cohomology module $\underline{H}^1_{(x_0,\ldots,x_3)}\big(K[x_0, \ldots, x_3]/I(C)\big)$ then this vector space has a dimension ≤ 1.

Proof: Lemma 1.12 and Lemma 1.13.

With the use of Corollary 1.18 we can construct irreducible curves in $\mathbf{P}^3$ which are not ideally the intersection of three hypersurfaces by applying the methods of Evans-Griffith [1] (see our Chapter V, § 4). We mention here still another explicit description of curves in $\mathbf{P}^3$ which lie in a liaison equivalence class corresponding to a finite-dimensional vector space of dimension > 1. The procedure to obtain them is based on an old geometric idea. For instance, we claim that there exist irreducible curves C_{42}^{145} in $\mathbf{P}^3$ of degree 42 and genus 145 which belong to the liaison class corresponding to a vector space of dimension 3. In order to prove this assertion we need to mention that K. Rohn [2] studied the residual intersection of special classes of space curves lying on any surfaces of degree 4 already in 1897. From these specific data we get that the rational twisted cubic curve C_3^0, counted with multiplicity 6, is linked to irreducible curves C_{42}^{145} by two hypersurfaces of degree 4 and 15. Therefore our claim follows if we show that the local ring

$$A := K[x_0, \ldots, x_3]_{(x_0,\ldots,x_3)}/\mathfrak{p}^3$$

is a Buchsbaum ring with invariant $i(A) = 3$, where $\mathfrak{p}$ is the defining prime ideal of C_3^0. We will prove for this the following lemma.

Lemma 1.19. *Let* $\mathfrak{p} = (x_0x_2 - x_1^2, x_0x_3 - x_1x_2, x_1x_3 - x_2^2) =: (f_1, f_2, f_3)$ *be the defining prime ideal of* C_3^0. *Put* $R := K[x_0, \ldots, x_3]_{(x_0,\ldots,x_3)}$. *Let* $n \geq 1$ *be an integer. Then* $R/\mathfrak{p}^n$ *is a (local) Buchsbaum ring if and only if* $n \leq 3$ *in which case the invariants of the Buchsbaum rings are given by* $i(R/\mathfrak{p}) = 0$, $i(R/\mathfrak{p}^2) = 1$, *and* $i(R/\mathfrak{p}^3) = 3$.

Furthermore, $\mathfrak{p}^2$ *and* $\mathfrak{p}^3$ *are the defining ideals of* C_3^0, *counted with multiplicity* 3 *and* 6, *respectively.*

Proof: First, we will show that $R/\mathfrak{p}^3$ is a Buchsbaum ring with $i(R/\mathfrak{p}^3) = 3$. The same argument, suitably modified, can be used to prove the statements for $R/\mathfrak{p}^2$. Note that $R/\mathfrak{p}$ is a Cohen-Macaulay ring. We look first at the ideal $\mathfrak{p}^3 + (x_2^2)$. Let $U(\mathfrak{p}^3, x_2^2)$ denote the intersection of the primary ideals of R belonging to $\mathfrak{p}^3 + (x_2^2)$ of dimension 1. Using localizations of $\mathfrak{p}^3 + (x_2^2)$ at the prime ideals belonging to the radical of $\mathfrak{p}^3 + (x_2^2)$ it is not too difficult to show that

$$\begin{aligned} U(\mathfrak{p}^3, x_2^2) = {} & (x_0^3, x_1^3, x_2^2, x_0^2x_1, x_0x_1^2) \\ & \cap (x_2^2, x_0x_3^3 - 3x_1x_2x_3^2, x_0^2x_2x_3^2 - x_0x_1^2x_3^2 + 2x_1^3x_2x_3, x_0x_1x_3^3 - 2x_1^2x_2x_3^2, \\ & f_1^3, f_3^3, f_1^2f_2, f_1^2f_3, f_3^2f_1, f_3^2f_2, f_1f_2f_3). \end{aligned}$$

Some nasty, but routine calculations show

$$(x_0, x_1, x_2, x_3) \cdot U(\mathfrak{p}^3, x_2^2) \subseteq (\mathfrak{p}^3, x_2^2).$$

Therefore it follows from Proposition I.1.10 that $R/\mathfrak{p}^3 + (x_2^2)$ is a Buchsbaum ring. Lifting x_2^2 (see Proposition I.2.19) then provides that $R/\mathfrak{p}^3$ is a Buchsbaum ring since $\mathfrak{p}^n$ is a primary ideal for all $n \geq 2$ (this follows from Achilles-Schenzel-Vogel [1], Example (6.9)). Secondly, we calculate the invariant $i(R/\mathfrak{p}^3)$. We know that for every primary ideal $\mathfrak{q}$ generated by a system of parameters of $R/\mathfrak{p}^3$ we have:

$$i(R/\mathfrak{p}^3) = L(R/\mathfrak{p}^3 + \mathfrak{q}) - e_0(\mathfrak{q}, R/\mathfrak{p}^3)$$

(see Theorem I.1.12). Take $\mathfrak{q} = (x_0, x_3)$ in $R/\mathfrak{p}^3$. Bezout's theorem provides that $e_0(\mathfrak{q}, R/\mathfrak{p}^3) = 18$. It is not difficult to show that $L(R/(\mathfrak{p}^3, x_0, x_3)) = 21$, i.e. $i(R/\mathfrak{p}^3) = 3$. Thirdly, we show that $R/\mathfrak{p}^n$ is not a Buchsbaum ring if $n \geq 4$. Let $U(\mathfrak{p}^n, x_2)$ denote the intersection of the primary ideals of R belonging to $\mathfrak{p}^n + (x_2)$ of dimension 1. Using localization we get

$$U(\mathfrak{p}^n, x_2) = (x_2, (x_0, x_1)^n) \cap (x_2, (x_1^2, x_3)^n)$$

and

$$(\mathfrak{p}^n, x_2) = (x_2, (x_1^2, x_0x_3, x_1x_3)^n).$$

Let $k \geq 2$ be an integer and put $n = 2k$. For $f := x_1^{2k}x_3^k \in U(\mathfrak{p}^n, x_2)$ we see that $x_1f \notin (\mathfrak{p}^n, x_2)$. Let $n = 2k + 1$. Choosing $g := x_1^{2k+2}x_3^k \in U(\mathfrak{p}^n, x_2)$ we have $x_1g \notin (\mathfrak{p}^n, x_2)$. We get from Proposition I.1.10 and Corollary I.1.11 that $R/\mathfrak{p}^n$ is not a Buchsbaum ring for $n \geq 4$, q.e.d.

Finally, we want to give an illustration of some of our investigations by studying the class of monomial space curves in $\mathbf{P}_K^3$, that is, curves $C \subset \mathbf{P}_K^3$ given parametrically by

$$(s^d, s^bt^{d-b}, s^at^{d-a}, t^d)$$

where $d > b > a \geq 1$ are integers with g.c.d. $(d, b, a) = 1$.

We set $S := K[x_0, x_1, x_2, x_3]$, and let $I(C) \subset S$ be the defining ideal of C. We set $R := S/I(C)$. Let $\mu(M)$ denote the number of elements in a minimal basis of the module M over a local ring. We have the following theorem:

Theorem 1.20. *Let C be a monomial curve in $\mathbf{P}^3_K$ over an algebraically closed field K. Then the following conditions are equivalent:*

(a) *C is an arithmetically non-Cohen-Macaulay Buchsbaum curve.*

(b) *$\mu(I(C)) = 4$ and C lies on a quadric.*

(c) $I(C) = (x_0x_3 - x_1x_2, x_0^2x_2^{2n-1} - x_1^{2n+1}, x_0x_2^{2n} - x_1^{2n}x_3, x_2^{2n+1} - x_1^{2n-1}x_3^2)$ *for an integer $n \geq 1$.*

(d) *C is given parametrically by $(s^{4n}, s^{2n+1}t^{2n-1}, s^{2n-1}t^{2n+1}, t^{4n})$ for an integer $n \geq 1$.*

(e) *There is an integer $n \geq 1$ such that a minimal finite resolution of $S/I(C)$ over S has the following form:*

$$0 \to S(-2n-3) \to S^4(-2n-2) \to S(-2) \oplus S^3(-2n-1) \to S \to S/I(C) \to 0.$$

(f) $\bigoplus_v H^1(\mathbf{P}^3, \mathcal{J}_C(v)) \cong K(-2n+1)$ *for an integer $n \geq 1$.*

(g) *C and the union of the 2 skew lines $x_0 = x_1 = 0 \cup x_2 = x_3 = 0$ are linked.*

We need some preliminary results for the proof of Theorem 1.20.

Remark 1.21. Assume C is an arbitrary monomial curve. Then the forms of the defining ideal $I(C)$ are of the following types:

1. There is exactly one form of the type

$$F_1 = x_1^\alpha - x_0^{\alpha-(\alpha_2+\alpha_3)}x_2^{\alpha_2}x_3^{\alpha_3}, \quad \text{resp.} \quad F_2 = x_2^\beta - x_0^{\beta_0}x_1^{\beta_1}x_3^{\beta-(\beta_0+\beta_1)}.$$

2. There are forms of the type

$$G = x_0^{\gamma_0}x_3^{\gamma_3} - x_1^{\gamma_1}x_2^{\gamma_2}, \quad \min(\gamma_i) > 0, \quad \gamma_1 < \alpha \quad \text{and} \quad \gamma_2 < \beta.$$

3. There are forms of type

$$H = x_0^{\delta_0}x_2^{\delta_2} - x_1^{\delta_1}x_3^{\delta_3}, \quad \min(\delta_i) > 0, \quad \delta_1 < \alpha \quad \text{and} \quad \delta_2 < \beta.$$

This follows easily by studying the parametric representation of C, see also Bresinsky-Renschuch [1], Lemma 2. We note that in Bresinsky-Renschuch [1] there is given an algorithm for computing a base of the defining ideal of an arbitrary monomial curve in $\mathbf{P}^3$.

Next we want to relate the perfectness of C to the number of generators $\mu(I(C))$.

Lemma 1.22. *Let C be a monomial curve in $\mathbf{P}^3_K$. Then C is arithmetically Cohen-Macaulay if and only if $\mu(I(C)) \leq 3$.*

Proof: First of all let us assume that C is arithmetically Cohen-Macaulay. We consider the homogeneous coordinate ring of C. This ring is a two-dimensional Cohen-Macaulay ring. In particular we have $R = R_{(x_0)} \cap R_{(x_3)}$, since $\{x_0, x_3\}$ is a homogeneous system of parameters of R and $R_{(x_0)} \cap R_{(x_3)}/R \cong \underline{H}^1_\mathfrak{m}(R) = 0$. Hence it follows that there is no quotient $\in R_{(x_0)} \cap R_{(x_3)}$ which is not contained in R. In other words, there does not exist a form of type H contained in $I(C)$. Now assume that there are at least two different forms $G = x_0^p x_3^q - x_1^m x_2^n$ and $G' = x_0^{p'} x_3^{q'} - x_1^{m'} x_2^{n'}$ of the second type. Without loss of generality we may assume $p > p'$ and $q < q'$. By multiplying with $x_3^{q'-q}$ resp. with $x_0^{p-p'}$ we get

$$x_1^m x_2^n x_3^{q'-q} - x_0^{p-p'} x_1^{m'} x_2^{n'} \in I(C).$$

Since $I(C)$ is a prime ideal it follows that there exists a form of type H. But this contradicts the Cohen-Macaulay property of R.

For the converse we remark that C is in particular ideally the intersection of three hypersurfaces. Applying Proposition 1.9 we see that $M(C) = 0$, that is, C is arithmetically Cohen-Macaulay, q.e.d.

According to Lemma 1.22 one may hope that there are similar results relating the Buchsbaum property to increased numbers of generators. The only result in this direction which is known to us is the following Corollary.

Corollary 1.23. *Let C be a monomial curve. Assume that $\mu(I(C)) = 4$. Then the minimal number of generators of $\underline{H}_{\mathfrak{m}}^1(R)$ is equal to 1, i.e. $\mu(\underline{H}_{\mathfrak{m}}^1(R)) = 1$. The converse does not hold.*

Proof: First of all two forms of $I(C)$ are defined to be F_1 and F_2, see Remark 1.21. Now assume there exist two forms of type G. Then the discussion given in the proof of Lemma 1.22 yields the existence of a form $H \in I(C)$ which contradicts $\mu(I(C)) = 4$. A similar discussion shows that the existence of two forms of type H also results in contradiction. Therefore there are exactly one form G and exactly one form H. Since $R_{(x_0)} \cap R_{(x_3)}/R \simeq \underline{H}_{\mathfrak{m}}^1(R)$ it now follows that $\underline{H}_{\mathfrak{m}}^1(R)$ has exactly one generator. For an example which disproves the converse statement see Example 1.27(ii) below, q.e.d.

Lemma 1.24. *If C is arithmetically non-Cohen-Macaulay and C lies on a quadric then the defining equation of this quadric is given by $x_0x_3 - x_1x_2 = 0$.*

Proof: Using the above-mentioned Remark 1.21 and the parametric representation of C we get the following possibilities for defining equations of the quadric:

(i) $x_0x_2 - x_1^2 = 0$, (ii) $x_0x_3 - x_1^2 = 0$, (iii) $x_0x_3 - x_2^2 = 0$,

(iv) $x_1x_3 - x_2^2 = 0$, (v) $x_0x_3 - x_1x_2 = 0$.

We will show that C is arithmetically Cohen-Macaulay if the defining equation of the quadric has the form (i), (ii), (iii) or (iv). There are several possible ways to prove this assertion. For instance, we get from Bresinsky-Renschuch [1] that $\mu(I(C)) = 2$ if we have cases (ii) and (iii). Having (i) or (iv) we obtain $\mu(I(C)) \leq 3$. Therefore Lemma 1.22 says that C is arithmetically Cohen-Macaulay. Another possibility is to apply the same methods used for proving Lemma 1.22 and the implication (a) $\Rightarrow$ (c) of Theorem 1.20. From this we get $\underline{H}_{\mathfrak{m}}^1(S/I(C)) = 0$ if we have the cases (i), ..., (iv), q.e.d.

The proof of Lemma 1.24 implies the following interesting fact.

Corollary 1.25. *Let C be a monomial curve in $\mathbf{P}^3$. If C lies on a quadric cone then C is arithmetically Cohen-Macaulay.*

Proof of Theorem 1.20: We will show the following implications:

$$\begin{array}{ccccccccccccc} (a) & \Longrightarrow & (c) & \Longrightarrow & (d) & \Longrightarrow & (g) & \Longrightarrow & (e) & \Longrightarrow & (b) & \Longrightarrow & (a) \\ & \nwarrow & \Downarrow & & & & & & & & & & \\ & & (f) & & & & & & & & & & \end{array}$$

The difficult part is to prove the implications (a) $\Rightarrow$ (c) and (b) $\Rightarrow$ (a). The other implications are more or less clear.

(a) $\Rightarrow$ (c): Since C is not arithmetically Cohen-Macaulay there exists at least one form $H = x_0^{\delta_0}x_2^{\delta_2} - x_1^{\delta_1}x_3^{\delta_3}$ contained in a minimal generating set of $I(C)$, see the discussion in the proof of Lemma 1.22. Therefore the following element is a non-zero element of the first local cohomology module:

$$\xi := x_1^{\delta_1}/x_0^{\delta_0} = x_2^{\delta_2}/x_3^{\delta_2} \in R_{(x_0)} \cap R_{(x_3)}/R \cong \underline{H}_{\mathfrak{m}}^1(R).$$

Since R is a Buchsbaum ring $H_{\mathfrak{m}}^1(R)$ is annihilated by $\mathfrak{m}$, i.e., $x_i \cdot \xi \in R$ for $i = 0, 1, 2, 3$. In particular it follows that

$$H = x_0x_2^{\gamma} - x_1^{\gamma}x_3 \quad \text{with } \gamma \geq 1.$$

We choose γ minimal with respect to the property $H \in I(C)$. Then it follows easily

$$F_1 = x_1^{\gamma+1} - x_0^{\gamma+1-(\alpha_2+\alpha_3)}x_2^{\alpha_2}x_3^{\alpha_3} \quad \text{with } \gamma \geq \alpha_2 + \alpha_3,$$

$$F_2 = x_2^{\gamma+1} - x_0^{\beta_0}x_1^{\beta_1}x_3^{\gamma+1-(\beta_0+\beta_1)} \quad \text{with } \gamma \geq \beta_0 + \beta_1.$$

Next we consider the quotients

$$x_1^{\gamma+1}/x_0 = x_0^{\gamma-(\alpha_2+\alpha_3)}x_2^{\alpha_2}x_3^{\alpha_3} = x_1x_2^{\gamma}/x_3,$$

$$x_2^{\gamma+1}/x_3 = x_0^{\beta_0}x_1^{\beta_1}x_3^{\gamma-(\beta_0+\beta_1)} = x_1^{\gamma}x_2/x_0.$$

This gives two forms

$$x_0^{\gamma-(\alpha_2+\alpha_3)}x_3^{\alpha_3+1} = x_1x_2^{\gamma-\alpha_2}, \qquad x_0^{\beta_0+1}x^{\gamma-(\beta_0+\beta_1)} = x_1^{\gamma-\beta_1}x_2 \tag{$*$}$$

contained in $I(C)$. By comparing both x_1, x_2-terms we get a form

$$x_0^{\beta_0+1}x_2^{\gamma-\alpha_2-1}x_3^{\gamma-(\beta_0+\beta_1)} - x_0^{\gamma-(\alpha_2+\alpha_3)}x_1^{\gamma-\beta_1-1}x_3^{\alpha_3+1}$$

contained in the prime ideal $I(C)$. By factoring out irrelevant terms we get a form of type H by an easy discussion. By the Buchsbaum property of R we see, as above, that it is of the following form:

$$H' = x_0x_2^{\gamma'} - x_1^{\gamma'}x_3 \quad \text{with } \gamma' < \gamma.$$

But this contradicts the minimality of γ. Therefore both forms in $(*)$ have to be equal. It follows $\gamma - \beta_1 = \gamma - \alpha_2 = 1$ and $\beta_0 = \alpha_3 = 0$, i.e. the form is given by $x_0x_3 - x_1x_2$. Because $x_0x_3 - x_1x_2 \in I(C)$ it now follows that there is no other form of type G. On the other hand, there is also no other form of the type H contained in $I(C)$ because it would be equal to $x_0x_2^{\alpha} - x_1^{\alpha}x_3$ with $\alpha > \gamma$ by virtue of the Buchsbaum property of R. But $\alpha > \gamma$ is not possible because this reduces modulo F_1 resp. F_2. Therefore we get the following forms:

$$F_1 = x_1^{\gamma+1} - x_0^2 x_2^{\gamma-1}, \quad F_2 = x_2^{\gamma+1} - x_1^{\gamma-1} x_3^2,$$
$$G = x_0x_3 - x_1x_2, \quad H = x_0x_2^{\gamma} - x_1^{\gamma}x_3$$

contained in $I(C)$. Since F_1 resp. F_2 is uniquely determined we get $\gamma = 2n$. This proves (a) $\Rightarrow$ (c).

(c) $\Rightarrow$ (d), trivial.

(d) $\Rightarrow$ (g): Applying Bezout's theorem we get that

$$I(C) \cap (x_0, x_1) \cap (x_2, x_3) = (x_0x_3 - x_1x_2, x_0x_2^{2n} - x_1^{2n}x_3) \quad \text{for all } n \geq 1$$

since the degree of C is given by $4n$, that is C and the skew lines $x_0 = x_1 = 0$ and $x_2 = x_3 = 0$ are linked.

(g) $\Rightarrow$ (e): From (g) follows (a) by applying Theorem 1.2(a). Hence from (d) we obtain that C and the union of the two skew lines are linked by two hypersurfaces of degree 2 and $2n + 1$. Therefore Lemma 1.7 implies assertion (e).

(e) $\Rightarrow$ (b): The assertion results immediately from the term $S(-2) \oplus S^3(-2n-1)$.

(b) $\Rightarrow$ (a): Lemma 1.22 and Lemma 1.24 show that C lies on the quadric with defining equation $x_0x_3 - x_1x_2 = 0$. Hence our Example 1.27(iii) yields the elements of a minimal base for $I(C)$. From these elements (a) follows from Theorem 1.11.

(c) $\Rightarrow$ (f): It is easily seen that C is ideally the intersection of the three hypersurfaces

$$x_0x_3 - x_1x_2 = x_0^2x_2^{2n-1} - x_1^{2n+1} = x_2^{2n+1} - x_1^{2n-1}x_3^2 = 0.$$

Hence Proposition 1.9 proves claim (f) for the Rao invariant of C.

(f) $\Rightarrow$ (a) is trivial since $\underline{H}^1_{\mathfrak{m}}(S/I(C))$ is a vector space of dimension 1.

This concludes the proof of Theorem 1.20, q.e.d.

Remark 1.26. We note that S. Goto in [2] proves a Buchsbaum criterion for an arbitrary affine semigroup ring using the main results of Goto-Watanabe [2]. In fact, S. Goto [2] computes the first local cohomology module $\underline{H}^1_{\mathfrak{m}}$ of such a Buchsbaum ring in terms of the underlying semigroup. But these methods do not provide a proof of Theorem 1.20.

We want to conlude by studying three examples and by adding some remarks.

Examples 1.27.

(i) The following example shows the usefulness of Theorem 1.2 in projective space $\mathbf{P}^n$ with $n \geq 4$. Take the surface F in $\mathbf{P}^5$ with defining ideal

$$\mathfrak{a} = (x_0, x_1, x_2) \cap (x_1, x_2, x_3) \cap (x_2, x_3, x_4) \cap (x_3, x_4, x_5) \cap (x_4, x_5, x_0) \cap (x_5, x_0, x_1).$$

By using Theorem 1.2 we will prove that F is arithmetically Buchsbaum which also follows from Proposition V.2.7 and Example V.2.9(ii). To accomplish this, let G be the surface in $\mathbf{P}^5$ with defining ideal

$$\mathfrak{b} = (x_0, x_2, x_4) \cap (x_1, x_3, x_5).$$

Then F is linked to G since $\mathfrak{a} \cap \mathfrak{b} = (x_0x_3, x_1x_4, x_2x_5)$. If follows immediately from Proposition I.2.25 that G is arithmetically Buchsbaum. Theorem 1.2 then finishes the proof of our claim.

(ii) In order to prove Corollary 1.23 of Lemma 1.22 we investigate the following example.

Take the curve C given parametrically by (s^8, s^7t, s^3t^5, t^8). It follows immediately from Renschuch [2] that

$$I(C) = (x_1^5 - x_0^4x_2, x_2^5 - x_0x_1x_3^3, x_1^3x_2 - x_0^3x_3, x_1x_2^3 - x_0^2x_3^2, x_0x_2^2 - x_1^2x_3),$$

that is, $I(C)$ is generated by precisely five elements. We get that

$$\underline{H}^1_{\mathfrak{m}}(R) \cong (x_1^2/x_0, 1)\, R/R \qquad \text{where } R = S/I(C).$$

Hence $\underline{H}^1_{\mathfrak{m}}(R)$ is generated by precisely one element. The computation of $\underline{H}^1_{\mathfrak{m}}(R)$ is accomplished by applying the same methods used in the proof of Theorem 1.20(a) $\Rightarrow$ (c).

(iii) In connection with Theorem 1.20(a) $\Rightarrow$ (c) and Lemma 1.24 we now consider the monomial curves lying on a non-singular quadric. We collect some properties of these curves. Their proof follows easily by applying methods of Renschuch [2] or of the proof of Theorem 1.20.

Let C be a curve given parametrically by

$$(s^d, s^bt^a, s^at^b, t^d), \qquad b > a.$$

Then we have the following properties:

(a) The defining ideal $I(C)$ has the following minimal basis:

$$I(C) = (x_0x_3 - x_1x_2, F_0, F_1, \ldots, F_{b-a})$$

where

$$F_i = x_0^{b-a-i}x_2^{a+i} - x_1^{b-i}x_3^i, \qquad 0 \leq i \leq b - a.$$

(b) C is ideally the intersection of the three hypersurfaces defined by $x_1^b - x_0^{b-a}x_2^a = 0$, $x_2^b - x_1^ax_3^{b-a} = 0$, and $x_0x_3 - x_1x_2 = 0$.

(c) $\dim_K[\underline{H}^1_{\mathfrak{m}}(R)]_i = \begin{cases} (i - a + 1)(b - 1 - i) & \text{for } a \leq i \leq b - 2, \\ 0 & \text{otherwise.} \end{cases}$

(d) $\mu\big(\underline{H}^1_{\mathfrak{m}}(R)\big) = b - a - 1.$

(e) $l_R\big(\underline{H}^1_{\mathfrak{m}}(R)\big) = \binom{b - a + 1}{3}.$

At the end of this paragraph we finally come to the elementary proof of Theorem 1.2 as remarked after the statement of Theorem 1.2.

Let M be a Noetherian A-module. As in the graded case (see Corollary 0.4.16) we can define a *Noetherian A-module* $M^\times$ in the following way (A a local ring):

Take a free A-module F and an epimorphism $\pi: F \to M$ such that $\operatorname{Ker} \pi \subseteq \mathfrak{m} \cdot F$. Set

$$M^\times := \operatorname{Coker}\big(\operatorname{Hom}(M, A) \to \operatorname{Hom}(F, A)\big) = \operatorname{Coker} \operatorname{Hom}(\pi, A).$$

We note that F and hence $M^\times$ are uniquely determined by M (up to an isomorphism). If M is free itself, $M^\times = 0$. If $\operatorname{Supp} M \cap \operatorname{Ass} A = \emptyset$ (which means that $\operatorname{Hom}(M, A) = 0$) we set $M^\times = \operatorname{Hom}(F, A) \cong F$.

Proposition 1.28. *Let A be a local Gorenstein ring with $d := \dim A \geq 2$. Let E denote the injective hull of the residue field $k = A/m$. Furthermore, let M denote a non-free Noetherian A-module. Then:*

(i) *If M is an equidimensional locally Cohen-Macaulay module then the same is true for $M^\times$. If* $\dim M = d$

$$H^i_{\mathfrak{m}}(M^\times) \cong \begin{cases} \operatorname{Hom}\big(H^{d-i}_{\mathfrak{m}}(M), E\big) & \text{for all } i = 1, \ldots, d-1, \\ 0 & \text{for } i = 0. \end{cases}$$

(ii) *If M is a d-dimensional Buchsbaum module with* $\operatorname{depth} M > 0$ *then $M^\times$ is also a (d-dimensional) Buchsbaum module with $I(M^\times) = I(M)$.*

Proof: (i) If $\dim M < d$ then $M^\times$ is a free A-module (since $\operatorname{Hom}(M, A) = 0$) and there is nothing to prove. If $\dim M = d$ (i) follows by the same argument as in the proof of the graded version of this statement, see Corollary 0.4.16.

(ii) Take a free A-module F and an epimorphism $\pi: F \to M$ with $\operatorname{Ker} \pi \subseteq \mathfrak{m} \cdot F$. Set $U := \operatorname{Ker} \pi$. Since M is a non-free A-module, $U \neq 0$ and since $\operatorname{Ass} U \subseteq \operatorname{Ass} F = \operatorname{Ass} A$ we obtain that $\dim U = d$. Let $\mathfrak{p} \in \operatorname{Spec} A$ be minimal. Then $A_\mathfrak{p}$ is an injective $A_\mathfrak{p}$-module and therefore $(M^\times)_\mathfrak{p} = \operatorname{Hom}_{A_\mathfrak{p}}(U_\mathfrak{p}, A_\mathfrak{p})$ ($\operatorname{Hom}_{A_\mathfrak{p}}(\ , A_\mathfrak{p})$ is an exact functor). But $\operatorname{Hom}_{A_\mathfrak{p}}(U_\mathfrak{p}, A_\mathfrak{p}) \neq 0$ for all $\mathfrak{p} \in \operatorname{Ass} U \neq \emptyset$, i.e. we have $\dim M^\times = d$. An easy computation shows that $\operatorname{depth} \operatorname{Hom}(M, A) \geq 2$, i.e. we obtain from the exact sequence $0 \to \operatorname{Hom}(M, A) \to \operatorname{Hom}(F, A) \to M^\times \to 0$:

$$H^0_{\mathfrak{m}}(M^\times) \cong H^1_{\mathfrak{m}}\big(\operatorname{Hom}(M, A)\big) = 0.$$

Therefore $\operatorname{depth} M^\times \geq 1$. Also, there is an exact sequence $0 \to M^\times \to \operatorname{Hom}(U, A) \to (U, A) \to \operatorname{Ext}^1(M, A) \to 0$. Since by Corollary 0.3.5 $\operatorname{Hom}\big(\operatorname{Ext}^1(M, A), E\big) = H^{d-1}_{\mathfrak{m}}(M)$ we see that $l_A\big(\operatorname{Ext}^1(M, A)\big) < \infty$, especially is $\mathfrak{m} \cdot \operatorname{Ext}^1(M, A) = 0$. Therefore we get (since $\operatorname{depth} \operatorname{Hom}(U, A) \geq 2$):

$$\operatorname{Ext}^1(M, A) = H^0_{\mathfrak{m}}\big(\operatorname{Ext}^1(M, A)\big) \cong H^1_{\mathfrak{m}}(M^\times).$$

If now $d = 2$, $M^\times$ is a Buchsbaum module by Proposition I.2.12. We next use induction on $d \geq 2$. If $d = 2$ the proof is already complete. Let $d \geq 3$.

Take now an element $x \in \mathfrak{m}$ with $\dim M^\times/x \cdot M^\times = d - 1$. We will show that $M^\times/x \cdot M^\times : \langle \mathfrak{m} \rangle$ is a Buchsbaum module. Then by Proposition I.2.23 $M^\times$ is a Buchsbaum module. Let $\operatorname{Ass} A := \{\mathfrak{p}_1, \ldots, \mathfrak{p}_r\}$ and assume $x \notin \mathfrak{p}_1 \cup \ldots \cup \mathfrak{p}_s$, $x \in \mathfrak{p}_{s+1} \cap \ldots \cap \mathfrak{p}_r$ where $1 \leq s \leq r$. If $s < r$, $\operatorname{Ann} M^\times \not\subseteq \mathfrak{p}_{s+1} \cup \ldots \cup \mathfrak{p}_r$ (since $\dim A/xA + \operatorname{Ann} M^\times = \dim M^\times/xM^\times = d - 1$). Take an element $y \in \mathfrak{p}_1 \cap \ldots \cap \mathfrak{p}_s \cap \operatorname{Ann} M^\times$ with $y \notin \mathfrak{p}_{s+1} \cup \ldots \cup \mathfrak{p}_r$ and set $z := x + y$. Then $zM^\times = xM^\times$ and $z \notin \mathfrak{p}_1 \cup \ldots \cup \mathfrak{p}_r$, i.e. $0 :_A z = 0$. Therefore we can assume without loss of generality that $0 :_A x = 0$. Using the exact sequence $0 \to A \xrightarrow{x} A \to A/xA \to 0$ and the standard fact that $\operatorname{Hom}_{A/xA}(N/xN, A/xA)$

$= \operatorname{Hom}_A(N, A/xA)$ for any A-module N we obtain a commutative diagram with exact rows and columns

$$\begin{array}{ccccccc}
 & 0 & & 0 & & & \\
 & \downarrow & & \downarrow & & & \\
0 \to & \operatorname{Hom}(M, A) & \to & \operatorname{Hom}(F, A) & & & \\
 & \downarrow & & \downarrow & & & \\
0 \to & \operatorname{Hom}(M, A) & \to & \operatorname{Hom}(F, A) & \to & M^\times & \to 0 \\
 & \downarrow & & \downarrow & & \downarrow & \\
0 \to & \operatorname{Hom}_{A/xA}(M/xM, A/xA) & \to & \operatorname{Hom}_{A/xA}(F/xF, A/xA) & \to & (M/xM)^\times & \to 0 \\
 & \downarrow & & \downarrow & & & \\
 & \operatorname{Ext}^1(M, A) & & 0 & & & \\
 & \downarrow & & & & & \\
 & 0 & & & & &
\end{array}$$

From this we obtain an epimorphism $\mu\colon M^\times \to (M/xM)^\times$ ($(M/xM)^\times$ is defined as an A/xA-module but is now considered as an A-module). Let $C := \operatorname{Ker} \mu$. Then the above diagram gives exact sequences

(1) $\quad 0 \to M^\times \to C \to \operatorname{Ext}^1(M, A) \to 0$ (snake-lemma),

(2) $\quad 0 \to C \to M^\times \xrightarrow{\mu} (M/xM)^\times \to 0.$

From (1) and (2) we obtain the following exact sequence since the composition $M^\times \to C \to M^\times$ from (1) and (2) is just the map $M^\times \xrightarrow{x} M^\times$):

(3) $\quad 0 \to \operatorname{Ext}^1(M, A) \to M^\times/xM^\times \to (M/xM)^\times \to 0.$

Since $d \geq 3$ we have $\operatorname{depth}(M/xM)^\times \geq 1$, i.e.

$$\operatorname{Ext}^1(M, A) = H^0_{\mathfrak{m}}\big(\operatorname{Ext}^1(M, A)\big) \cong H^0_{\mathfrak{m}}(M^\times/xM^\times) = xM^\times : \langle \mathfrak{m} \rangle / xM^\times .$$

Therefore (3) gives rise to an isomorphism

$$M^\times/xM^\times : \langle \mathfrak{m} \rangle \cong (M/xM)^\times .$$

By the induction hypothesis (M/xM is a Buchsbaum module over A and hence over A/xA of dimension $d - 1 = \dim A/xA$) we know that $(M/xM)^\times$ is a Buchsbaum module over A/xA and hence over A. Therefore $M^\times/xM^\times : \langle \mathfrak{m} \rangle$ is a Buchsbaum module. It is $I(M^\times) = I(M)$ by (i), q.e.d.

Corollary 1.29. *Let A be a local Gorenstein ring with $d := \dim A \geq 2$. Let M be a d-dimensional Noetherian A-module with $\dim A/\mathfrak{p} = d$ for all $\mathfrak{p} \in \operatorname{Ass} M$ which has no free direct summand. Then M is a Buchsbaum module if and only if $M^\times$ is a Buchsbaum module.*

Proof: It is easy to show that by the assumptions made there is an isomorphism $M \xrightarrow{\sim} (M^\times)^\times$. The statement follows from the above proposition, q.e.d.

The following corollary establishes Theorem 1.2.

Corollary 1.30. *Let A be a local Gorenstein ring of dimension $d \geq 2$. Let $\mathfrak{a}, \mathfrak{b} \subset A$ be unmixed ideals without common components such that $\mathfrak{a} \cap \mathfrak{b} = 0$, $\mathfrak{a} \not\subseteq \mathfrak{b}$, $\mathfrak{b} \not\subseteq \mathfrak{a}$. Then $A/\mathfrak{a}$ is a Buchsbaum ring if and only if $A/\mathfrak{b}$ is a Buchsbaum ring. In this case $I(A/\mathfrak{a}) = I(A/\mathfrak{b})$.*

Proof: Set $M := A/\mathfrak{a}$. Then $M^\times = A/\mathfrak{b}$ and $M = (M^\times)^\times = A/\mathfrak{a}$, q.e.d.

§ 2. On liaison addition and applications

In this paragraph we will solve the liaison addition problem by following P. Schwartau's [1] work. We are able here to use this liaison addition to prove new theorems about liaison realization and self-linkage. Since the liaison procedure is explicit, we are able to solve a long-standing problem of liaison: to find explicit curve in $\mathbf{P}^3_K$ corresponding to the liaison invariant $\bigoplus_n \underline{K}$ for all $n \geq 1$; that is we will construct arithmetical Buchsbaum curves C in $\mathbf{P}^3$ with an invariant $i(C) = n$. Essentially the only well-understood case previously was the case $n = 1$, represented by the example of two skew lines above. Using this example and the liaison addition, we produce explicit examples for all $n \geq 1$. The ideal of the curve C with $i(C) = n$ will be the monomial ideal

$$(x_0, x_1)^n \cap (x_1, x_2)^{n-1} \cap (x_2, x_3)^n \cap (x_3, x_0)^{n-1};$$

note that this reduces to the skew lines when $n = 1$.

It is then surprising to discover that a slight change in the exponents above can make all the liaison invariants vanish. For example all the monomial ideals

$$(x_0, x_1)^n \cap (x_1, x_2)^n \cap (x_2, x_3)^n \cap (x_3, x_0)^{n-1}$$

have liaison invariant 0; that is they define arithmetically Cohen-Macaulay curves in $\mathbf{P}^3_K$.

Liaison addition in its most general form is an addition of resolutions of ideals. Of course one already knows that the direct sum of two such resolutions does provide a resolution, but not of an ideal. Nevertheless, our discovery is that for ideals of grade ≥ 2 the direct sum is very nearly the answer. The key result is the following

Theorem 2.1. *Let I, I' be homogeneous ideals of $S = K[x_0, \ldots, x_n]$ of grade ≥ 2. Choose any homogeneous elements $f \in I$ and $f' \in I'$ such that $\{f, f'\}$ is an S-sequence (such a choice is possible). Then $f' \cdot I + f \cdot I'$ is a homogeneous ideal of grade exactly 2, and there exist graded free resolutions $\mathbf{F}$, $\mathbf{F}'$, $\mathbf{G}$ of $I, I', f' \cdot I + f \cdot I'$ over S such that*

$$\mathbf{G} = \mathbf{F}(-\deg f') \oplus \mathbf{F}'(-\deg f)$$

up to relations and generators.

Remark 2.2. The theorem is stated only for the case which interests us: for homogeneous ideals of $S = K[x_0, \ldots, x_n]$. However, the first proof, given by P. Schwartau, applies to any commutative Noetherian ring, as long as we make the extra assumption that $\operatorname{pd} I$ and $\operatorname{pd} I' < \infty$. Another proof we shall give, due to David Buchsbaum, shows that even this assumption is unnecessary. In either proof, the grading of S plays no helpful role.

The existence of f, f': We may obtain the first non-zero-divisor f by choosing any non-zero homogeneous element of I. We next need a homogeneous element $f' \in I'$ which is a non-zero-divisor on $S/(f)$, which we try to obtain by advoiding the associated primes $\mathfrak{p}_1, \ldots, \mathfrak{p}_n$ of (f). If this is impossible, then for every degree d we get: $I'_d \subset \mathfrak{p}_1 \cup \ldots \cup \mathfrak{p}_n$. Therefore $I' \subset \mathfrak{p}_1 \cup \ldots \cup \mathfrak{p}_n$; that is $I' \subset \mathfrak{p}_i$ for some i. This is a contradiction, since grade $I' \geq 2$ but grade $\mathfrak{p}_i \leq 1$. Therefore the desired f' exists.

Proof of Theorem 2.1*:* David Buchsbaum has pointed out that once we known $f' \cdot I + f \cdot I'$ is the ideal to add resolutions, a simple "coordinate-free" explanation is possible. Namely, use the standard exact sequence:

$$0 \to f' \cdot I \cap f \cdot I' \xrightarrow{\alpha} f' \cdot I \oplus f \cdot I' \to f' \cdot I + f \cdot I' \to 0$$

where α is defined by $s \mapsto (s, -s)$.

Now observe that since f, f' are relatively prime, $f' \cdot I \cap f \cdot I' = (f \cdot f')$. Thus we can represent the ideal $f' \cdot I + f \cdot I'$ as:

$$\operatorname{coker}\left((f \cdot f') \xrightarrow{\alpha} f' \cdot I \oplus f \cdot I'\right)$$

where

$$f \cdot f' \mapsto (f \cdot f', -f \cdot f').$$

We now use this to write the resolution of $f' \cdot I + f \cdot I'$ as a cokernel. Namely, the Comparison Theorem guarantees that the map α (homogeneous of degree 0) lifts to a homogeneous chain map $\bar{\alpha}$ of resolutions. The sequence of cokernels will then form a graded complex $\mathbf{G}$, ending in the ideal $f' \cdot I + f \cdot I'$. The resolution $\mathbf{G}$ will be exact if the lifts can be chosen as inclusions, and will be free if the lifts are split inclusions. Under certain weak assumptions, this can be done as follows: Choose graded free resolutions $\mathbf{F}$, $\mathbf{F}'$, of I, I' over S:

$$\mathbf{F}\colon \ldots \to \bigoplus_1^c S(-c_k) \xrightarrow{\Phi_3} \bigoplus_1^b S(-b_j) \xrightarrow{\Phi_2} \bigoplus_1^a S(-a_i) \xrightarrow{\Phi_1} S,$$

$$\mathbf{F}'\colon \ldots \to \bigoplus_1^{c'} S(-c'_k) \xrightarrow{\Phi'_3} \bigoplus_1^{b'} S(-b'_j) \xrightarrow{\Phi'_2} \bigoplus_1^{a'} S(-a'_i) \xrightarrow{\Phi'_1} S.$$

We now demand that these resolutions select the elements f', f of the hypothesis; that is, that the original resolution $\mathbf{F}$ of I selects $f \in I$ as the image of $(0, \ldots, 0, 1)$ and that $\mathbf{F}'$ selects $f' \in I'$ as the image of $(1, 0, \ldots, 0)$. Then from the cokernels one gets a graded free resolution $\mathbf{G}$ of the ideal $f' \cdot I + f \cdot I'$, this can be done as follows:

$$\begin{array}{llllll}
\mathbf{F}(-\deg f')\colon \ldots & \bigoplus^{c} S(-c_k - \deg f') & \bigoplus^{b} S(-b_j - \deg f') & \bigoplus^{a} S(-a_i - \deg f') & & \\
\oplus & \oplus \longrightarrow & \oplus \longrightarrow & \oplus \longrightarrow & \begin{array}{c} f'\cdot I \\ \oplus \\ f \cdot I' \end{array} & \to 0 \\
\mathbf{F}'(-\deg f)\colon \ldots & \bigoplus^{c'} S(-c'_k - \deg f) \to & \bigoplus^{b'} S(-b'_j - \deg f) \to & \bigoplus^{a'} S(-a'_i - \deg f) & \uparrow \alpha & \\
 & \uparrow & \uparrow & \uparrow & & \\
\ldots \to 0 & \longrightarrow & 0 \longrightarrow & S(-\deg f - \deg f') & \xrightarrow{f \cdot f'} (f \cdot f') & \to 0
\end{array}$$

But since the resolution of $(f \cdot f')$ is so short, we see that $\mathbf{G}$ will be equal to $\mathbf{F}(-\deg f') \oplus \mathbf{F}'(-\deg f)$ up to relations and generators, as claimed in the theorem.

It now only remains to prove that $f' \cdot I + f \cdot I'$ has grade exactly 2. But this follows from the inclusions:

$$(f, f') \cdot (I \cap I') \subseteq f' \cdot I + f \cdot I' \subseteq (f, f') \cap I \cap I',$$

q.e.d.

The following principle fact is a direct consequence of the proof of Theorem 2.1.

Corollary 2.3. *For all $i \geq 2$ we have:*

$$\begin{aligned} &\underline{\operatorname{Ext}}_S^i(f' \cdot I + f \cdot I', S) \\ &= \underline{\operatorname{Ext}}_S^i(I, S)\,(\deg f') \oplus \underline{\operatorname{Ext}}_S^i(I', S)\,(\deg f), \end{aligned}$$

and therefore for all $i \geq 3$:

$$\begin{aligned} &\underline{\operatorname{Ext}}_S^i(S/f' + f \cdot I', S) \\ &= \underline{\operatorname{Ext}}_S^i(S/I, S)\,(\deg f') \oplus \underline{\operatorname{Ext}}_S^i(S/I', S)\,(\deg f). \end{aligned}$$

Now we will show how to solve the liaison problem. As is known Rao has proven that curves $C \subset \mathbf{P}_K^3$ are classified up to the liaison by the graded module $\bigoplus_v H^1(\mathbf{P}_K^3, \mathcal{J}_C(v))$. In the following we intend to use Schwartau's alternate formulation of the liaison invariant by use of $\underline{\operatorname{Ext}}_S^3(S/I(C), S)$. This allows us to investigate the liaison invariant via the theory of free resolutions.

Definition 2.4. Let C be a curve in $\mathbf{P}_K^3$. Define as above $M(C)$ to be the graded S-module $\bigoplus_v H^1(\mathbf{P}^3, \mathcal{J}_C(v))$ and let $E(C)$ be the graded S-module $\underline{\operatorname{Ext}}^3(S/I(C), S)$, where $I(C)$ is the defining ideal of C and $S = K[x_0, x_1, x_2, x_3]$.

Using our notations and results from Chapter 0, § 4 (see the section on dualization) we will first show how Rao's result may be stated in terms of Schwartau's module $E(C)$ rather than $M(C)$. This follows immediately from

Lemma 2.5. *For any curve $C \subset \mathbf{P}_K^3$,*

$$E(C) = \big((M(C))\big)^v\,(4).$$

Thus $E(C)$ determines the same liaison class of modules as $M(C)$.

Additionally we have the following transformation law: Suppose, C, C' are curves in $\mathbf{P}_K^3$ such that C and C' are algebraically linked by f, f'. Then:

$$\begin{aligned} M(C') &= \big(M(C)\big)^v\,(4 - \deg f - \deg f'), \\ E(C') &= \big(E(C)\big)^v\,(\deg f + \deg f' + 4). \end{aligned}$$

Proof: First we note that the proof of Corollary 0.4.7 provides the following property:

$$M(C) \cong \underline{H}^1_{(x_0,\ldots,x_3)}\big(K[x_0, \ldots, x_3]/I(C)\big).$$

Hence our duality theorem (see Theorem 0.4.14) yields the first assertion. The transformation law follows from Corollary 0.4.17, q.e.d.

Now, recapitulating the liaison addition problem was the following: Is there an addition of closed subschemes in $\mathbf{P}_K^n$ which induces an addition of their liaison classes? On might think that in the case of curves in $\mathbf{P}_K^3$ (without isolated or embedded points) this problem is already answered by Rao's theory of the liaison invariant. Each such curve determines a specific graded S-module of finite length. Thus the direct sum determines a unique algebraic liaison class which may be defined as the liaison sum. The reader will note, however, that this method is not only unsatisfactory but invalid.

It is unsatisfactory because the addition has no description in terms of the curves alone. It is invalid because of the following fact:

The direct sum of graded S-modules does not induce a sum on liaison classes of graded S-modules.

For example, E, $E(1)$, and $\underline{\mathrm{Hom}}_S(E, S)$ are all in the same liaison class, but $E \oplus E$, $E(1) \oplus E$, and $\underline{\mathrm{Hom}}_S(E, S) \oplus E$ all determine different liaison classes. Therefore the direct sum of liaison invariants does not induce an addition in liaison classes of curves. So the Rao theory does not solve the problem for curves in $\mathbf{P}^3_K$, but instead complicates the issue by raising another question:

Is there an addition of curves in $\mathbf{P}^3_K$ which induces the direct sum on their liaison invariants?

On might at first try the scheme-theoretic union as a means of adding curves, but the union of one line with a skew line shows that this idea is not very promising since scheme-theoretic union does not preserve the 0 class. The geometric formulation of our liaison addition will reveal just why scheme-theoretic union is insufficient. Here now is Schwartau's solution of the liaison addition problem:

Definition 2.6. Let V, V' be closed subschemes in $\mathbf{P}^n_K$ of codimension 2 with defining ideal $I(V) =: I$ and $I(V') =: I'$ (note that $(x_0, \ldots, x_n)$ is not associated to I and I'). Choose any hypersurfaces f containing V, f' containing V', such that (f, f') is a complete intersection in $\mathbf{P}^n_K$. We then define $Vf + f'V'$ to be the *closed subscheme* of $\mathbf{P}^n_K$ defined by the homogeneous ideal $f' \cdot I + f \cdot I'$.

Theorem 2.7. *Let V, V' be closed subschemes in $\mathbf{P}^n_K$ of codimension 2, and let $Vf + f'V'$ be defined as above. Then:*

1. *$(x_0, \ldots, x_n)$ is not associated to $f' \cdot I + f \cdot I'$; that is*
$$I(Vf + f'V') = f' \cdot I + f \cdot I'.$$
2. *$Vf + f'V'$ is again of codimension 2 in $\mathbf{P}^n_K$.*
3. *For all $i \geq 3$ we have*
$$\underline{\mathrm{Ext}}^i_S(S/I(Vf + f'V)) \cong \underline{\mathrm{Ext}}^i_S(S/I, S)\,(\deg f') \oplus \underline{\mathrm{Ext}}^i_S(S/I', S)\,(\deg f).$$
4. *$Vf + f'V'$ is locally Cohen-Macaulay and equi-dimensional if and only if V, V' are locally Cohen-Macaulay and equi-dimensional.*
5. *$Vf + f'V'$ is arithmetically Cohen-Macaulay if and only if V, V' are arithmetically Cohen-Macaulay.*
6. *If f is a non-zero divisor modulo I' and f' is a non-zero-divisor modulo I, then $Vf + f'V' = V \cup V' \cup (f, f')$ as closed subschemes of $\mathbf{P}^n_K$; that is, $f' \cdot I + f \cdot I' = I \cap I' \cap (f, f')$.*

The last equation is always true set-theoretically, even without the hypothesis of 6.

Corollary 2.8. *Let C, C' be any two curves in $\mathbf{P}^3_K$ with defining ideals I, I' and liaison invariants E, E'. Choose any surfaces f containing C, f' containing C', such that (f, f') is a complete intersection curve in $\mathbf{P}^3_K$. Then*

1. *$Cf + f'C'$ is a curve in $\mathbf{P}^3_K$, with liaison invariant $E(\deg f') \oplus E'(\deg f)$ and defining ideal $f' \cdot I + f \cdot I'$.*
2. *If C, C' are without isolated or embedded points, the same is true of $Cf + f'C'$.*

3. *$Cf + f'C' = C \cup C' \cup (f, f')$ set-theoretically.*
4. *If the surface f does not contain any component of C', and f' does not contain any component of C, then $Cf + f'C = C \cup C' \cup (f, f')$ scheme-theoretically; that is,*

$$f' \cdot I + f \cdot I' = I \cap I' \cap (f, f').$$

Proof: Theorem 2.7 we only need to remind the reader that the property of "no imbedded points" is for curves equivalent to the "locally Cohen-Macaulay" property stated in the earlier theorem.

Remark 2.9. We may always choose f, f' of the same degree d. Then the liaison addition procedure will give us a curve $Cf + f'C'$ with liaison invariant $E(d) \oplus E'(d) \cong (E \oplus E')(d)$. Thus the added curve in this case lies in the liaison class associated to the module $E \oplus E'$. However, for Buchsbaum curves we get the following statement:

Corollary 2.10. *$Cf + f'C'$ is arithmetically Buchsbaum if and only if C, C' are arithmetically Buchsbaum. (Note that f and f' have not necessarily the same degree.)*

Proof of Theorem 2.7: 1. Assume that $(x_0, \ldots, x_n)$ is associated to $f' \cdot I + f \cdot I'$. By the Auslander-Buchsbaum theorem we have $\operatorname{pd}(S/f' \cdot I + f \cdot I') = n + 1$; that is

$$\underline{\operatorname{Ext}}_S^{n+1}(S/f' \cdot I + f \cdot I', S) \neq 0.$$

Corollary 2.3 provides

$$\text{either } \underline{\operatorname{Ext}}_S^{n+1}(S/I, S) \neq 0 \quad \text{or} \quad \underline{\operatorname{Ext}}_S^{n+1}(S/I', S) \neq 0.$$

Hence

$$\text{either } \operatorname{pd} S/I \geq n + 1 \quad \text{or} \quad \operatorname{pd} S/I' \geq n + 1;$$

that is $(x_0, \ldots, x_n)$ is associated to I or I' which contradicts our assumptions. This proves 1.

2. We are to show $\operatorname{ht} I(Vf + f'V') = 2$. But since S is Cohen-Macaulay it is enough to show grade of $I(Vf + f'V') = 2$. This fact follows from 1. and Theorem 2.1.

3. This follows from 1. and Corollary 2.3.

4. and 5.: Since V, V', and $Vf + f'V'$ all have codimension 2, the properties at issue in 4., 5. depend only on the vanishing or finite length of $\underline{\operatorname{Ext}}^i(\ ,\ )$ (with suitable arguments) for $3 \leq i \leq n$. But then assertion 3. implies 4. and 5.

6. The reader will easily see that the hypothesis of 6. forces

$$f' \cdot I + f \cdot I' = I \cap I' \cap (f, f').$$

Note that even without the hypothesis of 6. we always have this equation up to radical. This follows from the inclusions:

$$(I \cap I') \cdot (f, f') \subseteq f' \cdot I + f \cdot I' \subseteq I \cap I' \cap (f, f'),$$

which always holds. This with our assertion 1. proves 6. and the theorem, q.e.d.

Next we discuss some applications of liaison addition and some new examples of curves in $\mathbf{P}_K^3$ defined by monomials. First of all, we need the so-called power formula.

Theorem 2.11 (Power Formula). *Let C be any curve in $\mathbf{P}_K^3$ with defining ideal I and liaison invariant E. Choose any S-sequence $f, f' \in I$ such that f, f' are homogeneous of the same degree d. Then $(f, f')^n \cdot I$ defines a curve in $\mathbf{P}_K^3$ with liaison invariant $\left(\overset{n+1}{\bigoplus} E\right)(nd)$.*

Proof: We use induction on n, the case $n = 0$ being obvious. Therefore assume $n > 0$ and that the theorem is true for $n - 1$. Thus $(f, f')^{n-1} \cdot I$ defines a curve $C^{(n-1)}$ with invariant $\left(\overset{n}{\bigoplus} E\right)\left((n-1)\, d\right)$. Notice that this ideal contains f^n, and that $(f^n, f') \cdot S$ is a complete intersection.

Thus we may consider the liaison sum $C^{(n-1)}f^n + f'C$. Corollary 2.8(1) then implies that this curve has defining ideal $f'\left((f, f')^{n-1} \cdot I\right) + f^n \cdot I = (f, f')^n \cdot I$, and liaison invariant

$$\left(\left(\overset{n}{\bigoplus} E\right)\left((n-1)\, d\right)\right)(d) \oplus E(nd) = \left(\overset{n}{\bigoplus} E\right)(nd) \oplus E(nd) = \left(\overset{n+1}{\bigoplus} E\right)(nd),$$

q.e.d.

Example 2.12. Let C be any complete intersection curve with defining ideal (f, f'). Then C is arithmetically Cohen-Macaulay and therefore has liaison invariant 0. It is clear that in the case of the trivial invariant the "same degree" hypothesis in the Power Formula is not needed. Thus we may apply the formula to the S-sequence $\{f, f'\} \in (f, f') \cdot S$. We deduce that $(f, f')^n \cdot (f, f') = (f, f')^{n+1}$ defines a curve in $\mathbf{P}_K^3$ of liaison invariant 0; that is an arithmetically Cohen-Macaulay curve. This is a special case of the well-known theorem that all powers of a complete intersection are perfect.

Furthermore, let I be any perfect ideal in $K[x_0, x_1, x_2, x_3]$. Then the Power Formula may be applied just as above to $\{f, f'\}$, an S-sequence in I, to conclude that the ideal $(f, f') \cdot I$ must be perfect.

Example 2.13. Let C be two skew lines with defining ideal $I = (x_0, x_1) \cap (x_2, x_3)$. Then I has generators $(x_0x_2, x_0x_3, x_1x_2, x_1x_3)$. Thus I contains the S-sequence x_0x_2, x_1x_3, and we may apply the Power Formula. Note that C has liaison invariant $\underline{K}(4)$ by use of Remark 1.5 and Lemma 2.5. Therefore we conclude that for all $n \geq 1$,

$$(x_0x_2, x_1x_3)^{n-1} \cdot \left((x_0, x_1) \cap (x_2, x_3)\right)$$

defines a curve in $\mathbf{P}_K^3$ of liaison invariant $\underline{K}^n(2 + 2n)$.

This gives us the first explicit examples of curves in $\mathbf{P}_K^3$ corresponding to the liaison invariants K^n for every $n \geq 1$; that is, we have equations defining arithmetically Buchsbaum curves C in $\mathbf{P}_K^3$ with Buchsbaum invariant $i(C) = n$. Also we have the following primary decomposition:

$$\begin{aligned}&(x_0x_2, x_1x_3)^{n-1} \cdot \left((x_0, x_1) \cap (x_2, x_3)\right)\\ &= (x_0, x_1)^n \cap (x_1, x_2)^{n-1} \cap (x_2, x_3)^n \cap (x_3, x_0)^{n-1}.\end{aligned}$$

Example 2.14. Consider the curve C in $\mathbf{P}_K^3$ with defining ideal

$$I = (x_0, x_1) \cap (x_1, x_2) \cap (x_2, x_3) = (x_0x_2, x_1x_2, x_1x_3).$$

Note that C is arithmetically Cohen-Macaulay. This follows immediately, for example from Proposition 1.9. Once again we may apply the Power Formula with respect to the S-sequence (x_0x_2, x_1x_3). Hence we conclude that for all $n \geq 1$

$$(x_0x_2, x_1x_3)^{n-1} \cdot \left((x_0, x_1) \cap (x_1, x_2) \cap (x_2, x_3)\right)$$

defines a curve in $\mathbf{P}_K^3$ of liaison invariant 0; that is these curves are arithmetically Cohen-Macaulay. We have the following primary decomposition:

$$(x_0x_2, x_1x_3)^{n-1} \cdot \big((x_0, x_1) \cap (x_1, x_2) \cap (x_2, x_3)\big)$$
$$= (x_0, x_1)^n \cap (x_1, x_2)^n \cap (x_2, x_3)^n \cap (x_3, x_0)^{n-1}.$$

Remark 2.15. Each of the above examples presents curves in $\mathbf{P}_K^3$ defined by monomials. We want to point out that not every liaison class in $\mathbf{P}_K^3$ contains such a curve. This assertion is a consequence of S. Goto's and K. Watanabe's paper [2] on $\mathbf{Z}^n$-graded rings. The main point here is following:

If C is a curve in $\mathbf{P}_K^3$ defined by monomials, then the quotient $S/I(C)$ is a $\mathbf{Z}^4$-graded S-module in the sense of Goto-Watanabe.

It follows that the liaison invariant $\underline{\mathrm{Ext}}_S^3\big(S/I(C), S\big)$ has also the structure of a $\mathbf{Z}^4$-graded S-module (see p. 243 of the just mentioned paper). Such a structure is preserved by K-duals and shifts, so we conclude that every module in the liaison class of $E(C)$ possesses a $\mathbf{Z}^4$-graded structure. But not every graded S-module of finite length has a $\mathbf{Z}^4$-graded structure, and therefore not every liaison class contains a curve defined by monomials.

Another application of liaison is the so-called liaison realization. Following Schwartau [1] we will explain this phenomenon. No complete proofs will be given here. The point of view discussed in what follows was introduced in the fundamental paper by Rao [1]:

There is a 1-1 correspondence between the liaison class of curves in $\mathbf{P}_K^3$ which are equidimensional without imbedded points and generic complete intersection and the liaison classes of graded S-modules of finite length. But this 1-1 correspondence of Rao does not produce a curve $C \subset \mathbf{P}_K^3$ of liaison invariant E for each graded S-module E of finite length. It only gives a curve for each liaison class of modules; that is the specific result of Rao says that for any graded S-module E of finite length there exists a curve $C \subset \mathbf{P}_K^3$ of liaison invariant $E(v)$ for some shift of E. It is in fact not true that an arbitrary graded S-module of finite length can be realized as the liaison invariant of a curve in $\mathbf{P}_K^3$. Indeed, we have the following lemma:

Lemma 2.16. *If $A \neq O$ is a graded S-module of finite length, then for $v \gg 0$ it is impossible to realize $A(-v)$ as the liaison invariant $E(C)$ of a curve $C \subset \mathbf{P}_K^3$.*

The proof of this lemma rests on the following

Claim 2.17: Any non-zero liaison invariant $E(C)$ "begins" in degree $= -4$.

After having established this claim it is easy to prove the lemma. Simply twist the given module A until it begins in degree -4. This would be impossible if A were to have infinitely many non-zero negative degrees, but this is precluded by finite length.

Therefore high negative twists of A cannot be realized as liaison invariants. However Rao's result quarentees that some twist $A(v)$ can be realized, and an examination of his proof shows that any twist $A(v + 2n)$, $n \geq 0$ can be realized. Unfortunately the bound for v is not proved to be sharp, and there is no way in general to obtain information about the twists $A(v + 2n + 1)$, $n \geq 0$. But this phenomenon may be clarified considerably through liaison addition:

Theorem 2.18 (Liaison Realization). *If a graded S-module E can be realized as the liaison invariant of some curve $C \subset \mathbf{P}_K^3$, then the same is true for each positive twist $E(\eta)$, $\eta \geq 1$.*

Proof: Choose any homogeneous element $f \in I(C)$ such that (x_0^η, f) is a complete intersection.

Now choose any curve C' lying in the plane $x_0 = 0$. Since any plane curve is a complete intersection, C' has liaison invariant 0. Therefore by liaison addition the curve $Cf + x_0^\eta C'$ has liaison invariant

$$E(\eta) + O(\deg f) = E(\eta).$$

A disadvantage in this method is that we can only produce curves without isolated points, whereas Rao produced non-singular curves.

§ 3. On curves linked to lines in $\mathbf{P}^3$ and applications

Consider the non-singular quadric surface $x_0x_3 - x_1x_2 = 0$ in $\mathbf{P}^3$. Such a surface has two rulings, and we will examine "m by n" configurations L of lines where L consists of m lines from one ruling and n from the other. We will show:

(i) The "m by n" configuration is arithmetically Cohen-Macaulay if and only if $|m - n| \leq 1$,

(ii) the configuration is arithmetically Buchsbaum if and only if $|n - m| \leq 2$.

Following Geramita-Maroscia-Vogel [1] we will outline an geometric approach to the content of (i) and (ii) by use of liaison. Therefore our proof is entirely different from the linear algebra proof of these results given by Geramita-Weibel [1]. Of course there are still different methods to study such configurations. By applying our method we are also able to produce equations defining connected curves in $\mathbf{P}^3$ which are set-theoretically complete intersections and not arithmetically Buchsbaum. Therefore we get counter-examples to an even stronger formulation of a conjecture of *R.* Hartshorne [3] (Conjecture 5.17 on p. 126). Our simplest counter-example to Hartshorne's conjecture is the curve $C \subset \mathbf{P}_K^3$ given by the following ideal $I(C)$ in $K[x_0, x_1, x_2, x_3]$:

$$I(C) = (x_0, x_1) \cap (x_2, x_3) \cap (x_0 - x_2, x_1 - x_3) \cap (x_0 - x_1, x_2 - x_3).$$

C is a set-theoretic complete intersection of the quadratic Q and cubic K defined by the equation $x_0x_3 - x_1x_2 = 0$ resp.

$$(x_0 - x_1)(x_2 - x_3)(x_0 - x_1 - x_2 + x_3) = 0.$$

The connected curve C is linked to the union of the following two skew lines:

$$x_0 = x_2 = 0 \quad \text{and} \quad x_1 = x_3 = 0.$$

Hence C is not arithmetically Cohen-Macaulay but arithmetically Buchsbaum. The same method suitably refined, can be used to construct curves which are set-theoretically complete intersections and not arithmetically Buchsbaum. The key here is that these curves are linked to the union of h skew lines which lie on a non-singular quadric in $\mathbf{P}^3$ with $h > 2$. Also we are able to give the equations for these examples (see Proposition 3.3 and Remark 3.4, below). From the point of view of commutative algebra we have the following fact:

Let F be an irreducible form in $S = K[x_0, x_1, x_2, x_3]$ and let $V(F) \subset \mathbf{P}_K^3$ be the surface defined by $F = O$. We let $R = S/(F)$ be the homogeneous coordinate ring of

$V(F)$. If $\mathfrak{p}$ is a homogeneous prime ideal in S having height 2 and containing F then $\mathfrak{p}$ describes an irreducible curve $V(\mathfrak{p})$ on $V(F)$. If $V(\mathfrak{p})$ contains at least one point which is not a singular point on $V(F)$, then the local ring $V(F)$ along the subvariety $V(\mathfrak{p})$ is a discrete valuation ring. Consequently, for each integer $n \geq 0$, there is a unique $\mathfrak{p}$-primary ideal of S which contains F and has length n.

Our aim in this paragraph is to investigate the special situation in which $\mathfrak{p}$ describes a line on $V(F)$. First we need the following lemma.

Lemma 3.1. *Let H be an irreducible form in $S := K[x_0, \ldots, x_r]$ (K any algebraically closed field, $r \geq 3$) which can be expressed as*

$$H = x_0 \cdot F - x_1 \cdot G,$$

where F an G are forms in $K[x_2, \ldots, x_r]$ of degree $\alpha \geq 1$. Let $A := K[x_0, \ldots, x_r]/(H)$. Set $\mathfrak{p} = (x_0, x_1) \cdot A$. Then, the local ring of the variety $V(H)$ along the subvariety $V(\mathfrak{p})$ is a discrete valuation ring.

Proof: It will suffice to show that there is a point on the linear subvariety $V(\mathfrak{p})$ of $V(H)$ which is a simple point on $V(H)$.

Suppose this is not the case, i.e. every point on $V(\mathfrak{p})$ is a singular point of $V(H)$. Consider the point $(1, 0, \ldots, 0) =: P$ on $V(H)$ and note that it is not on $V(\mathfrak{p})$. Let Q be a point on $V(\mathfrak{p})$. The line connecting P and Q meets P with multiplicity $\geq \alpha$ and meets Q with multiplicity ≥ 2. Since $\alpha + 2 > \deg H$, this line lies on $V(H)$. Since this was true for all points of $V(\mathfrak{p})$ we have that the hyperplane formed by these lines is contained in $V(H)$. This implies that $V(H)$ is a hyperplane, which is a contradiction, q.e.d.

We will say again that a homogeneous ideal $I \subset S := K[x_0, \ldots, x_r]$ is a *Cohen-Macaulay* (respectively, *Buchsbaum*) ideal if the local ring at the maximal homogeneous ideal of $K[x_0, \ldots, x_r]/I$ is a Cohen-Macaulay (respectively, Buchsbaum) ring.

In the following we will examine some properties of "multiple" lines which are on a non-singular quadric surface in $\mathbf{P}^3$.

Theorem 3.2. *Let $H = x_0F - x_1G$ be as in Lemma 3.1. Let $\mathfrak{p} = (x_0, x_1) \subset S$. Let $\mathfrak{q}$ be a $\mathfrak{p}$-primary ideal in S of length n (≥ 2) which contains H. Then:*

(i) $\mathfrak{q} = (\mathfrak{p}^n, H)$.

(ii) *If, moreover, $F \in (x_3)$ and $G \in (x_2)$ then if $\mathfrak{q}$ is a Buchsbaum ideal we must have $n = 2$ and $H = \alpha x_0x_3 - \beta x_1x_2$ with $\alpha, \beta \in K$.*

(iii) *If $n = 2$, $r = 3$ and $H = \alpha x_0x_3 - \beta x_1x_2$, $\alpha, \beta \in K$ ($\alpha \cdot \beta \neq 0$), then $\mathfrak{q}$ is a Buchsbaum ideal with invariant* 1.

(iv) *If $r > 3$ and $n > 1$, then $\mathfrak{q}$ is not a Buchsbaum ideal.*

Proof: From Lemma 3.1 we have that there is a unique $\mathfrak{p}$-primary ideal in S of length n which contains H. Thus, to prove (i), it suffices to show that $\mathfrak{a}_n := (\mathfrak{p}^n, H)$ is a primary ideal of length n.

Claim 1. $\mathfrak{a}_n := (\mathfrak{p}^n, H)$ is $\mathfrak{p}$-primary.

We know of several ways to prove this claim. We will give an elementary proof following Geramita-Maroscia-Vogel [1].

Since $\operatorname{Rad}(\mathfrak{a}_n) = \mathfrak{p}$, it will be sufficient to show that $\mathfrak{a}_n$ is unmixed. Suppose this is not the case and write $\mathfrak{a}_n = \mathfrak{q} \cap \mathfrak{q}_1 \ldots \cap q_s$, where $\mathfrak{q}$ is $\mathfrak{p}$-primary and $\mathfrak{q}_1, \ldots, \mathfrak{q}_s$ denote the embedded primary components of $\mathfrak{a}_n$. Hence $\operatorname{Rad}(\mathfrak{q}_i) =: \mathfrak{p}_i \supset (x_0, x_1)$. Thus, there is a form A, no monomial of which is divisible by x_0 or x_1, such that $\mathfrak{a}_n : A \supsetneqq \mathfrak{a}_n$. Let $B \in K[x_0, \ldots, x_r]$ be a form such that $B \notin \mathfrak{a}_n$ and $B \cdot A \in \mathfrak{a}_n$. Then $B \in \mathfrak{q} \subseteq (x_0, x_1)$. We may consider B as an element in $(K[x_2, \ldots, x_r])\,[x_0, x_1]$ and write B as a sum of forms in this ring; $B = B_1 + B_2 + \ldots + B_t$ where $\deg B_i = i$ (as a form in x_0, x_1). Since $\mathfrak{p}^n \subseteq \mathfrak{a}_n \subseteq \mathfrak{q}$ we may assume $t \leq n - 1$. Note that $\mathfrak{a}_n$ may also be considered as a homogeneous ideal in $(K[x_2, \ldots, x_r])\,[x_0, x_1]$. Thus we may write

$$B \cdot A = B_1 A + B_2 A + \ldots + B_t A\,.$$

Since $\deg A = 0$ (in x_0, x_1), $\deg B_i A = i$. Therefore, $B_i \cdot A \in \mathfrak{a}_n$ for $1 \leq i \leq t$. But, the only forms in $\mathfrak{a}_n$ of degree $< n$ (in x_0, x_1) are the multiples of H. Thus $H \mid B_i A$, $1 \leq i \leq t$. Since H is irreducible and $H \nmid A$ we obtain that $H \mid B_i$, $1 \leq i \leq t$, and so $B \in \mathfrak{a}_n$. This contraction establishes Claim 1.

Claim 2. $\mathfrak{a}_n$ is $\mathfrak{p}$-primary of length n.

Proof: Consider the following chain of length n of ideals:

$$\mathfrak{a}_n \subsetneqq (\mathfrak{a}_n, x_0^{n-1}) \subsetneqq \ldots \subsetneqq (\mathfrak{a}_n, x_0^2) \subsetneqq (x_0, x_1) = \mathfrak{p}\,.$$

It is easy to see that even after localizing at $\mathfrak{p}$, the inclusions in this chain remain proper. (One argues by degree in x_0, x_1 as in claim 1.) Thus $\mathfrak{a}_n$ has length at least n. Using the well-known criterion from Zariski-Samuel [1], Vol. 1, p. 237, Cor. 2, we see, after localizing at $\mathfrak{p}$, that this is a saturated chain of primary ideals and so $\mathfrak{a}_n$ has length n.

This completes the proof of claim 2 and hence the proof of (i) of the theorem.

Proof of (ii): Suppose H has the required form and $\mathfrak{q}$ is a Buchsbaum ideal. Then, since x_3 is not a zero-divisor modulo $\mathfrak{q}$, $(\mathfrak{q}, x_3)$ is also a Buchsbaum ideal (see Corollary I.1.11). Let $U(\mathfrak{q}, x_3)$ denote the primary component of $(\mathfrak{q}, x_3)$ belonging to $(x_0, x_1, x_3) = \operatorname{Rad}(\mathfrak{q}, x_3)$. By the special form of H we obtain that $U(\mathfrak{q}, x_3) = (x_0^n, x_1, x_3)$. By Proposition I.1.10 we must then have

$$(x_0, \ldots, x_r) \cdot U(\mathfrak{q}, x_3) \subseteq (\mathfrak{q}, x_3)\,.$$

In particular we obtain that $x_1 x_2 \in (\mathfrak{q}, x_3) = (\mathfrak{p}^n, H, x_3)$. Hence simple degree considerations (and the special form of H) give that $\deg H = 2$ and that $H = \alpha \cdot x_0 x_3 - \beta \cdot x_1 x_2$. Also, we must have $x_1^2 \in (\mathfrak{q}, x_3)$ and this shows that $n \leq 2$ since $(\mathfrak{q}, x_3) = (\mathfrak{p}^n, H, x_3)$. This completes the proof of (ii).

Proof of (iii): It follows immediately from the assumptions that

$$\mathfrak{q} = (x_0^2, x_1^2, x_0 x_1, \alpha x_0 x_3 - \beta x_1 x_2)\,.$$

The assertion (iii) now results since the subvariety defined by $\mathfrak{q}$ is linked to two skew lines given by the following equations:

$$(\alpha x_0 x_3 - \beta x_1 x_2, x_0 x_1) = (x_0, x_2) \cap (x_1, x_3) \cap \mathfrak{q}\,.$$

Assertion (i) and Bezout's Theorem provide this equality. Hence Theorem 1.2 completes the proof of (iii).

Proof of (iv): This is clear from (iii) since in a local Buchsbaum ring every localization at a non-maximal prime ideal must be Cohen-Macaulay, see Corollary I.1.11. This concludes the proof of the theorem.

We now apply Theorem 3.2 to the study of a special class of reducible curves lying on a non-singular quadratic surface in $\mathbf{P}^3$. These observations also yield the above-mentioned applications to set-theoretic intersections of algebraic subvarieties in n-space (see, for example, Kunz's book [1] or the report in Stückrad and Vogel [6]).

Let K be an algebraically closed field of arbitrary characteristic. Let V be an algebraic subvariety in n-space; that is V is a subvariety of affine space $\mathbf{A}_K^n$ or projective space $\mathbf{P}_K^n$. Then we have the following classical problem which is a major open problem in algebraic geometry today:

What is the smallest number s of equations defining V? We denote by ara V this number s. Looking at it from the point of view of commutative algebra we then get the following fact: The radical of the defining ideal $I(V)$ of V is the radical of an ideal generated by ara V (homogeneous) polynomials.

Let d be the dimension of the algebraic subvariety V in n-space. Then we have always the following bounds:

$$n - d \leq \operatorname{ara} V \leq n.$$

We also say that V is a *set-theoretic intersection of* ara V *hypersurfaces*. If $n - d = \operatorname{ara} V$ then V is called a *set-theoretic complete intersection*. We recall that V is called an (*ideal-theoretic*) *complete intersection* if $I(V)$ is generated by $n - d$ polynomials. Nowadays, we have the following major question in this area:

Classical problem. Is every (connected) curve in 3-space the set-theoretic intersection of two hypersurfaces?

There are some recent results in $\mathbf{A}_K^3$ which pertain to this problem. Nothing of real interest is known about general results on set-theoretic intersections of curves in $\mathbf{P}_K^3$ where K is a field of characteristic zero.

Considering the cone over a curve in $\mathbf{P}_K^3$ leads R. Hartshorne [3], p. 126, to the following conjecture in local algebra:

Let A be a regular local ring containing a field of characteristic zero. Let B be a local ring which is a finitely generated flat A-module. Then the reduced local ring B_{red} of B is also a flat A-module.

One important special case of this conjecture is the following question: Let C be a curve in $\mathbf{P}_K^3$ over a field K of characteristic zero. Is then C arithmetically Cohen-Macaulay if C is a set-theoretic complete intersection?

As it turns out this is not the case, however. The first counter-example was given by C. Peskine and L. Szpiro in [3]. Their curve is not irreducible but arithmetically Buchsbaum (see our Remark 3.5, below). As R. Hartshorne pointed out to us there are irreducible curves of genus 5 and degree 8 in $\mathbf{P}_K^3$ which are set-theoretic complete intersection and also not arithmetically Cohen-Macaulay. We note that these curves are again arithmetically Buchsbaum because their liaison invariants are given by $\underline{K}^2$ (see Rao [2]).

Motivated by the preceding we investigate the following stronger formulation of Hartshorne's question: Let C be a curve in $\mathbf{P}_K^3$ over a field K of characteristic 0. Is C then arithmetically Buchsbaum if C is a set-theoretic complete intersection?

However, this also is not true. The assertion results immediately from the following proposition (see our Remark 3.4).

Proposition 3.3. *Let $C = C_{m,n}$ be a reducible curve lying on a nonsingular quadric Q in $\mathbf{P}^3$, where $C_{m,n}$ is of the form*

$$C_{m,n} = (L_1 \cup \ldots \cup L_m) \cup (L'_1 \cup \ldots \cup L'_n)$$

where the L_i, $1 \leq i \leq m$ are lines in one ruling of Q and the L'_j, $1 \leq j \leq n$, are lines in the other ruling of Q. Assume $m \leq n$. Then

(a) *C is always a set-theoretic complete intersection, and a complete intersection when $n = m$.*

(b) *C is arithmetically Cohen-Macaulay if and only if $n - m \leq 1$.*

(c) *C is arithmetically Buchsbaum if and only if $n - m \leq 2$.*

Note the following basic facts of the ruled surface Q in $\mathbf{P}^3$: First, Q is equal to the Segre embedding of $\mathbf{P}^1 \times \mathbf{P}^1$ in $\mathbf{P}^3$, for a suitable choice of coordinates. Therefore Q contains two families of lines $\{L_i\}$, $\{L'_i\}$, each parametrized by $t \in \mathbf{P}^1$, with the properties: if $L_t \neq L_u$, then $L_t \cap L_u = \emptyset$; if $L'_t \neq L'_u$, then $L'_t \cap L'_u = \emptyset$, and for all t, u $L_t \cap L'_u =$ one point; that is, the quadric surface in $\mathbf{P}^3$ is a ruled surface in two different ways.

Proof: (a) Let $\Pi_{i,j}$ denote the plane containing the lines L_i and L'_j $(1 \leq i \leq m,\ 1 \leq j \leq n)$ Then, it is easy to check that the curve $C = C_{m,n}$ is the set-theoretic complete intersection of the quadric Q and the reducible surface, say F, of degree n given by:

$$F = \Pi_{1,1}\Pi_{2,2} \cdot \ldots \cdot \Pi_{m,m}\Pi_{m,(m+1)} \cdot \ldots \cdot \Pi_{m,(n-m)} .$$

By Bezout's Theorem, when $m = n$, $C_{m,m}$ is a complete intersection of F and Q.

(b) We first show that if C is arithmetically Cohen-Macaulay then $n - m \leq 1$. In fact, if $n - m =: h \geq 2$, $C_{m,n}$ is linked to the union of h skew lines lying on Q. This follows immediately from assertion (a) since $C_{n,n}$ is a complete intersection. But the union of h skew lines on Q is linked to a multiple line lying on Q whose defining ideal is precisely the primary ideal $\mathfrak{q} = (\mathfrak{p}^h, Q)$ we studied in Theorem 3.2. This assertion results immediately from the proof of (a) and Theorem 3.2(i). Since $h \geq 2$ it follows from theorem 3.2(ii) and (iii) that $\mathfrak{q}$ is not a Cohen-Macaulay ideal. Since we have the liaison of $C_{m,n}$ and the curve with defining ideal $\mathfrak{q}$ the conclusion follows from Remark 1.5(i). We now assume that $n - m \leq 1$. If $n = m$ then $C_{n,n}$ is a complete intersection by (a) and therefore is clearly arithmetically Cohen-Macaulay. If $n - m = 1$ we obtain from (a) that the ideal of the configuration is linked to the ideal (x_0, x_1) and thus again is Cohen-Macaulay by liaison (see Remark 1.5(i)).

(c) This is proven exactly as (b) by again using Theorem 3.2 and of course Theorem 1.2. This concludes the proof of Proposition 3.3, q.e.d.

Remark 3.4. Note that the configurations of lines in Proposition 3.3 for which $n - m > 1$ give an infinite family of counter-examples to the above-mentioned conjecture of Hartshorne. In particular when $n - m > 2$, none of these examples is arithmetically Buchsbaum. The simplest example $m = 1$, $n = 3$ was examined at the beginning of this section. The equations follow immediately from the proof of Proposition 3.3.

Remark 3.5. We now discuss the previously mentioned counter-example given by Peskine and Szpiro [3], 1.7: They exhibited a connected curve of degree 6 in $\mathbf{P}^3_K$ which is a set-theoretic complete intersection and not arithmetically Cohen-Macaulay. We will show that this curve is arithmetically Buchsbaum. More precisely, their example is the union of the irreducible quintic, say C, given parametrically by (s^5, s^4t, st^4, t^5) and the line, say L, defined by (x_0, x_1). This quintic has the following prime ideal, say $\mathfrak{p}$:

$$\mathfrak{p} = (x_0x_3 - x_1x_2, x_0^3x_2 - x_1^4, x_0^2x_2^2 - x_1^3x_3, x_0x_2^3 - x_1^2x_3^2, x_2^4 - x_1x_3^3).$$

Notice that the sextic $C \cup L$ lies on the quadric surface $Q = x_0x_3 - x_1x_2 = 0$. It is linked to a multiple line lying on Q. The liaison is given by the following equation:

$$(x_0x_3 - x_1x_2, x_0x_2^3 - x_1^2x_3^2) = \mathfrak{p} \cap (x_0, x_1) \cap (x_2^2, x_3^2, x_2x_3, x_0x_3 - x_1x_2).$$

Bezout's theorem and Theorem 3.2(i) imply this equality.

Therefore we can apply our observations. By Theorem 3.2(i) and (iii), $(x_2^2, x_3^2, x_2x_3, x_0x_3 - x_1x_2)$ is an (x_2, x_3)-primary ideal of length 2 and therefore is a Buchsbaum non-Cohen-Macaulay ideal. Hence, by Theorem 1.2, the sextic $C \cup L$ is not arithmetically Cohen-Macaulay but arithmetically Buchsbaum. By using another liaison we will see that $C \cup L$ is a set-theoretic complete intersection. The point here is that the quintic C is linked to another multiple line lying again on Q. The liaison is defined by the following equations:

$$(x_0x_3 - x_1x_2, x_1^4 - x_0^3x_2) = \mathfrak{p} \cap \big((x_0, x_1)^3, x_0x_3 - x_1x_2\big).$$

Bezout's theorem and Theorem 3.2(i) establish this equality. Therefore the property that the sextic $C \cup L$ is a set-theoretic complete intersection follows immediately.

Another application of our observations about multiple lines lying on a quadric surface in $\mathbf{P}^3$ is given in the following section on self-linked curves. We want to conclude this section with some work done by Migliore in [1] and [2].

Let $L_1, \ldots, L_m$ be a set of mutually skew lines in $\mathbf{P}^3$ and let $L'_1, \ldots, L'_n$ be another such set (we may assume $m \leq n$). We assume also that L_i meets L'_j in a point, for all i, j. We shall call the configuration in $\mathbf{P}^3$ formed by the union of these lines a $C_{m,n}$-configuration. Note that if $m \geq 3$ then $C_{m,n}$ lies on a smooth quadric surface. Hence for most choices of m and n our Proposition 3.3 tells us when the curve $C_{m,n}$ is arithmetically Cohen-Macaulay and when it is arithmetically Buchsbaum, leaving open only the case $C_{2,n}$. Since the Buchsbaum property of a space curve is reflected in the Hartshorne-Rao module in a simple way (see Corollary 0.4.7), the new techniques of J. Migliore [1, 2] can be used to complete Proposition 3.3.

First, the results of Proposition 3.3 and Geramita-Weibel [1] are summarized in the following theorem.

Theorem 3.6.

(a) *A $C_{0,1}$ is arithmetically Cohen-Macaulay, a $C_{0,2}$ is arithmetically Buchsbaum but not Cohen-Macaulay, and a $C_{0,n}$ is not arithmetically Buchsbaum for $n \geq 3$.*

(b) *A $C_{1,1}$ and a $C_{1,2}$ are arithmetically Cohen-Macaulay, a $C_{1,3}$ is arithmetically Buchsbaum but not Cohen-Macaulay, and a $C_{1,n}$ is not arithmetically Buchsbaum for $n \geq 4$.*

(c) *If a $C_{m,n}$ lies on a smooth quadric surface then it is arithmetically Cohen-Macaulay if and only if $n - m \leq 1$, and it is arithmetically Buchsbaum if and only if $n - m \leq 2$.*

(d) *A $C_{2,3}$ and a $C_{2,4}$ are arithmetically Cohen-Macaulay if they do not lie on a quadric surface.*

Proof: Proposition 3.3 provides (c). Note that this proof is entirely different from the linear algebra proof given by Geramita-Weibel [1], § 10. By using local cohomology it is not too difficult to show (a), (b) and (d). A more geometric approach to the content of (d) is given in the following proof:

Consider a $C_{2,4}$-configuration not lying on a quadric surface. Let Q_1 denote the unique nonsingular quadric surface containing L_1, L_2 and L_1', L_2', L_3' and let Q_2 denote the unique non-singular quadric surface containing L_1, L_2 and L_2', L_3' and L_4'. Select any point $P \in L_1'$ such that $P \notin L_1$, $P \notin L_2$ and let $\tilde{L}_1$ denote the line on Q_1 which meets L_1' at P. Also, pick any point $Q \in L_4'$ such that $Q \notin L_1$, $Q \notin L_2$ and let $\tilde{L}_2$ denote the line on Q_2 which meets L_4' at Q. Clearly $\tilde{L}_1 \neq \tilde{L}_2$ since we already have four lines (L_1, L_2, L_2', L_3') in the intersection of Q_1 and Q_2. Let Π_1, respectively Π_2, denote the tangent planes to Q_1 at P and to Q_2 at Q. Clearly, $\Pi_1 \neq \Pi_2$ since $L_1' \subseteq \Pi_1$ and $L_4 \subseteq \Pi_2$ and $L_1' \cap L_4' = \emptyset$.

Claim. The $C_{2,4}$-configuration above is linked to a Cohen-Macaulay curve τ, by the cubic surfaces $F = Q_1\Pi_2$ and $G = Q_2\Pi_1$ and therefore such a $C_{2,4}$-configuration is arithmetically Cohen-Macaulay.

Proof: We have

$$F \cap G = (L_1 \cup L_2 \cup L_2' \cup L_3') \cup (L_1' \cup \tilde{L}_1) \cup (L_4' \cup \tilde{L}_2) \cup (\Pi_1 \cap \Pi_2) \qquad (*)$$

and it remains to identify $\tilde{L}$ with $(\Pi_1 \cap \Pi_2)$.

First observe that $\tilde{L}$ meets $\tilde{L}_1$ and $\tilde{L}_2$ and also meets L_1' and L_4', hence $\tilde{L}$ is distinct from the other eight lines of (*).

Let $\tilde{C} = \tilde{L}_1 \cup \tilde{L}_2 \cup \tilde{L}$. We have only two possibilities:

(1) $\tilde{L}_1 \cap \tilde{L}_2 \neq \emptyset$, in which case $\tilde{C}$ is a plane curve and hence Cohen-Macaulay;

(2) $\tilde{L}_1 \cap \tilde{L}_2 = \emptyset$, in which case $\tilde{C}$ is still Cohen-Macaulay since $\tilde{L}$, $\tilde{L}_1$, $\tilde{L}_2$ form a $C_{1,2}$ which must lie on some non-singular quadric surface and the result now follows from Proposition 3.3(b), q.e.d.

Thus the Cohen-Macaulay and Buchsbaum properties of $C_{m,n}$-configurations are fully understood, expect for the case of $C_{2,n}$ for $n \geq 5$ if the configuration does not lie on a quadric surface. Following Migliore [2] we now answer this question.

Theorem 3.7. *Consider $C_{2,n}$-configurations not lying on a quadric surface.*

(a) *A $C_{2,5}$ is arithmetically Buchsbaum but not Cohen-Macaulay.*

(b) *A $C_{2,6}$ is arithmetically Buchsbaum but not Cohen-Macaulay provided it does not lie on a cubic surface.*

A $C_{2,6}$ on a cubic surface is not arithmetically Buchsbaum.

(c) *A $C_{2,n}$ is not arithmetically Buchsbaum for $n \geq 7$.*

We want to describe the basic idea of Migliore's proof since this proof requires a new technique: namely to examine the module structure of the Hartshorne-Rao module in some detail.

Put $S := K[x_0, x_1, x_2, x_3]$ for an algebraically closed field K. Given a graded S-module $M = \bigoplus_{n \in \mathbf{Z}} M_n$ of finite length, the action of $S_1 = H^0(\mathbf{P}^3, \mathcal{O}(1))$ between two consecutive components (i.e. $\Phi_n : S_1 \to \operatorname{Hom}(M_n, M_{n+1})$) gives rise to a degeneracy locus, which can be thought of as lying in $\mathbf{P}S_1 := (\mathbf{P}^3)^*$. Namely, let $V_{n,r} = \mathbf{P}\{L \in S_1 \mid \dim \Phi_n(L) \leq r\}$ and let $V_n = V_{n,s}$ where $s = \max\{r \mid V_{n,r} \subset (\mathbf{P}^3)^*\}$ (or else $V_n = \emptyset$). These loci are isomorphism invariants and are preserved under duals and shifts.

Let C be a curve of $\mathbf{P}^3$. For the Hartshorne-Rao module these loci are related to the geometry of C itself. The main philosophy that emerges from Migliore [1] is that the degeneracy locus generally corresponds to those planes H in $\mathbf{P}^3$ which meet C non-generically, either containing a component of C or having $C \cap H$ impose an unusually small number of conditions on some plane curves on H. By applying these ideas Migliore [1] also derives some necessary conditions for $M(C)$ to have components in negative degrees (see Lemma 3.8 below).

The main point to be made here is that although the dimensions of the various components of $M(C)$ are relatively easy to compute, the module structure requires much more careful study. The object of the following is to examine this structure in some detail:

Let $M = \bigoplus_{n \in \mathbf{Z}} M_n$ be a graded S-module of finite length, and let $S_1 = H^0(\mathbf{P}^3, \mathcal{O}(1))$. The S-module structure of M is given by the collection of vector space homomorphisms:

$$\Phi_n : S_1 \to \operatorname{Hom}_K(M_n, M_{n+1}).$$

Since Φ_n is trivial if either M_n or M_{n+1} is zero, assume that this is not the case for some choice of n. If we choose bases for M_n and for M_{n+1}, and if $L = \alpha_0 x_0 + \alpha_1 x_1 + \alpha_2 x_2 + \alpha_3 x_3 \in S_1$, then Φ_n can be viewed as a $(\dim M_{n+1}) \times (\dim M_n)$ matrix A_n whose entries are linear polynomials in the α_i.

Also, we must understand how the S-multiplication

$$\Phi_n : S_1 \to \operatorname{Hom}_K(M_n(C), M_{n+1}(C))$$

is induced for $M = M(C)$. Let $L \in S_1$. Then L gives a map of sheaves

$$\mathcal{J}_C(n) \xrightarrow{\times L} \mathcal{J}_C(n+1)$$

by usual multiplication, and this induces the homomorphism $\Phi_n(L)$ on the first cohomology.

Observe that the map $\times L$ is injective. A natural way to understand $\Phi_n(L)$ is then to find the cokernel of the map $\times L$ and study the associated long exact cohomology sequence. We will use this idea by considering the following exact sequences:

We have the exact sequence

$$0 \to \mathcal{J}_C(n) \xrightarrow{\times L} \mathcal{J}_C(n+1) \to \mathcal{J}_{H_L \cap C \mid H_L}(n+1) \to 0$$

since the cokernel of the map $\times L$ comes by restricting to the hyperplane H_L defined by $L = 0$. Taking cohomology, one gets

$$\begin{aligned} 0 &\to H^0(\mathbf{P}^3, \mathcal{J}_C(n)) \xrightarrow{\times L} H^0(\mathbf{P}^3, \mathcal{J}_C(n+1)) \\ &\to H^0(H_L, \mathcal{J}_{H_L \cap C \mid H_L}(n+1)) \to M_n(C) \xrightarrow{\Phi_n(L)} M_{n+1}(C) \to \ldots \end{aligned}$$

Now, recall that a curve C in $\mathbf{P}^3$ is arithmetically Buchsbaum if and only if the Hartshorne-Rao module $M(C) = \bigoplus_{n\in\mathbf{Z}} H^1_1(\mathbf{P}^3_K, \mathcal{J}_C(n))$ as a graded S-module is annihilated by the irrelevant ideal (x_0, x_1, x_2, x_3) of S (see Corollary 0.4.7).

Equivalently, C is arithmetically Buchsbaum if and only if the maps

$$\Phi_n(L) : M_n(C) \to M_{n+1}(C)$$

are zero for all $n \in \mathbf{Z}$ and all $L \in S_1$, thus our techniques are applicable to prove Theorem 3.7.

Applying these techniques we can draw some simple conclusions about curves C whose Hartshorne-Rao modules $M(C)$ are shifted sufficiently far to the left. Specifically we want to say something when $M(C)$ has components in negative degree. We will state without proof the following striking lemma of Migliore, Proposition 2.11 in [1]:

Lemma 3.8. *Suppose* $n \leq -1$. *Then*

$$\dim M_n(C) < \dim M_{n+1}(C).$$

In some sense this statement is as good as we can hope to obtain since there are examples where $\dim M_0(C) = \dim M_1(C)$. For example, let C be the disjoint union of a conic and a line. Then $\dim M_0(C) = \dim M_1(C) = 1$ and $\dim M_n(C) = 0$ otherwise. An interesting generalization of this is, to let C be the disjoint union of a line and a plane curve of degree $t + 1$. Then calculations show that $\dim M_n(C) = 1$ *for* $0 \leq n \leq t$ and $\dim M_n(C) = 0$ otherwise.

Proof of Theorem 3.7. We shall make frequent use of the following exact sequence:

$$0 \to \mathcal{J}_C(n) \to \mathcal{O}_{\mathbf{P}^3}(n) \to \mathcal{O}_C(n) \to 0.$$

Furthermore, we will apply the following basic fact of our $C_{2,n}$-configuration:

$$h^0(\mathcal{O}_{C_{2,n}}(d)) = (n+2)(d+1) - 2n \quad \text{for } d \geq 1.$$

To see this, note that no more than two lines in the $C_{2,n}$-configuration meet at any point, and that the restriction of $H^0(\mathcal{O}_{C_{2,n}}(d))$ to any line L_i or L'_j is isomorphic to a subspace of $H^0(\mathcal{O}_{\mathbf{P}^1}(d))$. We then carefully construct $H^0(\mathcal{O}_{C_{2,n}}(d))$ by patching together sections of the L_i and L'_j.

Consider first the curve $L_1 \cup L'_1$. We can find embeddings $\psi_1 : \mathbf{P}^1 \to \mathbf{P}^3$ and $\psi'_1 : \mathbf{P}^1 \to \mathbf{P}^3$ whose images are (respectively) L_1 and L'_1, subject to the condition that for some $P, Q \in \mathbf{P}^1$, $\psi_1(P) = \psi'_1(Q)$. Let $U = \mathbf{A}^1$ be the complement of any point other than P and Q in $\mathbf{P}^1$. There is a natural bijection between linear forms $F \in H^0(\mathcal{O}_{\mathbf{P}^1}(d))$ and polynomials f of degree $\leq d$ on $\mathbf{A}^1$. Then the elements of $H^0(\mathcal{O}_{L_1 \cup L'_1}(d))$ are in one-to-one correspondence with the pairs $(F_1, F_2) \in H^0(\mathcal{O}_{\mathbf{P}^1}(d)) \times H^0(\mathcal{O}_{\mathbf{P}^1}(d))$ such that $f_1(P) = f_2(Q)$. To find all such elements we may choose F_1 arbitrarily, then the condition $f_1(P) = f_2(Q)$ for F_2 gives a hyperplane (not necessarily through 0) in $H^0(\mathcal{O}_{\mathbf{P}^1}(d))$. Thus $h^0(\mathcal{O}_{L_1 \cup L'_1}(d)) = d + 1 + d = 2d + 1$.

To complete of proof, one then "attaches" the lines $L_2, L'_2, L'_3, \ldots, L'_n$ (in that order!), always using the same (well-chosen) copy of $\mathbf{A}^1 \subset \mathbf{P}^1$. Note that specifying the value of a polynomial at two points is always a codimension-two condition for $d \geq 1$. Therefore $h^0(\mathcal{O}_{C_{2,n}}(d)) = (d+1) + (d) + (d) + (n-1)(d-1) = (n+2)(d+1) - 2n$ as desired.

Proof of (a): Note first that $\dim M_1(C_{2,5}) = 0$, $\dim M_2(C_{2,5}) = 1$, and $\dim M_3(C_{2,5}) = h^0(\mathcal{J}_{C_{2,5}}(3)) - 2$. We will show that $\dim M_r(C_{2,5}) = 0$ for $r \geq 3$.

First suppose that $\dim M_3(C_{2,5}) \geq 1$. Then $h^0(\mathcal{J}_{C_{2,5}}(3)) \geq 3$. Note that even in the "worst" case, when $L_1, L_2, L_1', L_2', L_3'$ and L_4' all lie on a quadric surface Q, we still have $h^0(\mathcal{J}_{C_{2,5}}(3)) = 2$. (Any cubic containing $C_{2,5}$ is the union of Q with a plane containing L_5'.) Hence if $h^0(\mathcal{J}_{C_{2,5}}(3)) \geq 3$ we must have two cubics containing $C_{2,5}$ and meeting properly. These link $C_{2,5}$ to a curve C'. By Lemma 2.5 we get $\dim M_3(C_{2,5}) = \dim M_{-1}(C')$. But since $\dim M_0(C') = 1$, Lemma 3.8 shows that this is impossible. Thus $\dim M_3(C_{2,5}) = 0$.

In order to show that $\dim M_r(C_{2,5}) = 0$ for $r \geq 4$ we link $C_{2,5}$ to a curve C' via a cubic and a quartic surface and apply the same reasoning as above. Therefore $M(C_{2,5})$ is non-zero in exactly one component, so $C_{2,5}$ must be Buchsbaum.

Proof of (b): If the $C_{2,6}$ does not lie on a cubic surface then we have $\dim M_1(C_{2,6}) = 0$, $\dim M_2(C_{2,6}) = 2$, $\dim M_3(C_{2,6}) = 0$, and $\dim M_4(C_{2,6}) = h^0(\mathcal{J}_{C_{2,6}}(4)) - 7$. If there exist two quartic surfaces containing $C_{2,6}$ and meeting properly then they link $C_{2,6}$ to a curve C' with $\dim M_r(C_{2,6}) = \dim M_{4-r}(C')$. In particular $\dim M_1(C') = 0$, and $0 = \dim M_0(C') = \dim M_{-1}(C') = \ldots$ (since $\deg C' = 8$ and not every hyperplane meets C' in eight collinear points). Therefore $M(C_{2,6})$ is non-zero in just one component and $C_{2,6}$ is therefore Buchsbaum.

If all quartics containing $C_{2,6}$ have a common component then the condition $h^0(\mathcal{J}_{C_{2,6}}(4)) \geq 7$ forces $C_{2,6}$ to have the property that seven of the eight lines, say $L_1, L_2, L_1', \ldots, L_5'$, lie on a quadric surface Q. (Note that any quartic surface containing $L_1', \ldots, L_5'$, must contain Q.) But then $C_{2,6}$ lies on a pencil of (reducible) cubic surfaces, contradicting our hypothesis and completing the first part of (b).

For the second part of (b) we first consider the case just mentioned. Then $\dim M_2(C_{2,6}) = 2$, $\dim M_3(C_{2,6}) = h^0(\mathcal{J}_{C_{2,6}}(3)) = 2$ and we have the above-mentioned exact sequence

$$0 \to H^0(\mathcal{J}_{C_{2,6}}(3)) \to H^0(H_L, \mathcal{J}_{C_{2,6} \cap H_L}(3)) \to M_2(C_{2,6}) \xrightarrow{\Phi_2(L)} M_3(C_{2,6}) \qquad (*)$$

For a general hyperplane H_L, $C_{2,6} \cap H_L$ consists of seven co-conical points and one additional point. Then $h^0(H_L, \mathcal{J}_{C_{2,6} \cap H_L}(3)) = 2$, so $\Phi_2(L)$ cannot be the zero map. Therefore $C_{2,6}$ is not Buchsbaum. The only remaining case is when $C_{2,6}$ lies on a unique cubic surface S. We first claim that S must be irreducible. Clearly S cannot be the union of three planes. If S is the union of a quadric Q and a plane H then Q must contain exactly five of the lines L_j'. But then Q also contains L_1 and L_2, therefore as above H (and hence S) is not unique. Therefore S is irreducible.

Now $\dim M_3(C_{2,6}) = h^0(\mathcal{J}_{C_{2,6}}(3)) = 1$ and by $(*)$ we only need to show that there exists an H_L for which $h^0(H_L, \mathcal{J}_{C_{2,6} \cap H_L}(3)) < 3$.

Equivalently, we will show that not every hyperplane section of $C_{2,6}$ lies on a conic (since clearly the general hyperplane section does not have five collinear points).

In Chapter 4 of Griffiths-Harris [1] there is a description of all cubic surfaces and the lines they contain. From it one can check that the only irreducible cubic surface S in $\mathbf{P}^3$ which contains a $C_{2,6}$ is the projection from $\mathbf{P}^4$ of a Steiner surface. Hence S is to be counted twice along L_1 (say) and it contains infinitely many lines, each meeting L_1 and L_2. Among these lines are $L_1', \ldots, L_6'$, and no four of them can lie on the same quadric surface. Next let Q_1 be the quadric containing L_1', L_2' and L_3' and let

Q_2 be the quadric containing L_4', L_5' and L_6'. Then $Q_1 \cap Q_2$ consists of L_1, L_2 and two other lines (or a double line). A general hyperplane H_L meets L_1', L_2', L_3', L_1 and L_2 in a unique conic, namely $H_L \cap Q_1$. Similarly, H_L meets L_4', L_5', L_6', L_1 and L_2 in a unique conic $H_L \cap Q_2$. Since $Q_1 \neq Q_2$, $H_L \cap C_{2,6}$ cannot lie on a conic which finishes the proof of (b).

Proof of (c): We have $\dim M_2(C_{2,n}) = n - 4$ and $\dim M_3(C_{2,n}) = 2n - 12 + h^0(\mathcal{J}_{C_{2,n}}(3))$. By (*) we have to show that $h^0(H_L, \mathcal{J}_{C_{2,n} \cap H_L}(3)) < n - 4 + h^0(\mathcal{J}_{C_{2,n}}(3))$ for some H_L. Some thought shows that this is true since $n \geq 7$, q.e.d.

Remark 3.9. We know that the technique of Liaison Addition introduced in the second paragraph of this chapter is useful in constructing examples of Buchsbaum curves. For example, one can form a $C_{2,6}$ not lying on a cubic surface by "adding" two pairs of skew lines. Similarly, one can construct a Buchsbaum configuration of lines C for which $M(C)$ is one-dimensional in two consecutive components.

We conclude this parapraph with two examples by considering again multiple lines. First let it be said that a totally different approach to Theorem 3.2 for double lines can be found in Geramita-Maroscia-Vogel [2].

Example 3.10. We would like to construct in this example a double line on a non-singular cubic surface in $\mathbf{P}^3$ which is not arithmetically Buchsbaum.

Consider $G = x_0x_3^2 - x_1x_2^2$ and note that by Theorem 3.2 $\mathfrak{q} = (x_0^2, x_1^2, x_0x_1, G)$ is (x_0, x_1)-primary of length 2 and is not arithmetically Buchsbaum. If we let $F = x_1^3 - x_0^2x_2$ then $F \in \mathfrak{q}$. If the characteristic of the base field is $\neq 2, 3$ then $F + G$ describes a non-singular cubic surface and therefore $\mathfrak{q}$ describes a double line on this surface which is not arithmetically Buchsbaum.

One may also show that $\mathfrak{q}$ is linked, by F and G, to the monomial curve with generic point (s^7, s^5t^2, st^6, t^7). This curve has been shown not to be arithmetically Buchsbaum (see Theorem 1.20).

Examples 3.11. In this example we show that Theorem 3.2(ii) does not extend to lines lying on a nonsingular quadric hypersurfaces in $\mathbf{P}^n$ $(n > 3)$. More precisely, consider the following ideal in $K[x_0, x_1, \ldots, x_4]$

$$\mathfrak{a} = (x_1^2 - x_0x_3, x_2^2 - x_1x_3, x_3^2 - x_2x_4).$$

Then $\mathfrak{a}$ defines a curve in $\mathbf{P}^4(K)$ and it is easy to check that

$$\mathfrak{a} = \mathfrak{p} \cap \mathfrak{q}$$

where $\mathfrak{p}$ is the defining prime ideal of the monomial curve, C, with generic point $(s^5, s^3t^2, s^2t^3, st^4, t^5)$ and $\mathfrak{q}$ is an (x_1, x_2, x_3)-primary ideal of length 3. Now $\mathfrak{a}$ contains $Q = x_1^2 - x_0x_3 + x_3^2 - x_2x_4$ which is non-singular for char $K \neq 2$. Note that C is arithmetically Cohen-Macaulay since the defining prime ideal $\mathfrak{p}$ of C is given by

$$\mathfrak{p} = (x_1^2 - x_0x_3, x_0x_4 - x_1x_2, x_2^2 - x_1x_3, x_1x_4 - x_2x_3, x_2x_4 - x_3^2).$$

It follows by liaison (see Remark 1.5(i)) that $\mathfrak{q}$ is also arithmetically Cohen-Macaulay. Note that here we may only say that $\mathfrak{q}$ describes "a" 3-fold line on the quadric hypersurface Q but that "the" 3-fold line on a nonsingular quadric surface in P^3 is not even arithmetically Buchsbaum.

§ 4. On self-linked curves in $\mathbf{P}^3$

Let C be a curve of $\mathbf{P}^3$. Following Rao [2], Schwartau [1] and Geramita-Maroscia-Vogel [1] we wish to study the geometric situation where twice C is the intersection of two hypersurfaces, i.e., where there exist two hypersurfaces which meet exactly on C and which make contact with each other with multiplicity two, along C (see Corollary 4.9 below). We will look at the phenomena algebraically using the techniques of liaison. C will be called *self-linked* or *linked to itself* if we can find a complete intersection X containing C such that C is residual to itself in X, i.e.

$$\mathcal{J}_C = \operatorname{Ann}(\mathcal{J}_C/\mathcal{J}_X),$$

where $\mathcal{J}_C$ is the ideal sheaf of C.

The simplest non-trivial example of self-linkage is the case of the twisted cubic curve in $\mathbf{P}^3$, which, even though not a complete intersection, is set-theoretically a complete intersection by means of a quadric and a cubic which touch along C. Catanese [1] ipresents a method of Gallarati where, starting from this example, more complicated examples are created by preserving self-linkage under liaison. This method thus creates many examples of self-linked curves which are arithmetically Cohen-Macaulay. The other commonly known example of self-linkage is the example of two Kummer surfaces touching along a smooth curve of degree 8 and genus 5. This example was also discovered by M. Noether [1] in 1883 and has been recently treated by W. Barth [1] in a study of the Mumford-Horrocks bundle on $\mathbf{P}^4$. It is a curve that is arithmetically Buchsbaum with Buchsbaum invariant 2. One question that arises immediately is to find those liaison equivalence classes of curves in $\mathbf{P}^3$ which contain a self-linked curve. By studying liaison we see that an immediate necessary condition for a curve C to be self-linked is that its liaison invariant be self-dual up to grading. Since the liaison invariant characterizes liaison classes, this is a restrictive condition on the liaison class in order for it to contain a self-linked curve. As an example, the liaison class in $\mathbf{P}^3$ of two skew lines, contains no self-linked curve in case the characteristic of the ground field is not equal two. Nowadays, it is still an open question to determine those liaison classes in $\mathbf{P}^3$ which contain a self-linked curve. Liaison addition from § 2 of this chapter provides one step toward the solution of this problem, in the form of the following theorem and its corollary. First we recall a definition.

Definition 4.1. Let C be a curve in $\mathbf{P}_K^3$, K an algebraically closed field. C is said to be *self-linked* if there is a complete intersection V containing C such that

$$\mathcal{J}_C = \operatorname{Ann}(\mathcal{J}_C/\mathcal{J}_V)$$

where $\mathcal{J}_C$ is the ideal sheaf of C.

Now, if C is defined by the ideal $I(C)$ in $S = K[x_0, x_1, x_2, x_3]$ and V is defined by the ideal $(F, G) \subset S$, where F and G are forms of $I(C)$, then it is easy to see that this definition is equivalent to

$$(F, G) : I(C) = I(C).$$

So, in particular, if C is self-linked then $I(C)^2 \subseteq (F, G) \subseteq I(C)$ and hence C is a set-theoretic complete intersection. Of course, the converse is not true in general.

Theorem 4.2. *Suppose two curves $C, C' \subset \mathbf{P}^3_K$ are algebraically linked* (see Definition 1.1) *by f, f'. Then the curve $Cf + f'C'$* (see Definition 2.6) *is self-linked by f^2, f'^2.*

Proof: Since C and C' are algebraically linked by f, f' we get:

a) $f, f' \in I(C) \cap I(C')$, b) (f, f') is a complete intersection,

c) $(f, f'):I(C) = I(C')$, d) $(f, f'):I(C') = I(C)$.

Now a) and b) guarantee that $Cf + f'C'$ makes sense. This new curve has defining ideal $f' \cdot I(C) + f \cdot I(C')$ by Corollary 2.8. We are to show it is self-linked via the complete intersection f^2, f'^2. Since $f^2, f'^2 \in f' \cdot I(C) + f \cdot I(C')$ we have to prove:

$$(f^2, f'^2):\big(f' \cdot I(C) + f \cdot I(C')\big) = f' \cdot I(C) + f \cdot I(C').$$

We show this property in two steps:

1. $\supseteqq$: We must show that for all $i, j \in I(C)$, $i', j' \in I(C')$ we obtain:

$$(f' \cdot i + f \cdot i')\,(f' \cdot j + f \cdot j') \in (f^2, f'^2).$$

Since $i \cdot j'$ and $i' \cdot j$ are in (f, f') by either c) or d) the assertion follows immediately.

2. $\subseteqq$: Suppose $\psi \in (f^2, f'^2):\big(f' \cdot I(C) + f \cdot I(C')\big)$. We have to show $\psi \in f' \cdot I(C) + f \cdot I(C')$. We have $\psi \cdot f' \cdot f \in (f^2, f'^2)$. But then the relative primeness of f, f' implies:

$$\psi \cdot f' \in (f^2, f'^2):f = (f, f'^2).$$

Hence $\psi \in (f, f'^2):f' = (f, f')$. Set

$$\psi = \gamma \cdot f + \delta \cdot f' \quad \text{for some } \gamma, \delta \in S.$$

We finish by showing that $\gamma \in I(C')$ and $\delta \in I(C)$. The fact that

$$\gamma \cdot f + \delta \cdot f' \in (f, f^2):\big(f' \cdot I(C) + f \cdot I(C')\big)$$

implies that for all elements $i \in I(C)$,

$$(\gamma \cdot f + \delta \cdot f') \cdot f' \cdot i \in (f^2, f'^2).$$

Hence $\gamma \cdot f \cdot f' \cdot i \in (f^2, f'^2)$. The relative primeness of f, f' again provides: $\gamma \cdot i \in (f, f')$. Therefore we obtain $\gamma \in (f, f'):I(C) = I(C')$ by c).

Similary, by using d) above we can show $\delta \in I(C)$ as desired, q.e.d.

Remark 4.3. It is not too difficult to show that this theorem has the following converse: Let C, C' be two curves in $\mathbf{P}^3_K$ and (f, f') a complete intersection containing both C and C'. Assume that the curve $Cf + f'C'$ is self-linked by f^2, f'^2 then C and C' are algebraically linked by f, f'. Hence the liaison addition yields a characterization of the property that two curves are algebraically linked via self-linkage.

We now concentrate on the promised corollary to Theorem 4.2.

Corollary 4.4. *Let E be a graded S-module of finite length. Then for $v \gg 0$ or $v \ll 0$ the liaison class associated to the module $E \oplus E^v(2v)$ contains a self-linked curve.*

Proof: By the theory of liaison realization (see Theorem 2.18), some twist $E(\eta)$ can be realized as the liaison invariant $E(C)$ of a curve $C \subset \mathbf{P}^3_K$. Since $E(\eta)$ is again of finite length it is possible to construct an algebraic link of C and a new curve C' via any homogeneous S-sequence $(f, f') \subseteq I(C)$ where C' is defined by the homogeneous ideal quotient $(f, f'):I(C)$. The equivalent Definition 1.1 shows that C and C' are algebraically linked. Then by Theorem 4.2, we know that $Cf + f'C'$ is a self-linked curve in $\mathbf{P}^3_K$. But this curve has liaison invariant by Corollary 2.8 given by

$$E(C)\,(\deg f') \oplus E(C')\,(\deg f)$$
$$= E(\eta)\,(\deg f') \oplus [E(\eta)^v\,(\deg f + \deg f' + 4)]\,(\deg f)$$

by the transformation law of Lemma 2.5

$$= E(\eta + \deg f') \oplus E^v(-\eta + 2\deg f + \deg f' + 4).$$

Now, this module determines the same liaison class as the module

$$E \oplus E^v(2\deg f - 2\eta + 4) =: E \oplus E^v(2v)$$

where $v := \deg f - \eta + 2$.

We now point out that in choosing the homogeneous S-sequence $(f, f') \cdot S \subseteq I(C)$, the first element f may be taken as any non-zero homogeneous element of $I(C)$. In particular we may replace f by $x_0 \cdot f, x_0^2 \cdot f, x_0^3 \cdot f, \ldots$ etc. Thus we obtain self-linked curves as above for all $v \gg 0$.

To obtain the desired curves for $v \ll 0$ simply replace E by E^v in the argument above. We then obtain self-linked curves in the liaison class of $E^v \oplus E(2v)$ for $v \gg 0$; but these are the same liaison classes as those of the k-duals $E \oplus E^v(-2v)$, which finishes the proof, q.e.d.

Remark 4.5. In characteristic $\neq 2$ this result has been proven (independently) by Rao [2] using some transversality theorems of S. Kleiman. Following Schwartau [1] advantages of our method are, that it applies to any characteristic, and that explicit bounds for v may be given as follows below, since we have produced specific bounds for v during the course of the proof. For "$v \gg 0$" it suffices that

$$v \geq \min\{\deg f - \eta + 2 \mid E(\eta) \text{ is realized by a curve on the surface } f\}.$$

If the module E itself can be realized as the liaison invariant of a curve in $\mathbf{P}^3_K$, we may write a bound for v as follows:

$$v \geq \operatorname{def} f + 2,$$

where f is the least degree surface in $\mathbf{P}^3_K$ on which E may be realized.

For "$v \ll 0$" it is enough to have

$$v \leq -\min\{\deg f - \eta + 2 \mid E^v(\eta) \text{ is realized by a curve on the surface } f\}.$$

If E^v itself may be realized, we may write a bound as follows:

$$v \leq -(\deg f + 2),$$

where f is the least degree surface in $\mathbf{P}^3_K$ on which E^v may be realized.

We now show that a full converse to Corollary 4.4 is impossible.

We will see that in characteristic 2 the liaison class of arithmetically Buchsbaum curves with Buchsbaum invariant 1 (i.e. the liaison class of two skew lines in $\mathbf{P}^3_K$) contains a self-linked curve. In fact, the following theorem gives the curve as $(x_0^2, x_1^2, x_0x_1, x_0x_3 - x_1x_2)$ with Schwartau's liaisbn invariant $K(4)$. At any rate, we have now demonstrated as promised that Corollary 4.4 has no full converse. Hence we come back to our observations of § 3 of this chapter.

Theorem 4.6. *Let C_n denote the curve in $\mathbf{P}_K^3$ defined by the ideal*

$$\mathfrak{a}_n = (\mathfrak{p}^n, Q)$$

where $\mathfrak{p} = (x_0, x_1)$, $Q = x_0x_3 - x_1x_2$ and $n \geq 2$. Then:

(i) *If $n > 2$ then C_n is not self-linked,*

(ii) *C_2 is self-linked if and only if* char $K = 2$.

Proof: We shall give the proof in three steps.

Step 1: $\mathfrak{a}_n^2 = (\mathfrak{p}^{2n}, \mathfrak{p}^nQ, Q^2)$ is $\mathfrak{p}$-primary for all $n \geq 2$.

To prove this we use an argument similar to the one we developed in the proof of Theorem 3.2 (see Claim 1 of that proof). More precisely, if $\mathfrak{a}_n^2$ is not primary then we can find a linear form $L = \alpha x_2 + \beta x_3 \big(\alpha, \beta \in K, (\alpha, \beta) \neq (0, 0)\big)$ such that $\mathfrak{a}_n^2 : (L) \supsetneqq \mathfrak{a}_n$.

Let H be a form in $K[x_0, x_1, x_2, x_3]$ such that $H \notin \mathfrak{a}_n^2$ and $HL \in \mathfrak{a}_n^2$. Now consider H as a polynomial in $K([x_2, x_3])\,[x_0, x_1]$ and write $H = H_1 + \ldots + H_t$ where $\deg H_i$ (in x_0, x_1) is i. Since $H \notin \mathfrak{a}_n^2$ and $\mathfrak{p}^{2n} \subseteq \mathfrak{a}_n^2$, we may assume that $t \leq 2n - 1$. Thus we have

$$HL = H_1L + \ldots + H_tL \in \mathfrak{a}_n^2$$

and since $\deg L = 0$ (in x_0, x_1), $\deg H_iL = i$.

Since $\mathfrak{a}_n^2$ is still a homogeneous ideal, when considered in $(K[x_2, x_3])\,[x_0, x_1]$, it follows that $H_iL \in \mathfrak{a}_n^2$, $1 \leq i \leq t$. Hence we obtain:

$$\text{If } i = 1,\ H_1L \in \mathfrak{a}^2 \Rightarrow H_1 \equiv 0;$$

$$\text{if } 2 \leq i \leq n,\ H_iL^n \in \mathfrak{a}_n^2 \Rightarrow Q^2 \mid H_i;$$

$$\text{if } n < i \leq t,\ H_iL \in \mathfrak{a}_n^2 \Rightarrow Q \mid H_i \text{ and therefore } H_i = QH'_i$$

where $\deg H'_i \geq n$ (in x_0, x_1) and therefore $H_i \in \mathfrak{p}^nQ$. Thus, $H \in \mathfrak{a}_n^2$, which is a contradiction.

Step 2: $\mathfrak{a}_n^2$ is a $\mathfrak{p}$-primary ideal of length $3n$, for all $n > 1$.

It is enough to check that in the localization $K[x_0, \ldots, x_3]_\mathfrak{p}$ we have the following saturated chain of primary ideals:

$$\begin{aligned}
\mathfrak{a}_n^2 &\subset (\mathfrak{a}_n^2, x_0^{2n-1}) \subset (\mathfrak{a}_n^2, x_0^{2n-1}, x_0^{2n-2}) \subset \ldots \subset (\mathfrak{a}_n^2, x_0^{2n-1}, \ldots, x_0^n)\\
&\subset (\mathfrak{a}_n^2, \ldots, x_0^n, x_0^{n-1}x_1) \subset (\mathfrak{a}_n^2, \ldots, x_0^n, x_0^{n-1}x_1, x_0^{n-1})\\
&\subset (\mathfrak{a}_n^2, \ldots, x_0^{n-1}, x_0^{n-2}x_1) \subset (\mathfrak{a}_n^2, \ldots, x_0^{n-1}, x_0^{n-2}x_1, x_0^{n-2})\\
&\subset (\mathfrak{a}_n^2, \ldots, x_0^{n-1}, x_0^{n-2}x_1, x_0^{n-2}, x_0^{n-1}x_1)\\
&\subset (\mathfrak{a}_n^2, \ldots, x_0^{n-2}, x_0^{n-1}x_1, x_0^{n-3}) \subset \ldots \subset (\mathfrak{a}_n^2, \ldots, x_0^2)\\
&\subset (\mathfrak{a}_n^2, \ldots, x_0^2, x_0x_1) \subset (\mathfrak{a}_n^2, \ldots, x_0^2, x_0x_1, x_0) \subset (\mathfrak{a}_n^2, \ldots, x_0^2, x_0x_1, x_0, x_1),
\end{aligned}$$

of length precisely $(n + 1) + 2(n - 1) + 1 = 3n$.

We are now ready to prove the theorem.

Suppose C_n is self-linked ($n \geq 2$). Then we must have two forms F, G in $\mathfrak{a}_n$, such that $(F, G) : \mathfrak{a}_n = \mathfrak{a}_n$ and consequently $\mathfrak{a}_n^2 \subseteq (F, G) \subseteq \mathfrak{a}_n$.

It is clear from Step 1 that $\mathfrak{a}_n^2 \subset (F, G)$. Let $\alpha = \deg F$, $\beta = \deg G$. Then from Step 2 we must have $\alpha \cdot \beta < 3n$. We may as well assume $\alpha \leq \beta$ and since $(F, G) \subseteq \mathfrak{a}_n$ that $\alpha \geq 2$, $\beta \geq 2$.

We distinguish two cases:

(i) $n > 2$: If $\alpha > 2$ and $\alpha \leq \beta$ we must have $\alpha < n$ and $\beta < n$. But then $Q \mid F$ and $Q \mid G$ which is a contradiction.

If $\alpha = 2$ then $F = Q$ and therefore (F, G) is a $\mathfrak{p}$-primary ideal of length 2β containing Q. By Theorem 3.2 we obtain that $(F, G) = (Q, G) = (\mathfrak{p}^{2\beta}, Q)$ which is impossible since $(\mathfrak{p}^{2\beta}, Q)$ is never a complete intersection.

(ii) $n = 2$: We first show that if char $K \neq 2$ then C_2 is not self-linked.

Suppose C_2 is self-linked by two forms F, G. Then, in order to have $\mathfrak{a}_2^2 \subset (F, G) \subset (\mathfrak{p}^2, Q)$ we must have $\deg F = \deg G = 2$. (This follows since, by Step 2, $\mathfrak{a}_2^2$ has length 6 and we know $(\mathfrak{p}^2, Q)$ has length 2. We also know that neither $\mathfrak{a}_2$ nor $\mathfrak{a}_2^2$ is itself a complete intersection.) We distinguish two cases:

(a) Either F or G is nonsingular.

In this case, (assuming F is nonsingular), we have that (F, G) is a primary ideal of length 4 whose associated prime defines a line on the nonsingular quadric surface $F = 0$. But from Theorem 3.2 we see that this ideal is not a complete intersection. Thus this case cannot occur.

(b) Both F and G are singular.

It is easy to see that the general quadric in

$$\mathfrak{a}_2, \alpha x_0^2 + \beta x_1^2 + \gamma x_0 x_1 + \delta(x_0 x_3 - x_1 x_2) \qquad (\alpha, \beta, \gamma, \delta \in K),$$

is singular if and only if $\delta = 0$. Thus both F and G have this form. Since both F and G cannot have a common factor we are reduced to consider two cases:

(i) $F = x_0^2 + \gamma x_0 x$, $\quad G = x_1^2 + \gamma' x_0 x_1$,

(ii) $F = x_0^2 + \beta x_1^2$, $\quad G = x_0 x_1$.

In either case, we must have $Q^2 \in (F, G)$.

In case (i) this gives that

$$x_0 x_1(\gamma x_3^2 + \gamma' x_2^2 + 2x_2 x_3) \in (F, G).$$

But $\gamma x_3^2 + \gamma' x_2^2 + 2x_2 x_3$ (char $K \neq 2$) $\notin (x_0, x_1)$, and therefore $x_0 x_1 \in (F, G)$. Thus $(F, G) = (x_0^2, x_1^2)$ in this case and hence $2x_0 x_1 x_2 x_3 \in (x_0^2, x_1^2)$ which is a contradiction.

In case (ii) if we use the fact that $Q^2 \in (F, G)$ we find that $(x_0^2 x_3^2 + x_1^2 x_2^2) \in (F, G)$ and therefore $x_1^2(\beta x_3^2 + x_2^2) \in (F, G)$. Since $\beta x_3^2 + x_2^2 \notin (x_0, x_1)$ we obtain $x_1^2 \in (F, G)$ from which $(F, G) = (x_0^2, x_0 x_1)$, a contradiction.

Thus, if char $K \neq 2$, C_2 is not self-linked.

To see that C_2 is self-linked in characteristic 2 it will suffice to establish the following:

Claim. $(x_0^2, x_1^2):(\mathfrak{p}^2, Q) = (\mathfrak{p}^2, x_0 x_3 + x_1 x_2)$ for any field K.

Proof: We have

$$\begin{aligned}(x_0^2, x_1^2):(\mathfrak{p}^2, Q) &= [(x_0^2, x_1^2):x_0 x_1] \cap [(x_0^2, x_1^2):Q] \\ &= (x_0, x_1) \cap [(x_0^2, x_1^2):(x_0 x_3 - x_1 x_2)] = (x_0^2, x_1^2):(x_0 x_3 - x_1 x_2).\end{aligned}$$

Clearly $(\mathfrak{p}^2, x_0x_3 + x_1x_2) \subseteq (x_0^2, x_1^2):(x_0x_3 - x_1x_2)$. Now $(\mathfrak{p}^2, x_0x_3 + x_1x_2)$ is an (x_0, x_1)-primary ideal of length 2 (see Theorem 3.2) and therefore if this inclusion of primary ideals were strict we would have

$$(x_0^2, x_1^2):(x_0x_3 - x_1x_2) = (x_0, x_1).$$

This is clearly false and the inclusion is an equality. This concludes the proof of Theorem 4.6.

Remark 4.7. The statement (ii) in Theorem 4.6 was also proved by Rao [2] (Example 10) and by Schwartau [1] (Corollary 7.7). The extra information we obtained in Theorem 3.2 allows our method of proof to proceed in a more direct fashion. Rao [2] shows that from (ii) of Theorem 4.6 it follows that the liaison class of arithmetical Buchsbaum curves with invariant 1 contains no self-linked curves, when char $K \neq 2$. Therefore we want to look at an example from this liaison class which shows that self-linkage is not preserved under liaison, when char $K = 2$.

Example 4.8. The curve C_2 is linked to the rational nonsingular quartic curve in $\mathbf{P}_K^3$ for arbitrary characteristic of K. The liaison in this case is given by the following surfaces:

$$F = x_0x_3 - x_1x_2 \quad \text{and} \quad G = x_1^3 - x_0^2x_2,$$

since the rational nonsingular quartic is given parametrically by

$$(s^4, s^3t, st^3, t^4).$$

Now, this quartic is not self-linked in any characteristic. If it were self-linked, it would be a set-theoretic complete intersection, one of the surfaces being the (unique) quadric containing it by Corollary 4.9 below. But it is well-known that the quartic is never a set-theoretic complete intersection of the quadric with any surface which contains the quartic. This fact is true in any characteristic (see, e.g., Roloff-Stückrad [1], Lemma 1). Hence this quartic is an arithmetically Buchsbaum curve which shows us that when char $K = 2$ self-linkage is not preserved under liaison.

It is possible to give more general example of this phenomenon which show that self-linkage is not preserved under liaison in any characteristic (see Geramita-Maroscia-Vogel [1], Example 2.4). We have used the following corollary:

Corollary 4.9. *Let C be any irreducible curve in $\mathbf{P}^3$. Assume that C is self-linked by two surfaces F and G. Then*

$$\deg F \cdot \deg G = 2 \cdot \deg C$$

Proof: Let $\mathfrak{p}$ be the defining prime ideal of C. If $S = K[x_0, \ldots, x_3]$, then $S_\mathfrak{p}$ is a two-dimensional regular local ring and therefore $\mathfrak{p}^2 \cdot S$ is never a complete intersection. Since C is self-linked by F and G we obtain:

$$\mathfrak{p}^2 \subsetneqq (F, G) \subset \mathfrak{p}.$$

Since $\mathfrak{p}^2 \cdot S_\mathfrak{p} \neq (F, G) \cdot S_\mathfrak{p}$ and $\mathfrak{p}^2 \cdot S_\mathfrak{p}$ is a $\mathfrak{p} \cdot S_\mathfrak{p}$-primary ideal of length 3, we get that (F, G) is a $\mathfrak{p}$-primary ideal of length 2; that is, $\deg F \cdot \deg G = \deg V = 2 \cdot \deg C$, where V is the curve with defining ideal (F, G), q.e.d.

We conclude this chapter with the following remark:

Remark 4.10. In 1983, Juan C. Migliore [1] has given a classification of sets of skew lines up to liaison. It is shown that if C and C' consist of $t \geq 3$ and t' skew lines, respectively, then their linkage properties depend primarily on whether or not they lie on a smooth quadric surface. If C lies on a quadric surface then C is linked to C' if and only if $t = t'$ and C' also lies on the same quadric; that is, it is shown how the quadric fully determines which sets of skew lines are in the liaison class of C.

If C does not lie on a quadric then it is not evenly linked to any other set of skew lines. If C is in general position then it is not oddly linked to any set of skew lines, but in special cases it can be oddly linked to at most one other set. Hence it is shown that if C does not lie on a quadric surface then it is essentially unique with respect to liaison.

Furthermore, there is a classification of "double lines" up to liaison. More precisely, given two schemes C and C' of degree 2, each supported on a line, a necessary and sufficient set of conditions is given for C' to be in the liaison class of C and a description how they can be linked.

The reader is referred to Migliore's thesis for more details (see also Migliore [2], [3]).

Chapter IV

Rees modules and associated graded modules of a Buchsbaum module

Hironaka [1], in his paper on desingularization of algebraic varieties over a field of characteristic 0, in order to deal with singular points develops the algebraic apparatus of the associated graded ring and introduces the theory of normal flatness. Such an approach necessitates a deep investigation of blowing-up and monoidal transformations. Let R be a Noetherian ring, $\mathfrak{q} \subset R$ an ideal. $\bigoplus_{n=0}^{\infty} \mathfrak{q}^n$ is a graded R-algebra of finite type. Hence $\operatorname{Proj}\left(\bigoplus_{n=0}^{\infty} \mathfrak{q}^n\right)$ is a finite type projective (Spec R)-scheme. We will set $\operatorname{Proj}\left(\bigoplus_{n=0}^{\infty} \mathfrak{q}^n\right) = \mathrm{Bl}_{\mathfrak{q}}R =$ the *blowing-up of R with center* $\mathfrak{q}$. We recall the terminology that $\mathrm{Bl}_{\mathfrak{q}}R$ is called a *quadratic transformation* of Spec R if $\mathfrak{q}$ is a maximal ideal in R. If $\mathfrak{q}$ is a prime ideal, $\mathrm{Bl}_{\mathfrak{q}}R$ is called a *monoidal transformation* of Spec R. Let f be the projection

$$\mathrm{Bl}_{\mathfrak{q}}R \to \operatorname{Spec} R.$$

The scheme-theoretic inverse image $f^{-1}(\operatorname{Spec} R/\mathfrak{q})$ in $\mathrm{Bl}_{\mathfrak{q}}R$ (i.e., the fibre above the subscheme defined by $\mathfrak{q}$) is isomorphic as $R/\mathfrak{q}$-scheme to $\operatorname{Proj}\left(\bigoplus_{n\geq 0} \mathfrak{q}^n/\mathfrak{q}^{n+1}\right)$. Then the result on the resolution of singularities of algebraic varieties can be stated as follows:

Let K be a field of characteristic 0. If X is an algebraic K-scheme, say reduced and irreducible, then there exists an algebraic subscheme D of X such that

(i) the set of points of D is exactly the singular locus of X, and

(ii) if $g: \tilde{X} \to X$ is the monoidal transformation of X with center D, then $\tilde{X}$ is non-singular.

In Hironaka's proof of this statement, the notion of normal flatness plays an important role. We say that X is *normally flat* along D if the stalk of $\mathrm{gr}_D^p(X)$ at every point of D is a free $\mathcal{O}_{D,x}$-module for all non-negative integers p.

This chapter now has its origin in an effort to extend Hironaka's observations to a general situation due to G. Faltings [1] and M. Brodmann [1]:

In an abstract manner of speaking, consider an affine scheme $(X, \mathcal{O}_X)$ and a point $p \in X$ where p is an isolated singularity. Let X be pure-dimensional of dimension d. Take d elements $f_1, \ldots, f_d \in \mathcal{O}_X(X)$ defining p set-theoretically, that is, $V(f_1, \ldots, f_d) = \{p\}$. Then we want to investigate the blowing-up $\mathrm{Bl}_{\mathfrak{q}}(X)$ of X with center Spec $\mathcal{O}_x(X)/\mathfrak{q}$, where $\mathfrak{q} := (f_1, \ldots, f_d)$. That is, we do not investigate the blowing-up $\mathrm{Bl}_{(p)}(X)$ at p but we consider the blowing-up at an "infinitesimal neighbourhood", say $\tilde{p}$, of p. The following two facts are well-known (see M. Brodmann [1]):

(i) $\mathrm{Bl}_{\mathfrak{q}}(X) = \operatorname{Proj}\left(\bigoplus_{n\geq 0} \mathfrak{q}^n\right)$, where $\bigoplus_{n\geq 0} \mathfrak{q}^n$ is the Rees algebra with respect to $\mathfrak{q}$.

(ii) There is the following commutative diagram:

$$\begin{array}{ccc} \mathrm{Bl}_{(p)}(X) = \mathrm{Proj}\left(\bigoplus_{n\geq 0} \mathfrak{m}_p^n\right) & \xrightarrow{\Phi} & X \\ \downarrow \varphi & & \downarrow \mathrm{id} \\ \mathrm{Bl}_{(\tilde{p})}(X) = \mathrm{Proj}\left(\bigoplus_{n\geq 0} \mathfrak{q}^n\right) & \xrightarrow{\Psi} & X \end{array}$$

Assume now that $\tilde{p}$ is "large enough". Then by results from Faltings [1], $\mathrm{Bl}_{(p)}(X)$ is Cohen-Macaulay. This observation leads to the theory on the *Macaulayfication* of X; that is, we only want to obtain from Hironaka's desingularization process that $\tilde{X}$ is (locally) Cohen-Macaulay. The first global statements were obtained by G. Faltings [1]. Using Brodmann's [1] arithmetical approach we have to investigate the affine cone over the blowing-up $\mathrm{Bl}_{\mathfrak{q}}(X)$, that is the Rees algebra $\bigoplus_{n\geq 0} \mathfrak{q}^n$, and the conormal cone defined by the *form ring* (associated graded ring) $\bigoplus_{n\geq 0} \mathfrak{q}^n/\mathfrak{q}^{n+1}$. Therefore, taking this notion locally, we can work throughout in the theory of local rings. Such an approach involves a deep study of Rees rings and form rings with respect to an ideal which is not the maximal ideal of a local ring. Hence our main object in the following is the study of the stability of the Rees rings and form rings of parameter ideals with respect to the Buchsbaum property (see, for example, Theorem 3.3). These investigations and the main results of this chapter include and improve the observations of M. Brodmann, S. Goto, Y. Shimoda and N. V. Trung.

Now, let A denote a local ring with maximal ideal $\mathfrak{m}$. In the sequel we will use the following notation:

Let $\mathfrak{q}$ be an ideal of A. We set

$$R_{\mathfrak{q}}(A) := \bigoplus_{n\geq 0} \mathfrak{q}^n T^n \subset A[T], \qquad \text{where } T \text{ is an indeterminate.}$$

Clearly, $R_{\mathfrak{q}}(A)$ is a graded A-algebra (containing A and contained in $A[T]$) of finite type over A, i.e. $R_{\mathfrak{q}}(A)$ is Noetherian and $[R_{\mathfrak{q}}(A)]_n = \mathfrak{q}^n T^n$ for all $n \geq 0$ and $[R_{\mathfrak{q}}(A)]_n = O$ for $n < 0$. (Note that for a graded ring R and a graded R-module M we denote by $[M]_n$ the Abelian group of homogeneous elements of degree n in M. In our situation $[M]_n$ is even an $[R]_0$-module, i.e. an A-module.) $R_{\mathfrak{q}}(A)$ possesses only one homogeneous maximal ideal, which is the ideal

$$\mathfrak{M} := \mathfrak{m} \oplus \bigoplus_{n\geq 1} \mathfrak{q}^n T^n.$$

Furthermore, we put $G_{\mathfrak{q}}(A) := R_{\mathfrak{q}}(A)/\mathfrak{q}R_{\mathfrak{q}}(A)$ which is again a Noetherian graded A-algebra (with induced grading). $R_{\mathfrak{q}}(A)$ is called the *Rees algebra* and $G_{\mathfrak{q}}(A)$ is said to be the *associated graded algebra* (or form ring) of A with respect to $\mathfrak{q}$.

Let M now be an A-module. There is no difficulty to define in a similar way as above the Rees module $R_{\mathfrak{q}}(M)$ and the associated graded module $G_{\mathfrak{q}}(M)$ of M with respect to $\mathfrak{q}$ (see § 1). Clearly, $R_{\mathfrak{q}}(M)$ is a graded $R_{\mathfrak{q}}(A)$-module and $G_{\mathfrak{q}}(M)$ is a graded $G_{\mathfrak{q}}(A)$-module. If M is a Noetherian A-module then it is easy to see that $R_{\mathfrak{q}}(M)$ and $G_{\mathfrak{q}}(M)$ are Noetherian modules. We also will make use of the following definitions in the sequel:

We say that $R_{\mathfrak{q}}(M)$ or $G_{\mathfrak{q}}(M)$ is a *Cohen-Macaulay* (*Buchsbaum*) *module* if $R_{\mathfrak{q}}(M)_{\mathfrak{M}}$ or $G_{\mathfrak{q}}(M)_{\mathfrak{M}}$ has this property. Furthermore, we say that $R_{\mathfrak{q}}(M)$ or $G_{\mathfrak{q}}(M)$ is a *locally*

Cohen-Macaulay module if $R_\mathfrak{q}(M)_{(\mathfrak{P})}$ or $G_\mathfrak{q}(M)_{(\mathfrak{P})}$ are Cohen-Macaulay modules for all homogeneous primes $\mathfrak{P} \neq \mathfrak{M}$ of $R_\mathfrak{q}(A)$ or $G_\mathfrak{q}(A)$.

The aim of this chapter is to discuss the relationship between the Buchsbaum property of a Noetherian A-module M and the Buchsbaum (Cohen-Macaulay, locally Cohen-Macaulay) property of $R_\mathfrak{q}(M)$ and $G_\mathfrak{q}(M)$. Since this is possible in a very satisfactory way for all parameter ideals $\mathfrak{q}$ of M (see Theorem 2.1, Theorem 2.10, Theorem 3.2 and, above all, Theorem 3.3) we restrict ourselves in § 2 and § 3 to this case.

§ 1. Some preliminary results

Here in the first paragraph we collect (and prove) some basic results concerning Rees rings and Rees modules and associated graded rings and modules.

Let M always denote a Noetherian A-module although many of the following facts remain true if M is not Noetherian. Let $\mathfrak{q} \subset A$ be an ideal. We set

$$R_\mathfrak{q}(M) := \bigoplus_{n\geq 0} \mathfrak{q}^n M T^n \subset M[T] \quad (T \text{ an indeterminate})$$

and

$$G_\mathfrak{q}(M) := R_\mathfrak{q}(M)/\mathfrak{q}R_\mathfrak{q}(M) \simeq \bigoplus_{n\geq 0} \mathfrak{q}^n M/\mathfrak{q}^{n+1}M.$$

Both, $R_\mathfrak{q}(M)$ and $G_\mathfrak{q}(M)$ are modules over $R_\mathfrak{q}(A)$. (Clearly, $G_\mathfrak{q}(M)$ is a $G_\mathfrak{q}(A)$-module, but we always consider it as an $R_\mathfrak{q}(A)$-module using the natural epimorphism $R_\mathfrak{q}(A) \to G_\mathfrak{q}(A)$.)

$R_\mathfrak{q}(M)$ is called the *Rees module* and $G_\mathfrak{q}(M)$ the *associated graded module* of M with respect to $\mathfrak{q}$. Obviously, $R_\mathfrak{q}(M)$ and $G_\mathfrak{q}(M)$ are graded $R_\mathfrak{q}(A)$-modules (Noetherian if M is Noetherian) with

$$[R_\mathfrak{q}(M)]_n = \begin{cases} O & \text{for } n < 0, \\ \mathfrak{q}^n M T^n & \text{for } n \geq 0. \end{cases}$$

Identifying for simplicity the modules $\mathfrak{q}^n M T^n/\mathfrak{q}^{n+1} M T^{n+1}$ and $\mathfrak{q}^n M/\mathfrak{q}^{n+1}M$ we have

$$[G_\mathfrak{q}(M)]_n = \begin{cases} O & \text{for } n < 0, \\ \mathfrak{q}^n M/\mathfrak{q}^{n+1} M & \text{for } n \geq 0. \end{cases}$$

If $\mathfrak{A} \subset R_\mathfrak{q}(A)$ is a (not necessary homogeneous) ideal then we set for each $i \in \mathbf{N}$:

$$[\mathfrak{A}]_i := \{\text{forms of degree } i \text{ in } \mathfrak{A}\} = \mathfrak{A} \cap [R_\mathfrak{q}(A)]_i = \mathfrak{A} \cap \mathfrak{q}^i T^i.$$

It is clear that $\mathfrak{A}_h := \bigoplus_{i\geq 0} [\mathfrak{A}]_i$ is a homogeneous ideal of $R_\mathfrak{q}(A)$. It is the biggest homogeneous ideal contained in $\mathfrak{A}$. If $\mathfrak{A}$ is a prime ideal then $\mathfrak{A}_h$ is prime as well and $[\mathfrak{A}]_0 = [\mathfrak{A}_h]_0$ is a prime ideal in A. (If $\mathfrak{P} \in \operatorname{Proj} R_\mathfrak{q}(A) = \operatorname{Bl}_\mathfrak{q}(A)$ and if $f\colon \operatorname{Bl}_\mathfrak{q} A \to \operatorname{Spec} A$ is the projection mentioned above then $f(\mathfrak{P}) = [\mathfrak{P}]_0$.) Conversely, if $\mathfrak{p}$ is a prime ideal in A, then

$$\tilde{\mathfrak{p}} := \bigoplus_{i\geq 0} (\mathfrak{p} \cap \mathfrak{q}^i)\, T^i \quad \text{and} \quad \mathfrak{p}^* := \mathfrak{p} \oplus \bigoplus_{i>0} \mathfrak{q}^i T^i$$

are homogeneous prime ideals in $R_\mathfrak{q}(A)$.

As in the local case we denote by $\operatorname{Supp}_{R_\mathfrak{q}(A)} R_\mathfrak{q}(M)$ or, if no confusion is possible by $\operatorname{Supp} R_\mathfrak{q}(M)$, the *set of all primes* $\mathfrak{P}$ *of* $R_\mathfrak{q}(A)$ *with* $R_\mathfrak{q}(M)_\mathfrak{P} \neq O$ and by $\operatorname{Ass}_{R_\mathfrak{q}(A)} R_\mathfrak{q}(M)$ the *set of all associated primes of* $R_\mathfrak{q}(M)$. Since $R_\mathfrak{q}(M)$ is a graded $R_\mathfrak{q}(A)$-module, $\operatorname{Ass} R_\mathfrak{q}(M)$ consists only of homogeneous primes (see Matsumura [1], Prop. (10B), p. 62). Also all minimal elements of Supp $R_{\mathfrak{q}}(M)$ belong to Ass $R_{\mathfrak{q}}(M)$. Clearly the same is true for $G_{\mathfrak{q}}(M)$.

Our first aim is to calculate Supp $R_{\mathfrak{q}}(M)$ and Supp $G_{\mathfrak{q}}(M)$. First we note that for every homogeneous prime ideal $\mathfrak{P}$ of $R_{\mathfrak{q}}(A)$ with $[\mathfrak{P}]_1 \subsetneqq \mathfrak{q}T$ we get by Lemma I.2.27:

$$R_{\mathfrak{q}}(M)_{\mathfrak{P}} \cong R_{\mathfrak{q}}(M)^*_{(\mathfrak{P})}$$

(for the notation see the remarks made in conjunction with Lemma I.2.26).

Similar to Lemma I.2.27 we can prove the following

Lemma 1.1. *Let* $\mathfrak{p} \in \operatorname{Spec} A \setminus V(\mathfrak{q})$. *Then*

(i) $R_{\mathfrak{q}}(A)_{\mathfrak{p}^*} \cong A[X]_{\mathfrak{e}} \cong A_{\mathfrak{p}}[X]_{\mathfrak{e}'}$ *(X a variable),*
where

$$\mathfrak{e} := \mathfrak{p}A[X] + X \cdot A[X] \text{ and } \mathfrak{e}' := \mathfrak{p}A_{\mathfrak{p}}[X] + X \cdot A_{\mathfrak{p}}[X].$$

(ii) *For any A-module M we have*

$$R_{\mathfrak{q}}(M)_{\mathfrak{p}^*} \cong M[X]_{\mathfrak{e}} \cong M_{\mathfrak{p}}[X]_{\mathfrak{e}'}.$$

Proof: It is sufficient to prove (i) since the same arguments work for (ii). The isomorphism $A[X]_{\mathfrak{e}} \cong A_{\mathfrak{p}}[X]_{\mathfrak{e}'}$ is obvious. We construct an isomorphism $g\colon A[X]_{\mathfrak{e}} \to R_{\mathfrak{q}}(A)_{\mathfrak{p}^*}$ in the same way as in the proof of Lemma I.2.27. We choose an element $x \in \mathfrak{q} \setminus \mathfrak{p}$ and define $g(X) := x \cdot T$. We leave the remainder of the proof to the reader, q.e.d.

Corollary 1.2. $R_{\mathfrak{q}}(M)_{\mathfrak{p}^*}$ *is a Cohen-Macaulay module if and only if* $M_{\mathfrak{p}}$ *is a Cohen-Macaulay module.*

We note that this is not true for the Buchsbaum property.

Let $\mathfrak{P} \subset R_{\mathfrak{q}}(A)$ be a prime ideal. It is easy to see that $R_{\mathfrak{q}}(M)_{\mathfrak{P}} \neq 0$ (or $R_{\mathfrak{q}}(M)_{(\mathfrak{P})} \neq 0$ if $\mathfrak{P}$ is homogeneous) implies $[\mathfrak{P}]_0 \in \operatorname{Supp} M$.

If $\mathfrak{q}T \subseteq \mathfrak{P}$ then $\mathfrak{P}$ is homogeneous and has the form $\mathfrak{P} = \mathfrak{p}^*$ with $\mathfrak{p} := [\mathfrak{P}]_0$. Then $R_{\mathfrak{q}}(M)_{(\mathfrak{P})} \cong M_{\mathfrak{p}} \neq 0$ if $\mathfrak{p} \in \operatorname{Supp} M$. The same is true for $G_{\mathfrak{q}}(M)$:

If $G_{\mathfrak{q}}(M)_{\mathfrak{P}} \neq 0$ (or $G_{\mathfrak{q}}(M)_{(\mathfrak{P})} \neq 0$ if $\mathfrak{P}$ is homogeneous) then $[\mathfrak{P}]_0 \in \operatorname{Supp} M \cap V(\mathfrak{q})$.

If $\mathfrak{P} = \mathfrak{p}^*$, $\mathfrak{p} \in \operatorname{Supp} M \cap V(\mathfrak{q})$ we obtain $G_{\mathfrak{q}}(M)_{\mathfrak{P}} \cong (M/\mathfrak{q}M)_{\mathfrak{p}} \neq 0$.

Collecting all these facts we get for every homogeneous prime ideal $\mathfrak{P} \subset R_{\mathfrak{q}}(A)$:

$$R_{\mathfrak{q}}(M)_{\mathfrak{P}} \neq 0 \quad \text{if and only if } R_{\mathfrak{q}}(M)_{(\mathfrak{P})} \neq 0,$$

$$G_{\mathfrak{q}}(M)_{\mathfrak{P}} \neq 0 \quad \text{if and only if } G_{\mathfrak{q}}(M)_{(\mathfrak{P})} \neq 0.$$

Furthermore, if $\mathfrak{P}$ is a homogeneous prime ideal of $R_{\mathfrak{q}}(A)$ with $\mathfrak{P} \notin \{\mathfrak{p}^* \mid \mathfrak{p} \in V(\mathfrak{q})\}$, Lemma I.2.27, I.2.26 and Lemma 1.1 show that $R_{\mathfrak{q}}(M)_{(\mathfrak{P})}$ $\big(G_{\mathfrak{q}}(M)_{(\mathfrak{P})}\big)$ is Cohen-Macaulay if and only if $R_{\mathfrak{q}}(M)_{\mathfrak{P}}$ $\big(G_{\mathfrak{q}}(M)_{\mathfrak{P}}\big)$ is Cohen-Macaulay.

We prove some further lemmata:

Lemma 1.3. *Let* $\mathfrak{P} \subset R_{\mathfrak{q}}(A)$ *be a homogeneous prime ideal.*

(i) *If* $[\mathfrak{P}]_i \subsetneqq \mathfrak{q}^i T^i$ *for some* $i \geq 1$ *then we have for* $x \in \mathfrak{q}^i$: *If* $xT^i \in \mathfrak{P}$ *then* $x \in \mathfrak{P}$. *The converse is true if* $\mathfrak{q} \not\subseteq [\mathfrak{P}]_0$.

(ii) *Assume that* $[\mathfrak{P}]_0 \in \operatorname{Supp} M$ *and* $[\mathfrak{P}]_1 \subsetneqq \mathfrak{q}T$. *Then we have for every* $p \in \mathbf{N}$:

$$\big(R_{\mathfrak{q}}(M)\,(p)\big)_{(\mathfrak{P})} = \big(\mathfrak{q}^p R_{\mathfrak{q}}(M)\big)_{(\mathfrak{P})} = x^p R_{\mathfrak{q}}(M)_{(\mathfrak{P})} \quad \text{for all } x \in \mathfrak{q}$$

with $xT \notin [\mathfrak{P}]_1$.

Every such x is a non-zero divisor with respect to $R_{\mathfrak{q}}(M)_{(\mathfrak{P})}$ and we have an exact sequence

$$0 \to R_{\mathfrak{q}}(M)_{(\mathfrak{P})} \xrightarrow{x} R_{\mathfrak{q}}(M)_{(\mathfrak{P})} \to G_{\mathfrak{q}}(M)_{(\mathfrak{P})} \to 0.$$

Proof: (i) Let $xT^i \in \mathfrak{P}$. We choose an element $y \in \mathfrak{q}^i$ with $yT^i \notin [\mathfrak{P}]_i$. Then $x(yT^i) = y(xT)^i \in \mathfrak{P}$, hence $x \in \mathfrak{P}$.

Let $\mathfrak{q} \subseteq [\mathfrak{P}]_0$ and assume $x \in \mathfrak{P}$. Then $\mathfrak{q}^i \setminus [\mathfrak{P}]_0 \neq \emptyset$ for all $i \geq 0$. Choose $y \in \mathfrak{q}^i \setminus [\mathfrak{P}]_0$. Then $y \notin \mathfrak{P}$ and $(xT^i)\ y = x(yT^i) \in \mathfrak{P}$, hence $xT^i \in \mathfrak{P}$.

(ii) Let $p \in \mathbf{N}$. Since in $\big(R_{\mathfrak{q}}(M)\ (p)\big)_{(\mathfrak{P})}$ only elements of degree ≥ 0 occur as numerator, we see that $\big(R_{\mathfrak{q}}(M)\ (p)\big)_{(\mathfrak{P})} = \big(\mathfrak{q}^p R_{\mathfrak{q}}(M)\big)_{(\mathfrak{P})}$.

Let $x \in \mathfrak{q}$, $xT \notin [\mathfrak{P}]_1$. Then we have for every $y \in \mathfrak{q}^p : y = x^p \cdot \dfrac{yT^p}{x^pT^p}$ in $R(A)_{(\mathfrak{P})}$. Therefore $\big(\mathfrak{q}^p R_{\mathfrak{q}}(M)\big)_{(\mathfrak{P})} = x^p R_{\mathfrak{q}}(M)_{(\mathfrak{P})}$.

Now a straightforward calculation shows that x is a non-zero divisor of $R_{\mathfrak{q}}(M)_{(\mathfrak{P})}$. If we localize the exact sequence

$$0 \to \mathfrak{q}R_{\mathfrak{q}}(M) \to R_{\mathfrak{q}}(M) \to G_{\mathfrak{q}}(M) \to 0$$

with $\mathfrak{P}$ and use $\big(\mathfrak{q}R_{\mathfrak{q}}(M)\big)_{(\mathfrak{P})} = xR_{\mathfrak{q}}(M)_{(\mathfrak{P})} \cong R_{\mathfrak{q}}(M)_{(\mathfrak{P})}$ we get the desired exact sequence, q.e.d.

Corollary 1.4. *Let $\mathfrak{P}$ be as in Lemma 1.3 but assume that $\mathfrak{p} := [\mathfrak{P}]_0 \not\supseteq \mathfrak{q}$. Then either $\mathfrak{P} = \mathfrak{p}^*$ (i.e. $[\mathfrak{P}]_i = \mathfrak{q}^iT^i$ for all $i \geq 1$) or $\mathfrak{P} = \tilde{\mathfrak{p}}$ (i.e. $[\mathfrak{P}]_i = (\mathfrak{p} \cap \mathfrak{q}^i)\ T^i$ for all $i \geq 0$).*

Proof: If $[\mathfrak{P}]_i = \mathfrak{q}^iT^i$ for some $i \geq 1$ then $(\mathfrak{q}T)^i = \mathfrak{q}^iT^i \subseteq \mathfrak{P}$, hence $\mathfrak{q}T \subseteq \mathfrak{P}$ and therefore $[\mathfrak{P}]_i = \mathfrak{q}^iT^i$ for all $i \geq 1$.

If $[\mathfrak{P}]_i \subsetneqq \mathfrak{q}^iT^i$ for all $i \geq 1$ then we get by Lemma 1.3: If $x \in \mathfrak{p} \cap \mathfrak{q}^i$ then $x \in \mathfrak{P}$ and therefore $xT^i \in [\mathfrak{P}]_i$. If $xT^i \in [\mathfrak{P}]_i$ then $x \in \mathfrak{P}$ and $x \in \mathfrak{q}^i$, i.e. $x \in \mathfrak{p} \cap \mathfrak{q}^i$, q.e.d.

Corollary 1.5. *Let $\mathfrak{P} \subset R_{\mathfrak{q}}(A)$ be a prime ideal of the form $\mathfrak{P} = \tilde{\mathfrak{p}}$ or $\mathfrak{P} = \mathfrak{p}^*$ with $\mathfrak{p} \in \operatorname{Spec} A$. Let N be an A-module and assume that $\mathfrak{N}$ is a graded $R_{\mathfrak{q}}(A)$-module which is a (graded) submodule of $N[T]$ with $[\mathfrak{N}]_0 = N$. Then $\mathfrak{N}_{(\mathfrak{P})} \cong N_{\mathfrak{p}}$.*

Proof: We consider the natural homomorphism (of $A_{\mathfrak{p}}$-modules)

$$f: N_{\mathfrak{p}} \to \mathfrak{N}_{(\mathfrak{P})}$$

which is defined by $f\left(\dfrac{n}{p}\right) = \dfrac{n}{p}$ for all $n \in N = [\mathfrak{N}]_0$ and all $p \in A \setminus \mathfrak{p} = [R_{\mathfrak{q}}(A)]_0 \setminus [\mathfrak{P}]_0$.

Let $\dfrac{v}{pT^i} \in \mathfrak{N}_{(\mathfrak{P})}$ with $v \in [\mathfrak{N}]_i \subseteq NT^i$, $pT^i \in [R_{\mathfrak{q}}(A)]_i \setminus [\mathfrak{P}]_i$. Then $v = nT^i$ with $n \in N$ and $p \in A \setminus \mathfrak{p}$ by our assumption (if $\mathfrak{P} = \tilde{\mathfrak{p}}$ then $p \in \mathfrak{q}^i \setminus (\mathfrak{p} \cap \mathfrak{q}^i) \subseteq A \setminus \mathfrak{p}$, if $\mathfrak{P} = \mathfrak{p}^*$ then $i = 0$). Hence $\dfrac{v}{pT^i} = f\left(\dfrac{n}{p}\right)$ and f is surjective.

If $\dfrac{n}{p} \in \operatorname{Ker} f$ $(n \in N,\ p \in A \setminus \mathfrak{p})$ then there is an $i \in \mathbf{N}$ and a $qT^i \in R_{\mathfrak{q}}(A) \setminus \mathfrak{P}$ $(q \in \mathfrak{q}^i)$ with $qT^in = 0$, i.e. $qn = 0$. But by the same argument as before $q \in A \setminus \mathfrak{p}$ and this implies $\dfrac{n}{p} = 0$, i.e. $\operatorname{Ker} f = 0$ and f is an isomorphism, q.e.d.

Lemma 1.6. *Let $\mathfrak{P}, \mathfrak{Q} \subset R_{\mathfrak{q}}(A)$ be homogeneous prime ideals with $\mathfrak{P} \subseteq \mathfrak{Q}$ and either $[\mathfrak{Q}]_1 \subsetneqq \mathfrak{q}T$ or $[\mathfrak{P}]_1 = \mathfrak{q}T$. Then there is a prime ideal $\mathfrak{v} \in \operatorname{Spec} R_{\mathfrak{q}}(A)_{\mathfrak{Q}}$ with $R_{\mathfrak{q}}(A)_{(\mathfrak{P})} \cong \big(R_{\mathfrak{q}}(A)_{(\mathfrak{Q})}\big)_{\mathfrak{v}}$ and therefore $R_{\mathfrak{q}}(M)_{(\mathfrak{P})} \cong \big(R_{\mathfrak{q}}(M)_{(\mathfrak{Q})}\big)_{\mathfrak{v}}$.*

Proof: If $[\mathfrak{P}]_1 = \mathfrak{q}T$ then $[\mathfrak{Q}]_1 = \mathfrak{q}T$ and $\mathfrak{P} = \mathfrak{p}_1^*$, $\mathfrak{Q} = \mathfrak{p}_2^*$ with $\mathfrak{p}_1, \mathfrak{p}_2 \in \operatorname{Spec} A$. Then $R_\mathfrak{q}(A)_{(\mathfrak{P})} \cong A_{\mathfrak{p}_1}$, $R_\mathfrak{q}(A)_{(\mathfrak{Q})} = A_{\mathfrak{p}_2}$ by Corollary 1.5 and our statement follows by a standard argument of local algebra.

Let $[\mathfrak{Q}]_1 \subsetneq \mathfrak{q}T$. Choose $x \in \mathfrak{q}$ with $xT \notin \mathfrak{Q}$. Then $xT \notin \mathfrak{P}$.

Let $\mathfrak{e}$ denote the maximal ideal of $R_\mathfrak{q}(A)_{(\mathfrak{P})}$. Then the natural homomorphism $f\colon R_\mathfrak{q}(A)_{(\mathfrak{Q})} \to R_\mathfrak{q}(A)_{(\mathfrak{P})}$ induces a homomorphism

$$g\colon \left(R_\mathfrak{q}(A)_{(\mathfrak{Q})}\right)_{f^{-1}(\mathfrak{e})} \to R_\mathfrak{q}(A)_{(\mathfrak{P})}.$$

Let $\dfrac{aT^i}{pT^i} \in R_\mathfrak{q}(A)_{(\mathfrak{P})}$ $(a, p \in \mathfrak{q}^i, pT^i \notin \mathfrak{P})$. Then $\dfrac{pT^i}{x^iT^i} \in R_\mathfrak{q}(A)_{(\mathfrak{Q})} \setminus f^{-1}(\mathfrak{e})$ and $\dfrac{aT^i}{pT^i} = g\left(\dfrac{aT^i}{x^iT^i}\left(\dfrac{pT^i}{x^iT^i}\right)^{-1}\right)$, i.e. g is surjective.

Let $\alpha \in \operatorname{Ker} g$. We may assume without loss of generality that $\alpha \in R_\mathfrak{q}(A)_{(\mathfrak{Q})}$. Write $\alpha = \dfrac{aT^i}{qT^i}$ with $a, q \in \mathfrak{q}^i$, $qT^i \notin \mathfrak{Q}$. Since $\dfrac{aT^i}{qT^i} = 0$ in $R_\mathfrak{q}(A)_{(\mathfrak{P})}$ there is a $j \in \mathbf{N}$ and a $p \in \mathfrak{q}^j$, $pT^j \notin \mathfrak{P}$ with $apT^{i+j} = pT^jaT^i = 0$, i.e. $ap = 0$. But $\dfrac{pT^j}{x^jT^j} \in R_\mathfrak{q}(A)_{(\mathfrak{Q})} \setminus f^{-1}(\mathfrak{e})$ and $\dfrac{pT^j}{x^jT^j} \cdot \dfrac{aT^i}{pT^i} = 0$, i.e. $\alpha = 0$. Hence g is injective and therefore an isomorphism, q.e.d.

Lemma 1.7. *Let $\mathfrak{P} \subset R_\mathfrak{q}(A)$ be a (not necessarily homogeneous) prime ideal. Then $R_\mathfrak{q}(M)_\mathfrak{P} \neq O$ if and only if there is a $\mathfrak{p} \in \operatorname{Supp} M$ with $\tilde{\mathfrak{p}} \subseteq \mathfrak{P}$.*

Proof: Assume $\tilde{\mathfrak{p}} \not\subseteq \mathfrak{P}$ for all $\mathfrak{p} \in \operatorname{Supp} M$. We may assume that $M \neq O$, i.e. $\operatorname{Supp} M \neq \emptyset$. Let $\{\mathfrak{p}_1, \ldots, \mathfrak{p}_s\}$ denote the set of minimal elements of $\operatorname{Supp} M$. Then there are $\eta_i \in \tilde{\mathfrak{p}}_i \setminus \mathfrak{P}$ for $i = 1, \ldots, s$ and, consequently, $\eta_1 \cdot \ldots \cdot \eta_s \in (\tilde{\mathfrak{p}}_1 \cap \ldots \cap \tilde{\mathfrak{p}}_s) \setminus \mathfrak{P}$. Since for all $i = 1, \ldots, s$ the coefficients of η_i (considered as a polynomial in T) lie in $\mathfrak{p}_i$, all coefficients of $\eta_1 \cdot \ldots \cdot \eta_s$ lie in $\mathfrak{p}_1 \cap \ldots \cap \mathfrak{p}_s$. Therefore there is a $h \in \mathbf{N}$ such that all coefficients of $\eta := (\eta_1 \cdot \ldots \cdot \eta_s)^h$ lie in $\operatorname{Ann} M$. Hence $\eta \cdot R_\mathfrak{q}(M) = O$, i.e. $R_\mathfrak{q}(M)_\mathfrak{P} = O$ since $\eta \notin \mathfrak{P}$. Assume there is a $\mathfrak{p} \in \operatorname{Supp} M$ with $\tilde{\mathfrak{p}} \subseteq \mathfrak{P}$. Then there is a $\mathfrak{v} \in \operatorname{Spec} R_\mathfrak{q}(A)_\mathfrak{P}$ with

$$\left(R_\mathfrak{q}(M)_\mathfrak{P}\right)_\mathfrak{v} \cong R_\mathfrak{q}(M)_{\tilde{\mathfrak{p}}} \cong R_\mathfrak{q}(M)^*_{\tilde{\mathfrak{p}}} \cong M^*_\mathfrak{p} \neq O$$

(see Lemma I.2.27 and Corollary 1.5). Hence $R_\mathfrak{q}(M)_\mathfrak{P} \neq O$, q.e.d.

Corollary 1.8. *Let $\mathfrak{P}$ be as before. Then $G_\mathfrak{q}(M)_\mathfrak{P} \neq O$ if and only if $R_\mathfrak{q}(M)_\mathfrak{P} \neq O$ and $\mathfrak{q} \subseteq [\mathfrak{P}]_0$.*

Proof: The "only-if-part" follows by the remarks made above.

Assume that $R_\mathfrak{q}(M)_\mathfrak{P} \neq O$ and $[\mathfrak{P}]_0 \in V(\mathfrak{q})$. Then there is a $\mathfrak{p} \in \operatorname{Supp} M$ with $\tilde{\mathfrak{p}} \subseteq \mathfrak{P}$. If we replace $\mathfrak{P}$ by $\mathfrak{P}_h$, the biggest homogeneous ideal contained in $\mathfrak{P}$, then $\tilde{\mathfrak{p}} \subseteq \mathfrak{P}_h$. Therefore $R_\mathfrak{q}(M)_{\mathfrak{P}_h} \neq O$ and $[\mathfrak{P}_h]_0 = [\mathfrak{P}]_0 \in V(\mathfrak{q})$. If $G_\mathfrak{q}(M)_{\mathfrak{P}_h} \neq O$ then $G_\mathfrak{q}(M)_\mathfrak{P} \neq O$ since there is a $\mathfrak{v} \in \operatorname{Spec} R_\mathfrak{q}(A)_\mathfrak{P}$ with $\left(G_\mathfrak{q}(M)_\mathfrak{P}\right)_\mathfrak{v} \cong G_\mathfrak{q}(M)_{\mathfrak{P}_h}$. Therefore we may assume without loss of generality that $\mathfrak{P}$ is homogeneous and it is now sufficient to show that $G_\mathfrak{q}(M)_{(\mathfrak{P})} \neq O$.

If $\mathfrak{P} = [\mathfrak{P}]_0^*$ then $G_\mathfrak{q}(M)_{(\mathfrak{P})} \cong (M/\mathfrak{q}M)_{[\mathfrak{P}]_0} \neq O$ since $[\mathfrak{P}]_0 \in V(\mathfrak{q}) \cap \operatorname{Supp} M = \operatorname{Supp} M/\mathfrak{q}M$.

If $\mathfrak{P} \neq [\mathfrak{P}]_0^*$ then $[\mathfrak{P}]_1 \subsetneq \mathfrak{q}T$. Now take the exact sequence of Lemma 1.3(ii). The element x used there belongs to $\mathfrak{q} \subseteq [\mathfrak{P}]_0$, hence x is not a unit in $R_\mathfrak{q}(A)_{(\mathfrak{P})}$. Therefore $G_\mathfrak{q}(M)_{(\mathfrak{P})} \cong R_\mathfrak{q}(M)_{(\mathfrak{P})}/xR_\mathfrak{q}(M)_{(\mathfrak{P})} \neq O$, q.e.d.

Proposition 1.9.

$$\dim_{R_\mathfrak{q}(A)} R_\mathfrak{q}(M) = \begin{cases} \dim_A M \textit{ if } \mathfrak{q} \subseteq \mathfrak{p} \textit{ for all } \mathfrak{p} \in \operatorname{Supp} M \textit{ with } \dim A/\mathfrak{p} = \dim M, \\ 1 + \dim_A M \textit{ otherwise}, \end{cases}$$

$$\dim_{R_\mathfrak{q}(A)} G_\mathfrak{q}(M) = \dim_A M.$$

Proof: Let $\mathfrak{p} \in \operatorname{Supp} M$. Then $\tilde{\mathfrak{p}} \in \operatorname{Supp} R_\mathfrak{q}(M)$ by Lemma 1.7 and $R_\mathfrak{q}(A)/\tilde{\mathfrak{p}} \cong R_\mathfrak{q}(A/\mathfrak{p})$ is true. Therefore we get by Lemma 1.7:

$$\dim R_\mathfrak{q}(M) = \sup \{\dim R_\mathfrak{q}(A/\mathfrak{p}) \mid \mathfrak{p} \in \operatorname{Supp} M\}.$$

Thus for the first statement it is sufficient to prove: Let A be an integral domain. Then we have for every ideal $\mathfrak{q} \subset A$:

$$\dim R_\mathfrak{q}(A) = \begin{cases} \dim A & \text{if } \mathfrak{q} = 0, \\ 1 + \dim A & \text{if } \mathfrak{q} \neq 0. \end{cases}$$

If $\mathfrak{q} = 0$, $R_\mathfrak{q}(A) = A$ (with the trivial grading) and we are done in this case.

If $\mathfrak{q} \neq 0$, write $\mathfrak{q} = (a_1, \ldots, a_t)A$ with $t \geq 1$. We use induction on t. If $t = 1$ there is an isomorphism

$$A[X] \cong R_\mathfrak{q}(A) \qquad (X \text{ a variable}),$$

which is given by $X \mapsto a_1 T$. This finishes the proof in this case since $\dim A[X] = 1 + \dim A$ (c.f. Matsumura [1], Theorem 22, p. 83).

Let $t \geq 2$ and set $\mathfrak{q}' := (a_1, \ldots, a_{t-1}) A$. Then there is an epimorphism (X a variable)

$$\pi\colon R_{\mathfrak{q}'}(A)\,[X] \to R_\mathfrak{q}(A),$$

which is given by $X \mapsto a_t T$. Now $R_{\mathfrak{q}'}(A)$ and $R_\mathfrak{q}(A)$ are integral domains and $\dim R_{\mathfrak{q}'}(A) = 1 + \dim A$ by our induction hypothesis. Also $a_1 X - a_t(a_1 T) \in \operatorname{Ker} \pi$, hence $\operatorname{Ker} \pi \neq 0$ and we get

$$\begin{aligned} \dim R_\mathfrak{q}(A) &= \dim R_{\mathfrak{q}'}(A)\,[X]/\operatorname{Ker} \pi \leq \dim R_{\mathfrak{q}'}(A)\,[X] - 1 \\ &= \dim R_{\mathfrak{q}'}(A) = 1 + \dim A. \end{aligned}$$

Next let $\mathfrak{p}_0 \subsetneqq \mathfrak{p}_1 \subsetneqq \ldots \subsetneqq \mathfrak{p}_d$, $d := \dim A$ be a chain of prime ideals in A. (Then $\mathfrak{p}_0 = 0$ and $\mathfrak{p}_d = \mathfrak{m}$.) Since $\mathfrak{q} \neq 0$, $\tilde{0} \subsetneqq 0^* \subsetneqq \mathfrak{p}_1^* \subsetneqq \ldots \subsetneqq \mathfrak{p}_{d-1}^* \subsetneqq \mathfrak{m}^*$ is a chain of prime ideals in $R_\mathfrak{q}(A)$, i.e. $\dim R_\mathfrak{q}(A) \geq 1 + \dim A$ and our first statement is therefore proven. Now let us calculate $\dim G_\mathfrak{q}(M)$.

If $\dim R_\mathfrak{q}(M) = \dim M$ we deduce from the following two epimorphisms (of $R_\mathfrak{q}(A)$-modules) $R_\mathfrak{q}(M) \to G_\mathfrak{q}(M) \to \underline{M/\mathfrak{q}M}$ (underlining an A-module N means consider N as an $R_\mathfrak{q}(A)$-module by the canonical epimorphism $R_\mathfrak{q}(A) \to A$ where N is equipped with the trivial grading):

$$\begin{aligned} \dim_A M &= \dim_A M/\mathfrak{q}M = \dim_{R_\mathfrak{q}(A)} \underline{M/\mathfrak{q}M} \leq \dim_{R_\mathfrak{q}(A)} G_\mathfrak{q}(M) \leq \dim_{R_\mathfrak{q}(A)} R_\mathfrak{q}(M) \\ &= \dim M \end{aligned}$$

and we are done in this case.

Assume now $\dim R_{\mathfrak{q}}(M) = 1 + \dim M$. Then there is a prime ideal $\mathfrak{p} \in \operatorname{Supp} M$ with $\dim A/\mathfrak{p} = \dim M$ and $\mathfrak{q} \not\subseteq \mathfrak{p}$. From the proof of the first statement we know that

$$\{\mathfrak{P} \in \operatorname{Supp} R_{\mathfrak{q}}(M) \mid \dim R_{\mathfrak{q}}(A)/\mathfrak{P} = 1 + \dim M\}$$
$$= \{\tilde{\mathfrak{p}} \mid \mathfrak{p} \in \operatorname{Supp} M,\ \dim A/\mathfrak{p} = \dim M,\ \mathfrak{q} \not\subseteq \mathfrak{p}\}.$$

Now choose an element $a \in \mathfrak{q}$ with $a \notin \mathfrak{p}$ for all $\mathfrak{p} \in \operatorname{Supp} M$, with $\dim A/\mathfrak{p} = \dim M$, $\mathfrak{q} \not\subseteq \mathfrak{p}$. Then $\dim R_{\mathfrak{q}}(M)/aR_{\mathfrak{q}}(M) = \dim M$ and hence the canonical epimorphism $R_{\mathfrak{q}}(M)/aR_{\mathfrak{q}}(M) \to G_{\mathfrak{q}}(M)$ shows that $\dim G_{\mathfrak{q}}(M) \leq \dim M$. Thus we have to prove that $\dim G_{\mathfrak{q}}(M) \geq \dim M$.

Let U denote the intersection of all those primary submodules of M belonging to prime ideals $\mathfrak{p}$ in $\operatorname{Supp} M$ with $\dim A/\mathfrak{p} = \dim M$ and $\mathfrak{q} \not\subseteq \mathfrak{p}$. Then by the natural epimorphism $G_{\mathfrak{q}}(M) \to G_{\mathfrak{q}}(M/U)$ we have $\dim G_{\mathfrak{q}}(M) \geq \dim G_{\mathfrak{q}}(M/U)$ and since $\dim M/U = \dim M$ we may assume without loss of generality that $\dim A/\mathfrak{p} = \dim M$, $\mathfrak{q} \not\subseteq \mathfrak{p}$ for each prime ideal $\mathfrak{p}$ in $\operatorname{Ass} M$. Now the set

$$\mathfrak{B} := \{\mathfrak{b} \mid \mathfrak{b} \text{ ideal in } A,\ \mathfrak{b} \subseteq \mathfrak{q} \text{ and there is an } n \in \mathbf{N} \text{ with } \mathfrak{b}^{n-1}M \subseteq \mathfrak{q}^n M\}$$

is not empty ($\mathfrak{q}^2 \in \mathfrak{B}$) and contains a unique maximal element, say $\mathfrak{q}_0$. (If $\mathfrak{b}_1, \mathfrak{b}_2 \in \mathfrak{B}$, $\mathfrak{b}_1^{n-1}M \subseteq \mathfrak{q}^n M$, $\mathfrak{b}_2^{m-1}M \subseteq \mathfrak{q}^m M$, then $(\mathfrak{b}_1 + \mathfrak{b}_2)^{nm-1} M \subseteq \mathfrak{q}^{nm}M$.) Clearly, $\mathfrak{q}_0 \subsetneqq \mathfrak{q}$ since $\mathfrak{q}^n M \neq 0$ for all $n \in \mathbf{N}$. Choose an $a \in \mathfrak{q}$ with $a \notin \mathfrak{q}_0$, $a \notin \mathfrak{p}$ for all $\gamma \in \operatorname{Ass} M$ (c.f. Matsumura [1], (1.B), p. 2). Then by the same argument as above $\dim R_{\mathfrak{q}}(M)/aR_{\mathfrak{q}}(M) = \dim M$ and we have a natural epimorphism $R_{\mathfrak{q}}(M)/aR_{\mathfrak{q}}(M) \to G_{\mathfrak{q}}(M)$. By Lemma 1.7 $\mathfrak{P}$ is minimal in $\operatorname{Supp} R_{\mathfrak{q}}(M)$ if and only if $\mathfrak{P} = \tilde{\mathfrak{p}}$ for some minimal $\mathfrak{p} \in \operatorname{Supp} M$. Hence all minimal elements of $\operatorname{Supp} R_{\mathfrak{q}}(M)$ have the same dimension. By Krull's Hauptidealsatz the same is true for $R_{\mathfrak{q}}(M)/aR_{\mathfrak{q}}(M)$.

If $aT \in \mathfrak{Q}$ for all $\mathfrak{Q} \in \operatorname{Supp} R_{\mathfrak{q}}(M)/aR_{\mathfrak{q}}(M)$ with $\dim R_{\mathfrak{q}}(A)/\mathfrak{Q} = \dim M$ then there is an integer $n \in \mathbf{N}$ with $(aT)^n\big(R_{\mathfrak{q}}(M)/aR_{\mathfrak{q}}(M)\big) = 0$. This implies $a^n M \subseteq a\mathfrak{q}^n M$ or $a^{n-1}M \subseteq \mathfrak{q}^n M + 0 :_M a = \mathfrak{q}^n M$ since $0 :_M a = 0$ by our assumption. But this is impossible since $a \notin \mathfrak{q}_0$. Hence there is a prime ideal $\mathfrak{Q} \in \operatorname{Supp} R_{\mathfrak{q}}(M)/aR_{\mathfrak{q}}(M)$ with $\dim R_{\mathfrak{q}}(A)/\mathfrak{Q} = \dim M$ and $aT \notin \mathfrak{Q}$. If $b \in \mathfrak{q}$ then $b(aT) = a(bT) \in \mathfrak{Q}$, hence $b \in \mathfrak{Q}$ and therefore $\mathfrak{q} \subseteq [\mathfrak{Q}]_0$, i.e. $\mathfrak{Q} \in \operatorname{Supp} G_{\mathfrak{q}}(M)$ by Corollary 1.8. This shows $\dim G_{\mathfrak{q}}(M) \geq \dim R_{\mathfrak{q}}(A)/\mathfrak{Q} = \dim M$, q.e.d.

Corollary 1.10.

$$\dim R_{\mathfrak{q}}(M) = \dim R_{\mathfrak{q}}(M)_{\mathfrak{M}} \quad \textit{and} \quad \dim G_{\mathfrak{q}}(M) = \dim G_{\mathfrak{q}}(M)_{\mathfrak{M}}.$$

The exact sequence of Lemma 1.3(ii) has an interesting consequence: Let $\mathfrak{P} \subset R_{\mathfrak{q}}(A)$ be a homogeneous prime ideal with $[\mathfrak{P}]_1 \neq \mathfrak{q}T$ and assume $G_{\mathfrak{q}}(M)_{(\mathfrak{P})} \neq 0$. Then $G_{\mathfrak{q}}(M)_{(\mathfrak{P})}$ is a Cohen-Macaulay module if and only if the same is true for $R_{\mathfrak{q}}(M)_{(\mathfrak{P})}$. We say $R_{\mathfrak{q}}(M)_{(\mathfrak{P})}$ and $G_{\mathfrak{q}}(M)_{(\mathfrak{P})}$ are *simultaneously Cohen-Macaulay.* Per definitionem, the zero module is Cohen-Macaulay. Hence the assumption $G_{\mathfrak{q}}(M)_{(\mathfrak{P})} \neq 0$ is necessary for this statement. The last lemma of this paragraph shows that we can omit this assumption if we allow $\mathfrak{P}$ to vary in the following way.

Lemma 1.11. *$R_{\mathfrak{q}}(M)_{(\mathfrak{P})}$ is Cohen-Macaulay for all* $\mathfrak{P} \in \operatorname{Proj} R_{\mathfrak{q}}(A) \setminus \{\mathfrak{p}^* \mid \mathfrak{p} \in V(\mathfrak{q})\}$ *if and only if $G_{\mathfrak{q}}(M)_{(\mathfrak{P})}$ is Cohen-Macaulay for all* $\mathfrak{P} \in \operatorname{Proj} R_{\mathfrak{q}}(A) \setminus \{\mathfrak{p}^* \mid \mathfrak{p} \in V(\mathfrak{q})\}$.

Proof: The "only-if-part" follows immediately from the above remark. Assume that $G_{\mathfrak{q}}(M)_{(\mathfrak{P})}$ is Cohen-Macaulay for all

$$\mathfrak{P} \in P(\mathfrak{q}) := \operatorname{Proj} R_{\mathfrak{q}}(A) \setminus \{\mathfrak{p}^* \mid \mathfrak{p} \in V(\mathfrak{q})\}.$$

Let $\mathfrak{P} \in P(\mathfrak{q})$. If $G_\mathfrak{q}(M)_{(\mathfrak{P})} \neq O$, $R_\mathfrak{q}(M)_{(\mathfrak{P})}$ is Cohen-Macaulay. Let $G_\mathfrak{q}(M)_{(\mathfrak{P})} = O$ and $R_\mathfrak{q}(M)_{(\mathfrak{P})} \neq O$. Then $[\mathfrak{P}]_0 \notin V(\mathfrak{q})$ by Corollary 1.8, i.e. $\mathfrak{P} = \mathfrak{p}^*$ or $\mathfrak{P} = \tilde{\mathfrak{p}}$ with $\mathfrak{p} \in \operatorname{Supp} M \setminus V(\mathfrak{q})$ and $R_\mathfrak{q}(M)_{(\mathfrak{P})} \cong M_\mathfrak{p}$ by Corollary 1.4 and Corollary 1.5. Take a minimal prime ideal $\mathfrak{Q}$ of $R_\mathfrak{q}(A)$ containing $\mathfrak{q}R_\mathfrak{q}(A) + \tilde{\mathfrak{p}}$. If all these $\mathfrak{Q}$ are of the form $\mathfrak{Q} = \mathfrak{v}^*$ with $\mathfrak{v} \in V(\mathfrak{q})$, then $[\operatorname{rad}(\mathfrak{q}R_\mathfrak{q}(A) + \tilde{\mathfrak{p}})]_1 = \mathfrak{q}T$, i.e. there is an integer n with $\mathfrak{q}^nT^n = [\mathfrak{q}R_\mathfrak{q}(A) + \tilde{\mathfrak{p}}]_n = (\mathfrak{q}^{n+1} + (\mathfrak{q}^n \cap \mathfrak{p}))\,T^n$. But this is impossible. Hence we can find a $\mathfrak{Q} \in P(\mathfrak{q})$ with $[\mathfrak{Q}]_0 \in V(\mathfrak{q})$ and $\tilde{\mathfrak{p}} \subset \mathfrak{Q}$. By Lemma 1.7 and Corollary 1.8 $G_\mathfrak{q}(M)_{(\mathfrak{Q})} \neq O$ and this implies that $R_\mathfrak{q}(M)_{(\mathfrak{Q})}$ is Cohen-Macaulay. Now $[\mathfrak{Q}]_1 \supsetneqq \mathfrak{q}T$ and Lemma 1.6 guarantees the existence of a prime ideal $\mathfrak{v} \in \operatorname{Spec} R_\mathfrak{q}(A)_{(\mathfrak{P})}$ with $M_\mathfrak{p} \cong R_\mathfrak{q}(M)_{(\mathfrak{P})} = (R_\mathfrak{q}(M)_{(\mathfrak{P})})_\mathfrak{v}$ and this is a Cohen-Macaulay module, q.e.d.

Corollary 1.12. *Assume* $\dim_A M/\mathfrak{q}M = 0$. *Then* $R_\mathfrak{q}(M)$ *is locally Cohen-Macaulay if and only if* $G_\mathfrak{q}(M)$ *is locally Cohen-Macaulay.*

Remark. The property "Cohen-Macaulay" in the statement of Lemma 1.11 can be replaced by every other local property which is stable under localizations, which is preserved if we factor a module having this property by a non-zero divisor and which is "discovered" in such factor modules where we have some possibilities of choosing "admissible" non-zero divisors. In particular we have

Corollary 1.13. $R_\mathfrak{q}(M)_{(\mathfrak{P})}$ *is a Buchsbaum module for all* $\mathfrak{P} \in \operatorname{Proj} R_\mathfrak{q}(A) \setminus \{\mathfrak{p}^* \mid \mathfrak{p} \in V(\mathfrak{q})\}$ *if and only if* $G_\mathfrak{q}(M)_{(\mathfrak{P})}$ *is a Buchsbaum module for all* $\mathfrak{P} \in \operatorname{Proj} R_\mathfrak{q}(A) \setminus \{\mathfrak{p}^* \mid \mathfrak{p} \in V(\mathfrak{q})\}$.

Proof: The only difficulty is to conclude from the Buchsbaum property of $G_\mathfrak{q}(M)_{(\mathfrak{P})}$, where $G_\mathfrak{q}(M)_{(\mathfrak{P})} \neq O$, the Buchsbaum property of $R_\mathfrak{q}(M)_{(\mathfrak{P})}$. But this is possible using the exact sequence of Lemma 1.3(ii) where the element x must be contained in the square of the maximal ideal of $R_\mathfrak{q}(A)_{(\mathfrak{P})}$ (c.f. Proposition I.2.19). For this replace x by x^2 if necessary, q.e.d.

§ 2. The Buchsbaum property of Rees modules and associated graded modules

The aim of this paragraph is to show that the Rees module $R_\mathfrak{q}(M)$ and the associated graded module $G_\mathfrak{q}(M)$ is Buchsbaum whenever M is Buchsbaum and $\mathfrak{q}$ is a parameter ideal of M. To this end we calculate the local cohomology modules $\underline{H}^i_\mathfrak{M}(R_\mathfrak{q}(M))$, $i \leq \dim M$, and $\underline{H}^i_\mathfrak{M}(G_\mathfrak{q}(M))$, $i < \dim M$.

Theorem 2.1. *Let M be a Noetherian A-module. M is a Buchsbaum module if and only if $G_\mathfrak{q}(M)$ is a Buchsbaum module for all parameter ideals $\mathfrak{q}$ of M.*
Moreover in this situation we have

$$\underline{H}^i_\mathfrak{M}(G_\mathfrak{q}(M)) \cong \underline{H^i_\mathfrak{m}(M)}\,(i) \quad \textit{for all } 0 \leq i < \dim M$$

and

$$I(G_\mathfrak{q}(M)) = I(M).$$

Proof: Let $d := \dim M$. If $d = 0$ there is nothing to prove. Let $d > 0$. If $G_\mathfrak{q}(M)$ is a Buchsbaum module for every parameter ideal $\mathfrak{q}$ of M, then we obtain for any parameter ideal $\mathfrak{q}$ of M:

$$G_\mathfrak{q}(M)/\mathfrak{q}TG_\mathfrak{q}(M) = \underline{M/\mathfrak{q}M}, \text{ i.e. } \mathfrak{q}TG_\mathfrak{q}(A) \text{ is a parameter ideal of } G_\mathfrak{q}(M).$$

Assume $\mathfrak{q} = (x_1, \ldots, x_d)\, A$. But $\mathfrak{q}' := (x_1, \ldots, x_{d-1})\, A$. Then

$$[(x_1T, \ldots, x_{d-1}T)\, G_{\mathfrak{q}}(M) :_{G_{\mathfrak{q}}(M)} x_dT]_0$$
$$= (\mathfrak{q}^2M + \mathfrak{q}'M) :_M x_d/\mathfrak{q}M = (x_d^2M + \mathfrak{q}'M) : x_d/\mathfrak{q}M = x_dM + (\mathfrak{q}'M : x_d)/\mathfrak{q}M.$$

Since $G_{\mathfrak{q}}(M)$ is a Buchsbaum module, this last module is annihilated by $\mathfrak{m}$, i.e. $(x_dM + (\mathfrak{q}'M : x_d)) \subseteq (x_dM + \mathfrak{q}'M) : \mathfrak{m}$.

If we replace x_d by x_d^n, $n \geq 1$, then for sufficiently large n

$$\mathfrak{q}'M : x_d^n = \mathfrak{q}'M : \langle \mathfrak{m} \rangle$$

and, using Krull's Intersection Theorem, we obtain

$$\mathfrak{q}'M : x_d \subseteq \mathfrak{q}'M : \langle \mathfrak{m} \rangle = \bigcap_{n>0} x_d^nM + (\mathfrak{q}'M : x_d^n) \subseteq \bigcap_{n>0} (x_d^nM + \mathfrak{q}'M) : \mathfrak{m}$$
$$= \mathfrak{q}'M : \mathfrak{m}.$$

Therefore M is a Buchsbaum module by Proposition I.1.10(i)'.

Now let M be a Buchsbaum module and $\mathfrak{q}$ a parameter ideal of M. Assume $\mathfrak{q}M = (x_1, \ldots, x_d)\, M$, $d = \dim M > 0$. We prove inductively on d:

(a) $[\underline{H}^i_{\mathfrak{M}}(G_{\mathfrak{q}}(M))]_n = 0$ for all $n > -i$ and all i and

$[\underline{H}^i_{\mathfrak{M}}(G_{\mathfrak{q}}(M))]_n = 0$ for all $n < -i$ and all $i < d$.

(b) For all i there is a natural (A-)monomorphism

$$\mu^i_M : [\underline{H}^i_{\mathfrak{M}}(G_{\mathfrak{q}}(M))]_{-i} \to H^i_{\mathfrak{m}}(M)$$

which is an isomorphism for $i < d$.

(c) The canonical maps

$$\lambda^i_M : \underline{H}^i(\mathfrak{M}; G_{\mathfrak{q}}(M)) \to \underline{H}^i_{\mathfrak{M}}(G_{\mathfrak{q}}(M))$$

are surjective for all $i < d$.

Our theorem will follow by Theorem I.2.15.

Let depth $M > 0$. Then

$$[0 :_{G_{\mathfrak{q}}(M)} \mathfrak{M}]_n = \mathfrak{q}^nM \cap (\mathfrak{q}^{n+1}M :_M \mathfrak{m}) \cap (\mathfrak{q}^{n+2}M :_M \mathfrak{q})/\mathfrak{q}^{n+1}M = \mathfrak{q}^{n+1}M/\mathfrak{q}^{n+1}M = 0$$

by Lemma I.1.15. Hence depth $G_{\mathfrak{q}}(M) > 0$. Now let M be an arbitrary Buchsbaum module. Since $(0 :_M \mathfrak{m}) \cap \mathfrak{q}M = 0$ (c.f. Lemma I.1.14) we have a monomorphism $H^0_{\mathfrak{m}}(M) = 0 :_M \mathfrak{m} \to M/\mathfrak{q}M$ which extends to an exact sequence of graded $R_{\mathfrak{q}}(A)$-modules

$$0 \to \underline{O :_M \mathfrak{m}} \to G_{\mathfrak{q}}(M) \to G_{\mathfrak{q}}(\overline{M}) \to 0,$$

where $\overline{M} := M/0 :_M \mathfrak{m}$. Since depth $\overline{M} > 0$, we obtain an isomorphism $\underline{H}^0_{\mathfrak{M}}(G_{\mathfrak{q}}(M)) \simeq \underline{H}^0_{\mathfrak{M}}(\underline{O :_M \mathfrak{m}}) = \underline{O :_M \mathfrak{m}} = \underline{H^0_{\mathfrak{m}}(M)}$ which gives rise to an isomorphism

$$\mu^0_M : [\underline{H}^0_{\mathfrak{M}}(G_{\mathfrak{q}}(M))]_0 \to H^0_{\mathfrak{m}}(M).$$

Clearly, μ^0_M is natural (i.e. if there is a Buchsbaum module N of (positive) dimension such that $\mathfrak{q}$ is again a parameter ideal of N and a map $f: M \to N$ then there is a commutative diagram

$$\begin{array}{ccc} [\underline{H}^0_{\mathfrak{M}}(G_{\mathfrak{q}}(M))]_0 & \xrightarrow{\mu^0_M} & H^0_{\mathfrak{m}}(M) \\ \downarrow & & \downarrow \\ [\underline{H}^0_{\mathfrak{M}}(G_{\mathfrak{q}}(N))]_0 & \xrightarrow{\mu^0_N} & H^0_{\mathfrak{m}}(N) \end{array}$$

where the vertical maps are induced by f).

Furthermore

$$\underline{H}^0(\mathfrak{M}; G_{\mathfrak{q}}(M)) = 0 :_{G_{\mathfrak{q}}(M)} \mathfrak{M} \cong 0 :_{\underline{H^0_{\mathfrak{m}}(M)}} \mathfrak{M} = \underline{H^0_{\mathfrak{m}}(M)} = \underline{H}^0_{\mathfrak{M}}(G_{\mathfrak{q}}(M))$$

and λ^0_M is an isomorphism. Therefore (a), (b) and (c) are correct for $i = 0$.

We note that for depth $M > 0$ we have an exact sequence

$$(\mathrm{E}) : 0 \to G_{\mathfrak{q}}(M)(-1) \xrightarrow{x_1 T} G_{\mathfrak{q}}(M) \to G_{\mathfrak{q}}(M/x_1M) \to 0.$$

(It is only necessary to show that x_1T is a non-zero divisor on $G_{\mathfrak{q}}(M)$. But this is easy to see by virtue of Lemma I.1.15.) If $d = 1$ and depth $M = 1$, we therefore have an exact sequence

$$0 \to \underline{H}^0_{\mathfrak{M}}(G_{\mathfrak{q}}(M/x_1M)) \to \underline{H}^1_{\mathfrak{M}}(G_{\mathfrak{q}}(M))(-1) \xrightarrow{x_1 T} \underline{H}^1_{\mathfrak{M}}(G_{\mathfrak{q}}(M)) \to 0,$$

where $\underline{H}^0_{\mathfrak{M}}(G_{\mathfrak{q}}(M/x_1M)) \cong \underline{M/x_1M}$.

Hence $[\underline{H}^1_{\mathfrak{M}}(G_{\mathfrak{q}}(M))]_n \cong [\underline{H}^1_{\mathfrak{M}}(G_{\mathfrak{q}}(M))]_{n+1}$ for all $n \geq 0$ and this shows that $[\underline{H}^1_{\mathfrak{M}}(G_{\mathfrak{q}}(M))]_n = 0$ for all $n \geq 0$ since $\underline{H}^1_{\mathfrak{M}}(G_{\mathfrak{q}}(M))$ is Artinian. We also have an isomorphism

$$[\underline{H}^1_{\mathfrak{M}}(G_{\mathfrak{q}}(M))]_{-1} \cong M/x_1M.$$

If depth $M = 0$, $\underline{H}^1_{\mathfrak{M}}(G_{\mathfrak{q}}(M)) \cong \underline{H}^1_{\mathfrak{M}}(G_{\mathfrak{q}}(\overline{M}))$ with $\overline{M} := M/H^0_{\mathfrak{m}}(M)$ and this proves (a). Furthermore, there is a natural isomorphism

$$[\underline{H}^1_{\mathfrak{M}}(G_{\mathfrak{q}}(M))]_{-1} \cong \overline{M}/x_1\overline{M}$$

and a natural monomorphism

$$\overline{M}/x_1\overline{M} \to H^1_{\mathfrak{m}}(\overline{M}) \cong H^1_{\mathfrak{m}}(M)$$

coming from the exact sequence

$$0 \to \overline{M} \xrightarrow{x_1} \overline{M} \to \overline{M}/x_1\overline{M} \to 0.$$

Thus (b) is proven and we are finished for $d = 1$.

Let $d > 1$. If we have proven (a), (b) and (c) for Buchsbaum modules of dimension d and positive depth, we get in the case depth $M = 0$:

Let $\overline{M} := M/H^0_{\mathfrak{m}}(M)$. Since $\underline{H}^i_{\mathfrak{M}}(G_{\mathfrak{q}}(M)) \cong \underline{H}^i_{\mathfrak{M}}(G_{\mathfrak{q}}(\overline{M}))$ for all $i > 0$ we obtain (a). If we set $\mu^i_M := \mu^i_{\overline{M}}$ for $i > 0$, (b) holds for all i. To verify (c) we again consider the exact sequence

$$0 \to \underline{H^0_{\mathfrak{m}}(M)} \to G_{\mathfrak{q}}(M) \to G_{\mathfrak{q}}(\overline{M}) \to 0.$$

From this we obtain for $0 < i < d$ commutative diagrams with exact rows

$$\begin{array}{ccccc} \underline{H}^i(\mathfrak{M}; \underline{H^0_{\mathfrak{m}}(M)}) & \to & \underline{H}^i(\mathfrak{M}; G_{\mathfrak{q}}(M)) & \xrightarrow{\pi^i} & \underline{H}^i(\mathfrak{M}; G_{\mathfrak{q}}(\overline{M})) \\ \downarrow & & \downarrow \lambda^i_M & & \downarrow \lambda^i_{\overline{M}} \\ 0 & \to & H^i_{\mathfrak{M}}(G_{\mathfrak{q}}(M)) & \longrightarrow & H^i_{\mathfrak{M}}(G_{\mathfrak{q}}(\overline{M})) \to 0. \end{array}$$

λ_M^i is surjective. Since by (a) $[\underline{H}_{\mathfrak{M}}^i(G_{\mathfrak{q}}(M))]_n = 0$ for all $n \neq -i$, the surjectivity of λ_M^i will follow if π^i is surjective in degree $-i$. Hence we show:

The monomorphism $\underline{0:_M \mathfrak{m}} \to G_{\mathfrak{q}}(M)$ induces for all $i \leq d$ monomorphisms

$$[\underline{H}^i(\mathfrak{M}; \underline{0:_M \mathfrak{m}})]_{-i+1} \to [\underline{H}^i(\mathfrak{M}; G_{\mathfrak{q}}(M))]_{-i+1}.$$

Choose a minimal generating set $\{a_1, \ldots, a_t\}$ of $\mathfrak{m}$. Then $\{a_1, \ldots, a_t, x_1T, \ldots, x_dT\}$ is a minimal generating set of $\mathfrak{M}$. Consider the Koszul complexes $K^{\cdot}(\mathfrak{M}; \underline{0:_M \mathfrak{m}})$ and $K^{\cdot}(\mathfrak{M}; G_{\mathfrak{q}}(M))$ and denote by $e_{i_1\ldots i_{p-q}}^{j_1\ldots j_q}$, $1 \leq j_1 < \ldots < j_q \leq d$, $1 \leq i_1 < \ldots < i_{p-q} \leq t$, $0 \leq q \leq p$ free generators of $K^p(\mathfrak{M}; R_{\mathfrak{q}}(A))$ of degree $-q$, where the upper indices are related to the elements $x_1T, \ldots, x_dT$ and the lower indices to $a_1, \ldots, a_t$.

If $d^{\cdot}$ is the differentiation in $K^{\cdot}(\mathfrak{M}; G_{\mathfrak{q}}(M))$, we have to show for all $i \leq d$:

$$[\operatorname{Im} d^{i-1}]_{-i+1} \cap [K^i(\mathfrak{M}; \underline{0:_M \mathfrak{m}})]_{-i+1} = 0$$

($d^{\cdot}$ acts trivially on $K^{\cdot}(\mathfrak{M}; \underline{0:_M \mathfrak{m}})$).

Let $x \in [K^{i-1}(\mathfrak{M}; G_{\mathfrak{q}}(M))]_{-i+1}$ with $d^{i-1}(x) \in [K^i(\mathfrak{M}; \underline{0:_M \mathfrak{m}})]_{-i+1}$. Write

$$x = \sum_{1 \leq j_1 < \ldots < j_{i-1} \leq d} e^{j_1\ldots j_{i-1}} \mu^{j_1\ldots j_{i-1}}$$

where $\mu^{j_1\ldots j_{i-1}} \in [G_{\mathfrak{q}}(M)]_0 = M/\mathfrak{q}M$ and where

$$d^{i-1}(x) = \sum_{\substack{1 \leq l \leq t \\ 1 \leq j_1 < \ldots < j_{i-1} \leq d}} e_l^{j_1\ldots j_{i-1}} (a_l \mu^{j_1\ldots j_{i-1}}) - \sum_{1 \leq j_1 < \ldots < j_i \leq d} e^{j_1\ldots j_i} \left(\sum_{h=1}^{i} (-1)^{h-1} x_{j_h} T \mu^{j_1\ldots \hat{j}_h \ldots j_i} \right).$$

$d^{i-1}(x) \in [K^i(\mathfrak{M}; \underline{0:_M \mathfrak{m}})]_{-i+1}$ implies

1. $a_l \mu^{j_1\ldots j_{i-1}} \in \underline{0:_M \mathfrak{m}}$ for all $1 \leq j_1 < \ldots < j_{i-1} \leq d$, $1 \leq l \leq t$ and

2. $\sum_{h=1}^{i} (-1)^{h-1} x_{j_h} T \mu^{j_1\ldots \hat{j}_h \ldots j_i} = 0$ for all $1 \leq j_1 < \ldots < j_i \leq d$. If we choose $m^{j_1\ldots j_{i-1}} \in M$, $1 \leq j_1 < \ldots < j_{i-1} \leq d$ with $\mu^{j_1\ldots j_{i-1}} \equiv m^{j_1\ldots j_{i-1}} \bmod \mathfrak{q}M$ then 2. means

$$x_p m^{j_1\ldots j_{i-1}} \in \sum_{q \in \{j_1\ldots j_{i-1}\}} x_q M + \mathfrak{q}^2 M$$

for all $p \in \{1, \ldots, d\} \setminus \{j_1, \ldots, j_{i-1}\}$ $(\neq \emptyset)$. Set (p fixed)

$$\mathfrak{q}' := (x_1, \ldots, x_{p-1}, x_{p+1}, \ldots, x_d)\, A.$$

Then

$$m^{j_1\ldots j_{i-1}} \in (\mathfrak{q}'M + \mathfrak{q}^2 M) : x_p = (\mathfrak{q}'M + x_p^2 M) : x_p = x_p M + (\mathfrak{q}'M : \mathfrak{m}) \subseteq \mathfrak{q}M : \mathfrak{m},$$

i.e.

$$\mu^{j_1\ldots j_{i-1}} \in \mathfrak{q}M : \mathfrak{m}/\mathfrak{q}M.$$

Therefore we have for all $l = 1, \ldots, t$: $a_l \mu^{j_1\ldots j_{i-1}} = 0$, i.e. $d^{i-1}(x) = 0$ and this proves (c) for this case.

Finally it remains to consider the case depth $M > 0$.

If we apply the functors $\underline{H}_{\mathfrak{M}}^i$ to the exact sequence (E) we obtain (using the induction hypothesis) for all i isomorphisms

$$[\underline{H}_{\mathfrak{M}}^i(G_{\mathfrak{q}}(M))]_n \xrightarrow[\sim]{x_1 T} [\underline{H}_{\mathfrak{M}}^i(G_{\mathfrak{q}}(M))]_{n+1} \quad \text{for all } n > -i$$

and for all $i < d$ (at least) monomorphisms

$$[\underline{H}^i_{\mathfrak{M}}(G_\mathfrak{q}(M))]_n \xrightarrow{x_1T} [\underline{H}^i_{\mathfrak{M}}(G_\mathfrak{q}(M))]_{n+1} \quad \text{for all } n < -i-1.$$

Hence $[\underline{H}^i_{\mathfrak{M}}(G_\mathfrak{q}(M))]_n = O$ for all $n > -i$ since $\underline{H}^i_{\mathfrak{M}}(G_\mathfrak{q}(M))$ is Artinian. Now the map μ^0_M is the zero map. We construct μ^i_M inductively on i, $i \leq d$. If μ^{i-1}_M is constructed, $i \leq d$, then we have a commutative diagram with exact rows

$$\begin{array}{ccccccccc} 0 \to & [\underline{H}^{i-1}_{\mathfrak{M}}(G_\mathfrak{q}(M))]_{-i+1} & \to & [\underline{H}^{i-1}_{\mathfrak{M}}(G_\mathfrak{q}(M/x_1M))]_{-i+1} & \to & [\underline{H}^i_{\mathfrak{M}}(G_\mathfrak{q}(M))]_{-i} & \to 0 \\ & \downarrow \mu^{i-1}_M & & \downarrow \mu^{i-1}_{M/x_1M} & & & \\ 0 \to & H^{i-1}_\mathfrak{m}(M) & \to & H^{i-1}_\mathfrak{m}(M/x_1M) & \xrightarrow{\varphi^i} & H^i_\mathfrak{m}(M) & \end{array}$$

Hence there is a unique map $\mu^i_M\colon [\underline{H}^i_{\mathfrak{M}}(G_\mathfrak{q}(M))]_{-1} \to H^i_\mathfrak{m}(M)$ making this diagram commutativ. Clearly, μ^i_M is a natural map. It is a monomorphism and an isomorphism whenever $i < d$ since φ^i is surjective and μ^{i-1}_{M/x_1M} is an isomorphism in this situation. This proves (b). Now for $i < d$, we consider the commutative diagram with exact rows:

$$\begin{array}{ccccccc} 0 \to & [\underline{H}^i_{\mathfrak{M}}(G_\mathfrak{q}(M))]_{-i-1} & \to & [\underline{H}^i_{\mathfrak{M}}(G_\mathfrak{q}(M))]_{-i} & \to & [\underline{H}^i_{\mathfrak{M}}(G_\mathfrak{q}(M/x_1M))]_{-i} \\ & & & \downarrow \mu^i_M & & \downarrow \mu^i_{M/x_1M} \\ & 0 & \to & H^i_\mathfrak{m}(M) & \to & H^i_\mathfrak{m}(M/x_1M) \end{array}$$

It shows that $[\underline{H}^i_{\mathfrak{M}}(G_\mathfrak{q}(M))]_{-i-1} = O$, hence $[\underline{H}^i_{\mathfrak{M}}(G_\mathfrak{q}(M))]_n = O$ for all $n < -i$ by the above monomorphisms. Therefore (a) is proven.

Now, from the exact sequence (E) result for each $i < d$ commutative diagrams with exact rows

$$\begin{array}{ccccccc} O \to & \underline{H}^{i-1}(\mathfrak{M}; G_\mathfrak{q}(M)) & \to & H^{i-1}(\mathfrak{M}; G_\mathfrak{q}(M/x_1M)) & \to & H^i(\mathfrak{M}; G_\mathfrak{q}(M))(-1) & \to 0 \\ & \downarrow \lambda^{i-1}_M & & \downarrow \lambda^{i-1}_{M/x_1M} & & \downarrow \lambda^i_M & \\ 0 \to & \underline{H}^{i-1}_{\mathfrak{M}}(G_\mathfrak{q}(M)) & \to & \underline{H}^{i-1}_{\mathfrak{M}}(G_\mathfrak{q}(M/x_1M)) & \to & \underline{H}^i_{\mathfrak{M}}(G_\mathfrak{q}(M))(-1) & \to 0 \end{array}$$

Since λ^{i-1}_{M/x_1M} is surjective by the induction hypothesis for $i < d$ $\lambda^i_M(-1)$ and hence λ^i_M is surjective for all $i < d$ and (c) is proven, q.e.d.

Corollary 2.2. *Let M be a Buchsbaum module of dimension $d > 0$ and $x_1, \ldots, x_d$ a system of parameters of M. Set $\mathfrak{q} := (x_1, \ldots, x_d)\,A$. Then for all integers $n, r_1, \ldots, r_d > 0$*

$$\mathfrak{q}^n M \cap (x_1^{r_1}, \ldots, x_d^{r_d})\,M = x_1^{r_1}\mathfrak{q}^{n-r_1}M + \ldots + x_d^{r_d}\mathfrak{q}^{n-r_d}M,$$

where $\mathfrak{q}^m M := M$ if $m \leq 0$.

Proof: It is sufficient to prove "$\subseteq$". Fix $r_1, \ldots, r_d$ and denote for notational convenience:

$$\mathfrak{q}' := (x_1^{r_1}T^{r_1}, \ldots, x_d^{r_d}T^{r_d})\,R_\mathfrak{q}(A), \qquad M' := M/(x_1^{r_1}, \ldots, x_d^{r_d})\,M,$$

$$U_n := (x_1^{r_1}\mathfrak{q}^{n-r_1}M + \ldots + x_d^{r_d}\mathfrak{q}^{n-r_d}M), \qquad V_n := \mathfrak{q}^n M \cap (x_1^{r_1}, \ldots, x_d^{r_d})\,M.$$

Consider the canonical epimorphism $\pi\colon G_{\mathfrak{q}}(M)/\mathfrak{q}'G_{\mathfrak{q}}(M) \to G_{\mathfrak{q}}(M')$. We get

$$\begin{aligned} l\big(G_{\mathfrak{q}}(M)/\mathfrak{q}'G_{\mathfrak{q}}(M)\big) &= e_0\big(\mathfrak{q}'; G_{\mathfrak{q}}(M)\big) + I\big(G_{\mathfrak{q}}(M)\big) \\ &= r_1 \cdot \ldots \cdot r_d e_0\big((x_1T, \ldots, x_dT); G_{\mathfrak{q}}(M)\big) + I(M) \\ &= r_1 \cdot \ldots \cdot r_d\big(l(G_{\mathfrak{q}}(M)/\mathfrak{q}TG_{\mathfrak{q}}(M)) - I(G_{\mathfrak{q}}(M))\big) + I(M) \\ &= r_1 \cdot \ldots \cdot r_d\big(l(M/\mathfrak{q}M) - I(M)\big) + I(M) \\ &= r_1 \cdot \ldots \cdot r_d e_0(\mathfrak{q}; M) + I(M) \\ &= e_0\big((x_1^{r_1}, \ldots, x_d^{r_d}); M\big) + I(M) = l(M') = l\big(G_{\mathfrak{q}}(M')\big), \end{aligned}$$

i.e. π is an isomorphism. Therefore

$$\begin{aligned} \mathfrak{q}^nM/(U_n + \mathfrak{q}^{n+1}M) &= [G_{\mathfrak{q}}(M)/\mathfrak{q}'G_{\mathfrak{q}}(M)]_n \cong [G_{\mathfrak{q}}(M')]_n \\ &= \big(\mathfrak{q}^nM + (x_1^{r_1}, \ldots, x_d^{r_d})\,M\big)/\big(\mathfrak{q}^{n+1}M + (x_1^{r_1}, \ldots, x_d^{r_d})\,M\big) \\ &\cong \mathfrak{q}^nM/(V_n + \mathfrak{q}^{n+1}M), \end{aligned}$$

hence $V_n \subseteq U_n + \mathfrak{q}^{n+1}M$ for all $n > 0$. This yields

$$V_n = (U_n + \mathfrak{q}^{n+1}M) \cap (x_1^{r_1}, \ldots, x_d^{r_d})\,M = U_n + V_{n+1} \subseteq U_n + \mathfrak{q}^{n+2}M$$

and so on. Hence

$$V_n \subseteq \bigcap_{m>n} (U_n + \mathfrak{q}^mM) = U_n$$

by Krull's Intersection Theorem, q.e.d.

Next we investigate the Rees module $R_{\mathfrak{q}}(M)$ where $\mathfrak{q}$ is a parameter ideal of M and M is a Buchsbaum module. First we have:

Proposition 2.3. *Let M be a Buchsbaum module and $\mathfrak{q}$ a parameter ideal of M. Assume that $d := \dim M > 0$. Then*

(i) $\underline{H}^0_{\mathfrak{M}}\big(R_{\mathfrak{q}}(M)\big) = \underline{O :_M \mathfrak{m}} = \underline{H^0_{\mathfrak{m}}(M)}$.

(ii) *For all $i = 1, \ldots, d$ we have*

$$[\underline{H}^i_{\mathfrak{M}}(R_{\mathfrak{q}}(M))]_n \cong \begin{cases} O & \text{for} \quad n \geq 0 \text{ and } n \leq -i+1, \\ H^{i-1}_{\mathfrak{m}}(M) & \text{for} \quad -i+2 \leq n \leq -1. \end{cases}$$

(iii) $\mathfrak{M} \cdot \underline{H}^i_{\mathfrak{M}}\big(R_{\mathfrak{q}}(M)\big) = O$ *for all* $i \leq d$.

Proof: Before embarking on the proof we note that "parameter ideal of M" means: There is an ideal $\mathfrak{q}$ in A and d elements $x_1, \ldots, x_d$ such that $\mathfrak{q}M = (x_1, \ldots, x_d)\,M$. This does not imply that $\mathfrak{q} = (x_1, \ldots, x_d)\,A$ although one can assume this in most cases.

Let $x_1, \ldots, x_d$ be elements of $\mathfrak{m}$ such that $\mathfrak{q}M = (x_1, \ldots, x_d)\,M$. An easy calculation shows that

$$\underline{H}^0_{\mathfrak{M}}\big(R_{\mathfrak{q}}(M)\big) \cong \bigoplus_{n\geq 0} \big(\mathfrak{q}^nM \cap H^0_{\mathfrak{m}}(M)\big)\,T^n = \underline{H^0_{\mathfrak{m}}(M)} = \underline{O :_M \mathfrak{m}},$$

since $\mathfrak{q}^nM \cap H^0_{\mathfrak{m}}(M) = \mathfrak{q}^nM \cap (O :_M \mathfrak{m}) = O$ for all $n > 0$ by Lemma I.1.14. This proves (i).

Let $\overline{M} := M/0:_M\mathfrak{m}$. Then the exact sequences

$$0 \to {}'0:_M\mathfrak{m} \to M \to \overline{M} \to 0 \qquad \text{and} \qquad 0 \to \underline{O:_M\mathfrak{m}} \to R_\mathfrak{q}(M) \to R_\mathfrak{q}(\overline{M}) \to 0$$

give rise for $i > 0$ to isomorphisms

$$H^i_\mathfrak{m}(M) \cong H^i_\mathfrak{m}(\overline{M}) \qquad \text{and} \qquad \underline{H}^i_\mathfrak{M}(R_\mathfrak{q}(M)) \cong \underline{H}^i_\mathfrak{M}(R(\overline{M}))$$

and we can assume without loss of generality that depth $M > 0$. Now we have two exact sequences

$$(\mathrm{E}_1)\colon\ 0 \to G_\mathfrak{q}(M)\,(-1) \xrightarrow{\iota} R_\mathfrak{q}(M)/x_1R_\mathfrak{q}(M) \xrightarrow{\sigma} R_\mathfrak{q}(M/x_1M) \to 0,$$

$$(\mathrm{E}_2)\colon\ 0 \to R_\mathfrak{q}(M) \xrightarrow{x_1} R_\mathfrak{q}(M) \to R_\mathfrak{q}(M)/x_1R_\mathfrak{q}(M) \to 0.$$

We use induction on d. If $d = 1$, $R_\mathfrak{q}(M)/x_1R_\mathfrak{q}(M) = G_\mathfrak{q}(M)$ and (E_2) and Theorem 2.1 imply:

$$0 = \underline{H}^0_\mathfrak{M}(G_\mathfrak{q}(M)) = \underline{H}^0_\mathfrak{M}(R_\mathfrak{q}(M)/x_1R_\mathfrak{q}(M)) = 0:_{\underline{H}^1_\mathfrak{M}(R_\mathfrak{q}(M))}x_1,$$

hence $\underline{H}^1_\mathfrak{M}(R_\mathfrak{q}(M)) = 0$ (every element of $\underline{H}^1_\mathfrak{M}(R_\mathfrak{q}(M))$ is annihilated by some power of x_1). Therefore we are finished with the proof in this case.

Let $d > 1$. By the proof of Theorem 2.1 (statement (a)), we have for all i and all $n \geq -i+1$: $[\underline{H}^{i+1}_\mathfrak{M}(G_\mathfrak{q}(M))\,(-1)]_n = 0$. Therefore (E_1) induces (with the induction hypothesis) for all $i < d$ exact sequences

$$0 \to \underline{H}^i_\mathfrak{M}(G_\mathfrak{q}(M))\,(-1) \xrightarrow{\iota^i} \underline{H}^i_\mathfrak{M}(R_\mathfrak{q}(M)/x_1R_\mathfrak{q}(M)) \xrightarrow{\sigma^i} \underline{H}^i_\mathfrak{M}(R_\mathfrak{q}(M/x_1M)) \to 0$$

and this implies $[\underline{H}^i_\mathfrak{M}(R_\mathfrak{q}(M)/x_1R_\mathfrak{q}(M))]_n = 0$ for all $n \geq 0$ and $n \leq -i$. Furthermore, $\mathfrak{M}\cdot\underline{H}^i_\mathfrak{M}(R_\mathfrak{q}(M)/x_1R_\mathfrak{q}(M)) = 0$ by this exact sequence, since the left term is concentrated in degree $-i+1$ and all elements of the right-hand term are of degree $\geq -i+2$.

From (E_2) result for all $i \leq d$ exact sequences

$$\underline{H}^{i-1}_\mathfrak{M}(R_\mathfrak{q}(M)/x_1R_\mathfrak{q}(M)) \to \underline{H}^i_\mathfrak{M}(R_\mathfrak{q}(M)) \xrightarrow{x_1} \underline{H}^i_\mathfrak{M}(R_\mathfrak{q}(M)).$$

As we will see later, $x_1\underline{H}^i_\mathfrak{M}(R_\mathfrak{q}(M)) = 0$ and this implies (iii).

But first of all we see that for $n \geq 0$ and $n \leq -i+1$:

$$0:_{[\underline{H}^i_\mathfrak{M}(R_\mathfrak{q}(M))]_n}x_1 = [0:_{\underline{H}^i_\mathfrak{M}(R_\mathfrak{q}(M))}x_1]_n = 0 \qquad (i \leq d),$$

and this implies $[\underline{H}^i_\mathfrak{M}(R_\mathfrak{q}(M))]_n = 0$ for $n \geq 0$ and $n \leq -i+1$.

We note (c.f. Sharp [1], Theorem 4.3) that for all i we have $\underline{H}^i_\mathfrak{M}(\underline{M}) \cong \underline{H^i_\mathfrak{M}(M)}$. Now we consider the exact sequences

$$(\mathrm{E}_3)\ O \to \mathfrak{q}R_\mathfrak{q}(M)\,(-1) \to R_\mathfrak{q}(M) \to \underline{M} \to 0,$$

$$(\mathrm{E}_4)\ O \to R_\mathfrak{q}(M) \to \mathfrak{q}R_\mathfrak{q}(M) \to G_\mathfrak{q}(M) \to 0.$$

Since $[\underline{H}^i_\mathfrak{M}(R_\mathfrak{q}(M))]_n = 0$ for $n \geq 0$, we get from (E_3) exact sequences

$$0 \to \underline{H^{i-1}_\mathfrak{m}(M)} \to \underline{H}^i_\mathfrak{M}(\mathfrak{q}R_\mathfrak{q}(M))\,(-1) \to \underline{H}^i_\mathfrak{M}(R_\mathfrak{q}(M)) \to 0$$

which yield isomorphisms

$$\begin{aligned} &[\underline{H}^i_\mathfrak{M}(\mathfrak{q}R_\mathfrak{q}(M))]_{-1} \cong H^{i-1}_\mathfrak{m}(M), \\ &[H^i_\mathfrak{M}(\mathfrak{q}R_\mathfrak{q}(M))]_n \cong [\underline{H}^i_\mathfrak{M}(R_\mathfrak{q}(M))]_{n+1} \quad \text{for all } n \leq -2. \end{aligned} \tag{*}$$

In particular $[\underline{H}^i_{\mathfrak{M}}(\mathfrak{q}R_\mathfrak{q}(M))]_n = O$ for $n \geq 0$ and $n \leq -i$. Therefore (E_4) together with Theorem 2.1 implies exact sequences ($1 \leq i \leq d$):

$$0 \to \underline{H}^{i-1}_{\mathfrak{M}}(G_\mathfrak{q}(M)) \to \underline{H}^i_{\mathfrak{M}}(\mathfrak{q}R_\mathfrak{q}(M)) \to \underline{H}^i_{\mathfrak{M}}(R_\mathfrak{q}(M)) \to O$$

and we obtain for all $n \geq -i + 2$:

$$[\underline{H}^i_{\mathfrak{M}}(\mathfrak{q}R_\mathfrak{q}(M))]_n = [\underline{H}^i_{\mathfrak{M}}(R_\mathfrak{q}(M))]_n.$$

Together with the isomorphisms (*) we obtain (ii). Since $x_1 \in R_\mathfrak{q}(A)$ is an element of degree zero and since $x_1 H^{i-1}_{\mathfrak{m}}(M) = 0$ for all $i \leq d$, we have $x_1 \underline{H}^i_{\mathfrak{M}}(R_\mathfrak{q}(M)) = O$ for all $i \leq d$ which concludes our proof, q.e.d.

To prove our next main result we need some technical lemmata which we will now prove. We note that some of them are of independent interest (e.g., Lemma 2.4, Lemma 2.6 or Lemma 2.8). First of all we have an easy consequence of Proposition I.1.9:

Lemma 2.4. *Let M be a Noetherian A-module of dimension d and let $\mathfrak{a}, \mathfrak{b} \subset A$ be ideals with* $\dim_A M/\mathfrak{a}M = \dim_A M/\mathfrak{b}M = 0$. *Assume we are given an M-basis $a_1, \ldots, a_t$ of $\mathfrak{a}$* (c.f. Definition I.1.7). *Then there is an M-basis $b_1, \ldots, b_s$ of $\mathfrak{b}$ having the following property:*

For all integers p, q with $0 \leq p \leq t$, $0 \leq q \leq s$, $p + q = d$ and all $1 \leq i_1 < \ldots < i_p \leq t$, $1 \leq j_1 < \ldots < j_q \leq s$ the elements $a_{i_1}, \ldots, a_{i_p}, b_{j_1}, \ldots, b_{j_q}$ form a system of parameters of M.

Proof: Choose a common $M/(a_{i_1}, \ldots, a_{i_p})$ M-basis of $\mathfrak{b}$ for all $0 \leq p \leq d$ and all $1 \leq i_1 < \ldots < i_p \leq d$ (see Proposition I.1.9), q.e.d.

Next we have

Lemma 2.5. *Let M be a Buchsbaum module of dimension $d \geq 2$ and let $x_1, \ldots, x_d$ be a system of parameters of M. If $z \in \mathfrak{m}$ is an element such that $x_1, \ldots, x_{i-1}, z, x_{i+1}, \ldots, x_d$ is again a system of parameters of M for all $i = 1, \ldots, d$ then we have for all p, q with $1 \leq p < q \leq d$:*

$$\big((x_1, \ldots, x_p)(x_{p+1}, \ldots, x_q) M :_M z\big) \subseteq O :_M \mathfrak{m} + (x_1, \ldots, x_p) M.$$

Proof: We use induction on p. If $p = 1$ and $m \in \big(x_1(x_2, \ldots, x_q) M\big) : z$ then there is an $m' \in (x_2, \ldots, x_q) M$ with $zm = x_1 m'$, hence $m' \in (zM : x_1) \cap (x_2, \ldots, x_q) M \subseteq (zM : \mathfrak{m}) \cap (z, x_2, \ldots, x_d) M = zM$ by Lemma I.1.14. We select an element m'' of M with $m' = zm''$ and obtain $zm = x_1 zm''$, hence $m \in x_1 M + O : z = O : \mathfrak{m} + x_1 M$.

Now let $p \geq 2$. Then we obtain using the induction hypothesis

$$\begin{aligned}
&\big((x_1, \ldots, x_p)(x_{p+1}, \ldots, x_q) M\big) : z \\
&= \big((x_1, \ldots, x_p)(x_{p+1}, \ldots, x_q) M : z\big) \\
&\quad \cap \big((x_p M + (x_1, \ldots, x_{p-1})(x_{p+1}, \ldots, x_q) M) : z\big) \\
&\subseteq \big((x_1, \ldots, x_{p-1}) M + (x_p(x_{p+1}, \ldots, x_q) M : z)\big) \cap \big(x_p M : \mathfrak{m} + (x_1, \ldots, x_{p-1}) M\big) \\
&\subseteq \big((x_1, \ldots, x_{p-1}) M + O : \mathfrak{m} + x_p M\big) \cap \big(x_p M : \mathfrak{m} + (x_1, \ldots, x_{p-1}) M\big) \\
&= O : \mathfrak{m} + (x_1, \ldots, x_p) M,
\end{aligned}$$

q.e.d.

Lemma 2.6. *Let M be a Buchsbaum module and $x_1, \ldots, x_i$, $1 \leq i \leq d$ a part of a system of parameters of M. Then the homomorphism*

$$H^j(\mathfrak{m}; (x_1, \ldots, x_i) M) \to H^j(\mathfrak{m}; M)$$

induced by the embedding $(x_1, \ldots, x_i) M \subset M$ is the zero homomorphism whenever $j \leq d - i$.

Proof: Let $i = 1$ and denote by $g^j: H^j(\mathfrak{m}; x_1M) \to H^j(\mathfrak{m}; M)$ the map induced by the embedding $x_1M \subset M$. For $j \leq d - 1$ let $\gamma \in H^j(\mathfrak{m}; x_1M)$. Choose a cocycle $c \in K^j(\mathfrak{m}; x_1M)$ representing γ. Then the image $\tilde{c}$ of c in $K^j(\mathfrak{m}; M)$ has the form $\tilde{c} = x_1c'$ with $c' \in K^j(\mathfrak{m}; M)$. An easy calculation shows that c' is also a cocycle (choose an M-basis of $\mathfrak{m}$, use $0 :_M x_1 = 0 :_M \mathfrak{m}$ and apply Lemma I.1.14). Therefore x_1c' is a coboundary (since $x_1H^j(\mathfrak{m}; M) = 0$), i.e. $g^j(\gamma) = 0$.

Now $g^j = 0$ is equivalent to the injectivity of the map $H^j(\mathfrak{m}; M) \to H^j(\mathfrak{m}; M/x_1M)$ obtained from the epimorphism $M \to M/x_1M$ by using the long exact cohomology sequence which results from the short exact sequence $0 \to x_1M \to M \to M/x_1M \to 0$. Repeating this argument again and again we get a chain of monomorphisms for $j \leq d - i$

$$H^j(\mathfrak{m}; M) \to H^j(\mathfrak{m}; M/x_1M) \to \ldots \to H^j(\mathfrak{m}; M/(x_1, \ldots, x_i) M),$$

i.e. a monomorphism $H^j(\mathfrak{m}; M) \to H^j(\mathfrak{m}; M/(x_1, \ldots, x_i)M)$ induced by the epimorphism $M \to M/(x_1, \ldots, x_i) M$. This proves the lemma, q.e.d.

Lemma 2.7. *Let M be a Buchsbaum module with* depth $M > 0$ *and let $\mathfrak{a} \subset A$ be an ideal with* $\dim_A M/\mathfrak{a}M = 0$. *Consider the double complex $(K^{\cdot\cdot}, d_1^{\cdot\cdot}, d_2^{\cdot\cdot})$ defined by $K^{p,q} = K^p(\mathfrak{a}; K^q(\mathfrak{m}; M))$ and differentiations $d_1^{p,q}: K^{p,q} \to K^{p+1,q}$, $d_2^{p,q}: K^{p,q} \to K^{p,q+1}$ given by the differentiation of the underlying Koszul complexes. Then we have for all $p, q \geq 0$ with $p + q \leq d$:*

$$\operatorname{Ker} d_1^{p,q} \cap \operatorname{Ker} d_2^{p,q} \subseteq \operatorname{Im} d_2^{p,q-1}.$$

Proof: We denote by $(K_1^{\cdot}, d_1^{\cdot})$ the Koszul complex $K^{\cdot}(\mathfrak{a}; \ldots)$ with differentiation $d_1^{\cdot}$ and by $(K_2^{\cdot}, d_2^{\cdot})$ the Koszul complex $K^{\cdot}(\mathfrak{m}; \ldots)$ with differentiation $d_2^{\cdot}$. For $l = 1, 2$ let $e^{(l)}_{i_1 \ldots i_j}$ be free generators of K_l^j, $1 \leq i_1 < \ldots < i_j \leq t_l$, where $t_1 := \operatorname{rank}_{A/\mathfrak{m}} \mathfrak{a}/\mathfrak{m}\mathfrak{a}$ and $t_2 := \operatorname{rank}_{A/\mathfrak{m}} \mathfrak{m}/\mathfrak{m}^2$. Assume that $a_1, \ldots, a_{t_1}$ and $b_1, \ldots, b_{t_2}$ are M-bases of $\mathfrak{a}$, resp. $\mathfrak{m}$, satisfying the statement of Lemma 2.4. Let $K^{\cdot} := \operatorname{Tot} K^{\cdot\cdot}$ with differentiation $d^{\cdot}$, i.e. we have

$$K^i = \bigoplus_{p+q=1} K^{p,q}$$

and for $\alpha \in K^{p,q}$, $p + q = i$

$$d^i(\alpha) = d_1^{p,q}(\alpha) + (-1)^p d_2^{p,q}(\alpha).$$

By $H_1^{\cdot}$, $H_2^{\cdot}$, $H^{\cdot}$ we denote the cohomology of these complexes. Note that a free basis of $K^{p,q}$ is given by the formal products

$$e^{(1)}_{i_1 \ldots i_p} \cdot e^{(2)}_{j_1 \ldots j_q}, \quad 1 \leq i_1 < \ldots < i_p \leq t_1,\ 1 \leq j_1 < \ldots < j_q \leq t_2.$$

In addition the map $K^{\cdot} \to K^{\cdot}(a_1, \ldots, a_{t_1}, b_1, \ldots, b_{t_2}; M)$ given by

$$e^{(1)}_{i_1 \ldots i_p} \cdot e^{(2)}_{j_1 \ldots j_q} m \mapsto e_{i_1 \ldots i_p, j_1 + t_1 \ldots j_q + t_1} m \qquad (m \in M)$$

is an isomorphism of complexes.

Now choose $x \in \mathfrak{a}$ such that $a_1, \ldots, a_{t_1}$ and $b_1, \ldots, b_{t_2}$ are again (M/xM)-bases of $\mathfrak{a}$ and $\mathfrak{m}$ (this is possible since $\dim_A M/\mathfrak{a}M = 0$).

Let $\xi \in \operatorname{Ker} d_1^{p,q} \cap \operatorname{Ker} d_2^{p,q}$, $p + q \leq d$. If $p = 0$ or $q = 0$ there is nothing to show since $\operatorname{Ker} d_1^{p,q} = 0$ or $\operatorname{Ker} d_2^{p,q} = 0$ in this case (depth $M > 0$). Thus we assume $p \geq 1$, $q \geq 1$. Since $xH_1^i = xH_2^i = O$ for all i ($x \in \mathfrak{a}$), there are $\alpha \in K^{p-1,q}$ and $\beta \in K^{p,q-1}$ with

$$d_1^{p-1,q}(\alpha) = d_2^{p,q-1}(\beta) = x\xi.$$

Now $d_2^{p+1,q-1} d_1^{p,q-1}(\beta) = d_1^{p,q} d_2^{p,q-1}(\beta) = d_1^{p,q} d_1^{p-1,q}(\alpha) = 0$, i.e. there is a $\gamma_1 \in K^{p+1,q-2}$ with $d_1^{p,q-1}(x\beta) = d_2^{p+1,q-2}(\gamma_1)$. Again we have $d_2^{p+2,q-2} d_1^{p+1,q-2}(\gamma_1) = d_1^{p+1,q-1} d_1^{p+1,q-2}(\gamma_1) = 0$, i.e. there is a $\gamma_2 \in K^{p+2,q-3}$ with $d_1^{p+1,q-2}(x\gamma_1) = d_2^{p+2,q-3}(\gamma_2)$, and so on. If we put $\gamma_0 := \beta$, we find elements $\gamma_i \in K^{p+i,q-i-1}$, $i \geq 1$, with $d_2^{p+i,q-i-1}(\gamma_i) = d_1^{p+i-1,q-i}(x\gamma_{i-1})$.

If we put $\gamma_0' := \alpha$, similar calculations show that there are $\gamma_i' \in K^{p-i-1,q+i}$, $i \geq 1$, with $d_2^{p-i-1,q+i}(x\gamma_i') = d_1^{p-i-2,q+i+1}(\gamma_{i+1}')$. We define

$$\gamma := \sum_{i \geq 0} x^{p+q-i-1}\big((-1)^{\varepsilon_i} \gamma_i + (-1)^{\varepsilon_i'} \gamma_i'\big) \in K^{p+q-1},$$

where $\varepsilon_i := \binom{p+i+1}{2}$, $\varepsilon_i' := \binom{p-i}{2}$. (Note $\gamma_i, \gamma_i' = 0$ if $i \geq p + q$.) Then it is easy to verify that $d^{p+q-1}(\gamma) = 0$. Now $xH^{p+q-1} = O$, i.e. there is a $\delta \in K^{p+q-2}$ with $x\gamma = d^{p+q-2}(\delta)$. Write

$$\delta = \sum_{r+s=p+q-2} \delta_{rs}, \ \delta_{rs} \in K^{r,s}.$$

But this implies

$$\begin{aligned} & d_2^{p,q-1} d_1^{p-1,q-1}\big((-1)^{\varepsilon_0} \delta_{p-1,q-1}\big) \\ &= d_2^{p,q-1}\big((-1)^{\varepsilon_0} d_1^{p-1,q-1}(\delta_{p-1,q-1}) + (-1)^{\varepsilon_0+p} d_2^{p,q-2}(\delta_{p,q-2})\big) \\ &= d_2^{p,q-1}\big((-1)^{\varepsilon_0} \cdot x \cdot x^{p+q-1}(-1)^{\varepsilon_0} \cdot \beta\big) = x^{p+q} d_2^{p,q-1}(\beta) = x^{p+q+1}\xi, \end{aligned}$$

i.e. there is an element $\eta \in K^{p-1,q-1}$ with $x^{p+q+1}\xi = d_2^{p,q-1} d_1^{p-1,q-1}(\eta)$.

Let

$$\xi = \sum e^{(1)}_{i_1 \ldots i_p} \cdot e^{(2)}_{j_1 \ldots j_q} \cdot m^{j_1 \ldots j_q}_{i_1 \ldots i_p},$$

where the sum is taken over all $i_1, \ldots, i_p, j_1, \ldots, j_q$ with $1 \leq i_1 < \ldots < i_p \leq t_1$, $1 \leq j_1 < \ldots < j_q \leq t_2$ and

$$\eta = \sum e^{(1)}_{l_1 \ldots l_{p-1}} \cdot e^{(2)}_{h_1 \ldots h_{q-1}} \cdot \mu^{h_1 \ldots h_{q-1}}_{l_1 \ldots l_{p-1}},$$

where the indices are defined similarly. Then

$$x^{p+q+1} m^{j_1 \ldots j_q}_{i_1 \ldots i_p} = \sum_{u=1}^{p} \sum_{v=1}^{q} (-1)^{u+v} a_{i_u} b_{j_v} \mu^{j_1 \ldots \hat{j}_v \ldots j_q}_{i_1 \ldots \hat{i}_u \ldots i_p} \in (a_{i_1}, \ldots, a_{i_p})(b_{j_1}, \ldots, b_{j_q}) M.$$

By Lemma 2.5 and the choice of x and the a_i's and b_j's,

$$m^{j_1 \ldots j_q}_{i_1 \ldots i_p} \in (a_{i_1}, \ldots, a_{i_p}) M.$$

Consider for all $1 \leq i_1 < \ldots < i_p \leq t_1$ the element

$$m_{i_1 \ldots i_p} := \sum_{1 \leq j_1 < \ldots < j_q \leq t_2} e^{(2)}_{j_1 \ldots j_q} m^{j_1 \ldots j_q}_{i_1 \ldots i_p} \in K_2^q = K^q(\mathfrak{m}; M).$$

$\xi \in \operatorname{Ker} d_2^{p,q}$ implies that $m_{i_1 \ldots i_p} \in \operatorname{Ker} d_2^q \cap K^q(\mathfrak{m}; (a_{i_1}, \ldots, a_{i_p}) M)$. But the embedding $(a_{i_1}, \ldots, a_{i_p}) M \subset M$ induces the zero homomorphism $H^j(\mathfrak{m}; (a_{i_1}, \ldots, a_{i_p}) M) \to H^j(\mathfrak{m}; M)$ by Lemma 2.6 for all $j \leq d - p$. Therefore there are $m'_{i_1 \ldots i_p} \in K^{q-1}(\mathfrak{m}; M)$ (since $q \leq d - p$) with $m_{i_1 \ldots i_p} = d_2^{p-1}(m'_{i_1 \ldots i_p})$, i.e.

$$\xi = d_2^{p,q-1} \Big(\sum_{1 \leq i_1 < \ldots < i_p \leq t_1} e^{(1)}_{i_1 \ldots i_p} m'_{i_1 \ldots i_p} \Big) \in \operatorname{Im} d_2^{p,q-1},$$

q.e.d.

Lemma 2.8. *Let M be a Buchsbaum module and $\mathfrak{q}$ a parameter ideal of M (for further reference to this lemma we note here that this does not imply that $\mathfrak{q}$ is generated by* dim M *elements). If $i, j \geq 0$, then*

$$d^j : K^j(\mathfrak{q}; \mathfrak{q}^j M) \to K^{j+1}(\mathfrak{q}; \mathfrak{q}^j M)$$

maps into $K^{j+1}(\mathfrak{q}; \mathfrak{q}^{j+1} M)$ and therefore we can define complexes

$$L^{\cdot}_h(M) : 0 \to K^0(\mathfrak{q}; \mathfrak{q}^{-h} M) \to K^1(\mathfrak{q}; \mathfrak{q}^{-h+1} M) \to \ldots,$$

where $h \in \mathbf{Z}$ and *and $\mathfrak{q}^h M := M$ for $h \leq 0$.*

We denote the differentiation by $\partial^{\cdot}_h$. Then

$$H^i(L^{\cdot}_h(M)) \cong \begin{cases} H^i(\mathfrak{q}; M) & \text{for } i \leq h, \\ 0 & \text{for } i > h. \end{cases}$$

Proof: It is clear that we only have to show $H^i(L^{\cdot}_h(M)) = 0$ for all $i > h$. Let $\overline{M} := M/0{:}\mathfrak{m}$. The map $M \to \overline{M}$ induces an epimorphism of complexes $L^{\cdot}_h(M) \to L^{\cdot}_h(\overline{M})$ which is an isomorphism for all degrees $i > h$ (since $\mathfrak{q}^r M \cong \mathfrak{q}^r \overline{M}$ for $r > 0$). Therefore $H^i(L^{\cdot}_h(M)) \cong H^i(L^{\cdot}_h(\overline{M}))$ for $i > h$ and we may assume that depth $M > 0$ if dim $M > 0$. We note that there is nothing to prove if dim $M = 0$, since, per definitionem, $\mathfrak{q}^r M = 0$ for all $r > 0$ in this case.

We use induction on $d :=$ dim M. For $d = 0$ we are finished. Let $d > 0$ and (without loss of generality) depth $M > 0$. Choose $x \in \mathfrak{q}$ such that dim $M/xM = d - 1$ and $\mathfrak{q}$ is again a parameter ideal of M/xM. Then there is an exact sequence of complexes (set $M' := M/xM$)

$$0 \to L^{\cdot}_{h+1}(M) \to L^{\cdot}_h(M) \to L^{\cdot}_h(M') \to 0,$$

where for each i the map $L^i_{h+1}(M) \to L^i_h(M)$ is obtained by multiplication with x (the exactness follows from $xM \cap \mathfrak{q}^r M = x\mathfrak{q}^{r-1} M \cong \mathfrak{q}^{r-1} M$ for all $r \in \mathbf{Z}$, see Corollary 2.2). The resulting long exact cohomology sequence gives for each $i > h$ an isomorphism

$$H^i(L^{\cdot}_h(M)) \cong H^i(L^{\cdot}_{i-1}(M))$$

(use induction on $i - h \geq 1$ and $H^i(L^{\cdot}_{h+1}(M)) \cong H^i(L^{\cdot}_h(M))$ for $i > h + 1$ by the above exact sequence and the induction hypothesis on d).

If we set $h = i - 1$, we get an exact sequence

$$(\mathrm{E}) : H^{i-1}(L^{\cdot}_{i-1}(M)) \to H^{i-1}(L^{\cdot}_{i-1}(M')) \to H^i(L^{\cdot}_i(M)) \to H^i(L^{\cdot}_{i-1}(M)) \to 0.$$

On the other hand, the exact sequence $0 \to M \xrightarrow{x} M \to M' \to 0$ induces an exact sequence $0 \to H^{i-1}(\mathfrak{q}; M) \to H^{i-1}(\mathfrak{q}; M') \to H^i(\mathfrak{q}; M) \to 0$, where $H^j(\mathfrak{q}; M^{(\prime)}) = H^j(L^{\cdot}_j(M^{(\prime)}))$ for all j. Calculating the lengths of the modules occuring in (E), we obtain

$$l\big(H^i(L^{\cdot}_{i-1}(M))\big) \leq l\big(H^i(L^{\cdot}_i(M))\big) - l\big(H^{i-1}(L^{\cdot}_{i-1}(M'))\big) + l\big(H^{i-1}(L^{\cdot}_{i-1}(M))\big) = 0,$$

i.e.

$$H^i\big(L^{\cdot}_{i-1}(M)\big) = 0,$$

q.e.d.

Lemma 2.9. *Let M be a Buchsbaum module and $x_1, \ldots, x_d$, $d = \dim M$ a system of parameters of M. Set $\mathfrak{q} := (x_1, \ldots, x_d)\, A$. Then:*

(i) *The embedding $\underline{O :_M \mathfrak{m}} \subset R_{\mathfrak{q}}(M)$ induces for all $i \leq d+1$ and all $q < 0$ monomorphisms*

$$[H^i(\mathfrak{M}; \underline{O :_M \mathfrak{m}})]_q \to \big[H^i(\mathfrak{M}; R_{\mathfrak{q}}(M))\big]_q.$$

(ii) *Let* depth $M > 0$. *Then the embedding $\underline{x_d M :_M \mathfrak{m}} \subset R_{\mathfrak{q}}(M)/x_dTR_{\mathfrak{q}}(M)$ (induced by $x_dM :_M \mathfrak{m} \subset M$) results for all $i \leq d$ in monomorphisms*

$$H^i(\mathfrak{M}; \underline{x_dM :_M \mathfrak{m}}) \to H^i\big(\mathfrak{M}; R_{\mathfrak{q}}(M)/x_dTR_{\mathfrak{q}}(M)\big).$$

Proof: For all $p \geq 0$, $\max(0, p-t) \leq h \leq \min(p, d)$, where $t = \operatorname{rank}_{A/\mathfrak{m}} \mathfrak{m}/\mathfrak{m}^2$ and all $i_1, \ldots, i_{p-h}, j_1, \ldots j_h$ with $1 \leq i_1 < \ldots < i_{p-h} \leq t$, $1 \leq j_1 < \ldots < j_h \leq d$ let $e^{j_1 \ldots j_h}_{i_1 \ldots i_{p-h}}$ be generators of the (free $R_{\mathfrak{q}}(A)$-)module $K^p\big(\mathfrak{M}; R_{\mathfrak{q}}(A)\big)$ with $\deg e^{j_1 \ldots j_h}_{i_1 \ldots i_{p-h}} = -h$. Let $d^{\cdot}$ denote the differentiation.

Note that $[K^p(\mathfrak{M}; N)]_q \neq 0$ for any A-module $N \neq 0$ and all $0 \leq p \leq t + d$ if and only if $\max(0, p-t) \leq -q \leq \min(p, d)$.

To prove (i) we have to show that for all $p \leq d$, $q < 0$

$$[K^{p+1}(\mathfrak{M}; \underline{O : \mathfrak{m}})]_q \cap [\operatorname{Im} d^p]_q = 0$$

(note that $d^{\cdot}$ acts trivially on $K^{\cdot}(\mathfrak{M}; \underline{O : \mathfrak{m}})$). We choose an M-basis $a_1, \ldots, a_t$ of $\mathfrak{m}$. Let

$$\xi = \sum_{h=-q}^{p} \sum_{\substack{1 \leq i_1 < \ldots < i_{p+q+1} \leq t \\ 1 \leq j_1 < \ldots < j_{-q} \leq d}} e^{j_1 \ldots j_h}_{i_1 \ldots i_{p-h}} \cdot m^{j_1 \ldots j_h}_{i_1 \ldots i_{p-h}} T^{q+h}$$

$$\in \big[K^p\big(\mathfrak{M}; R_{\mathfrak{q}}(M)\big)\big]_q \; (m^{j_1 \ldots j_h}_{i_1 \ldots i_{p-h}} \in \mathfrak{q}^{q+h} M)$$

with $d^p(\xi) \in [K^{p+1}(\mathfrak{M}; \underline{O : \mathfrak{m}})]_q$, i.e. we have for all $h = -q, \ldots, \min(p+1, d)$, $1 \leq i_1 < \ldots < i_{p+1-h} \leq t$, $1 \leq j_1 < \ldots < j_h \leq d$:

$$\sum_{l=1}^{p+1-h} (-1)^{l-1} a_{i_l} m^{j_1 \ldots j_h}_{i_1 \ldots \hat{i}_l \ldots i_{p+1-h}} + (-1)^{p+1-h} \sum_{n=1}^{h} (-1)^{n-1} x_{j_n} m^{j_1 \ldots \hat{j}_n \ldots j_h}_{i_1 \ldots i_{p+1-h}} \begin{cases} = 0 & \text{for} \quad h > -q, \\ \in 0 : \mathfrak{m} & \text{for} \quad h = -q. \end{cases}$$

But $q \leq -1$, hence we get for $h = -q$:

$$\sum_{l=1}^{p+1+q} (-1)^{l-1} a_{i_l} m^{j_1 \ldots j_{-q}}_{i_1 \ldots \hat{i}_l \ldots i_{p+1+q}} \in 0 : \mathfrak{m} \cap (a_{i_1}, \ldots, a_{i_{p+1+q}})\, M = 0,$$

since $p + 1 + q \leq p \leq d$ (note that $m^{j_1 \ldots \hat{j}_n \ldots j_{-q}}_{i_1 \ldots\, p+1+q} = 0$), i.e. $d^p(\xi) = 0$ and this proves (i).

To prove (ii) let us denote by $K^{\cdot}$ the Koszul complex $K^{\cdot}(\mathfrak{M}; R_{\mathfrak{q}}(M)/x_d TR_{\mathfrak{q}}(M))$ with differentiation $d^{\cdot}$ and by $\tilde{K}^{\cdot}$ the Koszul complex $K^{\cdot}(\mathfrak{M}; \overline{x_d M : \mathfrak{m}})$ with differentiation $\tilde{d}^{\cdot}$. We have to show for all $p \le d$ and all q with $0 \le q \le p-1$:

$$[\mathrm{Im}\, d^{p-1}]_{-q} \cap [\tilde{K}^p]_{-q} \subseteq [\mathrm{Im}\, d^{p-1}]_{-q} \qquad ([\tilde{K}^q]_n = 0 \text{ for } n < -p \text{ and } n > 0).$$

Let

$$\xi = \sum_{h=q}^{p-1} \sum_{\substack{1 \le i_1 < \ldots < i_{p-h-1} \le t \\ 1 \le j_1 < \ldots < j_h \le d}} e^{j_1 \ldots j_h}_{i_1 \ldots i_{p-h-1}} \overline{m^{j_1 \ldots j_h}_{i_1 \ldots i_{p-h-1}} T^{h-q}} \in [K^{p-1}]_{-q}$$

($\overline{}$ denotes the residue class modulo $x_d TR_{\mathfrak{q}}(M)$),

$$m^{j_1 \ldots j_h}_{i_1 \ldots i_{p-h-1}} \in \mathfrak{q}^{h-q} M, \text{ with } d^{p-1}(\xi) \in [\tilde{K}^p]_{-q}.$$

First of all we prove inductively on s, $1 \le s \le p-q$, that there are elements

$$\xi^{(s)} = \sum_{h=q}^{p-s} \sum_{\substack{1 \le i_1 < \ldots < i_{p-h-1} \le t \\ 1 \le j_1 \ldots < j_h \le d}} e^{j_1 \ldots j_h}_{i_1 \ldots i_{p-h-1}} \overline{m^{(s) j_1 \ldots j_h}_{i_1 \ldots i_{p-h-1}} T^{h-q}} \in [K^{p-1}]_{-q}, \text{ with } d^{p-1}(\xi^{(s)}) = d^{p-1}(\xi).$$

For $s = 1$ there is nothing to prove (put $\xi^{(1)} := \xi$). Assume we have determined $\xi^{(s)}$ for $1 \le s < p-q$. Then

$$\sum_{l=1}^{p-s+1} (-1)^{l-1} x_{j_l} m^{(s) j_1 \ldots \hat{j}_l \ldots j_{p-s+1}}_{i_1 \ldots i_{s-1}} \in x_d \mathfrak{q}^{p-s-q} M$$

for all $1 \le i_1 < \ldots < i_{s-1} \le t$, $1 \le j_1 < \ldots < j_{p-s+1} \le d$.

Let

$$\mu^{(s) j_1 \ldots j_{p-s}}_{i_1 \ldots i_{s-1}} := m^{(s) j_1 \ldots j_{p-s}}_{i_1 \ldots i_{s-1}} \bmod x_d M, \qquad M' := M/x_d M$$

and — using the notation of Lemma 2.8 —

$$\mu_{i_1 \ldots i_{s-1}} := \sum_{1 \le j_1 < \ldots < j_{p-s} \le d} e_{j_1 \ldots j_{p-s}} \mu^{(s) j_1 \ldots j_{p-s}}_{i_1 \ldots i_{s-1}} \in K^{p-s}(\mathfrak{q}; \mathfrak{q}^{p-s-q} M') = L^{p-s}_q(M').$$

Then $\partial^{p-s}(\mu_{i_1 \ldots i_{s-1}}) = 0$.

By using Lemma 2.8 we find elements

$$\mu'_{i_1 \ldots i_{s-1}} = \sum_{1 \le j_1 < \ldots < j_{p-s-1} \le d} e_{j_1 \ldots j_{p-s-1}} \mu'^{j_1 \ldots j_{p-s-1}}_{i_1 \ldots i_{s-1}} \in L^{p-s-1}_q(M')$$

with $\mu_{i_1 \ldots i_{s-1}} = \partial^{p-s-1}(\mu'_{i_1 \ldots i_{s-1}})$ (since $p - s > q$).

Let $m'^{j_1 \ldots j_{p-s-1}}_{i_1 \ldots i_{s-1}} \in \mathfrak{q}^{p-s-q-1} M$ be elements with $\mu'^{j_1 \ldots j_{p-s-1}}_{i_1 \ldots i_{s-1}} \equiv m'^{j_1 \ldots j_{p-s-1}}_{i_1 \ldots i_{s-1}} \bmod x_d M$. For $q \le h \le p-s-1$ we put

$$m^{(s+1) j_1 \ldots j_h}_{i_1 \ldots i_{p-h-1}} := \begin{cases} (0 & \text{for } h < q, h \ge p-s) \\ m^{(s) j_1 \ldots j_h}_{i_1 \ldots i_{p-h-1}} & \text{for } q \le h < p-s-1, \\ m^{(s) j_1 \ldots j_{p-s-1}}_{i_1 \ldots i_s} & \\ \quad + (-1)^s \sum_{l=1}^{s} (-1)^{l-1} a_{i_l} m'^{j_1 \ldots j_{p-s-1}}_{i_1 \ldots \hat{i}_l \ldots i_s} & \text{for } h = p-s-1. \end{cases}$$

Then an easy calculation shows that we have with

$$\xi^{(s+1)} := \sum_{h=q}^{p-s-1} \sum_{\substack{1\le i_1<\ldots<i_{p-h-1}\le t\\ 1\le j_1<\ldots<j_h\le d}} e^{j_1\ldots j_h}_{i_1\ldots i_{p-h-1}} \overline{m^{(s+1)j_1\ldots j_h}_{i_1\ldots i_{p-h-1}}T^{h-q}} \in [K^{p-1}]_{-q}:$$

$d^{p-1}(\xi^{(s+1)}) = d^{p-1}(\xi^{(s)}) = d^{p-1}(\xi)$ and we are done.

Using $\xi^{(p-q)}$ we may assume without loss of generality that

$$\xi = \sum_{\substack{1\le i_1<\ldots<i_{p-q-1}\le t\\ 1\le j_1<\ldots<j_q\le d}} e^{j_1\ldots j_q}_{i_1\ldots i_{p-q-1}} \; m^{j_1\ldots j_q}_{i_1\ldots i_{p-q-1}} \in [K^{p-1}]_{-q},$$

where $m^{j_1\ldots j_q}_{i_1\ldots i_{p-q-1}} \in M = [R_{\mathfrak{q}}(M)/x_dTR_{\mathfrak{q}}(M)]_0$.

$d^{p-1}(\xi) \in [\tilde{K}^p]_{-q}$ means that

$$\sum_{n=1}^{q+1} (-1)^{n-1} x_{j_n} m^{j_1\ldots \hat{j}_n \ldots j_{q+1}}_{i_1\ldots i_{p-q-1}} \in x_dM$$

for all $i_1, \ldots, i_{p-q-1}, j_1, \ldots, j_{q+1}$ with $1 \le i_1 < \ldots < i_{p-q-1} \le t$, $1 \le j_1 < \ldots < j_{q+1} \le d$ and

$$\sum_{l=1}^{p-q} (-1)^{l-1} a_{i_l} m^{j_1\ldots j_q}_{i_1\ldots \hat{i}_l \ldots i_{p-q}} \in x_dM:\mathfrak{m}$$

for all $i_1, \ldots, i_{p-q}, j_1, \ldots, j_q$ with $1 \le i_1 < \ldots < i_{p-q} \le t$, $1 \le j_1 < \ldots < j_q \le d$.

Set $\overline{M} := M/x_dM:\mathfrak{m}$ and for $m \in M$ let $\overline{m}$ denote the residue class of m in $\overline{M}$.

We consider the double complex $K^{\cdot}(\mathfrak{q}; K^{\cdot}(\mathfrak{m}; \overline{M}))$ introduced in Lemma 2.7, with $\mathfrak{a} = \mathfrak{q}$ and $\overline{M}$ instead of M. We set

$$\overline{\eta} := \sum_{\substack{1\le i_1<\ldots<i_{p-q-1}\le t\\ 1\le j_1<\ldots<j_q\le d}} e^{(1)}_{j_1\ldots j_q} \cdot e^{(2)}_{i_1\ldots i_{p-q-1}} \overline{m}^{j_1\ldots j_q}_{i_1\ldots i_{p-q-1}} \in K^q(\mathfrak{q}; K^{p-q-1}(\mathfrak{m}; \overline{M})).$$

Then $\overline{\eta} \in \operatorname{Ker} d_1^{q,p-q-1} \cap \operatorname{Ker} d_2^{q,p-q-1}$ in the notation of Lemma 2.7. But then $\overline{\eta} \in \operatorname{Im} d_2^{q,p-q-2}$ by Lemma 2.7 $(q + (p-q-1) = p-1 \le d-1 = \dim \overline{M})$, i.e. there are $m'^{j_1\ldots j_q}_{i_1\ldots i_{p-q-2}} \in M$ with

$$\overline{m}^{j_1\ldots j_q}_{i_1\ldots i_{p-q-1}} = \sum_{l=1}^{p-q-1} (-1)^{l-1} a_{i_l} \overline{m}'^{j_1\ldots j_q}_{i_1\ldots \hat{i}_l \ldots i_{p-q-1}}$$

for all $1 \le i_1 < \ldots < i_{p-q-1} \le t$, $1 \le j_1 < \ldots < j_q \le d$, i.e.,

$$\tilde{m}^{j_1\ldots j_q}_{i_1\ldots i_{p-q-1}} := m^{j_1\ldots j_q}_{i_1\ldots i_{p-q-1}} - \sum_{l=1}^{p-q-1} (-1)^{l-1} a_{j_l} m'^{j_1\ldots j_q}_{i_1\ldots \hat{i}_l \ldots i_{p-q-1}} \in x_dM:\mathfrak{m}$$

for all $1 \le i_1 < \ldots < i_{p-q-1} \le t$, $1 \le j_1 < \ldots < j_q \le d$.

Let

$$\tilde{\xi} := \sum_{\substack{1\le i_1<\ldots<i_{p-q-1}\le t\\ 1\le j_1<\ldots<j_q\le d}} e^{j_1\ldots j_q}_{i_1\ldots i_{p-q-1}} \; \tilde{m}^{j_1\ldots j_q}_{i_1\ldots i_{p-q-1}} \in [K^{p-1}(\mathfrak{M}; \underline{x_dM:\mathfrak{m}})]_{-q}.$$

Then $d^{p-1}(\xi) = \tilde{d}^{p-1}(\tilde{\xi}) \in \operatorname{Im} \tilde{d}^{-1}$, q.e.d.

Now we are able to prove:

Theorem 2.10. *Let M be a Buchsbaum module of positive dimension. Then $R_{\mathfrak{q}}(M)$ is a Buchsbaum module for all parameter ideals $\mathfrak{q}$ of M.*

Proof: Let $\mathfrak{q}$ be a parameter ideal of M and $x_1, \ldots, x_d$, $d := \dim M$, a system of parameters of M with $\mathfrak{q}M = (x_1, \ldots, x_d)\, M$. Let $\mathfrak{q}' := (x_1, \ldots, x_d)\, A$. Then $R_\mathfrak{q}(M) = R_{\mathfrak{q}'}(M)$ is a Buchsbaum module over $R_\mathfrak{q}(A)$ if and only if it is a Buchsbaum module over $R_{\mathfrak{q}'}(A)$, i.e. we may assume that $\mathfrak{q}$ is generated by a system of parameters of M.

We use induction on d. If $d = 1$ or $d = 2$ then by Proposition 2.3 $\underline{H}^i_{\mathfrak{M}}(R_\mathfrak{q}(M)) = 0$ for $i \neq 0$, $d + 1$ $(= \dim R_\mathfrak{q}(M))$ and $\mathfrak{M}\underline{H}^0_{\mathfrak{M}}(R_\mathfrak{q}(M)) = 0$. Hence $R_\mathfrak{q}(M)$ is a Buchsbaum module by Proposition I.2.12.

Let $d \geq 3$ and let $\overline{M} := M/0{:}\mathfrak{m}$. Consider the exact sequence

$$0 \to \underline{O{:}\mathfrak{m}} \to R_\mathfrak{q}(M) \to R_\mathfrak{q}(\overline{M}) \to 0.$$

Using Lemma 2.9(i) and the fact that $\underline{H}^i_{\mathfrak{M}}(\underline{O{:}\mathfrak{m}}) = 0$ for $i > 0$, we get for all $0 < i \leq d$ and all $p < 0$ commutative diagrams with exact rows (the vertical maps are the canonical ones)

$$\begin{array}{ccccccc} O \to & [\underline{H}^i(\mathfrak{M}; \underline{O{:}\mathfrak{m}})]_p & \to & [\underline{H}^i(\mathfrak{M}; R_\mathfrak{q}(M))]_p & \to & [\underline{H}^i(\mathfrak{M}; R_\mathfrak{q}(\overline{M}))]_p & \to 0 \\ & \downarrow & & \downarrow{\scriptstyle [\lambda^i]_p} & & \downarrow{\scriptstyle [\bar\lambda^i]_p} & \\ 0 \to & 0 & \to & [\underline{H}^i_{\mathfrak{M}}(R_\mathfrak{q}(M))]_p & \to & [\underline{H}^i_{\mathfrak{M}}(R_\mathfrak{q}(\overline{M}))]_p & \to O. \end{array}$$

If $R_\mathfrak{q}(\overline{M})$ is a Buchsbaum module, then $[\bar\lambda^i]_p$ is surjective for all $1 \leq i \leq d$ by Theorem I.2.15, hence $[\lambda^i]_p$ is surjective for all $1 \leq i \leq d$ and all $p < 0$. But then λ^i is surjective since $[\underline{H}^i_{\mathfrak{M}}(R_\mathfrak{q}(M))]_p = 0$ for $p \geq 0$ by Proposition 2.3. Since $\mathfrak{M}\underline{H}^0_{\mathfrak{M}}(R_\mathfrak{q}(M)) = 0$ (Proposition 2.3), λ^0 is surjective too and $R_\mathfrak{q}(M)$ is a Buchsbaum module by Theorem I.2.15. Therefore we can assume that depth $M > 0$.

Set $\tilde{M} := M/x_dM{:}\mathfrak{m}$. Then by the induction hypothesis, $R_\mathfrak{q}(\tilde{M})$ is a Buchsbaum module, i.e. the canonical maps

$$\tilde\lambda^i\colon \underline{H}^i(\mathfrak{M}; R_\mathfrak{q}(\tilde{M})) \to \underline{H}^i_{\mathfrak{M}}(R_\mathfrak{q}(\tilde{M}))$$

are surjective for all $i \leq d$ by Theorem I.2.15.

Now the kernel of the natural epimorphism $R_\mathfrak{q}(M) \to R_\mathfrak{q}(\tilde{M})$ contains $x_dTR_\mathfrak{q}(M)$, i.e. we have an epimorphism

$$\pi\colon R_\mathfrak{q}(M)/x_dTR_\mathfrak{q}(M) \to R_\mathfrak{q}(\tilde{M}).$$

By Lemma I.1.14 we have for all $p > 0$ with $\mathfrak{q}' := (x_1, \ldots, x_{d-1})\, A$

$$\begin{aligned}[\operatorname{Ker} \pi]_p &= \overline{(\mathfrak{q}^pM \cap (x_dM{:}\mathfrak{m}))\, T^p} = \overline{(x_d\mathfrak{q}'^{p-1}M + (\mathfrak{q}'^pM \cap (x_dM{:}\mathfrak{m})))\, T^p} \\ &= \overline{(x_d\mathfrak{q}'^{p-1}M)\, T^p} = O\end{aligned}$$

($\overline{}$ means here: residue class modulo $x_dTR_\mathfrak{q}(M)$) and for $p = 0$ we see that

$$[\operatorname{Ker} \pi]_0 = x_dM{:}\mathfrak{m},$$

i.e. we obtain an exact sequence

$$0 \to \underline{x_dM{:}\mathfrak{m}} \to R_\mathfrak{q}(M)/x_dTR_\mathfrak{q}(M) \to R_\mathfrak{q}(\tilde{M}) \to 0.$$

From Sharp [1], 4.3., it follows readily that $\underline{H}^i_{\mathfrak{M}}(x_dM{:}\mathfrak{m}) \simeq \underline{H^i_{\mathfrak{M}}(x_dM{:}\mathfrak{m})}$, i.e. $[\underline{H}^i_{\mathfrak{M}}(\underline{x_dM{:}\mathfrak{m}})]_p = 0$ for $p \neq 0$. But $[\underline{H}^i_{\mathfrak{M}}(R_\mathfrak{q}(\tilde{M}))]_p = 0$ for $p \geq 0$ (Proposition 2.3). By using this and Lemma 2.9(ii) we get for all $i < d$ a commutative diagram with exact rows (the vertical homomorphisms are again the canonical ones):

$$\begin{array}{ccccccc} O \to \underline{H}^i(\mathfrak{M}; \underline{x_d M : \mathfrak{m}}) & \to & \underline{H}^i(\mathfrak{M}; R_\mathfrak{q}(M)/x_d T R_\mathfrak{q}(M)) & \to & \underline{H}^i(\mathfrak{M}; R_\mathfrak{q}(\tilde{M})) \to 0 \\ \downarrow \lambda^i_* & & \downarrow \bar{\lambda}^i & & \downarrow \tilde{\lambda}^i \\ O \to \underline{H}^i_\mathfrak{M}(\underline{x_d M : \mathfrak{m}}) & \to & \underline{H}^i_\mathfrak{M}(R_\mathfrak{q}(M)/x_d T R_\mathfrak{q}(M)) & \to & \underline{H}^i_\mathfrak{M}(R_\mathfrak{q}(\tilde{M})) \to O \end{array}$$

We know that $\tilde{\lambda}^i$ is surjective for $i < d$ and we now show that λ^i_* also is surjective for $i < d$. Then $\bar{\lambda}^i$ is surjective for $i < d$, i.e. $R_\mathfrak{q}(M)/x_d T R_\mathfrak{q}(M)$ is a Buchsbaum module. But then $R_\mathfrak{q}(M)$ is a Buchsbaum module by Proposition I.2.19, since $x_d T$ is a non-zero divisor with respect to $R_\mathfrak{q}(M)$ and $x_d T \underline{H}^i_\mathfrak{M}(R_\mathfrak{q}(M)) = 0$ for all $i \leq d$ by Proposition 2.3. Since $\dim_{R_\mathfrak{q}(A)} \underline{x_d M : \mathfrak{m}} = \dim_A(x_d M : \mathfrak{m}) = \dim M$, the surjectivity of λ^i_* for all $i < d$ is equivalent to the Buchsbaum property of $\underline{x_d M : \mathfrak{m}}$ (over $R_\mathfrak{q}(A)$), c.f. Theorem I.2.15. But $\underline{x_d M : \mathfrak{m}}$ is a Buchsbaum module over $R_\mathfrak{q}(A)$ if and only if $x_d M : \mathfrak{m}$ is a Buchsbaum module over A (by virtue of the canonical epimorphism $R_\mathfrak{q}(A) \to A$). Consider the exact sequence $0 \to M \to x_d M : \mathfrak{m} \to (x_d M : \mathfrak{m})/x_d M \to 0$, where the first map is given by $m \mapsto x_d m$. Now $H^i_\mathfrak{m}((x_d M : \mathfrak{m})/x_d M) = 0$ (since $\dim_A (x_d M : \mathfrak{m})/x_d M = 0$) for all $i > 0$, i.e. we have epimorphisms $H^i_\mathfrak{m}(M) \to H^i(x_d M : \mathfrak{m})$ for all $i \geq 0$ $(H^0(x_d M : \mathfrak{m}) = 0)$. Therefore we get commutative diagrams

$$\begin{array}{ccc} H^i(\mathfrak{m}; M) & \to & H^i(\mathfrak{m}; x_d M : \mathfrak{m}) \\ \downarrow \lambda^i_M & & \downarrow \lambda^i_{x_d M : \mathfrak{m}} \\ H^i_\mathfrak{m}(M) & \to & H^i_\mathfrak{m}(x_d M : \mathfrak{m}) \end{array}$$

(where the vertical homomorphisms are the canonical ones). But the surjectivity of λ^i_M for $i < d$ implies the surjectivity of $\lambda^i_{x_d M : \mathfrak{m}}$ for $i < d$, i.e. $x_d M : \mathfrak{m}$ is a Buchsbaum module, q.e.d.

Finally, we will give a characterization of the Cohen-Macaulay property of $R_\mathfrak{q}(M)$ where $\mathfrak{q}$ runs through all parameter ideals of M. We have

Theorem 2.11. *Let M be a Noetherian A-module of dimension $d \geq 2$. Let S denote the set of non-zero divisors in A with respect to M. The following conditions are equivalent:*

(i) *M is a Buchsbaum module with $H^i_\mathfrak{m}(M) = 0$ for all $i \neq 1, d$.*

(ii) *$R_\mathfrak{q}(M)$ is a Cohen-Macaulay module for all parameter ideals $\mathfrak{q}$ of M.*

(iii) *There is a d-dimensional Noetherian Cohen-Macaulay A-module N with $M \subseteq N \subseteq M_S$ and $\mathfrak{m} \cdot N \subseteq M$.*

Proof: (i) ⇒ (ii) is an immediate consequence of Proposition 2.3.

(ii) ⇒ (i): Since $H^0_\mathfrak{m}(M) = [\underline{H}^0_\mathfrak{M}(R_\mathfrak{q}(M))]_0 = 0$, we have depth $M > 0$. By Theorem 3.3, which we will prove in the next section, we get that M is a Buchsbaum module. By using again Proposition 2.3 we find $H^i_\mathfrak{m}(M) = 0$ for all $i \neq 1, d$.

(i) ⇒ (iii): Since $H^0_\mathfrak{m}(M) = 0$, we have an exact sequence

$$0 \to M \to H^0(M) \to H^1_\mathfrak{m}(M) \to 0,$$

where $H^0(M) := \varinjlim_n \operatorname{Hom}_A(\mathfrak{m}^n; M)$ (c.f. Lemma 0.1.8 for the notation). Since $H^1_\mathfrak{m}(M)$ is Noetherian $H^0(M)$ is Noetherian as well and $\mathfrak{m} \cdot H^1_\mathfrak{m}(M) = 0$ implies $\mathfrak{m} \cdot H^0(M) \subseteq M$.

Also by Lemma 0.1.8 there is a natural embedding $H^0(M) \subseteq M_S$ (Ass M consists only of minimal primes) and we have $H^0_{\mathfrak{m}}(H^0(M)) = H^1_{\mathfrak{m}}(H^0(M)) = 0$ and $H^i_{\mathfrak{m}}(H^0(M)) \cong H^i_{\mathfrak{m}}(M) = 0$ for $2 \leq i < d$, i.e. $H^0(M) =: N$ is a Cohen-Macaulay module.

(iii) $\Rightarrow$ (i): We consider the exact sequence $0 \to M \to N \to N/M \to 0$. Since $\mathfrak{m}(N/M) = 0$, $\dim_A(N/M) = 0$, i.e. $H^i_{\mathfrak{m}}(N/M) = 0$ for $i > 0$ and $H^0_{\mathfrak{m}}(N/M) = N/M$. Therefore the long exact cohomology sequence gives isomorphisms $H^i_{\mathfrak{m}}(M) \cong H^i_{\mathfrak{m}}(N)$ for $i \geq 2$, $H^0_{\mathfrak{m}}(M) = 0$ (since $H^0_{\mathfrak{m}}(N) = 0$) and an exact sequence $0 \to N/M \to H^1_{\mathfrak{m}}(M) \to H^1_{\mathfrak{m}}(N) \to 0$. But N is a Cohen-Macaulay module and therefore $H^i_{\mathfrak{m}}(M) = 0$ for all $i \neq 1, d$ and $\mathfrak{m}H^i_{\mathfrak{m}}(M) \cong \mathfrak{m}(N/M) = 0$, q.e.d.

§ 3. Blowing-up characterization of Buchsbaum modules

Troughout this paragraph M denotes a non-zero Noetherian A-module. First of all we need the following

Lemma 3.1. *Let $\mathfrak{q}$ be a parameter ideal with respect to M. Assume that $\mathfrak{q}$ is generated by a system of parameters of M. Let $\mathfrak{P}$ be a prime ideal of $R_{\mathfrak{q}}(A)$. Then*

(i) *$\mathfrak{m}R_{\mathfrak{q}}(A)$ is a prime ideal in $R_{\mathfrak{q}}(A)$.*

(ii) *$R_{\mathfrak{q}}(M)_{\mathfrak{P}} \neq 0$ if and only if $[\mathfrak{P}]_0 \in \operatorname{Supp} M$.*

(iii) *$G_{\mathfrak{q}}(M)_{\mathfrak{P}} \neq 0$ if and only if $[\mathfrak{P}]_0 = \mathfrak{m}$.*

Proof: (i): There is an $n \in \mathbf{N}$ with $\mathfrak{m}^n(M/\mathfrak{q}M) = 0$. Therefore $\mathfrak{m}^n R_{\mathfrak{q}}(M) \subseteq \mathfrak{q}R_{\mathfrak{q}}(M) \subseteq \mathfrak{m}R_{\mathfrak{q}}(M)$, hence $\dim R_{\mathfrak{q}}(M)/\mathfrak{m}R_{\mathfrak{q}}(M) = \dim G_{\mathfrak{q}}(M) = \dim M$.

Let $\bar{A} := A/\operatorname{Ann} M$, $\bar{\mathfrak{q}} := \mathfrak{q}\bar{A}$, $\bar{\mathfrak{m}} := \mathfrak{m}\bar{A}$. Then $\bar{\mathfrak{q}}$ is a parameter ideal of $\bar{A}$ and we have an epimorphism $\pi\colon R_{\mathfrak{q}}(A)/\mathfrak{m}R_{\mathfrak{q}}(A) \to R_{\bar{\mathfrak{q}}}(\bar{A})/\bar{\mathfrak{m}}R_{\bar{\mathfrak{q}}}(\bar{A})$. If $\mathfrak{q} = (x_1, \ldots, x_d)\,A$ ($d = \dim M = \dim \bar{A}$) then there are additional natural epimorphisms ($X_1, \ldots, X_d$ indeterminates):

$$\varphi\colon (A/\mathfrak{m})\,[X_1, \ldots, X_d] \to R_{\mathfrak{q}}(A)/\mathfrak{m}R_{\mathfrak{q}}(A),$$

defined by $X_i \mapsto \overline{x_i T}$ and

$$\bar{\varphi}\colon (\bar{A}/\bar{\mathfrak{m}})\,[X_1, \ldots, X_d] \to R_{\bar{\mathfrak{q}}}(\bar{A})/\bar{\mathfrak{m}}R_{\bar{\mathfrak{q}}}(\bar{A})$$

defined in the same way. Clearly, $\bar{A}/\bar{\mathfrak{m}} \cong A/\mathfrak{m}$ and we have a commutative diagram

$$\begin{array}{ccc} (A/\mathfrak{m})\,[X_1, \ldots, X_d] & \xrightarrow{\varphi} & R_{\mathfrak{q}}(A)/\mathfrak{m}R_{\mathfrak{q}}(A) \\ \downarrow\wr & & \downarrow\pi \\ (\bar{A}/\bar{\mathfrak{m}})\,[X_1, \ldots, X_d] & \xrightarrow{\bar{\varphi}} & R_{\bar{\mathfrak{q}}}(\bar{A})/\bar{\mathfrak{m}}R_{\bar{\mathfrak{q}}}(\bar{A}). \end{array}$$

Now $(\bar{A}/\bar{\mathfrak{m}})\,[X_1, \ldots, X_d]$ is an integral domain of dimension d. As we have seen in the beginning $R_{\bar{\mathfrak{q}}}(\bar{A})/\bar{\mathfrak{m}}R_{\bar{\mathfrak{q}}}(\bar{A})$ is a (graded) ring of dimension d. Hence $\bar{\varphi}$ must be an isomorphism and therefore φ is an isomorphism, i.e. $R_{\mathfrak{q}}(A)/\mathfrak{m}R_{\mathfrak{q}}(A)$ is an integral domain. This proves (i).

(ii) and (iii): The "only-if-parts" are clear. By Lemma 1.7 and Corollary 1.8 it is enough to prove: $[\mathfrak{P}]_0 \in \operatorname{Supp} M \Rightarrow R_{\mathfrak{q}}(M)_{\mathfrak{P}} \neq 0$.

Using the same argument as in the proof of Corollary 1.8 we may assume without loss of generality that $\mathfrak{P}$ is homogeneous. It is sufficient to prove $R_{\mathfrak{q}}(M)_{(\mathfrak{P})} \neq O$. If $[\mathfrak{P}]_0 \notin V(\mathfrak{q})$ then $R_{\mathfrak{q}}(M)_{(\mathfrak{P})} \cong M_{[\mathfrak{P}]_0} \neq O$ by Corollary 1.4 and Corollary 1.5. If $[\mathfrak{P}]_0 \in V(\mathfrak{q})$ then $[\mathfrak{P}]_0 \in V(\mathfrak{q}) \cap \operatorname{Supp} M = \{\mathfrak{m}\}$, i.e. $[\mathfrak{P}]_0 = \mathfrak{m}$. Take a prime ideal $\mathfrak{p} \in \operatorname{Supp} M$ with $\dim A/\mathfrak{p} = \dim M$. Then $\mathfrak{p} \in \operatorname{Ass} M$ and we have a monomorphism $A/\mathfrak{p} \to M$ which induces a monomorphism $R_{\mathfrak{q}}(A/\mathfrak{p}) \to R_{\mathfrak{q}}(M)$. Note that $\mathfrak{q}$ is a parameter ideal of $A/\mathfrak{p}$. Hence it is enough to prove $R_{\mathfrak{q}}(A/\mathfrak{p})_{(\mathfrak{P})} \neq O$.

If $R_{\mathfrak{q}}(A/\mathfrak{p})_{(\mathfrak{P})} = O$, there is an $n \in \mathbf{N}$ and a $p \in \mathfrak{q}^n$ such that $pT^n \notin \mathfrak{P}$ but $pT^n(A/\mathfrak{p}) = O$, i.e. $p \in \mathfrak{p}$. Now write $\mathfrak{q} = (x_1, \ldots, x_d)\, A$ $(d = \dim M)$, take indeterminates $X_1, \ldots, X_d$ and an element $f \in A[X_1, \ldots, X_d]$, homogeneous of degree n with $p = f(x_1, \ldots, x_d)$. Let $\bar{f}$ denote the image of f in $(A/\mathfrak{p})\,[X_1, \ldots, X_d]$. Then $\bar{f}(x_1, \ldots, x_d) = 0$ and Theorem 21 of Zariski-Samuel [1], p. 292, shows, that all coefficients of $\bar{f}$ belong to the maximal ideal of $A/\mathfrak{p}$, i.e. all coefficients of f belong to $\mathfrak{m}$. But this means $p = f(x_1, \ldots, x_d) \in \mathfrak{m}\mathfrak{q}^n$ hence $pT^n \in \mathfrak{m}\mathfrak{q}^n T^n = [\mathfrak{P}]_0\, [R_{\mathfrak{q}}(A)]_n \subseteq \mathfrak{P}$, a contradiction. This concludes our proof, q.e.d.

Now we can prove:

Theorem 3.2. *Let M be a Noetherian A-module of positive dimension. Then the following statements are equivalent:*

(i) *$R_{\mathfrak{q}}(M)$ is a locally Cohen-Macaulay module for all parameter ideals $\mathfrak{q}$ of M.*

(ii) *$G_{\mathfrak{q}}(M)$ is a locally Cohen-Macaulay module for all parameter ideals $\mathfrak{q}$ of M.*

(iii) *$M/H^0_{\mathfrak{m}}(M)$ is a Buchsbaum module.*

Proof: The equivalence of (i) and (ii) is a consequence of Corollary 1.12.

To prove the equivalence of (ii) and (iii) we first note that we can assume without loss of generality that A is complete and that depth $M > 0$. This is clear for (iii) (c.f. Lemma I.1.13). On the other hand, $G_{\hat{\mathfrak{q}}}(\hat{M}) \cong G_{\mathfrak{q}}(M)$ for all parameter ideals $\mathfrak{q}$ of M (as $R_{\mathfrak{q}}(A)$-modules, hence as $R_{\hat{\mathfrak{q}}}(\hat{A})$-modules by using the canonical map $R_{\mathfrak{q}}(A) \to R_{\hat{\mathfrak{q}}}(\hat{A})$ of graded rings). Conversely, for a parameter ideal $\mathfrak{v}$ of $\hat{M}$ (in $\hat{A}$) there is a parameter ideal $\mathfrak{q}$ of M (in A) with $\mathfrak{v}^n\hat{M}/\mathfrak{v}^{n+1}\hat{M} \cong \mathfrak{q}^n M/\mathfrak{q}^{n+1}M$, i.e. $G_{\mathfrak{v}}(\hat{M}) \cong G_{\mathfrak{q}}(M)$ and therefore we can assume that A is complete. Furthermore, let π denote the canonical epimorphism $G_{\mathfrak{q}}(M) \to G_{\mathfrak{q}}(M/H^0_{\mathfrak{m}}(M))$. Since for $n \geq 0$

$$\operatorname{Ker}[\pi]_n = \mathfrak{q}^n M \cap \left(\mathfrak{q}^{n+1}M + H^0_{\mathfrak{m}}(M)\right)/\mathfrak{q}^{n+1}M = O$$

for all sufficiently large n, there is an integer $t \geq 0$ with $\mathfrak{M}^t \operatorname{Ker} \pi = 0$. Therefore $G_{\mathfrak{q}}(M)_{(\mathfrak{P})} \cong G_{\mathfrak{q}}(M/H^0_{\mathfrak{m}}(M))_{(\mathfrak{P})}$ for all $\mathfrak{P} \in \operatorname{Proj} R_{\mathfrak{q}}(A)$ and we can assume that depth $M > 0$.

First we prove the implication (iii) $\Rightarrow$ (ii). But this is a consequence of Theorem 2.1, since $G_{\mathfrak{q}}(M)_{\mathfrak{M}}$ is a Buchsbaum module, hence $G_{\mathfrak{q}}(M)_{\mathfrak{P}}$ is a Cohen-Macaulay-module for all $\mathfrak{P} \in \operatorname{Proj} R_{\mathfrak{q}}(A)$ and all parameter ideals $\mathfrak{q}$ of M (c.f. Corollary I.1.11). Therefore $G_{\mathfrak{q}}(M)_{(\mathfrak{P})}$ is a Cohen-Macaulay-module for all $\mathfrak{P} \in \operatorname{Proj} R_{\mathfrak{q}}(A)$ by the remarks made in § 1.

Now we prove the implication (ii) $\Rightarrow$ (iii). We use induction on $d := \dim M$. If $d = 1$, there is nothing to prove. Let $d \geq 2$ and $x_1, \ldots, x_d$ a system of parameters of M. Set $\mathfrak{q} := (x_1, \ldots, x_d)\, A$. If $\mathfrak{P} \in \operatorname{Proj} R_{\mathfrak{q}}(A)$ with $G_{\mathfrak{q}}(M)_{(\mathfrak{P})} \neq O$ then $[\mathfrak{P}]_0 = \mathfrak{m}$ by Lemma 3.1 and therefore the prime ideal $\mathfrak{m}R_{\mathfrak{q}}(A)$ (Lemma 3.1) is contained in $\mathfrak{P}$. Hence $\mathfrak{m}R_{\mathfrak{q}}(A)$

is the only minimal element in $\operatorname{Supp} G_{\mathfrak{q}}(M)$. Since $\operatorname{depth} G_{\mathfrak{q}}(M)_{(\mathfrak{P})} > 0$ for every $\mathfrak{P} \in \operatorname{Proj} R_{\mathfrak{q}}(A)$ with $\mathfrak{m}R_{\mathfrak{q}}(A) \subsetneqq \mathfrak{P} \subsetneqq \mathfrak{M}$, $\operatorname{Ass} G_{\mathfrak{q}}(M) \subseteq \{\mathfrak{m}R_{\mathfrak{q}}(A), \mathfrak{M}\}$. Since $x_1T \notin \mathfrak{m}R_{\mathfrak{q}}(A)$, $0 :_{G_{\mathfrak{q}}(M)} x_1T$ is annihilated by some power of $\mathfrak{M}$, i.e. there is a $c > 0$ with $(\mathfrak{q}^n M :_M x_1) \cap \mathfrak{q}^c M = \mathfrak{q}^{n-1}M$ for all $n > c$. Hence by Krull's Intersection Theorem $\mathfrak{q}^c(0 :_M x_1) \subseteq (0 :_M x_1) \cap \mathfrak{q}^c M = \bigcap_{n>c} (\mathfrak{q}^n M :_M x_1) \cap \mathfrak{q}^c M = \bigcap_{n>c} \mathfrak{q}^{n-1}M = 0$. Therefore $0 :_M x_1 \subseteq 0 :_M \mathfrak{q}^c = 0$ since depth $M > 0$ and x_1 is a non-zero divisor with respect to M.

By the Lemma of Artin-Rees (c.f. Matsumura [1] (11.E), p. 70) there is an integer $r > 0$ with $\mathfrak{q}^n M :_M x_1 = \mathfrak{q}^{n-r}(\mathfrak{q}^r M :_M x_1) \subseteq \mathfrak{q}^c M$ for all $n \geq r + c$. This implies for $n \geq r + c$:

$$\mathfrak{q}^n M :_M x_1 = (\mathfrak{q}^n M :_M x_1) \cap \mathfrak{q}^c M = \mathfrak{q}^{n-1} M.$$

We let $M' := M/x_1M$. The canonical epimorphism $G_{\mathfrak{q}}(M) \to G_{\mathfrak{q}}(M')$ factors through $G_{\mathfrak{q}}(M)/x_1TG_{\mathfrak{q}}(M)$, i.e. we have an epimorphism

$$\pi\colon G_{\mathfrak{q}}(M)/x_1TG_{\mathfrak{q}}(M) \to G_{\mathfrak{q}}(M').$$

Let $\mathfrak{q}' := (x_2, \ldots, x_d)\, A$. Then for all $n \geq r + c$:

$$\begin{aligned}\operatorname{Ker}[\pi]_n &= \big(\mathfrak{q}^n M \cap (x_1 M + \mathfrak{q}'^{n+1} M)\big)/(\mathfrak{q}^{n+1} M + x_1 \mathfrak{q}^{n-1} M) \\ &= \big(\mathfrak{q}'^{n+1} M + (\mathfrak{q}^n M \cap x_1 M)\big)/(\mathfrak{q}'^{n+1} M + x_1 \mathfrak{q}^{n-1} M) = O.\end{aligned}$$

For all $n < r + c$ we have $\mathfrak{m}^t \operatorname{Ker}[\pi]_n = 0$ for sufficiently large t. Therefore a power of $\mathfrak{M}$ annihilates $\operatorname{Ker} \pi$, i.e.

$$\big(G_{\mathfrak{q}}(M)/x_1TG_{\mathfrak{q}}(M)\big)_{(\mathfrak{P})} \cong G_{\mathfrak{q}}(M')_{(\mathfrak{P})} \quad \text{for all } \mathfrak{P} \in \operatorname{Proj} R_{\mathfrak{q}}(A).$$

Hence $G_{\mathfrak{q}}(M')$ is a locally Cohen-Macaulay module. Assume we have proved (iii) in case $d = 2$. Then for $d \geq 3$ we know that $M/x_1M :_M \langle\mathfrak{m}\rangle \cong (M/x_1M)/H^0_{\mathfrak{m}}(M/x_1M) = M'/H^0_{\mathfrak{m}}(M')$ is a Buchsbaum module. Therefore M is a Buchsbaum module by Corollary I.2.24, since A is complete and hence an epimorphic image of a local Gorenstein ring. Thus it remains to show (iii) in case $d = 2$. Let x_1, x_2 be an arbitrary system of parameters of M. Since $0 :_M x_1 = 0$ as demonstrated above, x_1, x_2 is a filter-regular sequence of M, see Definition 1 of the Appendix. Then by Proposition 16 of the Appendix, $l\big(H^1_{\mathfrak{m}}(M)\big) < \infty$. Therefore we can find an $x \in \mathfrak{m}$ with $\dim M/xM = 1$ and $xH^1_{\mathfrak{m}}(M) = 0$. Take $y \in \mathfrak{m}$ with $\dim M/(x, y)\, M = 0$ and let $\mathfrak{q} := (x, y)\, A$. We set $\mathfrak{P} := \mathfrak{m}R_{\mathfrak{q}}(A) + xTR_{\mathfrak{q}}(A)$. If Y is an indeterminate, we have an epimorphism $\varphi\colon (A/\mathfrak{m})\,[Y] \to R_{\mathfrak{q}}(A)/\mathfrak{P}$ defined by $Y \mapsto \overline{yT}$. Now $\mathfrak{P}$ is not $\mathfrak{M}$-primary, hence $\dim R_{\mathfrak{q}}(A)/\mathfrak{P} \geq 1$. But $(A/\mathfrak{m})\,[Y]$ is a one-dimensional integral domain and therefore φ is an isomorphism. Hence $\mathfrak{P} \in \operatorname{Proj} R_{\mathfrak{q}}(A)$.

Let $g := \dfrac{xT}{yT} \in R_{\mathfrak{q}}(A)_{(\mathfrak{P})}$. The natural embedding $A \subset R_{\mathfrak{q}}(A)$ (of rings) induces a homomorphism of rings

$$\lambda\colon A \to R_{\mathfrak{q}}(A)_{(\mathfrak{P})}/gR_{\mathfrak{q}}(A)_{(\mathfrak{P})}.$$

Let $\mu = \dfrac{aT^n}{pT^n} \in R_{\mathfrak{q}}(A)_{(\mathfrak{P})}$, where $a, p \in \mathfrak{q}^n$, $pT^n \notin \mathfrak{P}$, i.e. $p \notin \mathfrak{m}\mathfrak{q}^n + x\mathfrak{q}^{n-1}$. We write $a = \sum_{i=0}^{n} a_i x^i y^{n-i}$, $p = \sum_{i=0}^{n} b_i x^i y^{n-i}$ $(a_j, b_j \in A,\ j = 0, \ldots, n)$. Then $b_0 \notin \mathfrak{m}$ and we have

$$\mu = \frac{a_0}{b_0} + g\mu' \quad \text{with } \mu' = \frac{y^nT^n}{b_0pT^n} \sum_{i=1}^{n} g^{i-1}(b_0a_i - b_ia_0) \in R_{\mathfrak{q}}(A)_{(\mathfrak{P})}.$$

This shows that λ is surjective. Replacing A by M we obtain by similar calculations an epimorphism of A-modules

$$\psi\colon M \to R_{\mathfrak{q}}(M)_{(\mathfrak{P})}/gR_{\mathfrak{q}}(M)_{(\mathfrak{P})}$$

(we consider $R_{\mathfrak{q}}(M)_{(\mathfrak{P})}/gR_{\mathfrak{q}}(M)_{(\mathfrak{P})}$ via λ as an A-module).

By Lemma 1.7 a prime ideal $\mathfrak{P}$ is minimal in Supp $R_{\mathfrak{q}}(M)$ if and only if $\mathfrak{P} = \tilde{\mathfrak{p}}$, where $\mathfrak{p}$ is minimal in Supp M. Let $\mathfrak{p}$ be minimal in Supp M. Then dim $A/\mathfrak{p} = 2$ and we have a chain of primes in $R_{\mathfrak{q}}(A)$: $\tilde{\mathfrak{p}} \subset \mathfrak{m}R_{\mathfrak{q}}(A) \subset \mathfrak{P} \subset \mathfrak{M}$ ($\mathfrak{p} \cap \mathfrak{q}^i \subseteq \mathfrak{m}\mathfrak{q}^i$ since $\mathfrak{q}$ is a parameter ideal of $A/\mathfrak{p}$, compare with the argument used in the proof of Lemma 3.1). Hence dim $R_{\mathfrak{q}}(M)_{(\mathfrak{P})} = 2$. Also it is easy to see that $g \notin \tilde{\mathfrak{p}}R_{\mathfrak{q}}(M)_{(\mathfrak{P})}$. This shows that g is a parameter element of $R_{\mathfrak{q}}(M)_{(\mathfrak{P})}$ and therefore $R_{\mathfrak{q}}(M)_{(\mathfrak{P})}/gR_{\mathfrak{q}}(M)_{(\mathfrak{P})}$ is a one-dimensional Cohen-Macaulay module. Hence $M/\mathrm{Ker}\,\psi$ is a Cohen-Macaulay module.

Claim: Ker $\psi = xM:y$.

If $m \in xM:y$ or $ym = xm'$, then $\psi(m) = \overline{gm'} = 0$, i.e. $xM:y \subseteq \mathrm{Ker}\,\psi$. Since $0:_M y = O$ we have an exact sequence $0 \to M \xrightarrow{y} M \to M/yM \to 0$. By applying the local cohomology functors we get a mononiorphism $(H^0_{\mathfrak{m}}(M) = 0)$: $H^0_{\mathfrak{m}}(M/yM) \to H^1_{\mathfrak{m}}(M)$. Now $H^0_{\mathfrak{m}}(M/yM) = (yM:\langle\mathfrak{m}\rangle)/yM$. Since $xH^1_{\mathfrak{m}}(M) = 0$, $x(yM:\langle\mathfrak{m}\rangle) \subseteq yM$, i.e. $yM:\langle\mathfrak{m}\rangle \subseteq yM:x \subseteq yM:x^2 \subseteq \ldots \subseteq yM:\langle\mathfrak{m}\rangle$. Therefore

$$yM:x = yM:x^2 = \ldots = yM:\langle\mathfrak{m}\rangle.$$

Let $m \in \mathrm{Ker}\,\psi$. Then there are $n \in \mathbf{N}$, $m' \in \mathfrak{q}^n M$, $p \in \mathfrak{q}^n$ with $pT^n \notin \mathfrak{P}$ such that $m = g\dfrac{m'T^n}{pT^n}$ in $R_{\mathfrak{q}}(M)_{(\mathfrak{P})}$. Hence there are a $j \in \mathbf{N}$ and a $q \in \mathfrak{q}^j$, $q^jT \notin \mathfrak{P}$ with $qT^j(mypT^{n+1} - xm'T^{n+1}) = 0$, i.e. $q(myp - xm') = 0$ (in M). Write $p = xp' + \alpha y^n$, $q = xq' + \beta y^j$ with $p' \in \mathfrak{q}^{n-1}$, $q' \in \mathfrak{q}^{j-1}$. Since $p, q \notin \mathfrak{P}$, $\alpha, \beta \notin \mathfrak{m}$. Furthermore $(x, q)\,A = (x, y^j)\,A$, i.e. x, q is a system of parameters of M and we have $0:_M q = 0$. Hence $0 = myp - xm' = \alpha my^{n+1} - xm''$ with $m'' := m' - myp' \in \mathfrak{q}^n M$. We write $m'' = \sum_{i=0}^{n} x^i y^{n-i} m_i$. If $n > 0$, $x^{n+1}m_n \in yM$, hence $m_n \in yM:x^{n+1} = yM:x$. Let $x^{n+1}m_n = x^n y m'_n$. Then we obtain

$$y\Big(\alpha my^n - x\sum_{i=0}^{n-1} x^i y^{n-i-1} m_i^{(1)}\Big) = 0$$

with $m_i^{(1)} := m_i$ for $0 \leq i \leq n-2$, $m_{n-1}^{(1)} := m_{n-1} + m'_n$. Since $0:_M y = 0$, we get

$$my - x\sum_{i=0}^{n-1} x^i y^{n-i-1} m_i^{(1)} = 0.$$

Repeating this procedure, we obtain after n steps

$$my - xm_0^{(n)} = 0, \quad \text{i.e.} \quad \alpha m \in xM:y.$$

Therefore $m \in xM:y$ (α is a unit in A) and this proves our claim.

Since $M/xM:y$ is a Cohen-Macaulay module, $xM:y$ is unmixed in M, i.e. $xM:y = xM:\langle\mathfrak{m}\rangle$. Therefore $yH^0_{\mathfrak{m}}(M/xM) = y(xM:\langle\mathfrak{m}\rangle/xM) = 0$.

Since $xH^1_{\mathfrak{m}}(M) = 0$, the exact sequence $0 \to M \xrightarrow{x} M \to M/xM \to 0$ results in an isomorphism $H^0_{\mathfrak{m}}(M/xM) \cong H^1_{\mathfrak{m}}(M)$ and therefore $yH^1_{\mathfrak{m}}(M) = 0$, where y is an arbitrary element of $\mathfrak{m}$ with $\dim_A M/(x, y)\, M = 0$. If we choose an (M/xM)-basis $y_1, \ldots, y_t$ of $\mathfrak{m}$ (c.f. Definition I.1.7 and Proposition I.1.9) it follows that $y_i H^1_{\mathfrak{m}}(M) = 0$ for all $i = 1, \ldots, t$, i.e. $\mathfrak{m}H^1_{\mathfrak{m}}(M) = 0$ and M is a Buchsbaum module by Proposition I.1.12, q.e.d.

In the last theorem of this paragraph we collect the main results of this chapter.

Theorem 3.3. *For a Noetherian A-module M of positive dimension with* depth $M > 0$ *the following conditions are equivalent:*

(i) *M is a Buchsbaum module.*

(ii) *$G_{\mathfrak{q}}(M)$ is a Buchsbaum module for all parameter ideals $\mathfrak{q}$ of M.*

(ii)′ *$G_{\mathfrak{q}}(M)$ is a locally Cohen-Macaulay module for all parameter ideals $\mathfrak{q}$ of M.*

(iii) *$R_{\mathfrak{q}}(M)$ is a Buchsbaum module for all parameter ideals $\mathfrak{q}$ of M.*

(iii)′ *$R_{\mathfrak{q}}(M)$ is a locally Cohen-Macaulay module for all parameter ideals $\mathfrak{q}$ of M.*

Proof: The equivalence of (i) and (ii) is the content of Theorem 2.1 and the equivalence of (i), (ii)′ and (iii)′ was proven previously in Theorem 3.2. The implications (ii) ⇒ (ii)′ and (iii) ⇒ (iii)′ are clear by the remarks of § 1. Finally, the implication (i) ⇒ (iii) follows from Theorem 2.10, q.e.d.

Remark 3.4. Without the assumption depth $M > 0$ (but $\dim M \geq 2$) we only have the following implications:

$$\begin{array}{ccccc} \text{(i)} \Leftrightarrow & \text{(ii)} & \Rightarrow & \text{(iii)} \\ & \Downarrow & & \Downarrow \\ & \text{(ii)}' & \Leftrightarrow & \text{(iii)}' \end{array}$$

The following example shows that none of the implications (ii) ⇒ (ii)′, (ii) ⇒ (iii) and (iii) ⇒ (iii)′ is generally reversible.

Example 3.5. Let M be a Noetherian A-module of dimension 2 and with $\mathfrak{m}H^1_{\mathfrak{m}}(M) = 0$. We prove for every parameter ideal $\mathfrak{q}$ of M:

(i) $\underline{H}^i_{\mathfrak{M}}(R_{\mathfrak{q}}(M)) = 0$ for $i = 1, 2$.

(ii) $[\underline{H}^0_{\mathfrak{M}}(R_{\mathfrak{q}}(M))]_p = \begin{cases} 0 & \text{for } p < 0, \\ (H^0_{\mathfrak{m}}(M) \cap \mathfrak{q}^p M)\, T^p & \text{for } p \geq 0. \end{cases}$

Proof: Let $\overline{M} := M/H^0_{\mathfrak{m}}(M)$. Then depth $\overline{M} > 0$ and $\mathfrak{m}H^1_{\mathfrak{m}}(\overline{M}) = \mathfrak{m}H^1_{\mathfrak{m}}(M) = 0$, i.e. $\overline{M}$ is a Buchsbaum module by Proposition I.2.12. Therefore $\underline{H}^i_{\mathfrak{M}}(R_{\mathfrak{q}}(\overline{M})) = 0$ for $i = 0, 1, 2$ by Proposition 2.3 for all parameter ideals $\mathfrak{q}$ of $\overline{M}$ (of M).

The kernel of the natural epimorphism $R_{\mathfrak{q}}(M) \to R_{\mathfrak{q}}(\overline{M})$ must be $\underline{H}^0_{\mathfrak{M}}(R_{\mathfrak{q}}(M))$ (since depth $R_{\mathfrak{q}}(\overline{M}) > 0$). An easy calculation shows that

$$[\underline{H}^0_{\mathfrak{M}}(R_{\mathfrak{q}}(M))]_p = \begin{cases} 0 & \text{for } p < 0, \\ (H^0_{\mathfrak{m}}(M) \cap \mathfrak{q}^p M)\, T^p & \text{for } p \geq 0 \end{cases}$$

and this proves (ii).

If we apply $\underline{H}^{\cdot}_{\mathfrak{M}}(\ )$ to the exact sequence

$$O \to \underline{H}^0_{\mathfrak{M}}(R_{\mathfrak{q}}(M)) \to R_{\mathfrak{q}}(M) \to R_{\mathfrak{q}}(\overline{M}) \to 0,$$

we find $\underline{H}^i_{\mathfrak{M}}(R_{\mathfrak{q}}(M)) \cong \underline{H}^i_{\mathfrak{M}}(R_{\mathfrak{q}}(\overline{M})) = 0$ for $i = 1, 2$ and this proves (i).

Now let us consider the following two cases:

(a) $\mathfrak{m}H^0_{\mathfrak{m}}(M) = O$. Then $\mathfrak{M}\underline{H}^0_{\mathfrak{M}}(R_{\mathfrak{q}}(M)) = 0$ by (ii) and $R_{\mathfrak{q}}(M)$ is a Buchsbaum module for all parameter ideals $\mathfrak{q}$ of M by Proposition I.2.12 although M need not be a Buchsbaum module (see, e.g., Example I.2.5).

(b) $\mathfrak{m}H^0_{\mathfrak{m}}(M) \neq O$. Then $\mathfrak{M}\underline{H}^0_{\mathfrak{M}}(R_{\mathfrak{q}}(M)) \neq 0$ and $R_{\mathfrak{q}}(M)$ is not a Buchsbaum module but a locally Cohen-Macaulay module for all parameter ideals $\mathfrak{q}$ of M.

(a) shows that (iii) $\Rightarrow$ (ii) and (ii)' $\Rightarrow$ (ii) cannot be true in general and (b) shows that (iii)' $\Rightarrow$ (iii) is not true in general.

Chapter V

Further applications and examples

§ 1. A Buchsbaum criterion for affine semigroup rings

We have seen that the ring

$$K[s^4, s^3 \cdot t, s \cdot t^3, t^4]$$

is a non-Cohen-Macaulay Buchsbaum ring. In fact it is a special case of an affine semigroup ring. More general, we will prove in the following a Buchsbaum criterion for affine semigroup rings. To accomplish this K will be a field, S a finitely generated (additive) submonoid of $\mathbf{N}^n$, and L the subgroup of $H := \mathbf{Z}^n$ generated by S.

We set $d :=$ rank L. Let $K[H]$ denote the group algebra of H over K and let X^a denote the image of $a \in H$ in $K[H]$. For every subset V of H we define $K[V] := \sum_{a \in V} K \cdot X^a$. The ring $K[S]$ of course coincides with the monoid algebra of S over K and may be considered as an H-graded subring of $K[H]$. For this and more technical results concerned the subject, compare the article [2] of S. Goto and K. Watanabe. Also we define

$$\bar{S} := \{a \in L \mid t \cdot a \in S \text{ for some integer } t > 0\}.$$

$\bar{S}$ is called the *normalization* of S. It is known that $K[\bar{S}]$ coincides with the normalization of the ring $K[S]$, compare M. Hochster [1], § 1. For simplicity we assume that $\bar{S} = L \cap \mathbf{N}^n$. In general, $\bar{S}$ is isomorphic to a monoid of this form, compare M. Hochster [1], § 2.

Next we define additional notation. For $i = 1, \ldots, n$ let

$$F_i := \{(a_1, \ldots, a_n) \in S \mid a_i = 0\} \quad \text{and} \quad S_i := S - F_i.$$

Let L_i denote the subgroup of L generated by F_i, $1 \leq i \leq n$. We assume $L_i \neq L_j$ for $i \neq j$. Set

$$\tilde{S} := \bigcap_{i=1}^{n} S_i.$$

Then $\tilde{S}$ is a finitely generated submonoid of $\bar{S}$ containing S (compare S. Goto and K. Watanabe [2], Proof of Lemma 3.3.8). S. Goto and K. Watanabe showed in [2], 3.3.3, that $K[\tilde{S}]$ is a d-dimensional Cohen-Macaulay ring.

Unfortunately, this result is not correct in general. A counterexample was given by Ngo Viet Trung and Le Tuan Hoa [1]. Moreover, this paper also contains a corrected version of the result of Goto and Watanabe. Especially, as a consequence of this new criterion, one sees that the property of $K[\tilde{S}]$ being Cohen-Macaulay is dependent upon the characteristic of K. Therefore we will suppose the Cohen-Macaulay property of $K[\tilde{S}]$ for the following statement.

Theorem 1.1 (S. Goto [2]). *Assume that $K[\tilde{S}]$ is a Cohen-Macaulay ring. Let $\mathfrak{m}$ be the unique H-graded maximal ideal of $K[S]$, i.e. $\mathfrak{m} = K[S \setminus \{0\}]$, and suppose that $d := \operatorname{rank} L \geq 2$. Then the following conditions are equivalent:*

(i) $A := K[S]_{\mathfrak{m}}$ *is a Buchsbaum ring.*

(ii) $\mathfrak{m} \cdot H^i_{\mathfrak{m}\cdot A}(A) = 0$ *for all* $i = 0, \ldots, d-1$.

(iii) $(S \setminus \{0\}) + \tilde{S} \subseteq S$.

In this case $H^i_{\mathfrak{m}\cdot A}(A) = 0$ *for all* $i \neq 1, d$ *and*

$$I(A) = (d-1) \cdot \#(\tilde{S} \setminus S),$$

where $\#(\tilde{S} \setminus S)$ *denotes the number of elements of* $\tilde{S} \setminus S$.

Proof: (iii) $\Rightarrow$ (i): Set $B := K[\tilde{S}]_{\mathfrak{m}}$. Then $A \subseteq B \subseteq Q(A)$ and B is an intermediate ring such that B is a d-dimensional Cohen-Macaulay ring which satisfies the condition (iii) of Theorem IV.2.11. Thus A is a Buchsbaum ring. The last assertion also follows from Theorem IV.2.11, since by Proposition I.2.6

$$\begin{aligned} I(A) &= \sum_{i=0}^{d-1} \binom{d-1}{i} \cdot l_A\big(H^i_{\mathfrak{m}\cdot A}(A)\big) = (d-1) \cdot l_A\big(H^1_{\mathfrak{m}\cdot A}(A)\big) \\ &= (d-1) \cdot \#(\tilde{S} \setminus S). \end{aligned}$$

The implication (i) $\Rightarrow$ (ii) is clear (see Corollary I.2.4).

(ii) $\Rightarrow$ (iii): By applying the functor $H^i_{\mathfrak{m}}(\ )$ to the exact sequence

$$0 \to K[S] \to K[\tilde{S}] \to K[\tilde{S}]/K[S] \to 0$$

we obtain $H^i_{\mathfrak{m}}(K[\tilde{S}]/K[S]) \cong H^{i+1}_{\mathfrak{m}}(K[S])$ for every $i = 0, \ldots, d-2$. Notice that $\operatorname{Supp}_{K[S]}(K[\tilde{S}]/K[S]) \subseteq \{\mathfrak{m}\}$.

Assume this not to be true and let $r := \dim_{K[S]} K[\tilde{S}]/K[S]$. Then $0 < r \leq d-2$ (compare Goto-Watanabe [2], Proof of 3.3.3) and (by our last result)

$$H^{r+1}_{\mathfrak{m}\cdot A}(A) \cong H^{r+1}_{\mathfrak{m}}(K[S]) \cong H^r_{\mathfrak{m}}(K[\tilde{S}]/K[S])$$

is a non-Noetherian A-module (see Chap. 0, § 1, 3.). But this contradicts the fact that all $H^i_{\mathfrak{m}\cdot A}(A)$, $i = 0, \ldots, d-1$ are finite-dimensional vector spaces over $A/\mathfrak{m} \cdot A$ by our assumption. Therefore $H^1_{\mathfrak{m}}(K[S]) \cong K[\tilde{S}]/K[S]$ which is annihilated by $\mathfrak{m}$. This implies $\mathfrak{m} \cdot K[\tilde{S}] \subseteq K[S]$. This is of course equivalent to $(S \setminus \{0\}) + \tilde{S} \subseteq S$ and we have completed the proof, q.e.d.

Corollary 1.2. *Under the same hypotheses as in Theorem 1.1 suppose that* $(S \setminus \{0\}) + \bar{S} \subset S$. *Then* $A := K[S]_{\mathfrak{m}}$ *is a Buchsbaum ring with* $I(A) = (d-1) \cdot \#(\bar{S} \setminus S)$ *and* $H^i_{\mathfrak{m}\cdot A}(A) = 0$ *for all* $i \neq 1, d$.

Proof: It suffices to show that $\tilde{S} = \bar{S}$. First notice that $\mathfrak{m} \cdot K[\bar{S}] \subseteq K[\tilde{S}]$ since $\mathfrak{m} \cdot K[\bar{S}] \subseteq K[S]$ and we have

$$l_{K[S]}(K[\bar{S}]/K[\tilde{S}]) < \infty.$$

Thus the assertion follows from the fact that $K[\tilde{S}]$ and $K[\bar{S}]$ are Cohen-Macaulay rings of dimension d. For the proof of the Cohen-Macaulayness of $K[\bar{S}]$ see M. Hochster, [1], Theorem 1, q.e.d.

Next we want to discuss some examples:

Examples 1.3. Let r, d be integers with $d \geq 2$ and $r \geq 1$ and let

$$T := \{(a_1, \ldots, a_d) \in \mathbf{N}^d \mid \sum_{i=1}^{d} a_i \equiv 0 \bmod r\}.$$

For a subset I of $\{(a_1, \ldots, a_d) \in \mathbf{N}^d \mid \sum_{i=1}^{d} a_i = r\}$ we set $S := T \setminus I$. Then S is a finitely generated submonoid of $\mathbf{N}^d$ with $T = \overline{S}$ and

$$(S \setminus \{0\}) + T \subset S.$$

Thus, by Corollary 1.2, $A := K[S]_{\mathfrak{m}}$ is a d-dimensional Buchsbaum ring with $I(A) = (d-1) \cdot \#I$ and $H^i_{\mathfrak{m}\cdot A}(A) = 0$ for $i \neq 1, d$.

In particular, the famous ring $K[s^4, s^3 \cdot t, s \cdot t^3, t^4] \cong K[x_0, x_1, x_2, x_3]/\mathfrak{p}$ is a special case of this example (take $r = 4$, $d = 2$ and $I = (2, 2)$). In his fundamental paper [2] W. Gröbner showed that the ideal $\mathfrak{p}$ is a special member of a large class of imperfect ideals. The result shows that projections of Veronesian ideals (perfect ideals) lead us to certain degenerations (imperfect ideals). In view of this phenomenon W. Gröbner posed the problem to classify simple projections of Veronesian ideals, compare [2], p. 263. In the sequel we give a complete solution of this problem. In particular, we show that the local rings at the vertices of affine cones over almost all projections of Veronesian varieties are Buchsbaum rings.

Let $M_{d,m}$ be the semigroup generated by 0 and all elements $(a_0, \ldots, a_d) \in \mathbf{N}^{d+1}$ with $a_0 + \ldots + a_d = m$ over $\mathbf{N}^{d+1}$.

For an arbitrary field K we denote by $R_{d,m} := K[M_{d,m}]$ the corresponding affine semigroup ring. $R_{d,m}$ is the coordinate ring of the image of $\mathbf{P}^d_K$ in $\mathbf{P}^N_K$, $N = \binom{m+d}{d} - 1$, by the Veronesian embedding. If $(i) := (i_0, \ldots, i_d)$, $i_0 + \ldots + i_d = m$, we denote by $M^{(i)}_{d,m}$ the subsemigroup of $M_{d,m}$ generated by all $(a_0, \ldots, a_d) \in M_{d,m}$ with $(a_0, \ldots, a_d) \neq (i)$.

The affine semigroup ring $R^{(i)}_{d,m}$ is then the coordinate ring of the projection of the Veronesian variety from infinity to the hyperplane $x_{(i)} = 0$. After changing the variables we can assume

$$m \geq i_0 \geq \ldots \geq i_d \geq 0.$$

For short we write (0) for $(m, 0, \ldots, 0)$ and (1) for $(m-1, 1, 0, \ldots, 0)$. In the following we assume $d \geq 1$ and $m \geq 2$.

Theorem 1.4. *Let $(i) \neq (0), (1)$. Then the affine semigroup ring $R^{(i)}_{d,m}$ has the following properties:*

(a) depth $R^{(i)}_{d,m} = 1$.

(b) *$R^{(i)}_{d,m}$ is a Buchsbaum ring with $I(R^{(i)}_{d,m}) = d$.*

(c) *For the cohomology of $X := V(\mathfrak{v}^{(i)}_{d,m}) \subset \mathbf{P}^{N-1}_K$ ($\mathfrak{v}^{(i)}_{d,m}$ is the defining ideal of the projection of the Veronesian variety in $\mathbf{P}^{N-1}_K$) we have*

$\dim_K \square$

n	$[R^{(i)}_{d,m}]_n$	$H^0(X, \mathcal{O}_X(n))$	$H^r(X, \mathcal{O}_X(n))$ $1 \leq r < d$	$H^d(X, \mathcal{O}_X(n))$
≤ -1	0	0	0	$(-1)^d \cdot \binom{n \cdot m + d}{d}$
0	1	1	0	0
1	$\binom{m+d}{d} - 1$	$\binom{m+d}{d}$	0	0
≥ 2	$\binom{n \cdot m + d}{d}$	$\binom{n \cdot m + d}{d}$	0	0

Proof: All statements of the theorem will follow from Example 1.3 by setting $M^{(i)}_{d,m} = M_{d,m} \setminus \{(i_0, \ldots, i_d)\}$. This equality follows from simple calculations, q.e.d.

In Theorem 1.4 only the cases $(i) \neq (0), (1)$ are handled. Now we add some results concerning the case $(i) = (0), (i) = (1)$.

Theorem 1.5.

(a) *The affine semigroup rings $R^{(0)}_{d,m}$, $R^{(1)}_{1,m}$ and $R^{(1)}_{2,2}$ are Cohen-Macaulay rings.*

(b) *For the local cohomology of $R := R^{(1)}_{d,m}$, $m \geq 3$, $d \geq 2$ we get:*

$\dim_K \square$

n	$[\underline{H}^1_{\mathfrak{m}}(R)]_n$	$[\underline{H}^2_{\mathfrak{m}}(R)]_n$	$[\underline{H}^r_{\mathfrak{m}}(R)]_n$, $2 < r \leq d$	$[\underline{H}^{d+1}_{\mathfrak{m}}(R)]_n$
≤ -1	0	1	0	$(-1)^d \cdot \binom{n \cdot m + d}{d}$
0	0	1	0	0
≥ 1	0	0	0	0

(c) *For the local cohomology of $R := R^{(1)}_{d,2}$, $d \geq 3$ we have:*

$\dim_K \square$

n	$[\underline{H}^r_{\mathfrak{m}}(R)]_n$, $0 \leq r < 3$	$[\underline{H}^3_{\mathfrak{m}}(R)]_n$	$[\underline{H}^r_{\mathfrak{m}}[R]]_n$, $3 < r \leq d$	$[\underline{H}^{d+1}_{\mathfrak{m}}(R)]_n$
≤ -1	0	$-n$	0	$(-1)^d \cdot \binom{2 \cdot n + d}{d}$
0	0	0	0	1
≥ 1	0	0	0	0

Proof: (a) According to Theorem 1 of M. Hochster [1] it is enough to show that the semigroup $M^{(0)}_{d,m}$ is normal, compare Corollary 1.2. In fact, this is easy to check. For the remaining two cases the proof follows readily.

(b) We consider the inclusion map $R^{(1)}_{d,m} \to R_{d,m}$. Let N be its cocernel. By an easy computation we obtain for the Hilbert function

$$H(n, R^{(1)}_{d,m}) = \begin{cases} 1 & \text{if } n = 0, \\ \binom{n \cdot m + d}{d} - 1 & \text{if } n > 0. \end{cases}$$

Hence for N it follows that

$$H(n, N) = \begin{cases} 0 & \text{for } n = 0, \\ 1 & \text{for } n > 0. \end{cases}$$

The next step is to show that N is a one-dimensional Cohen-Macaulay module over $R^{(1)}_{d,m}$. But this follows because $e_0(\mathfrak{m}, N) = 1$. Since $R_{d,m}$ is also a Cohen-Macaulay module over $R^{(1)}_{d,m}$, we get $\underline{H}^2_{\mathfrak{m}}(R) = \underline{H}^1_{\mathfrak{m}}(N)$ and $\underline{H}^r_{\mathfrak{m}}(R) = 0$ for $r \neq 2, d + 1$.

Now, comparing the Hilbert function and the Hilbert-Samuel function of $R^{(1)}_{d,m}$ we get

$$[\underline{H}^2_{\mathfrak{m}}(R)]_n \cong \begin{cases} K & \text{if } n \leq 0, \\ 0 & \text{if } n > 0. \end{cases}$$

Thus, the statement has been proven.

(c) The proof of this statement is obtained similarly. We remark only, that the cokernel N of the embedding $R^{(1)}_{d,2} \hookrightarrow R_{d,2}$ is a two-dimensional Cohen-Macaulay module over $R^{(1)}_{d,2}$. Furthermore,

$$H(n, R^{(1)}_{d,2}) = \binom{2 \cdot n + d}{d} - n, \quad n \geq 0,$$

q.e.d.

We also can consider multiple projections of Veronesian ideals. Here the situation is more complicated. We illustrate this by two examples.

Examples 1.6. We consider $R_1 := R^{((i),(j))}_{1,6}$ with $(i) = (4, 2)$, $(j) = (3, 3)$ and $R_2 := R^{((k),(l))}_{1,6}$ with $(k) = (4, 2)$, $(l) = (2, 4)$. Then we get for the local cohomology:

$\dim_K \square$

n	$[R_1]_n$	$[R_2]_n$	$[\underline{H}^1_{\mathfrak{m}_1}(R_1)]_n$	$[\underline{H}^1_{\mathfrak{m}_2}(R_2)]_n$	$[\underline{H}^2_{\mathfrak{m}_i}(R_i)]_n$, $i = 1, 2$
≤ -1	0	0	0	0	$-(6 \cdot n + 1)$
0	1	1	0	0	0
1	5	5	2	2	0
2	12	13	1	0	0
≥ 3	$6 \cdot n + 1$	$6 \cdot n + 1$	0	0	0

Hence, R_2 is a Buchsbaum ring while R_1 is not a Buchsbaum ring.

Furthermore, the new results of Bresinsky [2] and Trung [8], [11] open the way to some deep lying applications of Goto's general Buchsbaum criterion in terms of semigroup ideals.

§ 2. Some examples related to problems of Hironaka and Seidenberg

First we need to mention a question posed by Hironaka in a discussion (at the University of Halle in 1974), who asked whether one can construct normal (and factorial) non-Cohen-Macaulay Buchsbaum singularities. By using Segre products of graded Cohen-Macaulay modules we have constructed normal (local) Buchsbaum rings A for every given dimension $d \geq 3$ and depth $t \geq 2$ such that $d > t$ (see Example I.4.14). We now give another example:

Let F be an arithmetically normal irregular surface. Let A be the local ring of the vertex of the affine cone over F. Then A is a normal non-Cohen-Macaulay ring. We claim that A is a Buchsbaum ring if $H^1(F, \mathcal{O}_F(p)) = O$ for all $p \neq 0$. Our assertion follows immediately from the following corollary of Proposition 2.4 below:

Corollary 2.1. *Let F be an arithmetically normal surface in projective space $\mathbf{P}^n_K$ over an arbitrary algebraically closed field K. Let A be the local ring of the vertex of the affine cone over F. If $H^1(F, \mathcal{O}_F(p)) \neq O$ for precisely one $p \in \mathbf{Z}$, then A is a non-Cohen-Macaulay Buchsbaum ring.*

We will give some examples of such surfaces.

Example 2.2. Take any non-singular irregular surface F, any ample line bundle $\mathcal{L}$ on F, and embed F in $\mathbf{P}^n_K$ by a sufficiently high power $\mathcal{L}^{\otimes l}$. Then F is arithmetically normal and we know that $H^1(F, \mathcal{O}_F(p)) \neq O$ only for $p = 0$. Corollary 2.1 shows that the local ring of the vertex of the affine cone over F is a normal (non-Cohen-Macaulay) Buchsbaum ring.

Example 2.3. G. Horrocks and D. Mumford [1] have constructed an indecomposable rank 2 vector bundle $\mathcal{F}$ on $P := \mathbf{P}^4_K$ with the following exact sequence:

$$0 \to \mathcal{O}_P(-5) \to \mathcal{F}(-5) \to \mathcal{O}_P \to \mathcal{O}_V \to O,$$

where V is an non-singular abelian surface. $\mathcal{F}$ is such that $H^2(\mathcal{F}(-5)) \cong K^2, H^2(\mathcal{F}(p-5)) = O$ for $p \neq 0$, $H^1(\mathcal{F}(p-5)) = O$ for $p \geq 6$ and $p \leq -6$. Therefore, if we embed P in

$$\mathbf{P}(\Gamma(\mathcal{O}_P(6))) = \mathbf{P}^{209}_K$$

by a Veronese morphism, we get an embedding of V in $\mathbf{P}^{209}_K$ which makes V arithmetically normal and such that $H^1(V, \mathcal{O}_V(p)) = O$ if and only if $p \neq 0$. Let A be the local ring of the vertex of the affine cone over $V \subset \mathbf{P}^{209}_K$. Then A is normal non-Cohen-Macaulay ring and Corollary 2.1 shows again that A is a Buchsbaum ring.

By proving our Corollary 2.1 we relate these questions to some global questions by considering affine cones over projective varieties in projective spaces. By studying the vertices of these cones we can give a proof of Corollary 2.1. In order to do this we make use of local cohomology, for which we obtain the following vanishing theorem:

Proposition 2.4. *Let V be a subvariety all of whose irreducible components are of dimension $d \geq 2$ in projective space $\mathbf{P}^n_K$ over an arbitrary algebraically closed field K. Let $(A, \mathfrak{m})$ be the local ring of the vertex of the affine cone over V. Let $t :=$ depth $A \geq 2$ and $H^i_{\mathfrak{m}}(A) = O$ for all $i \neq t$, $d + 1$. Then we have:*

(i) *A is a Buchsbaum ring with invariant $I(A) = \binom{d}{t}$ if and only if $H^{t-1}(V, \mathcal{O}_V(p)) \neq O$ for precisely one $p \in \mathbf{Z}$ and if for this p we have* dim $H^{t-1}(V, \mathcal{O}_V(p)) = 1$.

(ii) *A is a Buchsbaum non-Cohen-Macaulay ring and there exists an integer $p \in \mathbf{Z}$ such that*

$$I(A) = \binom{d}{t} \dim H^{t-1}(V, \mathcal{O}_V(p))$$

if and only if

$$H^{t-1}(V, \mathcal{O}_V(j)) = 0 \quad \textit{for all } j \neq p.$$

Proof: For $d > i \geq 1$ there are isomorphisms (of A-modules)

$$\bigoplus_{p \in \mathbf{Z}} H^i(V, \mathcal{O}_V(p)) \simeq H^{i+1}_{\mathfrak{m}}(A)$$

(see Proposition 0.2.3 and the arguments used in the proof of Corollary 0.4.7).

If A is a Buchsbaum ring of Krull dimension $d + 1$, then we have by Proposition I.2.6

$$I(A) = \sum_{i=0}^{d} \binom{d}{i} \dim_{A/\mathfrak{m}}(H^i_{\mathfrak{m}}(A)).$$

By hypothesis we obtain

$$I(A) = \binom{d}{t} \dim_{A/\mathfrak{m}}(H^t_{\mathfrak{m}}(A)).$$

This invariant for the Buchsbaum ring A, the above isomorphism, and Proposition I.2.12 yield our assertions (i) and (ii), q.e.d.

Proof of Corollary 2.1. This assertion follows easily from Proposition 2.4(ii).

For the construction of factorial non-Cohen-Macaulay-Buchsbaum singularities we will apply the theory of abelian varieties. First we will show that an abelian variety with a very ample invertible sheaf defines always a Buchbaum ring. To be more precise:

Let k denote any field and X an abelian k-variety with $g = \dim X$. Let $\mathcal{L}$ be a very ample invertible sheaf on X. Then we consider the graded k-algebra

$$R := \bigoplus_{n \in \mathbf{Z}} H^0(X, \mathcal{L}^{\otimes n})$$

with the naturally induced ring structure and

$$R_n := H^0(X, \mathcal{L}^{\otimes n}) \quad \text{for every } n \in \mathbf{Z}.$$

In fact R is a finitely generated, non-negatively graded k-algebra. We put $\mathfrak{m} = \bigoplus_{n \geq 1} R_n$. Following Schenzel [1], 5.4, we will prove the following theorem. The key is our Proposition I.3.10 and Mumford's [3] observations regarding abelian varieties.

Theorem 2.5. *Let X be an abelian variety with $g = \dim X \geq 1$. Let $\mathcal{L}$ be a very ample invertible sheaf on X. Then $R_{\mathfrak{m}}$ is a Buchsbaum ring with* $\dim R_{\mathfrak{m}} = g + 1$, depth $R_{\mathfrak{m}} = 2$, *and* $\dim_k H^i_{\mathfrak{m}}(R_{\mathfrak{m}}) = \binom{g}{i-1}$, $2 \leq i \leq g$, *i.e.*

$$I(R_{\mathfrak{m}}) = \binom{2g}{g-1} - g.$$

Proof: By using the results of D. Mumford [3], Section 16, we get the vanishing of the cohomology

$$H^i(X, \mathcal{L}^{\otimes n}) = 0 \quad \text{for } 0 < i < \dim X \text{ and all } n \neq 0.$$

From this follows

$$[\underline{H}_{\mathfrak{m}}^{i+1}(R)]_n \cong H^i(X, \mathcal{L}^{\otimes n}) = 0 \quad \text{for all } n \neq 0 \text{ and } 0 < i < \dim X.$$

Also we have $\underline{H}_{\mathfrak{m}}^i(R) = 0$ for $i = 0, 1$.

It follows immediately from Proposition I.3.10 that $R_{\mathfrak{m}}$ is a $(g+1)$-dimensional Buchsbaum ring.

Moreover we get from Mumford [3], Section 13, Corollary 2, that for $n = 0$ and $0 \leq i \leq \dim X$

$$\dim_k H^i(X, \mathcal{L}^{\otimes n}) = \binom{g}{i}.$$

Therefore the results follow for the local cohomology of $R_{\mathfrak{m}}$, the depth of $R_{\mathfrak{m}}$ and for $I(R_{\mathfrak{m}})$, q.e.d.

In particular, let C be a smooth irreducible curve over a field k with genus $g \geq 1$. Then the Jacobian variety J_C of C is a g-dimensional abelian variety (see, for example, Mumford's [4] lectures on curves and their Jacobians).

Example 2.6. Let k be a field of characteristic 0. Let $k(t)$ be the field of rational functions in one variable t over k. S. Mori [1] has constructed smooth and geometrically irreducible hyperelliptic curves C of genus $g \geq 1$ over $k(t)$ such that for the Jacobian variety J of C and its theta divisor Θ the completion $\hat{R}$ of

$$R = \bigoplus_{n \in \mathbf{Z}} H^0(J, \Theta^{\otimes n})$$

with respect to $\mathfrak{m} = \bigoplus_{n \geq 1} H^0(J, \Theta^{\otimes n})$ is a factorial domain. Hence we get non-Cohen-Macaulay factorial Buchsbaum rings $R_{\mathfrak{m}}$ for $g \geq 2$ with $\dim R_{\mathfrak{m}} = g + 1$, depth $R_{\mathfrak{m}} = 2$ and $I(R_{\mathfrak{m}}) = \binom{2g}{g-1} - g$.

By using the Theorem 5.8 of Freitag-Kiehl [1], R. Kiehl [1] has constructed other examples of non-Cohen-Macaulay factorial Buchsbaum rings of depth = 3 and, for example, of dimension 60. Moreover, there are non-Cohen-Macaulay factorial rings which are not Buchsbaum rings. This fact was mentioned by R. Kiehl [2] in a private letter:

„... Mit Hilfe Ihres Theorems 2 (see our Proposition 2.4) kann Herr Freitag ein Beispiel eines faktoriellen Ringes konstruieren, der nicht zur Klasse der Buchsbaum-Ringe gehört, nämlich: Der Kegelspitzenring des graduierten Ringes aller Modulformen zur vollen Siegelschen Modulgruppe genügend hohen Grades ist faktoriell (see Freitag [1], Theorem 3.6), aber nicht Buchsbaum. Es ist aber nicht klar, ob die Komplettierung dieses Ringes faktoriell ist, dazu müßte zumindest die Tiefe größer oder gleich 3 sein (s. Bericht von Lipman in den AMS)."

Today, the deep lying reason for this observation follows from results of Tsuchihashi [1] and Ishida [1, 2].

Other examples of factorial rings which are not Buchsbaum rings follow from Mori [1], Example 2.4. Also, Bertin's [1] four-dimensional non-Cohen-Macaulay factorial domain is not a Buchsbaum ring.

Next we discuss again Seidenberg's problem. A. Seidenberg [2], p. 620, has considered the difference between the dimension and depth of homogeneous polynomial ideals taking this difference as a measure of the deviation from the Cohen-Macaulay property. Our Example I.4.14 shows that this measure does not describe the non-Cohen-Macaulay property well. In the following we describe another such class of projective varieties by given their defining equations. For this we examine our ideals of type (r, d) studied in Definition I.2.13 and Lemma I.2.14. We get the following class of projective varieties:

Proposition 2.7. *Let d, r be arbitrary integers with $1 \leq r < d$. Then there exist projective varieties $V \subset \mathbf{P}_k^{2d-1}$ over an arbitrary field k such that the local ring A of the vertex of the affine cone over V is a Buchsbaum ring with*

$$\dim A = d \quad \text{and} \quad \operatorname{depth} A = r.$$

Here A is given by the quotient $A_d/\mathfrak{a}$ where $A_d = k[x_1, \ldots, x_{2d}]_{(x_1,\ldots,x_{2d})}$ and $\mathfrak{a}$ is the following ideal of type (r, d) if r is odd:

$$\begin{aligned}\mathfrak{a} &= (x_1x_{d+1}, x_2x_{d+2}, \ldots, x_{d-1}x_{2d-1}, x_d, x_{i_1}\bar{x}_{i_2}x_{i_3}\bar{x}_{i_4} \cdot \ldots \cdot \bar{x}_{i_{r-1}}x_{i_r}) \\ &\quad \cap (x_1x_{d+1}, \ldots, x_{d-1}x_{2d-1}, x_{2d}, \bar{x}_{i_1}x_{i_2}\bar{x}_{i_3} \cdot \ldots \cdot \bar{x}_{i_r}) \\ &=: \mathfrak{a}_1 \cap \mathfrak{a}_2\end{aligned}$$

where $\bar{x}_j$ is obtained by applying the permutation

$$\begin{pmatrix} x_1 & \ldots & x_d & x_{d+1} & \ldots & x_{2d} \\ x_{d+1} & \ldots & x_{2d} & x_1 & \ldots & x_d \end{pmatrix}$$

to x_j and the system of indices $(i_1, i_2, \ldots, i_r)$ runs through all r-tuples with $0 < i_1 < i_2 < \ldots < i_r < d$.

If r is even, then $\mathfrak{a}$ is given by

$$\begin{aligned}\mathfrak{a} &= (x_1x_{d+1}, \ldots, x_{d-1}x_{2d-1}, x_{2d}, \bar{x}_{i_1}x_{i_2}\bar{x}_{i_3} \cdot \ldots \cdot x_{i_r}) \\ &\quad \cap (x_1x_{d+1}, \ldots, x_{d-1}x_{2d-1}, x_d, x_{i_1}\bar{x}_{i_2}x_{i_3} \cdot \ldots \cdot \bar{x}_{i_r}).\end{aligned}$$

Remark 2.8. This class of projective varieties provides another interesting application to canonical modules as given by Yoichi Aoyama in [1], Theorem 1.

Example 2.9. (i) The ideal of type $(3, 5)$:

$$\begin{aligned}\mathfrak{a} &= (x_5, x_1x_6, x_2x_7, x_3x_8, x_4x_9, x_1x_3x_7, x_1x_4x_7, x_1x_4x_8, x_2x_4x_8) \\ &\quad \cap (x_{10}, x_1x_6, x_2x_7, x_3x_8, x_4x_9, x_2x_6x_8, x_2x_6x_9, x_3x_6x_9, x_3x_7x_9).\end{aligned}$$

(ii) The ideal of type $(2, d)$:

$$\begin{aligned}\mathfrak{a} &= (x_1, \ldots, x_d) \cap (x_2, \ldots, x_{d+1}) \cap \ldots \cap (x_{d+1}, \ldots, x_{2d}) \\ &\quad \cap (x_{d+2}, \ldots, x_{2d}, x_1) \cap \ldots \cap (x_{2d}, x_1, \ldots, x_{d-1}).\end{aligned}$$

Proof of Proposition 2.7: Lemma I.2.14 shows that we only have to prove that the ideal $\mathfrak{a} \subset A_d$ is of type (r, d). Let $r \geq 3$ be odd. By Definition I.2.13 we have to show that $A_d/\mathfrak{a}_1$ is a Cohen-Macaulay ring. It is not difficult to show that the elements $x_1 + x_{d+1}$, $x_2 + x_{d+2}, \ldots, x_{d-1} + x_{2d-1}, x_{2d}$ define an $A_d/\mathfrak{a}_1$-sequence of the d-dimensional ring $A_d/\mathfrak{a}_1$. Hence $A_d/\mathfrak{a}_1$ is Cohen-Macaulay. If r is even, the proof is analogous, q.e.d.

§ 3. On Buchsbaum rings obtained by glueing

We know already from the Introduction how a Buchsbaum ring is obtained by glueing, see Example 3 (3c). Following Shiro Goto [4] we now describe this procedure. Originally the concept of abstract glueing was introduced by C. Traverso [1] in order to clarify the structure of semi-normal rings. We also will study a key example involving this construction originally introduced by S. Goto [4] and G. Tamone [1].

In order to state our main result we need some definitions. Let S be a Cohen-Macaulay (not necessary local) ring with $\dim S_{\mathfrak{Q}} = \dim S$ for every maximal ideal $\mathfrak{Q}$ of S. Let R be a subring of S and assume that S is module-finite over R. Hence R is again a Noetherian ring. For every prime ideal $\mathfrak{p}$ (resp. $\mathfrak{P}$) of R (resp. S) we denote by $k(\mathfrak{p})$ (resp. $k(\mathfrak{P})$) the field $R_\mathfrak{p}/\mathfrak{p}R_\mathfrak{p}$ (resp. $S_\mathfrak{P}/\mathfrak{P}S_\mathfrak{P}$).

Let $\mathfrak{p}$ be a prime ideal of R. We put

$$W(\mathfrak{p}) = \{\mathfrak{P} \in \operatorname{Spec} S \mid \mathfrak{P} \cap R = \mathfrak{p}\}.$$

Notice that $W(\mathfrak{p})$ is a finite subset of Spec S. For every $\mathfrak{P} \in W(\mathfrak{p})$ let $i_\mathfrak{P}: k(\mathfrak{p}) \to k(\mathfrak{P})$ be the canonical monomorphism. We denote by $f(\mathfrak{P})$ the value of f at $\mathfrak{P}$ in $k(\mathfrak{P})$ for $f \in S$ and $\mathfrak{P} \in W(\mathfrak{p})$. Then we have the following definition by C. Traverso [1].

Definition. We put $\mathfrak{p}' = \bigcap_{\mathfrak{P} \in W(\mathfrak{p})} \mathfrak{P}$ and

$$R' = \{f \in S \mid \exists\, c \in k(\mathfrak{p}) \text{ such that } f(\mathfrak{P}) = i_\mathfrak{P}(c) \text{ for every } \mathfrak{P} \in W(\mathfrak{p})\}.$$

Then R' is a subring of S containing R and $\mathfrak{p}'$ is a prime ideal of R'. We call R' the *glueing* over $\mathfrak{p}$.

For an arbitrary local ring A we denote by emb (A) (resp. $e_0(A)$) the embedding dimension of A (resp. the multiplicity of A relative to the maximal ideal of A).

Now we are prepared to state the main result of this paragraph.

Theorem 3.1. *With the stated hypothesis let* $d = \dim R_\mathfrak{p}$. *Then* $A := R'_{\mathfrak{p}'}$ *is a Buchsbaum local ring of dimension* d. *Further, suppose* $d > 0$ *and let* $\mathfrak{m}$ *denote the maximal ideal of* A. *Then we have:*

(1) $H^i_\mathfrak{m}(A) = 0$ *for* $i \neq 1, d$.

(2) $I(A) = (d-1) \cdot \left[\sum_{\mathfrak{P} \in W(\mathfrak{p})} [k(\mathfrak{P}) : k(\mathfrak{p})] - 1\right]$.

(3) $\operatorname{emb}(A) = \sum_{P \in W(\mathfrak{p})} \operatorname{emb}(S_\mathfrak{P}) \cdot [k(\mathfrak{P}) : k(\mathfrak{p})]$.

(4) $e_0(A) = \sum_{P \in W(\mathfrak{p})} e_0(S_\mathfrak{P}) \cdot [k(\mathfrak{P}) : k(\mathfrak{p})]$.

If $d \geq 2$, *the Cohen-Macaulayfication of* A *coincides with* $S_{\mathfrak{P}'}$, *i.e.,* $\tilde{A} = S_{\mathfrak{P}'}$.

We will illustrate this theorem with the previously mentioned example from the Introduction. Let k be a field with char $k \neq 2$ and $S = k[x, y]$ a polynomial ring. Let $R = \{f \in S \mid f(-1, 0) = f(1, 0)\}$ and $\mathfrak{p} = \{f \in S \mid f(-1, 0) = f(1, 0) = 0\}$. Then R is a subring of S and S is module-finite over R. Moreover $\mathfrak{p}$ is a maximal ideal of R with $W(\mathfrak{p}) = \{\mathfrak{P}, \mathfrak{Q}\}$ where $\mathfrak{P} = (x + 1, y)$ and $\mathfrak{Q} = (x - 1, y)$. In this situation the glueing R' over $\mathfrak{p}$ coincides with the ring R itself, and the two points $\mathfrak{P}$ and $\mathfrak{Q}$ of Spec S are glued together into a single point $\mathfrak{p}$ of Spec R via the morphism Spec S $\to$ Spec R. Therefore we get that $A = R_{\mathfrak{p}}$ is a Buchsbaum (local) domain of dimension 2 with $I(A) = 1$. Clearly emb $(A) = 4$ and $e_0(A) = 2$.

In order to prove the theorem we need some lemmas.

Lemma 3.2. *With the same hypothesis as stated in the theorem let $B = S_{\mathfrak{p}'}$. Then we have*

(1) $\mathfrak{p}'$ *is also an ideal of S and* $\mathfrak{m} = \mathfrak{m}B \subset A$.

(2) $\mathfrak{p}'$ *is a unique prime ideal of R'' such that $\mathfrak{p}' \cap R = \mathfrak{p}$, and therefore* $\dim A = d$.

(3) $k(\mathfrak{p}) = k(\mathfrak{p}')$, *where $k(\mathfrak{p}')$ denotes the field $R'_{\mathfrak{p}'}/\mathfrak{p}'R'_{\mathfrak{p}'}$.*

(4) $\operatorname{Max} B = \{\mathfrak{P}B \mid \mathfrak{P} \in W(\mathfrak{p})\}$.

(5) $\dim S_{\mathfrak{P}} = d$ *for every* $\mathfrak{P} \in W(\mathfrak{p})$.

Proof:

(1) This is trivial since $\mathfrak{p}' = \bigcap_{\mathfrak{P} \in W(\mathfrak{p})} \mathfrak{P}$ by definition.

(2) It is again a trivial matter to verify the first assertion. The second one follows from the first, since $A = R'_{\mathfrak{p}'}$ and $d = \dim R_{\mathfrak{p}}$ by definition.

(3) This follows immediately from (2).

(4) Let $\mathfrak{P}$ be an element of $W(\mathfrak{p})$. Then $\mathfrak{P} \cap R' = \mathfrak{p}'$ by (2) since $(\mathfrak{P} \cap R') \cap R = \mathfrak{p}$. Hence $\mathfrak{P}B$ is a maximal ideal of $B = S_{\mathfrak{p}'}$. Every maximal ideal $\mathfrak{n}$ of B of course may be expressed as $\mathfrak{n} = \mathfrak{P}B$ for some $\mathfrak{P} \in W(\mathfrak{p})$ since $\mathfrak{p}' \cap R = \mathfrak{p}$, which again follows by (2).

(5) Let $\mathfrak{P}$ be an element of $W(\mathfrak{p})$ and choose a maximal ideal $\mathfrak{Q}$ of S containing $\mathfrak{P}$ such that $\dim S/\mathfrak{P} = \dim S_{\mathfrak{Q}}/\mathfrak{P}S_{\mathfrak{Q}}$. Then $S_{\mathfrak{Q}}$ is a Cohen-Macaulay local ring with $\dim S_{\mathfrak{Q}} = \dim S$ by the initial assumption on S. Hence we have

$$\dim S_{\mathfrak{P}} = \dim S_{\mathfrak{Q}} - \dim S_{\mathfrak{Q}}/\mathfrak{P}S_{\mathfrak{Q}} = \dim S - \dim S/\mathfrak{P} = \dim R - \dim R/\mathfrak{p},$$

and thus we see that $\dim S_{\mathfrak{P}}$ does not depend on $\mathfrak{P}$. Also $\dim B = d$ by (2), since B is module-finite over A. Thus we conclude by (4) that $\dim S_{\mathfrak{P}} = d$ for every $\mathfrak{P} \in W(\mathfrak{p})$, q.e.d.

Corollary 3.3. *Let $Q(A)$ denote the total quotient ring of A and suppose that $d > 0$. Then* depth $A > 0$ *and* $Q(A) \supset B$.

Proof: First note that $B = S_{\mathfrak{p}'}$ is a Cohen-Macaulay ring as is S by the initial assumption on S. Hence the A-module B is Cohen-Macaulay and of maximal dimension d, because $\dim B_{\mathfrak{n}} = d$ for every maximal ideal $\mathfrak{n}$ of B by Lemma 3.2 (see (4) and (5)). In particular $\operatorname{depth}_A B > 0$ as $d > 0$.

Let a be an element of $\mathfrak{m}$ and suppose that a is B-regular. Then clearly a is A-regular and so we have that depth $A > 0$. Also $\mathfrak{m}B \subset A$ by Lemma 3.2(1). Thus $B \subset A[a^{-1}]$ and therefore we have the inclusion $B \subset Q(A)$ as required.

Lemma 3.4..

(1) $\mathfrak{m} = \bigcap_{\mathfrak{n} \in \operatorname{Max} B} \mathfrak{n}$.

(2) $l_A(B/\mathfrak{m}^i) = \sum_{\mathfrak{P} \in W(\mathfrak{p})} l_{S_{\mathfrak{P}}}(S_{\mathfrak{P}}/\mathfrak{P}^i S_{\mathfrak{P}}) \cdot [k(\mathfrak{P}):k(\mathfrak{p})]$ *for every integer* $i \geq 0$.

Proof:

(1) Since $\mathfrak{m} = \mathfrak{p}'B$ by (1) of Lemma 3.2, we have

$$\mathfrak{m} = \Big(\bigcap_{\mathfrak{P} \in W(\mathfrak{p})} \mathfrak{P}\Big) \cdot B = \bigcap_{\mathfrak{P} \in W(\mathfrak{p})} \mathfrak{P}B = \bigcap_{\mathfrak{n} \in \operatorname{Max} B} \mathfrak{n}.$$

(2) Let $i \geq 0$ be an integer. It then follows from (1) that

$$\mathfrak{m}^i = \bigcap_{\mathfrak{n} \in \operatorname{Max} B} \mathfrak{n}^i \quad \text{and} \quad B/\mathfrak{m}^i = \bigoplus_{\mathfrak{n} \in \operatorname{Max} B} B/\mathfrak{n}^i$$

by virtue of the Chinese Remainder Theorem. Hence

$$l_A(B/\mathfrak{m}^i) = \sum_{\mathfrak{n} \in \operatorname{Max} B} l_B(B/\mathfrak{n}^i) \cdot [B/\mathfrak{n} : A/\mathfrak{m}]$$

and therefore recalling $A/\mathfrak{m} = k(\mathfrak{p})$ by (3) of Lemma 3.2, we conclude by (4) of Lemma 3.2 that

$$l_A(B/\mathfrak{m}^i) = \sum_{\mathfrak{P} \in W(\mathfrak{p})} l_{S_{\mathfrak{P}}}(S_{\mathfrak{P}}/\mathfrak{P}^i S_{\mathfrak{P}}) \cdot [k(\mathfrak{P}):k(\mathfrak{p})]$$

as claimed.

Proof of Theorem 3.1: We know that $\dim A = d$ by Lemma 3.2(2). Note that if $d \leq 1$, then A is a Cohen-Macaulay ring (see Corollary 3.3). Hence A is a Buchsbaum local ring with $I(A) = 0$ in this case. Now suppose that $d \geq 2$. Then, since B is a Cohen-Macaulay ring with $\dim B_{\mathfrak{n}} = d$ for every maximal ideal $\mathfrak{n}$ of B by (4) and (5) of Lemma 3.2, A is a Buchsbaum local ring with Cohen-Macaulayfication B and with $H^i_{\mathfrak{m}}(A) = (0)$ for $i \neq 1, d$ (see (1) of Lemma 3.2, Corollary 3.3 and Theorem IV.2.11. Moreover, by Proposition I.2.6, we see that

$$I(A) = (d-1) \dim_{A/\mathfrak{m}} B/A \quad \big(\text{since } B/A \cong H^1_{\mathfrak{m}}(A)\big).$$

Also

$$\dim_{A/\mathfrak{m}} B/A = \dim_{A/\mathfrak{m}} B/\mathfrak{m} - 1 \quad \text{and} \quad \dim_{A/\mathfrak{m}} B/\mathfrak{m} = \sum_{\mathfrak{P} \in W(\mathfrak{p})} [k(\mathfrak{P}):k(\mathfrak{p})]$$

by Lemma 3.4. Hence we have the required equation

$$I(A) = (d-1) \cdot \Big\{\sum_{\mathfrak{P} \in W(\mathfrak{p})} [k(\mathfrak{P}):k(\mathfrak{p})] - 1\Big\}.$$

Now consider the assertions (3) and (4). Suppose that $d > 0$ and let $i \geq 0$ be an integer. Then

$$l_A(B/\mathfrak{m}^i) = \sum_{\mathfrak{P} \in W(\mathfrak{p})} l_{S_{\mathfrak{P}}}(S_{\mathfrak{P}}/\mathfrak{P}^i S_{\mathfrak{P}}) \cdot [k(\mathfrak{P}):k(\mathfrak{p})]$$

by Lemma 3.4, and therefore we get

$$l_A(A/\mathfrak{m}^i) = \sum_{\mathfrak{P} \in W(\mathfrak{p})} l_{S_\mathfrak{P}}(S_\mathfrak{P}/\mathfrak{P}^i S_\mathfrak{P}) \cdot [k(\mathfrak{P}):k(\mathfrak{p})] - l_A(B/A).$$

This implies the required equation

$$e_0(A) = \sum_{\mathfrak{P} \in W(\mathfrak{p})} e_0(S_\mathfrak{P}) \cdot [k(\mathfrak{P}):k(\mathfrak{p})]$$

since $\dim A = \dim S_\mathfrak{P} = d > 0$ for every $\mathfrak{P} \in W(\mathfrak{p})$ (see Lemma 3.2(5)). Again by Lemma 3.4 we see that

$$\begin{aligned} \operatorname{emb}(A) &= l_A(\mathfrak{m}/\mathfrak{m}^2) = l_A(B/\mathfrak{m}^2) - l_A(B/\mathfrak{m}) \\ &= \sum_{\mathfrak{P} \in W(\mathfrak{p})} \left(l_{S_\mathfrak{P}}(S_\mathfrak{P}/\mathfrak{P}^2 S_\mathfrak{P}) - l_{S_\mathfrak{P}}(S_\mathfrak{P}/\mathfrak{P} S_\mathfrak{P})\right) \cdot [k(\mathfrak{P}):k(\mathfrak{p})] \\ &= \sum_{\mathfrak{P} \in W(\mathfrak{p})} \operatorname{emb}(S_\mathfrak{P}) \cdot [k(\mathfrak{P}):k(\mathfrak{p})]. \end{aligned}$$

This completes the proof of the theorem, q.e.d.

Following S. Goto [4] and G. Tamone [1] we next will examine a key example:

Let K/k be a field extension of fields and S a finitely generated K-algebra of dimension s. Suppose that S is a Cohen-Macaulay ring with $\dim S/\mathfrak{P} = s$ for every minimal prime ideal $\mathfrak{P}$ of S. Let W be a non-empty finite subset of Spec S and assume that $\dim S_\mathfrak{P} = d$ for every $\mathfrak{P} \in W$. We put

$$\mathfrak{p} = \bigcap_{\mathfrak{P} \in W} \mathfrak{P}.$$

Then, by virtue of the Normalization Lemma, one may find elements $x_1, x_2, \ldots, x_s$ of S such that S is module-finite over $K[x_1, x_2, \ldots, x_s]$ and

$$\mathfrak{p} \cap K[x_1, x_2, \ldots, x_s] = (x_{s-d+1}, x_{s-d+2}, \ldots, x_s).$$

Now let

$$R = k[x_1, x_2, \ldots, x_{s-d}] + \mathfrak{p}.$$

Then R is a subring of S and S is module-finite over R. Moreover $\mathfrak{p}$ is a prime ideal of R with

$$W(\mathfrak{p}) = W \quad \text{and} \quad R/\mathfrak{p} = k[x_1, x_2, \ldots, x_{s-d}].$$

It now follows that the glueing R' over $\mathfrak{p}$ coincides with the ring R itself (see Tamone [1], Theorem 1.3, for the details). Thus, by applying the above theorem to this situation, we obtain immediately the following results:

(1) $A = R_\mathfrak{p}$ is a Buchsbaum local ring of dimension d and with $H^i_\mathfrak{m}(A) = 0$ for $i \neq 1$, d. (Here $\mathfrak{m}$ denotes the maximal ideal of A.)

(2) Suppose that $d > 0$ and set $d(\mathfrak{P}) = [S_\mathfrak{P}/\mathfrak{P} S_\mathfrak{P} : k(x_1, \ldots, x_{s-d})]$ for every $\mathfrak{P} \in W(\mathfrak{p})$. Then

$$I(A) = (d-1) \cdot \left\{ \sum_{\mathfrak{P} \in W(\mathfrak{p})} d(\mathfrak{P}) - 1 \right\},$$

$$\operatorname{emb}(A) = \sum_{\mathfrak{P} \in W(\mathfrak{p})} \operatorname{emb}(S_\mathfrak{P}) \cdot d(\mathfrak{P}),$$

and

$$e_0(A) = \sum_{\mathfrak{P} \in W(\mathfrak{p})} e_0(S_{\mathfrak{P}}) \cdot d(\mathfrak{P}).$$

If $d \geq 2$ and $\# W > 2$, then A of course is not a Cohen-Macaulay ring.

(3) Suppose $d \geq 2$. Then the Cohen-Macaulayfication of A coincides with $S_{\mathfrak{P}}$, i.e., $\tilde{A} = S_{\mathfrak{P}}$.

For further examples see S. Goto [4].

We conclude by studying the following well-known example:

Let k be a field and $P = k[\![X_1, \ldots, X_d, Y_1, \ldots, Y_d]\!]$, $d \geq 2$, a formal power serie sring. Let

$$A = P/\mathfrak{p} \cap \mathfrak{q} \quad \text{and} \quad B = P/\mathfrak{p} \oplus P/\mathfrak{q},$$

where $\mathfrak{p} = (X_1, \ldots, X_d)\, P$ and $\mathfrak{q} = (Y_1, \ldots, Y_d)\, P$. Then A of course is a Buchsbaum ring. One may deduce this fact by using the above Theorem 3.1, since A is obtained from B by glueing over the maximal ideal.

Clearly $\tilde{A} = B$ in this case, $\tilde{A}$ coincides with the normalization of A and A is seminormal. We have obviously

$$e_0(A) = 2, \quad \operatorname{emb}(A) = 2d, \quad \text{and} \quad I(A) = d - 1.$$

§ 4. Construction of Buchsbaum rings with given local cohomology

By using results found in the papers of Evans-Griffith [1] and Goto [3] we prove the following theorem.

Theorem 4.1. *Let $d > 0$ and $h_0, h_1, \ldots, h_{d-1} \geq 0$ be integers. Then there is a Buchsbaum local ring A with maximal ideal $\mathfrak{m}$ such that* $\dim A = d$ *and* $\dim_{A/\mathfrak{m}} H^i_{\mathfrak{m}}(A) = h_i$ *for every $0 \leq i \leq d - 1$.*

Moreover, if $h_0 = 0$ (resp. $d \geq 2$ and $h_0 = h_1 = 0$), then A may also be taken to be an integral domain (resp. a normal domain).

Remark 4.2. In case $d \geq 2$ and $h_0 = 0$ the method of construction of these examples is due to E. G. Evans Jr. and P. A. Griffith [1]. The Buchsbaum property for their results follows immediately by Lemma I.3.10. Additional, the case $h_0 \neq 0$ was studied by S. Goto [3] and the examples are obtained by idealization.

First we will consider the case $d \geq 2$ and $h_0 = 0$.

Suppose that A is a regular local ring of dimension $d + 2$. We take a minimal resolution of $A/\mathfrak{m}$:

$$0 \to F_{d+2} \to \ldots \to F_1 \to F_0 = A \to A/\mathfrak{m} \to 0.$$

We set $Z_s := \operatorname{im}(F_{s+1} \to F_s)$ for $0 \leq s \leq d$.

Lemma 4.3.

(1) *Z_s is a Buchsbaum A-module with* $\dim_A Z_s = d + 2$.

(2) $H^i_{\mathfrak{m}}(Z_s) = \begin{cases} A/\mathfrak{m} & \textit{for } i = s + 1, \\ 0 & \textit{for all } i \neq s + 1, d + 2. \end{cases}$

(3) *The invariant* $I(Z_s) = \dbinom{d+1}{s+1}$.

Proof: The assertions (1) and (3) follow immediately from (2) by Propositions I.2.6 and I.2.12. For the claim (2), consider the exact sequence for $0 \leq s \leq d$:

$$0 \to Z_s \to F_s \to Z_{s-1} \to 0$$

where we set $Z_{-1} := A/\mathfrak{m}$. Then, after applying the functor $H^i_{\mathfrak{m}}(\ )$ to these sequences, the result follows by induction on s, q.e.d.

We now set $M_s := Z_s^{(h_s)}$, i.e. the direct sum of h_s copies of Z_s, for every $1 \leq s \leq d-1$, and we set $M := \bigoplus_{s=1}^{d-1} M_s$. Note that we may assume $M \neq 0$. In fact if $M = 0$, then $h_s = 0$ for all $0 \leq s \leq d-1$ and every regular local ring of dimension d satisfies all the conditions required in our theorem.

Lemma 4.4.

(1) *M is a Buchsbaum A-module with* $\dim_A M = d+2$.

(2) $H^i_{\mathfrak{m}}(M) = \begin{cases} (A/\mathfrak{m})^{(h_{i-1})} & \textit{for } i = 2, \ldots, d \\ (0) & \textit{for } i = 0, 1 \textit{ and } d+1. \end{cases}$

(3) *The invariant* $I(M) = \sum_{s=1}^{d-1} h_s \cdot \binom{d+1}{s+1}$.

(4) *M is a non-free and reflexive A-module.*

(5) *$M_{\mathfrak{p}}$ is a free A_p-module for every* $\mathfrak{p} \in \operatorname{Spec} A \setminus \mathfrak{m}$.

Proof: The assertion (2) follows from Lemma 4.3. Clearly $\dim_A M = d+2$. Since we have the exact sequence $0 \to Z_s \to F_s \to F_{s-1}$ for all $s = 1, \ldots, d-1$ we get that M is a reflexive A-module as is Z_s. Now M is not free since $\operatorname{depth}_A M \leq d < \dim_A M = d+2$. The assertion (5) is trivial since $(Z_s)_{\mathfrak{p}}$ is a free $A_{\mathfrak{p}}$-module for every $\mathfrak{p} \in \operatorname{Spec} A \setminus \{\mathfrak{m}\}$ and for every $1 \leq s \leq d-1$.

Now we prove the assertions (1) and (3). Let $\mathfrak{q}$ be a parameter ideal of A. Since $\dim_A M = \dim A$, $\mathfrak{q}$ is also a parameter ideal of Z_s for every $s = 1, \ldots, d-1$. Also since we have

$$l_A(M/\mathfrak{q}M) = \sum_{s=1}^{d-1} h_s \cdot l_A(Z_s/\mathfrak{q} \cdot Z_s)$$

and

$$e_0(\mathfrak{q}, M) = \sum_{s=1}^{d-1} h_s \cdot e_0(\mathfrak{q}, Z_s)$$

we get for the difference:

$$\begin{aligned} l_A(M/\mathfrak{q}M) - e_0(\mathfrak{q}, M) &= \sum_{s=1}^{d-1} h_s \cdot \big(l_A(Z_s/\mathfrak{q} \cdot Z_s) - e_0(\mathfrak{q}, Z_s)\big) \\ &= \sum_{s=1}^{d-1} h_s \cdot I(Z_s), \quad \text{by Lemma 4.3} \\ &= \sum_{s=1}^{d-1} h_s \cdot \binom{d+1}{s+1}; \end{aligned}$$

thus the difference does not depend on $\mathfrak{q}$. Therefore we have proven the assertions (1) and (3). (Note that (1) also follows from Theorem I.2.15, q.e.d.)

Since M is a non-free and reflexive A-module we get the following lemma from the results of M. Auslander [1], Theorem B and Proposition 4.4:

Lemma 4.5. *There is an exact sequence*

$$0 \to F \to M \to I \to 0$$

of A-modules such that F is a free A-module and I is an unmixed ideal of A of height 2.

In Lemma 4.5 we set $\bar{A} = A/I$ and $\bar{\mathfrak{m}} = \mathfrak{m}/I$. Then the following lemma proves our theorem in case $d \geq 2$ and $h_0 = 0$.

Lemma 4.6. *$\bar{A}$ is a Buchsbaum local ring with* $\dim \bar{A} = d$ *and* $\dim_{\bar{A}/\bar{\mathfrak{m}}} H^i_{\bar{\mathfrak{m}}}(\bar{A}) = h_i$ *for* $0 \leq i \leq d-1$.

Proof: Clearly $\dim \bar{A} = d$ and $H^0_{\bar{\mathfrak{m}}}(\bar{A}) = 0$ since I is unmixed. Consider the following two exact sequences (exact by Lemma 4.5):

$$0 \to F \to M \to I \to 0 \quad \text{and} \quad 0 \to I \to A \to \bar{A} \to 0.$$

By applying the functor $H^i_{\mathfrak{m}}(\ )$ we therefore get:

$$H^{i+1}_{\mathfrak{m}}(M) \simeq H^{i+1}_{\mathfrak{m}}(I) \quad \text{and} \quad H^i_{\mathfrak{m}}(\bar{A}) \simeq H^{i+1}_{\mathfrak{m}}(I) \quad \text{for every } i \leq d-1.$$

Thus we see by Lemma 4.4 that

$$H^i_{\bar{\mathfrak{m}}}(\bar{A}) \simeq H^i_{\mathfrak{m}}(\bar{A}) \simeq H^{i+1}_{\mathfrak{m}}(M) \simeq (\bar{A}/\bar{\mathfrak{m}})^{(h_i)} \quad \text{for all } i = 1, \ldots, d-1.$$

Now we prove that $\bar{A}$ is a Buchsbaum local ring. For this purpose it suffices to show that $\bar{A}$ is a Buchsbaum A-module. From the above two exact sequences we obtain the following commutative diagrams of A-modules for every $i \leq d-1$:

$$\begin{array}{ccc} \mathrm{Ext}^{i+1}_A(A/\mathfrak{m}, M) & \xrightarrow{\sim} & \mathrm{Ext}^{i+1}_A(A/\mathfrak{m}, I) \\ \downarrow{\scriptstyle \varphi^{i+1}_M} & & \downarrow{\scriptstyle \varphi^{i+1}_I} \\ H^{i+1}_{\mathfrak{m}}(M) & \xrightarrow{\sim} & H^{i+1}_{\mathfrak{m}}(I) \end{array} \qquad \begin{array}{ccc} \mathrm{Ext}^{i}_A(A/\mathfrak{m}, \bar{A}) & \xrightarrow{\sim} & \mathrm{Ext}^{i+1}_A(A/\mathfrak{m}, I) \\ \downarrow{\scriptstyle \varphi^{i}_{\bar{A}}} & & \downarrow{\scriptstyle \varphi^{i+1}_I} \\ H^{i}_{\mathfrak{m}}(\bar{A}) & \xrightarrow{\sim} & H^{i+1}_{\mathfrak{m}}(I) \end{array}$$

where the vertical maps are canonical homomorphisms. Since M is a Buchsbaum A-module by Lemma 4.4 we have that φ^{i+1}_M is surjective (see Corollary I.2.16) and therefore φ^{i+1}_M is also surjective. Hence we get that $\varphi^i_{\bar{A}}$ is a surjection. This property implies that A is a Buchsbaum A-module by Theorem I.2.10, q.e.d.

Remark 4.7. In case A is the localization of a polynomial ring $k[x_1, \ldots, x_{d+2}]$ (k an infinite field) by the irrelevant maximal ideal $(x_1, \ldots, x_{d+2})$, Evans and Griffith [1] have shown that I can be taken to be a prime ideal. Moreover they have proved that A/I may also be taken to be a normal domain if $h_1 = 0$.

The assertion of our theorem is trivial in case $d = 1$ and $h_0 = 0$.

We now will consider the case $h_0 > 0$. The examples are then obtained by idealization. We need some lemmas in order to describe the method of this construction. In the following we denote by A any local ring with maximal ideal $\mathfrak{m}$. Let V be a vector space over $A/\mathfrak{m}$ of dimension t and let $B := A \times V$ denote the idealization of V over A; that is the underlying additive group of B coincides with that of the direct sum of A and V and the multiplication in B is defined by $(a, x) \cdot (b, y) = (ab, ay + bx)$. Note that B is again a Noetherian local ring of dimension $d = \dim A$ and with maximal ideal $\mathfrak{e} = \mathfrak{m} \times V$. Let $p: B \to A$ be the canonical projection, i.e. $p\big((a, x)\big) = a$ for $(a, x) \in B$. Then clearly $\ker p = \{0\} \times V$ and, applying the functor $H^i_{\mathfrak{e}}(\ )$ to the exact sequence $0 \to \{0\} \times V \to B \xrightarrow{p} A \to 0$, we get the following lemma:

Lemma 4.8. *$H^i_{\mathfrak{e}}(B) = H^i_{\mathfrak{m}}(A)$ for $i \neq 0$ and $H^0_{\mathfrak{e}}(B) = H^0_{\mathfrak{m}}(A) \times V$.*

Lemma 4.9. *B is a Buchsbaum ring if and only if A is a Buchsbaum ring. In this case we have $I(B) = I(A) + t$.*

Proof: Let $\mathfrak{Q}$ be a parameter ideal of B. We set $\mathfrak{q} = p(\mathfrak{Q})$. Then $\mathfrak{q}$ is also a parameter ideal of A since $A/\mathfrak{q}$ is a homomorphic image of $B/\mathfrak{Q}$ via p. For every integer $s > 0$ there is an exact sequence

$$0 \to \{0\} \times V \to B/\mathfrak{Q}^s \xrightarrow{p} A/\mathfrak{q}^s \to 0,$$

which implies $l_B(B/\mathfrak{Q}^s) = l_A(A/\mathfrak{q}^s) + t$, in particular $e_0(\mathfrak{Q}, B) = e_0(\mathfrak{q}, A)$. Hence we get that $l_B(B/\mathfrak{Q}) - e_0(\mathfrak{Q}, B) = l_A(A/\mathfrak{q}) - e_0(\mathfrak{q}, A) + t$; that is B is Buchsbaum ring if A is a Buchsbaum ring. Conversely assume that B is a Buchsbaum ring and let $a_1, \ldots, a_d$ be a system of parameters for A. We set $\mathfrak{q} = (a_1, \ldots, a_d)\, A$ and $\mathfrak{Q} = (f_1, \ldots, f_d)\, B$ where $f_i = (a_i, 0)$ for $1 \leq i \leq d$. Then clearly $\mathfrak{Q}$ is a parameter ideal of B and $p(\mathfrak{Q}) = \mathfrak{q}$. Therefore we again obtain that $l_A(A/\mathfrak{q}) - e_0(\mathfrak{q}, A) = I(B) - t$, that is A is also a Buchsbaum ring, q.e.d.

Now we prove the assertion of our theorem in case $h_0 > 0$. If $d = 1$ we take A to be a Cohen-Macaulay ring of dimension 1. If $d \geq 2$ we take a Buchsbaum ring A with $\dim A = d$ and $\dim_{A/\mathfrak{m}} H^i_{\mathfrak{m}}(A) = h_i$ for $i = 1, \ldots, d-1$, and $H^0_{\mathfrak{m}}(A) = 0$. This is possible by Lemma 4.6. Let V be an h_0-dimensional vector space over $A/\mathfrak{m}$. Take the idealization $B = A \times V$ of V over A. Then Lemma 4.8 and 4.9 imply that the local ring B has all the properties required in our theorem for the case $h_0 > 0$. This completes the proof of the theorem, q.e.d.

§ 5. Some examples of Segre products

1. In view of Corollary I.4.15 we want to investigate Segre products of Buchsbaum varieties in $\mathbf{P}^n$.

Proposition 5.1. *Let $V \subseteq \mathbf{P}^n_K$ and $W \subseteq \mathbf{P}^n_K$ (K any field) be locally Buchsbaum varieties. In general, $S(V \times W)$, the Segre embedding of $V \times W$ in $\mathbf{P}^{n \cdot m + n + m}_K$, will not be a locally Buchsbaum variety.*

Before proving this by giving an example we describe an algorithm which enables one to calculate the defining equations of the Segre embedding of the product of two varieties using the equations of these varieties.

To this end let $V \subseteq \mathbf{P}^n_K$, $W \subseteq \mathbf{P}^m_K$ be projective varieties given by homogeneous ideals $\mathfrak{a} \subseteq R_1 := K[X_0, \ldots, X_n]$, resp. $\mathfrak{b} \subseteq R_2 := K[Y_0, \ldots, Y_m]$. Assume as given the generators of $\mathfrak{a}$ and $\mathfrak{b}$. Our aim is to obtain generators for the defining ideal $\mathfrak{c}$ of $S(V \times W) \subset \mathbf{P}^{n \cdot m + n + m}_K$.

Let $R := \sigma(R_1, R_2) = K[X_0 \cdot Y_0, \ldots, X_n \cdot Y_m]$, $S := K[T_{00}, \ldots, T_{nm}]$, T_{ij} variables. Then R is an epimorphic image of S given by the mapping $T_{ij} \mapsto X_i \cdot Y_j$. The kernel of this epimorphism is the prime ideal $\mathfrak{p}$ generated by $T_{ij} \cdot T_{rs} - T_{is} \cdot T_{rj}$, $0 \leq i < r \leq n$, $0 \leq j < s \leq m$. Also there is an ideal $\mathfrak{c}' \subset R$ such that

$$\sigma(R_1/\mathfrak{a} \cdot R_2/\mathfrak{b}) \simeq R/\mathfrak{c}' \simeq S/\mathfrak{c} \quad \text{where } \mathfrak{c} \text{ is the required ideal.}$$

By Cartan-Eilenberg [1], II, Prop. 4.3(c), we obtain an exact sequence

$$0 \to \big(\sigma(\mathfrak{a}, R_2) + \sigma(R_1, \mathfrak{b})\big) \to R \to R/\mathfrak{c}' \to 0,$$

i.e. $\mathfrak{c}' = \sigma(\mathfrak{a}, R_2) + \sigma(R_1, \mathfrak{b})$. Therefore

$$\{f \cdot Y_0^{j_0} \cdot \ldots \cdot Y_0^{j_m} \mid f \text{ a generating form of } \mathfrak{a},\ j_0 + \ldots + j_m = \deg f\}$$
$$\cup \{X_0^{i_0} \cdot \ldots \cdot X_n^{i_n} \cdot g \mid g \text{ a generating form of } \mathfrak{b},\ i_0 + \ldots + i_n = \deg g\}$$

will be a set of generators of $\mathfrak{c}'$. Now $\mathfrak{c}$ will be generated by the generators of the ideal $\mathfrak{p}$ and the inverse images of the generators of $\mathfrak{c}'$ (in S). (See also the remarks following Definition 0.2.5.) Now we are able to give our example which proves Proposition 5.1:

Example 5.2. Let $V \subset \mathbf{P}_K^4$ be the surface given parametrically by

$$\{(t_0^3, t_0^2 t_1, t_0 t_1 t_2, t_0 t_2(t_2 - t_0), t_2^2(t_2 - t_0))\}.$$

It was first studied by R. Hartshorne [1].

It is not difficult to verify that the ideal $\mathfrak{p}_V$ of V is

$$\mathfrak{p}_V = (X_1X_4 - X_2X_3, X_0X_1X_2 - X_0X_2^2 + X_1^2X_3, X_0X_2X_3 - X_0X_2X_4 + X_1X_3^2,$$
$$X_0X_3X_4 - X_0X_4^2 + X_3^3) \subset K[X_0, \ldots, X_4] =: R$$

(see Renschuch [2], p. 334).

First we show that V is a locally Buchsbaum variety. We know (see Hartshorne [1]) that $\operatorname{sing} V = \{(1, 0, 0, 0, 0)\}$. Therefore we only need to verify the Buchsbaum property of the local ring A of V at the origin of $\mathbf{P}_K^4$. Since $\dim A = 2$ and depth $A \geq 1$ (A is an integral domain), we can apply Proposition I.2.12. In order to use the proposition we have to find one non-zero $x \in \mathfrak{m}_A^2$ with $\mathfrak{m}_A \cdot U(x \cdot A) \subseteq x \cdot A$.

Choose $x = \overline{X}_3^2$. We obtain

$$\begin{aligned} U(X_3^2 \cdot A) &= (X_3^2, X_4, X_0X_2 + X_1X_3) \cdot A \\ &\qquad \cap \big(X_4^2, X_3 - X_4, X_0(X_1 - X_2) + X_1X_3\big) \cdot A \\ &= (X_3^2, X_4^2) \cdot A \\ &\qquad + (X_4, X_0X_2 + X_1X_3) \cdot \big(X_3 - X_4, X_0(X_1 - X_2) + X_1X_3\big) \cdot A \\ &= (X_3^2, X_4^2) \cdot A \end{aligned}$$

and therefore we see that $\mathfrak{m}_A \cdot U(X_3^2 \cdot A) \subseteq X_3^2 \cdot A$; that is V is a locally Buchsbaum variety.

Now let $X := S(V \times \mathbf{P}^1_K) \subset \mathbf{P}^9_K$. By the previous algorithm the defining ideal $\mathfrak{p}_X$ of X is given by (we set $T_{ij} =: X_{2i+j}$, $0 \leq i \leq 4$, $0 \leq j \leq 1$):

$$\begin{aligned}\mathfrak{p}_X = (&X_0X_3 - X_1X_2, X_0X_5 - X_1X_4, X_0X_7 - X_1X_6, X_0X_9 - X_1X_8,\\ &X_2X_5 - X_3X_4, X_2X_7 - X_3X_6, X_2X_9 - X_3X_8, X_4X_7 - X_5X_6,\\ &X_4X_9 - X_5X_8, X_6X_9 - X_7X_8, X_2X_8 - X_4X_6, X_2X_9 - X_4X_7,\\ &X_3X_9 - X_5X_7,\\ &X_0X_2X_4 - X_0X_4^2 + X_2X_6, X_0X_2X_5 - X_0X_4X_5 + X_2^2X_7,\\ &X_0X_3X_5 - X_0X_5^2 + X_2X_3X_7, X_1X_3X_5 - X_1X_5^2 + X_3^2X_7,\\ &X_0X_4X_6 - X_0X_4X_8 + X_2X_6^2, X_0X_4X_7 - X_0X_4X_9 + X_2X_6X_7,\\ &X_0X_5X_7 - X_0X_5X_9 + X_2X_7^2, X_1X_5X_7 - X_1X_5X_9 + X_3X_7^2,\\ &X_0X_6X_8 - X_0X_8^2 + X_6^3, X_0X_6X_9 - X_0X_8X_9 + X_6^2X_7,\\ &X_0X_7X_9 - X_0X_9^2 + X_6X_7^2, X_1X_7X_9 - X_1X_9^2 + X_7^3).\end{aligned}$$

Let B be the local ring of X at the line $X_2 = \ldots = X_9 = 0$. We claim that B is not a Cohen-Macaulay ring. Then by Corollary I.1.11 the local ring of X at the origin $X_1 = X_2 = \ldots = X_9 = 0$ cannot be a Buchsbaum ring.

It is now easy to see that rad $X_2 \cdot B = (X_2, X_3, X_4, X_5) \cdot B$ and this last ideal is a prime ideal in B, since $(\mathfrak{p}_X, X_2, X_3, X_4, X_5)$ is a prime ideal in $K[X_0, \ldots, X_9]$. $((\mathfrak{p}_X, X_2, X_3, X_4, X_5) = (\mathfrak{p}_W, X_2, X_3, X_4, X_5)$ where $\mathfrak{p}_W \subset K[X_0, X_1, X_6, X_7, X_8, X_9]$ may be considered as the ideal of the Segre embedding $W := S(C \times \mathbf{P}^1_K)$ of the product of the plane curve C given by $Z_0Z_1Z_2 - Z_0Z_2^2 + Z_1^3 = 0$ with $\mathbf{P}^1_K$ in $\mathbf{P}^5_K$. But $\mathfrak{p}_W$ is indeed a prime ideal.) Now $\overline{X_4(X_6 - X_8)} \in X_2 \cdot B$. Therefore $X_2 \cdot B$ must have an embedded component and B is not a Cohen-Macaulay ring.

Remark 5.3. By studying a primary decomposition of $X_2^2 \cdot B$ we get from Proposition I.2.12 that B is a Buchsbaum ring.

2. We now give examples of two arithmetically Buchsbaum varieties (non Cohen-Macaulay) such that their Segre product is again an arithmetically Buchsbaum variety.

Let $R := K[X_0, \ldots, X_3]/\mathfrak{p}$ (K a field), where

$$\mathfrak{p} := (X_0X_3 - X_1X_2, X_1^3 - X_0^2X_2, X_2^3 - X_1X_3^2, X_0X_2^2 - X_1^2X_3)$$

is the well-known ideal of Macaulay, and set $\mathfrak{m} := (X_0, X_1, X_2, X_3) \cdot R$. We have that R is a graded Buchsbaum ring, see the Introduction, Example 3 (3b). First we investigate the local cohomology modules of R. By using the exact sequence

$$0 \to R(-1) \xrightarrow{X_2} R \to R/X_2 \cdot R \to 0$$

we obtain an isomorphism

(i) $\underline{H}^0_{\mathfrak{m}}(R/X_2 \cdot R) \cong \underline{H}^1_{\mathfrak{m}}(R)(-1)$

and an exact sequence

(ii) $0 \to \underline{H}^1_{\mathfrak{m}}(R) \to \underline{H}^1_{\mathfrak{m}}(R/X_2 \cdot R) \to \underline{H}^2_{\mathfrak{m}}(R)(-1) \to \underline{H}^2_{\mathfrak{m}}(R) \to 0.$

Since $(\mathfrak{p}, X_2) = (X_0, X_1, X_2) \cap (X_1^3, X_2, X_3) \cap (X_0, X_1^2, X_2, X_3^2)$ we see that

$$\underline{H}^0_{\mathfrak{m}}(R/X_2 \cdot R) \cong \underline{K}(-2) \qquad \text{(generated by the coset of } X_1X_3)$$

and this gives $\underline{H}^1_{\mathfrak{m}}(R) \cong \underline{K}(-1)$.

Since $\underline{H}^1_{\mathfrak{m}}(R/X_2 \cdot R) \cong \underline{H}^1_{\mathfrak{m}}(R/U(X_2 \cdot R))$ and since $R/U(X_2 \cdot R)$ is isomorphic to a quotient of a polynomial ring by an ideal generated by monomials it is not difficult to calculate the Hilbert function and the Hilbert polynomial of this Cohen-Macaulay algebra. Hence we get by Lemma I.4.3 and Corollary I.4.4

$$[\underline{H}^1_{\mathfrak{m}}(R/X_2 \cdot R)]_n \cong [\underline{H}^1_{\mathfrak{m}}(R/U(X_2 \cdot R))]_n = 0 \qquad \text{for all } n \geq 2$$

and

$$[\underline{H}^1_{\mathfrak{m}}(R/X_2 \cdot R)]_1 \cong K.$$

Since $[\underline{H}^1_{\mathfrak{m}}(R)]_1 \cong K$ the exact sequence (ii) yields for each $n \geq 1$ isomorphisms

$$[\underline{H}^2_{\mathfrak{m}}(R)]_{n-1} \cong [\underline{H}^2_{\mathfrak{m}}(R)]_n.$$

Hence $[\underline{H}^2_{\mathfrak{m}}(R)]_q = 0$ for all $q \geq 0$, i.e. $e(\underline{H}^2_{\mathfrak{m}}(R)) = -1$.

Now let S be another graded K-algebra of dimension 2 with

$$\underline{H}^0_{\mathfrak{n}}(S) = 0, \quad [\underline{H}^1_{\mathfrak{n}}(S)]_p = 0 \quad \text{for all } p \neq 1 \quad \text{and} \quad e(\underline{H}^2_{\mathfrak{n}}(S)) < 0$$

($\mathfrak{n}$:= maximal ideal of S). Then S is a Buchsbaum algebra as well.

We define $T := \sigma(R, S)$ and get from our Künneth relations (Proposition 0.2.10):

$$\underline{H}^2_{\mathfrak{m}_T}(T) \cong \underline{H}^1(T) = \sigma(\underline{H}^1(R), \underline{H}^0(S)) \oplus \sigma(\underline{H}^0(R), \underline{H}^1(S)) = 0,$$

since $e(\underline{H}^1(R)) < 0$, $e(\underline{H}^1(S)) < 0$, $a(\underline{H}^0(R)) \geq 0$ and $a(\underline{H}^0(S)) \geq 0$.

Since $[R]_n \cong [\underline{H}^0(R)]_n$, $[S]_n \cong [\underline{H}^0(S)]_n$ for all $n \neq 1$ the same must be true for T, i.e. we have $[\underline{H}^1_{\mathfrak{m}_T}(T)]_n = 0$ for all $n \neq 1$. Therefore T is a Buchsbaum algebra by Corollary I.3.6. For instance, we can have $S = R$ or $S = K[X_0, X_1, X_2, X_3]/(X_0, X_1) \cap (X_2, X_3)$ since $e(\underline{H}^2_{\mathfrak{n}}(S)) < 0$ (see the following Example 3).

3. In Chapter I, § 2, we gave an example (Example 2.5) of a local non-Buchsbaum ring A, of depth 0, dimension 2, $\mathfrak{m} \cdot H^i_{\mathfrak{m}}(A) = 0$ for $i = 0, 1$. (See also Corollary I.2.4).

By using Segre products we will give now an example of a graded module M with $\mathfrak{m} \cdot \underline{H}^i_{\mathfrak{m}}(M) = 0$ for $i < \dim M$, depth $M > 0$ which is not a Buchsbaum module.

Example 5.4. Let $R := K[X_1, X_2, X_3, X_4]$, $S := K[Y_1, Y_2]$ ($X_1, \ldots, X_4, Y_1, Y_2$ indeterminates, K a field) and $\mathfrak{a} := (X_1, X_2) \cap (X_3, X_4) \subset R$. We set $M := \sigma(R/\mathfrak{a}(-2), S)$. It is a finitely generated $\sigma(R, S)$-module.

Let $\mathfrak{v} := \sigma(\mathfrak{m}_R, \mathfrak{m}_S)$ (the maximal ideal of $\sigma(R, S)$).

Since depth $R/\mathfrak{a}(-2)$ = depth $R/\mathfrak{a} = 1$, depth $S = 2$, Corollary 0.1.12 implies depth $M \geq 1$. Clearly, dim $M = 3$. We know (see Proposition I.2.25) that $R/\mathfrak{a}$ is a Buchsbaum module with $\underline{H}^1_{\mathfrak{m}_R}(R/\mathfrak{a}) \cong \underline{K}$ and the exact sequence in the proof of Lemma I.2.14 give rise to an isomorphism $\underline{H}^2_{\mathfrak{m}_R}(R/\mathfrak{a}) \cong \underline{H}^2_{\mathfrak{m}_R}(R/(X_1, X_2)) \oplus \underline{H}^2_{\mathfrak{m}_R}(R/(X_3, X_4))$. Therefore $e(\underline{H}^2_{\mathfrak{m}_R}(R/\mathfrak{a})) = -2$ and $[\underline{H}^2_{\mathfrak{m}_R}(R/\mathfrak{a})]_{-2} \cong K \oplus K$ $(= K^2)$.

Therefore by our Künneth relations (Proposition 0.2.10)

$$\begin{aligned} \underline{H}^2_{\mathfrak{v}}(M) &\cong \underline{H}^1(M) \cong \sigma(\underline{H}^1(R/\mathfrak{a})(-2), \underline{H}^0(S)) \oplus \sigma(\underline{H}^0(R/\mathfrak{a})(-2), \underline{H}^1(S)) \\ &\cong \underline{K}^2 \oplus \underline{0} = \underline{K}^2 \end{aligned}$$

and

$$\underline{H}^1_{\mathfrak{v}}(M) \cong \operatorname{coker}(M \to \underline{H}^0(M)) = \operatorname{coker}\big(\sigma(R/\mathfrak{a}(-2), S) \to \underline{H}^0(\sigma(R/\mathfrak{a}(-2), S))\big)$$
$$\cong \sigma\big(\underline{H}^1_{\mathfrak{m}_R}(R/\mathfrak{a})(-2), S\big) \cong \underline{K}^3(-2).$$

Hence depth $M = 1$ and $\mathfrak{v} \cdot \underline{H}^i_{\mathfrak{v}}(M) = O$ for all $i < 3 = \dim M$.

We claim that M is not a Buchsbaum module.

Let $x_i := X_i \bmod \mathfrak{a}$. Then $\mathfrak{v}$ is generated by the elements $x_i \otimes Y_j$ with $1 \leq i \leq 4$, $1 \leq j \leq 2$.

$a_1 := X_1 + X_4$, $a_2 := X_2 + X_3$ is a system of parameters of $R/\mathfrak{a}(-2)$ and Y_1, Y_2 is a system of parameters of S. Hence $s_1 := a_1 \otimes Y_1$, $s_2 := a_2 \otimes Y_2$ is a part of a system of parameters of M. (It is easy to see that we obtain a system of parameters of M if we add the element $a_1 \otimes Y_2 - a_2 \otimes Y_1$.) Let $m := 1 \otimes Y_1 Y_2 \in M$. We show that $m \in (s_1, s_2) \cdot M : \langle \mathfrak{v} \rangle$.

Since in $R/\mathfrak{a}(-2)$ we have $x_1^2 = x_1 a_1$, $x_2^2 = x_2 a_2$, $x_3^2 = x_3 a_2$, $x_4^2 = x_4 a_1$, we obtain for $j = 1, 2$:

$$(x_1^2 \otimes Y_j^2) \cdot m = x_1^2 \otimes Y_1 Y_2 Y_j^2 = x_1 a_1 \otimes Y_1 Y_2 Y_j^2 = s_1(x_1 \otimes Y_2 Y_j^2) \in s_1 \cdot M,$$
$$(x_2^2 \otimes Y_j^2) \cdot m = x_2^2 \otimes Y_1 Y_2 Y_j^2 = x_2 a_2 \otimes Y_1 Y_2 Y_j^2 = s_2(x_2 \otimes Y_1 Y_j^2) \in s_2 \cdot M$$

and so on.

Thus $m \in (s_1, s_2) \cdot M : (x_1^2 \otimes Y_1^2, \ldots, x_4^2 \otimes Y_2^2) \subseteq (s_1, s_2) \cdot M : \langle \mathfrak{v} \rangle$.

If M were a Buchsbaum module this would imply $(x_1 \otimes Y_1) \cdot m \in (s_1, s_2) \cdot M$ since $(s_1, s_2) \cdot M : \langle \mathfrak{v} \rangle = (s_1, s_2) \cdot M : \mathfrak{v}$, i.e. there are elements $m_1, m_2 \in M$ with $x_1 \otimes Y_1^2 Y_2 = s_1 m_1 + s_2 m_2$. Comparing degrees it is easy to see that without loss of generality $m_1 = 1 \otimes u_1$, $m_2 = 1 \otimes u_2$ with $u_1, u_2 \in [S]_2$. This yields

$$x_1 \otimes Y_1^2 Y_2 = a_1 \otimes u_1 Y_1 + a_2 \otimes u_2 Y_2.$$

But this contradicts the linear independence of the images of the elements X_1, a_1, a_2 in $[R/\mathfrak{a}(-2)]_3 = [R/\mathfrak{a}]_1$, i.e. M is not a Buchsbaum module.

Note that for this example the assumptions of Proposition I.3.10 are not fulfilled.

4. Now we will give an example of a graded Buchsbaum module not satisfying the assumptions of Proposition I.3.10. It shows that the criterion is not a necessary condition.

Example 5.5. Let $R := K[X_1, X_2, X_3]$, $S := K[Y_1, Y_2, Y_3, Y_4]$ ($X_1, X_2, X_3, Y_1, \ldots, Y_4$ indeterminates, K a field). Choose $0 \neq f \in R$, $\deg f = 5$, $0 \neq g \in S$, $\deg g = 2$. Let $M := R/f \cdot R(2)$, $N := S/g \cdot S$, $P := \sigma(M, N)$ (as a $\sigma(R, S)$-module) and $\mathfrak{m} :=$ maximal ideal of $\sigma(R, S)$.

Now $\dim M = 2$, $\dim N = 3$, $a(M) = -2$, $a(N) = 0$, $r(M) = 1$, $r(N) = -1$, i.e. the assumptions of Theorem I.4.6 are fulfilled and P is a Buchsbaum module.

By Proposition I.2.10 $\underline{H}^i_{\mathfrak{m}}(P) \neq O$ precisely for $i = 2, 3, 4$. Also $[\underline{H}^2_{\mathfrak{m}}(P)]_n = O$ for all $n \neq 0$ and $[\underline{H}^3_{\mathfrak{m}}(P)]_m = O$ for all $m \neq -2$. Therefore the assumptions of Proposition I.3.10 are not fulfilled since in this case $(2 + 0) - (3 - 2) = 1$.

5. We now give an example of two graded injective modules such that their Segre product is not an injective module (see Lemma 0.2.9).

Example 5.6. Let $R := K[X, Y]$ (X, Y indeterminates, K a field) and $I := Q =$ injective enveloppe of R and $J :=$ injective enveloppe of $\underline{K} = R/(X, Y)R$.

We know that $Q := \left\{\frac{p}{q} \,\middle|\, p, q \text{ homogeneous in } R,\ q \neq 0\right\}$ is the (homogeneous) field of quotients of R. By Corollary 0.4.10 and Theorem 0.4.14 we know that $J \cong \underline{H}^2_{\mathfrak{m}_R}(R)\,(-2)$.

We claim that $\underline{\mathrm{Ext}}^1_{\sigma(R,R)}(\underline{K}, \sigma(I, J)) \neq O$.

$\sigma(R, R)$ is generated (as a K-algebra) by the elements $z_1 := X \otimes X$, $z_2 := X \otimes Y$, $z_3 := Y \otimes X$, $z_4 := Y \otimes Y \in [\sigma(R, R)]_1$. Let $\mathfrak{m}$ denote the maximal ideal of $\sigma(R, R)$. We choose four elements $v_1, \ldots, v_4$ in $[Q]_0$ with $v_1v_4 - v_2v_3 \neq 0$.

Now $e(\underline{H}^2_{\mathfrak{m}_R}(R)) = -2$, i.e. $e(J) = 0$, $[J]_0 \cong K$ and $[J]_{-1} \cong K \oplus K$. Since $[\sigma(I, J)]_0 \cong [Q]_0 \otimes_K K = [Q]_0$, by $z_i \mapsto v_i \otimes 1$ a $\sigma(R,R)$-homomorphism $f\colon \mathfrak{m} \to \sigma(I, J)$ of degree -1 is given (notice that $[\sigma(I, J)]_p = O$ for all $p > 0$).

We suppose $\underline{\mathrm{Ext}}^1_{\sigma(R,R)}(\underline{K}, \sigma(I, J)) = O$ and we try to obtain a contradiction. We have an epimorphism

$$\sigma(I, J) \cong \underline{\mathrm{Hom}}_{\sigma(R,R)}(\sigma(R, R), \sigma(I, J)) \to \underline{\mathrm{Hom}}_{\sigma(R,R)}(\mathfrak{m}, \sigma(I, J)),$$

i.e. there is an $x \in [\sigma(I, J)]_{-1}$ with $f(z_i) = z_i x$ for $i = 1, \ldots, 4$. By studying the local cohomology module $\underline{H}^2_{\mathfrak{m}_R}(R)$ we see that $[J]_{-1} \cong K \oplus K$ possesses a basis e_1, e_2 with $Xe_1 = Ye_2 = 1$, $Xe_2 = Ye_1 = 0$ (in $K \cong [J]_0$). Using this we can write:

$$x = w_1 \otimes e_1 + w_2 \otimes e_2 \qquad \text{with } w_1, w_2 \in [Q]_{-1}.$$

Then $z_1x = Xw_1 \otimes 1$, $z_2x = Xw_2 \otimes 1$, $z_3x = Yw_1 \otimes 1$, $z_4x = Yw_2 \otimes 1$. Consequently $v_1 = Xw_1$, $v_2 = Xw_2$, $v_3 = Yw_1$, $v_4 = Yw_2$ and we have in Q: $v_1v_4 - v_2v_3 = 0$, a contradiction.

This example also shows that the homomorphisms τ^n of our generalized Künneth relations (Proposition 0.2.10(i)) are not generally isomorphisms. Namely, for $M_1 = I$, $M_2 = J$ on the left-hand side of this formula we always get zero for all $n > 0$. But the term on the right-hand side is not equal to zero (take for example $n = 1$, $N_1 = N_2 = \underline{K}$).

6. Finally, we give an example which shows that the sufficient criterion for Buchsbaum modules given by Theorem I.2.10 is in general not necessary.

Example 5.7. Let K be a field and $X_1, \ldots, X_{2n}$, $n \geq 3$ indeterminates and set $R := K[[X_1, \ldots, X_{2n}]]$, $\mathfrak{n} := (X_1, \ldots, X_{2n}) \cdot R$, $F := X_1^2 + \ldots + X_{2n}^2$. Let

$$\mathfrak{a} := (X_1, \ldots, X_n) \cdot R \cap (X_{n+1}, \ldots, X_{2n}) \cdot R.$$

By Lemma I.2.14, $R/\mathfrak{a}$ is a Buchsbaum-module over R with $\dim R/\mathfrak{a} = n$, depth $R/\mathfrak{a} = 1$. Therefore $M := R/\mathfrak{a} + F \cdot R$ is a Buchsbaum module over R with $\dim M = n - 1$, depth $M = 0$. Let $A := R/F \cdot R$, $\mathfrak{m} := \mathfrak{n} \cdot A$, $x_i := X_i \bmod F$, $i = 1, \ldots, 2n$. Then $x_1, \ldots, x_{2n}$ is a minimal basis of $\mathfrak{m}$. Since $F \cdot M = O$, M is also a Buchsbaum module over A (see, e.g. Corollary I.1.11 and Lemma I.1.6).

If the canonical maps

$$\varphi^i_M\colon \mathrm{Ext}^i_A(K, M) \to H^i_{\mathfrak{m}}(M)$$

were surjective for all $i = 0, \ldots, n - 2$, we then would obtain from the proof of Theorem I.2.10, that the homomorphisms

$$f_i\colon \mathrm{Ext}^i_A(K, O :_M \mathfrak{m}) \to \mathrm{Ext}^i_A(K, M) \qquad (\text{induced by } O :_M \mathfrak{m} \subset M)$$

are injective for all $i \leq n-1$. Let now

$$\ldots \to A^{n_2} \xrightarrow{g_2} A^{n_1} \xrightarrow{g_1} A \to K \to 0$$

be a minimal free resolution of $K = A/\mathfrak{m}$ over A. Obviously, we have $n_1 = 2n$ and $\ker g_1 = \operatorname{im} g_2$ is generated by the $\binom{2n}{n}$ syzygies $s_{ij} := (0, \ldots, 0, x_j, 0, \ldots, 0, -x_i, 0, \ldots, 0)$ for all i, j with $1 \leq i < j \leq 2n$ (x_j is in the ith place and $-x_i$ in the jth place) and the syzygy $s := (x_1, \ldots, x_{2n})$.

These syzygies form a minimal basis, i.e. we have $n_2 = \binom{2n}{n} + 1$. By applying the functors $\operatorname{Hom}_A(\ , O :_M \mathfrak{m})$, $\operatorname{Hom}_A(\ , M)$ to the above free resolution we obtain complexes (all squares are commutative):

$$\begin{array}{ccccccccc} O \to & O :_M \mathfrak{m} & \longrightarrow & (O :_M \mathfrak{m})^{n_1} & \longrightarrow & (O :_M \mathfrak{m})^{n_2} & \longrightarrow & \ldots \\ & \cap & & \cap & & \cap & & \\ 0 \to & M & \xrightarrow{g^0} & M^{n_1} & \xrightarrow{g^1} & M^{n_2} & \xrightarrow{g^2} & \ldots \end{array}$$

The boundary homomorphisms of the above complex are zero, i.e. $\operatorname{Ext}_A^i(K, O :_M \mathfrak{m}) \cong (O :_M \mathfrak{m})^{n_i}$. Let $Z := \ker g^2$, $B := \operatorname{im} g^1$. Then the following diagram with exact rows is commutative:

$$\begin{array}{ccccccc} O \to & & (O :_M \mathfrak{m})^{n_2} & \to & \operatorname{Ext}_A^2(K, O :_M \mathfrak{m}) & \to 0 \\ \downarrow & & \downarrow & & \downarrow f_2 & \\ 0 \to B \to & & Z & \to & \operatorname{Ext}_A^2(K, M) & \to O. \end{array}$$

Let $m := g^1(\bar{x}_1, \ldots, \bar{x}_n, 0, \ldots, 0) \in B$ ($\bar{x}_i$ is the coset of x_i in M). Set $\gamma := \bar{x}_1^2 + \ldots + \bar{x}_n^2 \neq 0$, then $m = (0, \ldots, 0, \gamma)$ (g^1 is obtained by scalare multiplication with the syzygies s_{ij} and s). Since $x_i \cdot \gamma = 0$ for all $i = 1, \ldots, 2n$, $\gamma \in O :_M \mathfrak{m}$, i.e. $m \in (O :_M \mathfrak{m})^{n_2} \cap B$. Therefore $f_2(m) = 0$, i.e. f_2 is not injective.

By using Lemma I.2.14 we see that $H^i_{\mathfrak{m}}(M) = O$ for all $i \neq 0, 1, n-1$. Since $n - 1 \geq 2$ and since f_0 is even an isomorphism, we obtain that φ^1_M is not surjective.

If we consider M as a ring, an easy computation shows that also in this case also the canonical homomorphism

$$\operatorname{Ext}^1_M(K, M) \to H^1_{\mathfrak{v}}(M) \quad (\mathfrak{v} \text{ maximal ideal of } M)$$

is not surjective.

Appendix
On generalizations of Buchsbaum modules

It is not difficult to see that there are various possibilities to generalize the concept of Buchsbaum ring and module. A first step in this direction is our Proposition I.2.1 where we have characterized those modules M for which $\mathfrak{m}H^i_\mathfrak{m}(M) = 0$ for all $i < \dim M$ (they are not Buchsbaum modules, in general, see, e.g., Example I.2.5 and Example V.5.4). From this point of view, we want to give in this Appendix an answer to the following two questions:

1. Which are the modules M (always Noetherian over the local ring A) such that $H^i_\mathfrak{m}(M)$, $i = 0, \ldots, \dim M - 1$, are Noetherian modules (and hence modules of finite length)?
   (See Proposition 13, Theorem 14, Lemma 15 and Proposition 16.)
2. With the hypothesis of 1.), what can we say about parameter ideals $\mathfrak{q}$ of M for which

$$l_A(M/\mathfrak{q}M) - e_0(\mathfrak{q}; M) = I(M),$$

where

$$I(M) := \sum_{i=0}^{d-1} \binom{d-1}{i} l_A(H^i_\mathfrak{m}(M)) \quad \text{with } d := \dim M > 0,$$

compare Proposition I.2.6?
(See Lemma 15, Theorem and Definition 14 and Theorem 20 as it relates to Definition 19.)

Let A always denote a local ring with maximal ideal $\mathfrak{m}$. Let M be a Noetherian A-module. If M is a Buchsbaum module, then for any part $x_1, \ldots, x_r$ of a system of parameters of M and any prime ideal $\mathfrak{p} \in \operatorname{Supp} M \setminus \{\mathfrak{m}\}$ containing $x_1, \ldots, x_r$ the images of $x_1, \ldots, x_r$ in $A_\mathfrak{p}$ form an $M_\mathfrak{p}$-sequence (see Chapter 0, § 2 for the definition and Corollary I.1.11). By virtue of this relation we generalize the notion of a weak M-sequence in the following manner. For this we recall first that for a submodule N of M:

$$N :_M \langle\mathfrak{m}\rangle = \{m \in M \mid \mathfrak{m}^n \cdot m \subseteq N \text{ for some } n \in \mathbf{N}\}.$$

Definition 1. Let $x_1, \ldots, x_r$ be a sequence of elements contained in $\mathfrak{m}$. It is called a *filter-regular* (f-regular) sequence with respect to M if for all $i = 1, \ldots, r$

$$(x_1, \ldots, x_{i-1}) \cdot M :_M x_i \subseteq (x_1, \ldots, x_{i-1}) \cdot M :_M \langle\mathfrak{m}\rangle.$$

It is immediately seen that a sequence of elements $x_1, \ldots, x_r$ is an f-regular sequence if and only if $\frac{x_1}{1}, \ldots, \frac{x_r}{1}$ in $A_\mathfrak{p}$ forms an $M_\mathfrak{p}$-sequence for all $\mathfrak{p} \in \operatorname{Supp} M \setminus \{\mathfrak{m}\}$ containing

$x_1, \ldots, x_r$. In particular an M-sequence or a weak M-sequence is an f-regular sequence. First we note the following easily provable facts:

Lemma 2.

(i) *A sequence of elements $x_1, \ldots, x_r$ is f-regular with respect to M if and only if it is f-regular with respect to $M/H^0_{\mathfrak{m}}(M)$.*

(ii) *If $x_1, \ldots, x_r$ is f-regular with respect to M, then for any $n \geq 1$ there is an element $y \in \mathfrak{m}^n$ such that $x_1, \ldots, x_r, y$ is f-regular with respect to M.*

(iii) *If $x_1, \ldots, x_r$ is f-regular with respect to M then*

$$\dim M/(x_1, \ldots, x_r) \cdot M = \sup\{\dim M - r, 0\}.$$

Proof: (i) is trivially true. For (ii) it suffices to show that $\mathfrak{m}$ contains an f-regular element y with respect to M. If $H^0_{\mathfrak{m}}(M) = M$, choose $y \in \mathfrak{m}^n$ arbitrarily. If $H^0_{\mathfrak{m}}(M) \neq M$ $H^0_{\mathfrak{m}}\big(M/H^0_{\mathfrak{m}}(M)\big) = 0$ implies depth $M/H^0_{\mathfrak{m}}(M) > 0$, hence there is an $M/H^0_{\mathfrak{m}}(M)$-regular element $y \in \mathfrak{m}^n$ and (ii) follows from (i). For the proof of (iii) we note that the case $r = 0$ is trivial. Thus let $r > 0$. If $\dim M = 0$, there is nothing to show. If $\dim M > 0$ then x_1 is $M/H^0_{\mathfrak{m}}(M)$-regular. Now $\operatorname{Ass} M/H^0_{\mathfrak{m}}(M) = \operatorname{Ass} M \setminus \{\mathfrak{m}\}$ shows that x_1 lies outside of all minimal primes of M. Let $M' := M/x_1 \cdot M$. Then $\dim M' = \dim M - 1$ and we get by an easy induction argument:

$$\begin{aligned}\dim M/(x_1, \ldots, x_r) \cdot M &= \dim M'/(x_2, \ldots, x_r) \cdot M = \sup\{\dim M' - (r-1), 0\} \\ &= \sup\{\dim M - r, 0\},\end{aligned}$$

q.e.d.

Let M be a Noetherian A-module of dimension $d \geq 1$. According to Lemma 2(iii) each f-regular sequence $x_1, \ldots, x_d$ with respect to M is a system of parameters with respect to M. The converse is not true, in general.

Proposition 3. *For a Noetherian A-module M of dimension $d \geq 1$ the following conditions are equivalent:*

(i) *Each system of parameters $x_1, \ldots, x_d$ of M is an f-regular sequence with respect to M.*

(ii) *Each part $x_1, \ldots, x_r$ of a system of parameters of M is unmixed up to an $\mathfrak{m}$-primary component, i.e. for all $\mathfrak{p} \in \operatorname{Ass}_A M/(x_1, \ldots, x_r) \cdot M \setminus \{\mathfrak{m}\}$ we have*

$$\dim A/\mathfrak{p} = d - r.$$

(iii) *For any $\mathfrak{p} \in \operatorname{Supp} M \setminus \{\mathfrak{m}\}$ we have*

$$\dim M = \operatorname{depth}_{A_\mathfrak{p}} M_\mathfrak{p} + \dim A/\mathfrak{p}.$$

(iv) *For all $\mathfrak{p} \in \operatorname{Supp} M \setminus \{\mathfrak{m}\}$*

$$\dim M = \dim_{A_\mathfrak{p}} M_\mathfrak{p} + \dim A/\mathfrak{p} \quad \text{and} \quad \dim_{A_\mathfrak{p}} M_\mathfrak{p} = \operatorname{depth}_{A_\mathfrak{p}} M_\mathfrak{p}.$$

(v) Supp M *is catenarian (i.e. $\operatorname{ht}_A \mathfrak{p} = \operatorname{ht}_A \mathfrak{q} + \operatorname{ht}_{A/\mathfrak{q}} \mathfrak{p}/\mathfrak{q}$ for all primes $\mathfrak{q} \subset \mathfrak{p}$ of* Supp M), *equidimensional (i.e. $\dim M = \dim A/\mathfrak{p}$ for all minimal primes $\mathfrak{p}$ of* Supp M) *and $M_\mathfrak{p}$ is a Cohen-Macaulay module for all $\mathfrak{p} \in \operatorname{Supp} M \setminus \{\mathfrak{m}\}$.*

Proof: (i) $\Rightarrow$ (ii): Suppose there exists a part of a system of parameters $x_1, \ldots, x_r$ such that there is a $\mathfrak{p} \in \operatorname{Ass}_A M/\mathfrak{x} \cdot M$ with $0 < \dim A/\mathfrak{p} < d - r$ (here $\mathfrak{x} := (x_1, \ldots, x_r) \cdot A$). Then we choose $y \in \mathfrak{p}$ such that $x_1, \ldots, x_r, y$ is again a part of a system of parameters.

By our assumption (i) we have

$$\mathfrak{x} \cdot M :_M y/\mathfrak{x} \cdot M \subseteq \mathfrak{x} \cdot M :_M \mathfrak{m}/\mathfrak{x} \cdot M$$

and

$$\operatorname{Ass} M/\mathfrak{x} \cdot M \cap \operatorname{Supp} A/y \cdot A \subseteq \operatorname{Ass} M/\mathfrak{x} \cdot M \cap \{\mathfrak{m}\} \subseteq \{\mathfrak{m}\}.$$

This is a contradiction since $\mathfrak{p} \in \operatorname{Ass} M/\mathfrak{x} \cdot M \cap \operatorname{Supp} A/y \cdot A$.

(ii) $\Rightarrow$ (iii): Let $\mathfrak{p} \in \operatorname{Supp} M \setminus \{\mathfrak{m}\}$. We choose a natural number r maximal with respect to the property that there are elements $x_1, \ldots, x_r$ in $\mathfrak{p}$ forming a part of a system of parameters of M. By the maximality of r we get $\mathfrak{p} \in \operatorname{Ass}_A M/(x_1, \ldots, x_r) \cdot M$.

By the unmixedness condition it follows that for all $i = 1, \ldots, r$

$$(x_1, \ldots, x_{i-1}) \cdot M_\mathfrak{p} :_{M_\mathfrak{p}} x_i = (x_1, \ldots, x_{i-1}) \cdot M_\mathfrak{p},$$

i.e. $r \leq \operatorname{depth}_{A_\mathfrak{p}} M_\mathfrak{p}$. But $\mathfrak{p}$ is minimal in $\operatorname{Supp} M/(x_1, \ldots, x_r) \cdot M$, i.e. $t \geq \dim_{A_\mathfrak{p}} M_\mathfrak{p}$. Therefore $r = \operatorname{depth}_{A_\mathfrak{p}} M_\mathfrak{p} = \dim_{A_\mathfrak{p}} M_\mathfrak{p}$. We also have $\dim A/\mathfrak{p} = \dim M - r$ and the conclusion is therefore proven.

It is now immediate that (iv) and (v) are equivalent formulations of (iii).

(iii) $\Rightarrow$ (ii): Suppose (ii) is not true. Then there is a part $x_1, \ldots, x_r$ of a system of parameters, r minimal, such that the assumption is not true. Take $\mathfrak{x} := (x_1, \ldots, x_r) \cdot A$ and $\mathfrak{p} \in \operatorname{Ass}_A M/\mathfrak{x} \cdot M$ with $0 < \dim A/\mathfrak{p} < \dim M - r$. Hence $r > \dim_{A_\mathfrak{p}} M_\mathfrak{p}$. By the minimality of r we get that $\frac{x_1}{1}, \ldots, \frac{x_r}{1}$ is an $M_\mathfrak{p}$-sequence of maximal length. Thus $\operatorname{depth}_{A_\mathfrak{p}} M_\mathfrak{p} = r$, which is a contradiction.

(ii) $\Rightarrow$ (i) is trivial and the proposition is thus proven, q.e.d.

Next we prove the following interesting fact.

Lemma 4. *Let M be a Noetherian A-module. If $x_1, \ldots, x_r$ is a part of a system of parameters with respect to M then there is an f-regular sequence $y_1, \ldots, y_r$ with respect to M such that*

$$(x_1, \ldots, x_r) \cdot A = (y_1, \ldots, y_r) \cdot A.$$

Proof: We use induction on r. The case $r = 0$ is trivial, therefore we assume $r \geq 1$ and $(x_1, \ldots, x_{r-1}) \cdot A = (y_1, \ldots, y_{r-1}) \cdot A$, where $y_1, \ldots, y_{r-1}$ denotes a suitable f-regular sequence.

Now we choose $y_r \in (x_1, \ldots, x_r) \cdot A \setminus \mathfrak{m} \cdot (x_1, \ldots, x_r) \cdot A$ such that $y_r \notin \mathfrak{p}$ for all $\mathfrak{p} \in \operatorname{Ass}_A M/(y_1, \ldots, y_{r-1}) \cdot M \setminus \{\mathfrak{m}\}$. From this we get the desired result, q.e.d.

Next we will show two technical results which will be useful to prove a relation between our f-regular sequences and reduced systems of parameters defined by M. Auslander and D. Buchsbaum in [1].

Proposition 5. *Let M be a Noetherian A-module and $x \in \mathfrak{m}$. Suppose the following conditions are satisfied:*

(i) *$x \notin \mathfrak{q}$ for all $\mathfrak{q} \in \operatorname{Ass}_A M \setminus \{\mathfrak{m}\}$.*

(ii) *Let $\mathfrak{p}$ be minimal in $\operatorname{Ass}_A A/(\mathfrak{q}' + x \cdot A)$ for some $\mathfrak{q}' \in \operatorname{Ass}_A M \setminus \{\mathfrak{m}\}$. Then we have $\mathfrak{p} \in \operatorname{Ass}_A M/x \cdot M$.*

Proof: Let $N := H^0_{\mathfrak{q}'}(M) = 0 :_M \langle \mathfrak{q}' \rangle \subseteq M$. Since $\operatorname{Supp} N/xN = \operatorname{Supp} N \cap V(xA) = V(\mathfrak{q}') \cap V(xA) = V(\mathfrak{q}' + xA)$ and $\mathfrak{p}$ is minimal in $V(\mathfrak{q}' + xA)$ it follows that $\mathfrak{p} \in \operatorname{Ass} N/xN$. Since $O :_{M/N} x = 0$, the embedding $N \subseteq M$ induces a monomorphism $N/(xN \to M/xM$. Therefore $\mathfrak{p} \in \operatorname{Ass} M/xM$, q.e.d.

Proposition 6. *Let M be a Noetherian A-module with* depth $M = 0$. *Suppose $x \in \mathfrak{m}$. Then there exists an integer n such that* depth $M/x^n M = 0$.

Proof: Let $N := H^0_{\mathfrak{m}}(M) \neq O$. The short exact sequence

$$O \to (N + x^n \cdot M)/x^n \cdot M \to M/x^n \cdot M$$

implies $\operatorname{depth}_A M/x^n \cdot M = 0$, since $x^n \cdot M \cap N = x^n(N :_M x^n) = O$ for sufficiently large n and

$$(N + x^n \cdot M)/x^n \cdot M \cong N/x^n \cdot M \cap N \cong N,$$

q.e.d.

Theorem 7. *For a Noetherian A-module M with $d :=$* dim $M \geq 1$ *the following conditions are equivalent:*

(i) *Each system of parameters of M is an f-regular sequence with respect to M.*

(ii) *Each system of parameters of M is reduced, i.e. for each system $x_1, \ldots, x_d$ of parameters we have $x_i \notin \mathfrak{p}$ for all $\mathfrak{p} \in \operatorname{Ass}_A M/(x_1, \ldots, x_{i-1}) \cdot M$ with* $\dim A/\mathfrak{p} \geq \dim M - i$ *for all $i = 1, \ldots, d$.*

(iii) *For each system of parameters $x_1, \ldots, x_d$ of M we have*

$$l_A\big(M/(x_1, \ldots, x_d) \cdot M\big) - e_0\big((x_1, \ldots, x_d) \cdot A, M\big)$$
$$= l_A\big((x_1, \ldots, x_{d-1}) \cdot M :_M x_d/(x_1, \ldots, x_{d-1}) \cdot M\big).$$

(iv) *Each part of a system of parameters of M consisting of $d - 2$ elements is unmixed up to $\mathfrak{m}$-primary components.*

Proof: (i) $\Rightarrow$ (ii) is clear by (ii) of Proposition 3.

(ii) $\Leftrightarrow$ (iii) is Corollary 4.8 of Auslander-Buchsbaum [1].

(ii) $\Rightarrow$ (iv) is trivial, thus it remains to prove (iv) $\Rightarrow$ (i).

Without loss of generality we may assume $d \geq 3$. According to Proposition 3 we have to show that a part $x_1, \ldots, x_r$ of a system of parameters of M is unmixed up to $\mathfrak{m}$-primary components. The statement is true for the cases $r = d - 2$, $d - 1$. Therefore let us assume $0 \leq r < d - 3$. Then $x_2, \ldots, x_r$ is a part of a system of parameters of the $(d-1)$-dimensional module M/x_1M. The condition (iv) remains valid for $M/x_1 \cdot M$ by passing to $M/x_1 \cdot M$, hence an easy induction argument yields that $(x_1, \ldots, x_r) \cdot M$ is unmixed up to an $\mathfrak{m}$-primary component. To complete the induction step we have to show the assertion for $r = 0$, i.e. for M itself. Assume there exists a $\mathfrak{p} \in \operatorname{Ass}_A M$ with $1 < \dim A/\mathfrak{p} < d$. Then we choose a parameter x for M such that $x \notin \mathfrak{q}$ for all $\mathfrak{q} \in \operatorname{Ass}_A M \setminus \{\mathfrak{m}\}$. Next we choose $\mathfrak{q}' \in \operatorname{Ass}_A A/(\mathfrak{p} + x \cdot A)$ with $\dim A/\mathfrak{q}' = \dim A/(\mathfrak{p} + x \cdot A) = \dim A/\mathfrak{p} - 1$. By Proposition 5 we have $\mathfrak{q}' \in \operatorname{Ass}_A M/x \cdot M$ with $0 < \dim A/\mathfrak{q}' < d - 1$ which contradicts the induction hypothesis for $M/x \cdot M$. Furthermore if there exists an $\mathfrak{p} \in \operatorname{Ass}_A M$ with $\dim A/\mathfrak{p} = 1$, we choose a parameter x for M in $\mathfrak{p}$. Then we have $\mathfrak{p} \cdot A_\mathfrak{p} \in \operatorname{Ass}_{A_\mathfrak{p}} M_\mathfrak{p}$ and $\operatorname{depth}_{A_\mathfrak{p}} M_\mathfrak{p} = 0$. For $n \gg 0$ it follows that $\operatorname{depth}_{A_\mathfrak{p}} M_\mathfrak{p}/x^n \cdot M_\mathfrak{p} = 0$ and therefore $\mathfrak{p} \in \operatorname{Ass}_A M/x^n \cdot M$, which contradicts the induction hypothesis for $M/x^n \cdot M$. Hence the inductive proof is complete, q.e.d.

Condition (iii) of the theorem is connected to our theory of Buchsbaum modules, more precisely, $l_A\big((x_1, \ldots, x_{d-1}) \cdot M :_M x_d/(x_1, \ldots, x_{d-1}) \cdot M\big)$ is independent of the choice of $x_1, \ldots, x_d$ if M is a Buchsbaum module.

Lemma 8. *Let M be a Noetherian A-module with $d := \dim M \geq 1$. Let $\hat{}$ denote the $\mathfrak{m}$-adic completion.*

(i) *If each system of parameters of $\hat{M}$ is an f-regular sequence of $\hat{M}$, then the same holds for M.*

(ii) *If A is a quotient of a Cohen-Macaulay ring the converse of* (i) *is true.*

Proof: Let $x_1, \ldots, x_r$ denote a part of a system of parameters of M. Then it is also a part of a system of parameters of $\hat{M}$. Since

$$\big((x_1, \ldots, x_{r-1}) \cdot \hat{M} :_{\hat{M}} x_r\big) \cap M = (x_1, \ldots, x_{r-1}) \cdot M :_M x_r$$

the result (i) follows immediately.

For the proof of (ii) let $\mathfrak{P} \in \operatorname{Spec} \hat{A} \setminus \{\hat{\mathfrak{m}}\}$ be a prime ideal and $\mathfrak{p} := \mathfrak{P} \cap A$. Since A is a quotient of a Cohen-Macaulay ring the fibre ring $k(\mathfrak{p}) \otimes \hat{A}_{\mathfrak{P}}$ of the canonical homomorphism

$$A_{\mathfrak{p}} \to \hat{A}_{\mathfrak{P}}$$

is a Cohen-Macaulay ring. (We may assume that A is a Cohen-Macaulay ring and apply Corollary (21.C) of Matsumura [1].) Since

$$\dim_{\hat{A}_{\mathfrak{P}}} \hat{M}_{\mathfrak{P}} = \dim_{A_{\mathfrak{p}}} M_{\mathfrak{p}} + \dim k(\mathfrak{p}) \otimes \hat{A}_{\mathfrak{P}},$$

resp.

$$\operatorname{depth}_{\hat{A}_{\mathfrak{P}}} \hat{M}_{\mathfrak{P}} = \operatorname{depth}_{A_{\mathfrak{p}}} M_{\mathfrak{p}} + \operatorname{depth} k(\mathfrak{p}) \otimes \hat{A}_{\mathfrak{P}},$$

it follows that

$$\dim_{\hat{A}_{\mathfrak{P}}} \hat{M}_{\mathfrak{P}} - \operatorname{depth}_{\hat{A}_{\mathfrak{P}}} \hat{M}_{\mathfrak{P}} = \dim_{A_{\mathfrak{p}}} M_{\mathfrak{p}} - \operatorname{depth}_{A_{\mathfrak{p}}} M_{\mathfrak{p}}.$$

By our assumptions we get that $\operatorname{Supp}_{\hat{A}} \hat{M} \setminus \{\hat{\mathfrak{m}}\}$ is a Cohen-Macaulay set. Since $\hat{A}$ is catenarian, by Proposition 3(v) it is enough to show that $\operatorname{Supp}_{\hat{A}} \hat{M}$ is equidimensional.

Suppose $\mathfrak{P} \in \operatorname{Supp}_{\hat{A}} \hat{M}$ is minimal. Then we get by

$$\operatorname{Ass}_{\hat{A}} \hat{M} = \bigcup_{\mathfrak{p} \in \operatorname{Ass}_A M} \operatorname{Ass}_{\hat{A}} \hat{A}/\mathfrak{p} \cdot \hat{A}$$

that $\mathfrak{P} \in \operatorname{Ass}_{\hat{A}} \hat{A}/\mathfrak{p} \cdot \hat{A}$ for $\mathfrak{p} = \mathfrak{P} \cap A$, which is minimal in $\operatorname{Supp} M$. Because A is a quotient of a Cohen-Macaulay ring, we have $\dim \hat{A}/\mathfrak{P} = \dim \hat{A}/\mathfrak{p} \cdot \hat{A}$ for all $\mathfrak{P} \in \operatorname{Ass}_{\hat{A}} \hat{A}/\mathfrak{p} \cdot \hat{A}$ by Nagata [1], Theorem (34.9). Since $\dim A/\mathfrak{p} = \dim M$ the equidimensionality follows, q.e.d.

Remark 9. Without additional assumptions on A the assertion (ii) in Lemma 8 does not remain valid. Ferrand and Raynaud [1] and Nagata [1] have constructed two-dimensional local integral domains R such that $\hat{R}$ admits a one-dimensional associated prime ideal. R is a local ring for which every system of parameters forms an f-regular sequence but $\hat{R}$ is not unmixed up to an $\mathfrak{m}$-primary component. Therefore it does not satisfy the equivalent conditions of Theorem 7.

A generalization of Buchsbaum modules is given by modules satisfying one of the four conditions of Theorem 7. We now give a cohomological characterization of these generalized Buchsbaum modules (see our Proposition 16). For this we first state another generalization of the notion of a weak M-sequence.

Definition 10. Let M be a Noetherian A-module with $d := \dim M > 0$ and $\mathfrak{q}$ an $\mathfrak{m}$-primary ideal. A system of elements $x_1, \ldots, x_r$ is called a $\mathfrak{q}$-*weak* M-*sequence* if

$$\mathfrak{q} \cdot \big((x_1, \ldots, x_{i-1}) \cdot M :_M x_i\big) \subseteq (x_1, \ldots, x_{i-1}) \cdot M$$

for all $i = 1, \ldots, r$.

Remark 11. Clearly, any $\mathfrak{q}$-weak M-sequence is an f-regular sequence. If there is an $\mathfrak{m}$-primary ideal $\mathfrak{q}$ such that every system of parameters is a $\mathfrak{q}$-weak M-sequence then every system of parameters is therefore an f-regular sequence with respect to M. But the converse is not true in general. (For an f-regular sequence of M there exists of course an $\mathfrak{m}$-primary ideal $\mathfrak{q}'$ such that this sequence is a $\mathfrak{q}'$-weak M-sequence.)

As a consequence of the following it will be true, that the converse holds if A is a "good" local ring.

Lemma 12. *Let M be a Noetherian A-module with $d := \dim M > 0$ and $\mathfrak{q}$ an $\mathfrak{m}$-primary ideal. If there exists a system of parameters $x_1, \ldots, x_d$ of M contained in $\mathfrak{m} \cdot \mathfrak{q}$ such that $x_1, \ldots, x_d$ is a $\mathfrak{q}$-weak M-sequence then*

$$\mathfrak{q} \cdot H^i_{\mathfrak{m}}(M) = 0 \quad \text{for all } i \neq d.$$

In particular, $H^i_{\mathfrak{m}}(M)$ is a finitely generated A-module for all $i \neq d$.

The proof of this lemma is obtained word by word from the proof of the implication (i) $\Rightarrow$ (iii) of Proposition I.2.1. Therefore we omit it. It also is a stronger version of one implication of the following proposition.

Proposition 13. *Let M be a Noetherian A-module with $d := \dim M > 0$ and $\mathfrak{q}$ an $\mathfrak{m}$-primary ideal. Then the following statements are equivalent:*

(i) *There is a system of parameters $x_1, \ldots, x_d$ of M contained in $\mathfrak{q}^2$ which is a $\mathfrak{q}$-weak M-sequence.*

(ii) *Each system of parameters of M contained in $\mathfrak{q}^2$ is a $\mathfrak{q}$-weak M-sequence.*

(iii) *$\mathfrak{q} \cdot H^i_{\mathfrak{m}}(M) = 0$ for all $i \neq \dim M$.*

The proof follows by using again exactly the same arguments as in the proof of Proposition I.2.1. Thus we again omit it.

Next we characterize those A-modules M for which all local cohomology modules $H^i_{\mathfrak{m}}(M)$, $i \neq \dim M$ are finitely generated, i.e. of finite length.

Theorem 14. *Let M denote a Noetherian A-module with $d := \dim M > 0$. Then the following conditions are equivalent:*

(i) *$l_A\big(H^i_{\mathfrak{m}}(M)\big) < \infty$ for all $i = 0, \ldots, d-1$.*

(ii) *There is an $\mathfrak{m}$-primary ideal $\mathfrak{q}$ such that every system of parameters of M is a $\mathfrak{q}$-weak M-sequence.*

(iii) *There is an* $\mathfrak{m}$*-primary ideal* $\mathfrak{r}$ *and a system of parameters* $x_1, \ldots, x_d$ *of* M *such that* $x_1^n, \ldots, x_d^n$ *is a* $\mathfrak{r}$*-weak* M*-sequence for all* $n \geq 1$.

(iv) *There is an integer* t *such that*

$$l_A(M/\mathfrak{x} \cdot M) - e_0(\mathfrak{x}, M) \leq t$$

for all parameter ideals $\mathfrak{x}$ *of* M.

(v) *There is an integer* s *and a system of parameters* $x_1, \ldots, x_d$ *of* M *such that for all* $n \geq 1$

$$l_A\big(M/(x_1^n, \ldots, x_d^n) \cdot M\big) - e_0\big((x_1^n, \ldots, x_d^n) \cdot A, M\big) \leq s.$$

Proof: (ii) $\Rightarrow$ (i) and (iii) $\Rightarrow$ (i) are consequences of Proposition 13 since in each case there is an $\mathfrak{m}$-primary ideal $\mathfrak{v}$ and a system of parameters in $\mathfrak{v}^2$ which is a $\mathfrak{v}$-weak M-sequence.

To show (i) $\Rightarrow$ (iv) we prove the following more general lemma:

Lemma 15. *Let* M *be a Noetherian* A*-module with* $d := \dim M > 0$. *For every parameter ideal* $\mathfrak{x}$ *of* M *we then have:*

(i) $l_A(M/\mathfrak{x} \cdot M) - e_0(\mathfrak{x}, M) \leq \sum_{j=0}^{d-1} \binom{d-1}{j} l_A\big(H^j_{\mathfrak{m}}(M)\big)$.

(ii) *If* $l_A\big(H^j_{\mathfrak{m}}(M)\big) < \infty$ *for all* $j = 0, \ldots, d-1$ *then there is an* $\mathfrak{m}$*-primary ideal* $\mathfrak{q}$ *such that equality holds in* (i) *for all parameter ideals* $\mathfrak{x} \subseteq \mathfrak{q}$.

Proof: If $l_A\big(H^j_{\mathfrak{m}}(M)\big) = \infty$ for some j, $0 \leq j < d$, then (i) is trivially true. Therefore assume $l_A\big(H^j_{\mathfrak{m}}(M)\big) < \infty$ for all $j = 0, \ldots, d-1$. Then there is an $\mathfrak{m}$-primary ideal $\mathfrak{v}$ such that $\mathfrak{v} \cdot H^j_{\mathfrak{m}}(M) = O$ for all $j < d$.

We show by induction on d that (i) holds with equality if $\mathfrak{x} \subseteq \mathfrak{v}^{2^{d-1}}$. If $d = 1$ then $\mathfrak{x} = x \cdot A$ with $x \in \mathfrak{m}$ and

$$0 :_M x \subseteq 0 :_M \langle \mathfrak{m} \rangle = 0 :_M \mathfrak{v} = H^0_{\mathfrak{m}}(M).$$

If $x \in \mathfrak{v}$, $0 :_M x = H^0_{\mathfrak{m}}(M)$. This gives (see Lemma 0.1.3(vi)!)

$$l_A(M/x \cdot M) - e_0(\mathfrak{x}, M) = l_A(O :_M x) \leq l_A\big(H^0_{\mathfrak{m}}(M)\big)$$

with equality if $x \in \mathfrak{v}$.

Let now $d \geq 2$ and $\mathfrak{x} = (x_1, \ldots, x_d) \cdot A$ be a parameter ideal of M. We have two exact sequences

$$0 \to 0 :_M x_1 \to M \xrightarrow{p} M/0 :_M x_1 \to 0$$

and

$$0 \to M/0 :_M x_1 \xrightarrow{i} M \to M/x_1 \cdot M \to 0.$$

By Lemma I.2.2 $0 :_M x_1 \subseteq H^0_{\mathfrak{m}}(M)$ with equality if $x_1 \in \mathfrak{v}$. Therefore $H^j_{\mathfrak{m}}(O :_M x_1) = 0$ for all $j > 0$ and $H^j_{\mathfrak{m}}(p)$ is an isomorphism. Also $H^j_{\mathfrak{m}}(i) \cdot H^j_{\mathfrak{m}}(p) = H^j_{\mathfrak{m}}(i \cdot p) = H^j_{\mathfrak{m}}(x_1) = x_1$. From the second exact sequence we obtain for all $j \geq 1$ (by using the isomorphisms $H^j_{\mathfrak{m}}(p)$)

(E_j): $H^{j-1}_{\mathfrak{m}}(M) \to H^{j-1}_{\mathfrak{m}}(M/x_1 \cdot M) \to H^j_{\mathfrak{m}}(M)$.

Thus for $j < d$ we have $\mathfrak{v}^2 \cdot H^{j-1}_{\mathfrak{m}}(M/x_1 \cdot M) = 0$ and

$$l_A\big(H^{j-1}_{\mathfrak{m}}(M/x_1 \cdot M)\big) \leq l_A\big(H^{j-1}_{\mathfrak{m}}(M)\big) + l_A\big(H^j_{\mathfrak{m}}(M)\big).$$

If $\mathfrak{x} \subseteq \mathfrak{v}^{2^{d-1}}$, $x_1 \cdot H^j_{\mathfrak{m}}(M) = 0$ for all $j < d$ and therefore the exact sequences ($\mathbf{E}_j$) give rise to short exact sequences

$$0 \to H^{j-1}_{\mathfrak{m}}(M) \to H^{j-1}_{\mathfrak{m}}(M/x_1 \cdot M) \to M^j_{\mathfrak{m}}(M) \to 0.$$

Hence in this case $l_A(H^{j-1}_{\mathfrak{m}}(M/x_1 \cdot M)) = l_A(H^{j-1}_{\mathfrak{m}}(M)) + l_A(H^j_{\mathfrak{m}}(M))$. Now by the induction hypothesis (applied to $M' := M/x_1 \cdot M$) we get for $\mathfrak{x}' := (x_2, \ldots, x_d) \cdot A$, by using Lemma 0.1.3(vi) (note $e_0(\mathfrak{x}', 0 :_M x_1) = 0$ by Lemma 0.1.4):

$$\begin{aligned} l_A(M/\mathfrak{x} \cdot M) - e_0(\mathfrak{x}, M) &= l_A(M'/\mathfrak{x}' \cdot M') - e_0(\mathfrak{x}', M') \\ &\leq \sum_{j=0}^{d-2} \binom{d-2}{j} l_A(H^j_{\mathfrak{m}}(M')) \quad \text{(with equality if } \mathfrak{x}' \subseteq \mathfrak{v}^{2^{d-1}}) \\ &\leq \sum_{j=0}^{d-2} \binom{d-2}{j} \left(l_A(H^j_{\mathfrak{m}}(M)) + l_A(H^{j+1}_{\mathfrak{m}}(M))\right) \\ &\quad \text{(with equality if } \mathfrak{x} \subseteq \mathfrak{v}^{2^{d-1}}) \\ &= \sum_{j=0}^{d-1} \binom{d-1}{j} l_A(H^j_{\mathfrak{m}}(M)), \end{aligned}$$

q.e.d.

We continue the proof of Theorem 14. (iv) $\Rightarrow$ (v) is trivial.

(v) $\Rightarrow$ (iii): Let $n_1, \ldots, n_d > 0$ denote integers and set

$$D(n_1, \ldots, n_d) := l_A\left(M/(x_1^{n_1}, \ldots, x_d^{n_d}) \cdot M\right) - e_0\left((x_1^{n_1}, \ldots, x_d^{n_d}) \cdot A, M\right).$$

First we prove that for integers $m_1, \ldots, m_d$ with $m_i \geq n_i$ for $i = 1, \ldots, d$:

$$D(n_1, \ldots, n_d) \leq D(m_1, \ldots, m_d).$$

Since $D(n_1, \ldots, n_d)$ does not depend on the order of the elements in the system of parameters it is sufficient to prove $D(1, \ldots, 1, n) \leq D(1, \ldots, 1, n+1)$ for all $n \geq 1$. For $i = 1, \ldots, d-1$ we write $M_i := (x_1, \ldots, x_{i-1}) \cdot M : x_1/(x_1, \ldots, x_{i-1}) \cdot M$. Then by Lemma 0.1.3(i), (v) and (vi) we get

$$\begin{aligned} &D(1, \ldots, 1, n+1) - D(1, \ldots, 1, n) \\ &= \sum_{i=1}^{d-1} e_0\left((x_{i+1}, \ldots, x_{d-1}, x_d^{n+1}) \cdot A, M_i\right) - \sum_{i=1}^{d-1} e_0\left((x_{i+1}, \ldots, x_{d-1}, x_d^n) \cdot A, M_i\right) \\ &\quad + l_A\left((x_1, \ldots, x_{d-1}) \cdot M : x_d^{n+1}/(x_1, \ldots, x_{d-1}) \cdot M\right) \\ &\quad - l_A\left((x_1, \ldots, x_{d-1}) \cdot M : x_d^n/(x_1, \ldots, x_{d-1}) \cdot M\right) \\ &= \sum_{i=1}^{d-1} e_0\left((x_{i+1}, \ldots, x_{d-1}, x_d) \cdot A, M\right) \\ &\quad + l_A\left((x_1, \ldots, x_{d-1}) \cdot M : x_d^{n+1}/(x_1, \ldots, x_{d-1}) \cdot M : x_d^n\right) \geq 0. \end{aligned}$$

Let now $n > 0$ be an integer and let $1 \leq i \leq d$. For all $m \geq n$ we get by Lemma 0.1.3(vi):

$$\begin{aligned} &l_A\left((x_1^n, \ldots, x_{i-1}^n, x_{i+1}^m, \ldots, x_d^m) \cdot M : x_i^n/(x_1^n, \ldots, x_{i-1}^n, x_{i+1}^m, \ldots, x_d^m) \cdot M\right) \\ &\leq D(n, \ldots, n, m, \ldots, m) \leq D(m, \ldots, m) \leq s \end{aligned}$$

by our assumption. Therefore

$$\mathfrak{m}^s\big((x_1^n, \ldots, x_{i-1}^n, x_{i+1}^m, \ldots, x_d^m) \cdot M : x_i^n/(x_1^n, \ldots, x_{i-1}^n, x_{i+1}^m, \ldots, x_d^m) \cdot M\big) = 0$$

or equivalently,

$$(x_1^n, \ldots, x_{i-1}^n, x_{i+1}^m, \ldots, x_d^m) \cdot M : x_i^n \subseteq (x_1^n, \ldots, x_{i-1}^n, x_{i+1}^m, \ldots, x_d^m) \cdot M : \mathfrak{m}^s.$$

By applying Krull's Intersection Theorem (the intersection is over all $m \geq n$) we obtain

$$(x_1^n, \ldots, x_{i-1}^n) \cdot M : x_i^n \subseteq (x_1^n, \ldots, x_{i-1}^n) \cdot M : \mathfrak{m}^s,$$

i.e. $x_1^n, \ldots, x_d^n$ is an $\mathfrak{m}^s$-weak M-sequence for all $n \geq 1$ and (ii) is proven.

(iv) $\Rightarrow$ (ii) follows from the validity of (v) $\Rightarrow$ (iii), q.e.d.

Finally we will now give an answer to the problem mentioned in Remark 11.

Proposition 16. *Let A be an epimorphic image of a local Cohen-Macaulay ring and let M be a Noetherian A-module with $d := \dim M \geq 1$. Then the following conditions are equivalent:*

(i) *$l_A(H_\mathfrak{m}^i(M)) < \infty$ for all $i \neq d$.*

(ii) *Every system of parameters of M is an f-regular sequence with respect to M.*

(iii) *There is an $\mathfrak{m}$-primary ideal $\mathfrak{q}$ such that every system of parameters of M is a $\mathfrak{q}$-weak M-sequence.*

(iv) *$M_\mathfrak{p}$ is a Cohen-Macaulay module (over $A_\mathfrak{p}$) for all $\mathfrak{p} \in \operatorname{Supp} M \setminus \{\mathfrak{m}\}$ and $\dim M = \dim A/\mathfrak{p}$ for all minimal primes $\mathfrak{p}$ in $\operatorname{Supp} M$.*

Proof: The equivalence of (i) and (iii) is clear by Theorem 14. The equivalence of (ii) and (iv) follows from Proposition 3(i) $\Leftrightarrow$ (v), since Spec A and hence Supp M is catenarian by the assumption on A. Remark 11 yields the implication (iii) $\Rightarrow$ (ii). Finally, we prove the implication (ii) $\Rightarrow$ (i). Let B denote a local Gorenstein ring with maximal ideal $\mathfrak{e}$ such that the $\mathfrak{m}$-adic completion $\hat{A}$ of A is a quotient of B. Since A is an epimorphic image of a Cohen-Macaulay ring, (ii) holds if we replace M by $\hat{M} = M \otimes_A \hat{A}$ and A by $\hat{A}$ and then by B (consider $\hat{M}$ as a B-module) by Lemma 8. Also (see, e.g. Sharp [2], Theorem 4.3) we have for all i that $H_\mathfrak{e}^i(M) \cong H_{\hat{\mathfrak{m}}}^i(\hat{M})$ (as B-modules) and $H_{\hat{\mathfrak{m}}}^i(\hat{M}) \cong H_\mathfrak{m}^i(M) \otimes_A \hat{A}$ and this results in

$$l_B(H_\mathfrak{e}^i(\hat{M})) = l_B(H_{\hat{\mathfrak{m}}}^i(\hat{M})) = l_{\hat{A}}(H_{\hat{\mathfrak{m}}}^i(\hat{M})) = l_A(H_\mathfrak{m}^i(M)).$$

Therefore we can assume without loss of generality that A is a local Gorenstein ring. Let $n := \dim A$. Let $\mathfrak{p} \in \operatorname{Supp} M$ with $\dim A/\mathfrak{p} > 0$. Assume there is an integer j with $n - d \leq j \leq n$ and $\mathfrak{p} \in \operatorname{Supp} \operatorname{Ext}_A^j(M, A)$, $\mathfrak{p} \notin \operatorname{Supp} \operatorname{Ext}_A^i(M, A)$ for all $i > j$. (Note that $\operatorname{Ext}_A^i(M, A) = 0$ for $i < n - d$ and $i > n$.) Then by local duality (see Corollary 0.3.5):

$$\begin{aligned} H_{\mathfrak{p}A_\mathfrak{p}}^{n-\dim(A/\mathfrak{p})-i}(M_\mathfrak{p}) &\cong \operatorname{Hom}_{A_\mathfrak{p}}(\operatorname{Ext}_{A_\mathfrak{p}}^i(M_\mathfrak{p}, A_\mathfrak{p}), I) \cong \operatorname{Hom}_{A_\mathfrak{p}}(\operatorname{Ext}_A^i(M, A)_\mathfrak{p}, I) \\ &\begin{cases} = 0 \text{ for } i > j, \\ \neq 0 \text{ for } i = j, \end{cases} \end{aligned}$$

where I denotes the injective hull of the residue field $A_\mathfrak{p}/\mathfrak{p} A_\mathfrak{p}$ of $A_\mathfrak{p}$. Therefore

$$n - \dim A/\mathfrak{p} - j = \operatorname{depth} M_\mathfrak{p} = \dim M_\mathfrak{p} = d - \dim A/\mathfrak{p},$$

since $M_{\mathfrak{p}}$ is a Cohen-Macaulay module by the equivalence of (ii) and (iv). Hence $j = n - d$, i.e. Supp $\mathrm{Ext}^i_A(M, A) \subseteq \{\mathfrak{m}\}$ for all $i > n - d$, and this means that $\mathrm{Ext}^i_A(M, A)$ is of finite length for all $i > n - d$. Therefore we obtain again by local duality, $H^i_{\mathfrak{m}}(M) \simeq \mathrm{Hom}_A\big(\mathrm{Ext}^{n-i}_A(M, A), E\big)$ (E denotes the injective hull of $A/\mathfrak{m}$) is of finite length for all $i < d$, q.e.d.

Reexamining our Lemma 15 we are lead to investigate yet another type of systems of parameters in a Noetherian A-module M. These observations go back originally to M. Brodmann [2] and N. V. Trung [6] and [10]. Following Trung [10], we prove

Theorem and Definition 17. *Let M denote a Noetherian A-module of dimension $d > 0$. Let $x_1, \ldots, x_d$ be a system of parameters of M and let $\mathfrak{q} := (x_1, \ldots, x_d) A$. The following conditions are equivalent:*

(i) $l_A\big(M/(x_1^2, \ldots, x_d^2) M\big) - e\,(x_1^2, \ldots, x_d^2 \,|M) = l_A\big(M/(x_1, \ldots, x_d) M\big) - e(x_1, \ldots, x_d|M)$.

(ii) $l_A\big(M/(x_1^{n_1}, \ldots, x_d^{n_d}) M\big) - e(x_1^{n_1}, \ldots, x_d^{n_d}|M)$
$= l_A\big(M/(x_1, \ldots, x_d) M\big) - e(x_1, \ldots, x_d|M)$ *for all* $n_1, \ldots, n_d \geq 1$.

(iii) $\big(l_A(H^i_{\mathfrak{m}}(M)\big) < \infty$ *for all* $i < d$ *and*$\big)$
$$l_A\big(M/(x_1, \ldots, x_d) M\big) - e(x_1, \ldots, x_d|M) = \sum_{i=0}^{d-1} \binom{d-1}{i} l_A\big(H^i_{\mathfrak{m}}(M)\big).$$

(iv) $x_1^{n_1}, \ldots, x_d^{n_d}$ *is a* $\mathfrak{q}$-*weak* M-*sequence for all* $n_1, \ldots, n_d \in \{1, 2\}$.

(v) $x_1^{n_1}, \ldots, x_d^{n_d}$ *is a* $\mathfrak{q}$-*weak* M-*sequence for all* $n_1, \ldots, n_d \geq 1$.

(vi) $\mathfrak{q}H^i_{\mathfrak{m}}\big(M/(x_1, \ldots, x_j) M\big) = 0$ *for all* $i, j \geq 0$ *with* $i + j < d$.

A system of parameters $x_1, \ldots, x_d$ of M fulfilling these equivalent conditions is called a *standard system of parameters* of M.

Proof: As in the proof of Theorem 14((v) $\Rightarrow$ (iii)) we let

$$D(n_1, \ldots, n_d) := l\big(M/(x_1^{n_1}, \ldots, x_d^{n_d}) M\big) - e(x_1^{n_1}, \ldots, x_d^{n_d}|M).$$

Clearly this number does not depend on the order of the elements $x_1, \ldots, x_d$ and for $1 \leq m_1 \leq n_1, \ldots, 1 \leq m_d \leq n_d$ we have by the proof of Theorem 14: $D(m_1, \ldots, m_d) \leq D(n_1, \ldots, n_d)$.

Hence Theorem 14 and Lemma 15 prove the equivalence of (ii) and (iii). The implication (ii) $\Rightarrow$ (i) is trivial. Now we prove (i) $\Rightarrow$ (ii) by induction on $N := \sum n_i \geq d$. If $N = d$ there is nothing to show. Let $N > d$ and without loss of generality assume $n_d = \max\{n_1, \ldots, n_d\}$. If $n_d = 2$, we then have by (i):

$$D(1, \ldots, 1) \leq D(n_1, \ldots, n_d) \leq D(2, \ldots, 2) = D(1, \ldots, 1)$$

which finishes the proof in this case. If $n_d \geq 3$, we have by the induction hypothesis

$$D(n_1, \ldots, n_{d-1}, n_d - 1) = D(1, \ldots, 1) = D(n_1, \ldots, n_{d-1}, 1).$$

Since $n_d - 1 \geq 2$, we find by using the proof of the implication (v) $\Rightarrow$ (iii) of Theorem 14 (for simplicity we set here $y_i := x_i^{n_i}$ for $i = 1, \ldots, d-1$, $n := n_d$):

$$0 = D(n_1, \ldots, n_{d-1}, n_d - 1) - D(n_1, \ldots, n_{d-1}, n_d - 2)$$
$$= \sum_{i=1}^{d-1} e\big(y_{i+1}, \ldots, y_{d-1}, x_d | (y_1, \ldots, y_{i-1}) M : y_i / (y_1, \ldots, y_{i-1}) M\big)$$
$$+ l\big((y_1, \ldots, y_{d-1}) M : x_d^{n-1} / (y_1, \ldots, y_{d-1}) M : x_d^{n-2}\big),$$

hence

$$e\big(y_{i+1}, \ldots, y_{d-1}, x_d | (y_1, \ldots, y_{i-1}) M : y_i / (y_1, \ldots, y_{i-1}) M\big) = 0$$

for $i = 1, \ldots, d-1$ and

$$(y_1, \ldots, y_{d-1}) M : x_d^{n-1} = (y_1, \ldots, y_{d-1}) M : x_d^{n-2}.$$

Therefore $(y_1, \ldots, y_{d-1}) M : x_d^n = (y_1, \ldots, y_{d-1}) M : x_d^{n-1}$ and this implies

$$D(n_1, \ldots, n_d) - D(1, \ldots, 1) = D(n_1, \ldots, n_d) - D(n_1, \ldots, n_{d-1}, n_d - 1) = 0.$$

Now the implication (v) $\Rightarrow$ (iv) is trivial and (iv) $\Rightarrow$ (vi) follows by Proposition 13 ((i) $\Rightarrow$ (iii)). Next we prove (vi) $\Rightarrow$ (iii). Clearly, $l\big(H^i_{\mathfrak{m}}(M)\big) < \infty$ for $i < d$ (let $j = 0$). We prove by induction on m, $1 \leq m \leq d$:

$$l(M/\mathfrak{q}M) - e_0(\mathfrak{q}; M) = \sum_{i=0}^{m-1} \binom{m-1}{i} l\big(H^i_{\mathfrak{m}}(M/(x_1, \ldots, x_{d-m}) M)\big)$$

(note: $e_0(\mathfrak{q}; M) = e(x_1, \ldots, x_d | M)$, c.f. Lemma 0.1.3). First by Lemma 0.2.2 and Theorem 7 we find

$$D(1, \ldots, 1) = l(M/\mathfrak{q}M) - e_0(\mathfrak{q}; M) = l(\mathfrak{q}'M : x_d / \mathfrak{q}'M),$$

where $\mathfrak{q}' := (x_1, \ldots, x_{d-1}) A$. Since $x_d H^0_{\mathfrak{m}}(M/\mathfrak{q}'M) = x_d(\mathfrak{q}'M : \langle\mathfrak{m}\rangle / \mathfrak{q}'M) = 0$, we have $\mathfrak{q}'M : \langle\mathfrak{m}\rangle = \mathfrak{q}'M : x_d$ and therefore $D(1, \ldots, 1) = l\big(H^0_{\mathfrak{m}}(M/\mathfrak{q}'M)\big)$ and we are done for $m = 1$. Let $1 < m \leq d$. Write $\mathfrak{q}_i := (x_1, \ldots, x_i) A$ for $0 \leq i < d$. Since for $1 \leq i \leq d$ we have $\dim_A \mathfrak{q}_{i-1}M : x_i / \mathfrak{q}_{i-1}M = 0$ the exact sequence

$$0 \to \mathfrak{q}_{d-m}M : x_{d-m+1} / \mathfrak{q}_{d-m}M \to M/\mathfrak{q}_{d-m}M \to M/\mathfrak{q}_{d-m}M : x_{d-m+1} \to 0$$

induces for all $i \geq 1$ isomorphisms

$$H^i_{\mathfrak{m}}(M/\mathfrak{q}_{d-m}M) \cong H^i_{\mathfrak{m}}(M/\mathfrak{q}_{d-m}M : x_{d-m+1}).$$

If we apply the functors $H^i_{\mathfrak{m}}(\ )$ to the exact sequence

$$0 \to M/\mathfrak{q}_{d-m}M : x_{d-m+1} \xrightarrow{g} M/\mathfrak{q}_{d-m}M \to M/\mathfrak{q}_{d-m+1}M \to 0,$$

where g is obtained by multiplication by x_{d-m+1}, the composition of the above isomorphisms with the maps induced by g in the resulting long exact cohomology sequence is just multiplication with x_{d-m+1} (on $H^i_{\mathfrak{m}}(M/\mathfrak{q}_{d-m}M)$) which is zero for $i \leq m-1$ by our assumption. Since $x_{d-m+1} H^0_{\mathfrak{m}}(M/\mathfrak{q}_{d-m}M) = 0$, we get $\mathfrak{q}_{d-m}M : x_{d-m+1} = \mathfrak{q}_{d-m}M : \langle\mathfrak{m}\rangle$, hence $H^0_{\mathfrak{m}}(M/\mathfrak{q}_{d-m}M : x_{d-m+1}) = 0$. Therefore the long exact cohomology sequence splits for all $i < m-1$ into exact sequences

$$0 \to H^i_{\mathfrak{m}}(M/\mathfrak{q}_{d-m}M) \to H^i_{\mathfrak{m}}(M/\mathfrak{q}_{d-m+1}M) \to H^{i+1}_{\mathfrak{m}}(M/\mathfrak{q}_{d-m}M) \to 0,$$

where we have replaced $H^{i+1}_{\mathfrak{m}}(M/\mathfrak{q}_{d-m}M:x_{d-m+1})$ by $H^{i+1}_{\mathfrak{m}}(M/\mathfrak{q}_{d-m}M)$ using the above isomorphism. This yields with the induction hypothesis

$$\begin{aligned} D(1, \ldots, 1) &= \sum_{i=0}^{m-2} \binom{m-2}{i} l\big(H^i_{\mathfrak{m}}(M/\mathfrak{q}_{d-m+1}M)\big) \\ &= \sum_{i=0}^{m-2} \binom{m-2}{i} \big[l\big(H^i_{\mathfrak{m}}(M/\mathfrak{q}_{d-m}M)\big) + l\big(H^{i+1}_{\mathfrak{m}}(M/\mathfrak{q}_{d-m}M)\big)\big] \\ &= \sum_{i=0}^{m-1} \binom{m-1}{i} l\big(H^i_{\mathfrak{m}}(M/\mathfrak{q}_{d-m}M)\big). \end{aligned}$$

(iii) follows for $m = d$.

Next we prove (ii) $\Rightarrow$ (vi). If (ii) holds then (iii) is also true.

We use induction on d. If $d = 1$, consider the embedding

$$0:_M x_1 \subseteq 0:_M \langle \mathfrak{m} \rangle = H^0_{\mathfrak{m}}(M).$$

Since by (iii) $l(0:x_1) = l(M/x_1M) - e(x_1|M) = l\big(H^0_{\mathfrak{m}}(M)\big)$, $O:x_1 = H^0_{\mathfrak{m}}(M)$ and this implies $x_1H^0_{\mathfrak{m}}(M) = O$.

Let $d > 1$. Since every system of parameters is f-regular (Theorem 7), we obtain with $\overline{M} := M/x_1^nM$, $n \geq 1$, $\mathfrak{q}' := (x_2, \ldots, x_d)\, A$:

$$\begin{aligned} &l\big(\overline{M}/(x_2^{n_2}, \ldots, x_d^{n_d})\, \overline{M}\big) - e(x_2^{n_2}, \ldots, x_d^{n_d}|\overline{M}) \\ &= D(n, n_2, \ldots, n_d) = D(n, 1, \ldots, 1) = l(\overline{M}/\mathfrak{q}'\overline{M}) - e_0(\mathfrak{q}'; \overline{M}). \end{aligned}$$

Hence $0 = \mathfrak{q}'H^i_{\mathfrak{m}}\big(\overline{M}/(x_2, \ldots, x_j)\, \overline{M}\big) \cong \mathfrak{q}'H^i_{\mathfrak{m}}\big(M/x_1^n, x_2, \ldots, x_j)\, M\big)$ for all $i \geq 0$, $j \geq 1$ with $i + j < d$ and it remains to consider the case $j = 0$ (set $n = 1$). The proof of the previous implication (vi) $\Rightarrow$ (iii) shows that there is an epimorphism for all $i \geq 0$ and all $n \geq 1$:

$$H^i_{\mathfrak{m}}(M/x_1^nM) \to 0:_{H^{i+1}_{\mathfrak{m}}(M)} x_1^n.$$

Since every element of $H^{i+1}_{\mathfrak{m}}(M)$ is annihilated by some power of x_1, we get $\mathfrak{q}'H^{i+1}_{\mathfrak{m}}(M) = O$ for all $0 \leq i < d - 1$. Since (ii) does not depend on the order of the elements $x_1, \ldots, x_d$, we find that $\mathfrak{q}H^i_{\mathfrak{m}}(M) = O$ for all $0 < i < d$. On the other hand $H^0_{\mathfrak{m}}(M) = O:x_1^n$ for sufficiently large n and the exact sequence

$$0 \to M/0:x_1^n \to M \to M/x_1^nM \to 0,$$

where the first map is induced by multiplication by x_1^n, gives rise to a monomorphism (since $H^0_{\mathfrak{m}}(M/O:x_1^n) = O$):

$$H^0_{\mathfrak{m}}(M) \to H^0_{\mathfrak{m}}(M/x_1^nM).$$

This shows that $\mathfrak{q}'H^0_{\mathfrak{m}}(M) = O$, hence $\mathfrak{q}H^0_{\mathfrak{m}}(M) = O$ and this concludes the proof of the statement.

Finally we prove (vi) $\Rightarrow$ (v). Let $n_1, \ldots, n_d \geq 1$. Since (vi) is equivalent to (ii) and (ii) does not depend on the order of the elements $x_1, \ldots, x_d$, we can assume without loss of generality that $1 \leq n_1 \leq \ldots \leq n_d$. If $n_1 \geq 2$, (v) follows from Proposition 13. If there is a j, $1 \leq j \leq d$ with $n_1 = \ldots = n_j = 1$, then we are finished if $j = d$.

If $j < d$, $x_{j+1}^{n_{j+1}}, \ldots, x_d^{n_d}$ is a system of parameters of $M/(x_1, \ldots, x_j)\, M$ contained in $(x_{j+1}, \ldots, x_d)^2 A$, i.e. $x_1, \ldots, x_j, x_{j+1}^{n_{j+1}}, \ldots, x_d^{n_d}$ is a $\mathfrak{q}$-weak M-sequence by Proposition 13 and (vi), q.e.d.

This theorem and Lemma 15 imply

Corollary 18. *Let M be a Noetherian A-module of positive dimension. There is a standard system of parameters of M if and only if $l\big(H^i_{\mathfrak{m}}(M)\big) < \infty$ for all $i < \dim M$. Moreover, in this situation there is an $\mathfrak{m}$-primary ideal $\mathfrak{q}$ of A such that every system of parameters of M contained in $\mathfrak{q}$ is a standard system of parameters.*

The corollary gives rise to the following definition:

Definition 19. Let M be a Noetherian A-module of dimension $d > 0$. An $\mathfrak{m}$-primary ideal $\mathfrak{q}$ of A is called a *standard ideal* with respect to M if every system of parameters of M contained in $\mathfrak{q}$ is a standard system of parameters of M.

Note that M is a Buchsbaum module if and only if $\mathfrak{m}$ is a standard ideal with respect to M. From this point of view our next (and last) theorem is not only a generalization of the important Theorem I.2.15 but it may also be considered as an alternative way of proving this main result (without using the "Ext-functors" and injective resolutions). We note that statement (ii), or better, the equivalence of (i) and (ii) essentially goes back to S. Goto.

Theorem 20. *Let M be a Noetherian A-module of dimension $d > 0$ and let $\mathfrak{a} \subset A$ be an ideal which is $\mathfrak{m}$-primary. Assume we are given an M-basis $x_1, \ldots, x_t$ of $\mathfrak{a}$. The following statements are equivalent:*

(i) *$\mathfrak{a}$ is a standard ideal with respect to M.*

(ii) *The canonical maps*

$$\lambda^i_M : H^i(\mathfrak{a}; M) \to H^i_{\mathfrak{m}}(M)$$

are surjective for all $i < d$.

(iii) *$x_{i_1}, \ldots, x_{i_d}$ is a standard system of parameters of M for all systems $i_1, \ldots, i_d$ of integers with $1 \leq i_1 < \ldots < i_d \leq t$.*

(iv) *Every system of parameters of M contained in $\mathfrak{a}$ is an $\mathfrak{a}$-weak M-sequence.*

Before proving this theorem we need some lemmata. But first of all we state:

Corollary 21. *Let $x_1, \ldots, x_d$ be a system of parameters of the Noetherian A-module M and assume that $d := \dim M > 0$. Then $x_1, \ldots, x_d$ is a standard system of parameters of M if and only if $\mathfrak{q} = (x_1, \ldots, x_d)\, A + \operatorname{Ann} M$ is a standard ideal of M.*

Now we prove:

Lemma 22. *Let M be a Noetherian A-module with $d := \dim M > 0$ and $l_A\big(H^i_{\mathfrak{m}}(M)\big) < \infty$ for all $i < d$. Let $x_1, \ldots, x_r$, $r \geq 1$ be a part of a system of parameters of M and let $\mathfrak{q} := (x_1, \ldots, x_r)\, A$. Then the maps*

$$H^i_{\mathfrak{m}}(M) \to H^i_{\mathfrak{q}}(M) \qquad (\textit{induced by } \mathfrak{q} \subseteq \mathfrak{m}, \text{ c.f. Chapter 0, § 1, 3.})$$

are isomorphisms for all $i < r$.

Proof: For $i = 0, \ldots, r$, $M_i := (x_1, \ldots, x_i)\, M$ is unmixed in M up to $\mathfrak{m}$-primary components. Hence we have for $i = 0$:

$$H^0_{\mathfrak{m}}(M) = 0 :_M \langle \mathfrak{m} \rangle = 0 :_M \langle \mathfrak{q} \rangle = H^0_{\mathfrak{q}}(M).$$

We have for all $i > 0$ a commutative diagram (c.f. Lemma 0.1.5(ii))

$$\begin{array}{ccc} H^i_{\mathfrak{m}}(M) & \xrightarrow{\sim} & H^i_{\mathfrak{m}}\big(M/H^0_{\mathfrak{m}}(M)\big) \\ \downarrow & & \downarrow \\ H^i_{\mathfrak{q}}(M) & \xrightarrow{\sim} & H^i_{\mathfrak{q}}\big(M/H^0_{\mathfrak{m}}(M)\big) \end{array}$$

Therefore we can assume that depth $M > 0$. Then $x := x_r$ is a non-zero divisor with respect to M. Set $\mathfrak{q}' := (x_1, \ldots, x_{r-1})\, A$. Then for all $n \geq 1$, $\mathfrak{q}'$ is a part of a system of parameters of M/x^nM and $l\big(H^i_{\mathfrak{m}}(M/x^nM)\big) \leq l\big(H^i_{\mathfrak{m}}(M)\big) + l\big(H^{i+1}_{\mathfrak{m}}(M)\big) < \infty$ for all $i < d - 1$. Consider the following commutative diagram

$$\begin{array}{ccccccccc} 0 & \to & M & \xrightarrow{x^n} & M & \to & M/x^nM & \to & 0 \\ & & \downarrow \| & & \downarrow x & & \downarrow & & \\ 0 & \to & M & \xrightarrow{x^{n+1}} & M & \to & M/x^{n+1}M & \to & 0. \end{array}$$

By applying $H^i_{\mathfrak{q}}(M)$ (resp. $H^i_{\mathfrak{m}}(M)$) and forming the direct limits of the corresponding long exact cohomology sequences we get for all $i > 0$ isomorphisms (since the limit of the direct system $H^i_{\mathfrak{q}}(M) \xrightarrow{x} H^i_{\mathfrak{q}}(M) \xrightarrow{x} \ldots$ is zero):

$$\varinjlim_n H^{i-1}_{\mathfrak{q}}(M/x^nM) \xrightarrow{\sim} H^i_{\mathfrak{q}}(M) \qquad \text{(the same is true if we replace } \mathfrak{q} \text{ by } \mathfrak{m}).$$

Hence we find for $i < r$ (use Corollary 0.1.7)

$$H^i_{\mathfrak{m}}(M) \cong \varinjlim_n H^{i-1}_{\mathfrak{m}}(M/x^nM) \cong \varinjlim_n H^{i-1}_{\mathfrak{q}'}(M/x^nM) = \varinjlim_n H^{i-1}_{\mathfrak{q}}(M/x^nM) \cong H^i_{\mathfrak{q}}(M),$$

q.e.d.

Lemma 23. *Let M be a Noetherian A-module of dimension $d > 0$ with $l_A\big(H^i_{\mathfrak{m}}(M)\big) < \infty$ for all $i < d$. Let $\mathfrak{a}$ be an $\mathfrak{m}$-primary ideal of A and assume that t is an integer with $1 \leq t \leq d$ such that for all $r = 0, \ldots, t - 1$*

$$\mathfrak{a}[(y_1, \ldots, y_r)\, M :_M \langle \mathfrak{m} \rangle] \subseteq (y_1, \ldots, y_r)\, M$$

for all parts $y_1, \ldots, y_r$ of systems of parameters of M contained in $\mathfrak{a}$. Let $x_1, \ldots, x_t$ be a part of a system of parameters of M contained in $\mathfrak{a}$. Then for all $n_1, \ldots, n_t \geq 1$ we have:

$$(x_1^{n_1+1}, \ldots, x_t^{n_t+1})\, M :_M x_1^{n_1} \cdot \ldots \cdot x_t^{n_t}$$
$$= (x_1, \ldots, x_t)\, M + \sum_{i=1}^{t} \big((x_1, \ldots, \hat{x}_i, \ldots, x_t)\, M :_M \mathfrak{a}\big).$$

Proof: It suffices to prove "$\subseteq$". We use induction on t. For $t = 1$ we have $(n := n_1)$: $x_1^{n+1}M : x_1^n = x_1M + 0 : x_1^n \subseteq x_1M + 0 : \langle \mathfrak{m} \rangle = x_1M + 0 : \mathfrak{a}$.

Let $t > 1$ and $x \in \mathfrak{a}$ be an element with $\dim M/xM = d - 1$. Then we have for all $n \geq 0$:

$$x^{n+1}M : \mathfrak{a} = (x^{n+1}M : x) \cap (x^{n+1}M : \mathfrak{a}) = (x^nM + O : x) \cap (x^{n+1}M : \mathfrak{a})$$
$$\subseteq (x^nM + O : \mathfrak{a}) \cap (x^{n+1}M : \mathfrak{a}) = 0 : \mathfrak{a} + x^n(x^{n+1}M : \mathfrak{a} \cdot x^n)$$
$$= 0 : \mathfrak{a} + x^n\big((xM + 0 : x^n) : \mathfrak{a}\big) \subseteq 0 : \mathfrak{a} + x^n(xM : \langle \mathfrak{m} \rangle) = 0 : \mathfrak{a} + x^n(xM : \mathfrak{a}).$$

Therefore for each $1 \leq i \leq t-1$:

$$(x_1, \ldots, \hat{x}_i, \ldots, x_{t-1}, x_t^{n_t+1})\, M : \mathfrak{a}$$
$$= (x_1, \ldots, \hat{x}_i, \ldots, x_{t-1})\, M : \mathfrak{a} + x_t^{n_t}\big((x_1, \ldots, \hat{x}_i, \ldots, x_{t-1}, x_t)\, M : \mathfrak{a}\big).$$

This with the induction hypothesis results in the following (set $M' := M/x_t^{n_t+1}M$):

$$[(x_1^{n_1+1}, \ldots, x_t^{n_t+1})\, M : x_1^{n_1} \cdot \ldots \cdot x_t^{n_t}]/x_t^{n_t+1}M$$
$$\cong \big((x_1^{n_1+1}, \ldots, x_{t-1}^{n_{t-1}+1})\, M' :_{M'} x_1^{n_1} \cdot \ldots \cdot x_{t-1}^{n_{t-1}}\big) :_{M'} x_t^{n_t}$$
$$= \left[\sum_{i=1}^{t-1} \big((x_1, \ldots, \hat{x}_i, \ldots, x_{t-1})\, M' : \mathfrak{a}\big) + (x_1, \ldots, x_{t-1})\, M'\right] :_{M'} x_t^{n_t}$$
$$= \left[\left(\sum_{i=1}^{t-1} (x_1, \ldots, \hat{x}_i, \ldots, x_{t-1})\, M : \mathfrak{a} + x_t^{n_t} \sum_{i=1}^{t-1} \big((x_1, \ldots, \hat{x}_i, \ldots, x_{t-1}, x_t)\, M : \mathfrak{a}\big)\right.\right.$$
$$\left.\left. + (x_1, \ldots, x_{t-1})\, M\right) : x_t^{n_t}\right] \Big/ x_t^{n_t+1}M$$
$$\subseteq \left[(x_1, \ldots, x_{t-1})\, M : \mathfrak{a} \cdot x_t^{n_t} + \sum_{i=1}^{t-1} \big((x_1, \ldots, \hat{x}_i, \ldots, x_{t-1}, x_t)\, M : \mathfrak{a}\big)\right] \Big/ x_t^{n_t+1}M$$
$$= \left[(x_1, \ldots, x_t)\, M + \sum_{i=1}^{t} \big((x_1, \ldots, \hat{x}_i, \ldots, x_t)\, M : \mathfrak{a}\big)\right] \Big/ x_t^{n_t+1}M,$$

since

$$(x_1, \ldots, x_t)\, M \subseteq \sum_{i=1}^{t} \big((x_1, \ldots, \hat{x}_i, \ldots, x_t)\, M : \mathfrak{a}\big) \quad \text{for } t \geq 2,$$

q.e.d.

Lemma 24. *Let M be a Noetherian A-module of dimension $d > 0$ with $l_A\big(H^i_{\mathfrak{m}}(M)\big) < \infty$ for all $i < d$ and let $\mathfrak{a} \subset A$ be an $\mathfrak{m}$-primary ideal such that the canonical maps*

$$\lambda^i_M : H^i(\mathfrak{a}; M) \to H^i_{\mathfrak{a}}(M) \big(= H^i_{\mathfrak{m}}(M)\big)$$

are surjective for all $i \leq r$, where $0 \leq r < d$. For every part $x_1, \ldots, x_r$ of a system of parameters of M contained in $\mathfrak{a}$ we have then

$$\mathfrak{a}[(x_1, \ldots, x_r)\, M : \langle\mathfrak{m}\rangle] \subseteq (x_1, \ldots, x_r)\, M.$$

Proof: Note that by our assumption for every part $x_1, \ldots, x_r$ of a system of parameters of M, $(x_1, \ldots, x_r)\, M$ is unmixed in M up to $\mathfrak{m}$-primary components. We use induction on r. Let $r = 0$. Since λ^0_M is the embedding $0 :_M \mathfrak{a} = H^0(\mathfrak{a}; M) \subseteq H^0_{\mathfrak{a}}(M) = 0 :_M \langle\mathfrak{a}\rangle = 0 :_M \langle\mathfrak{m}\rangle$, hence $\mathfrak{a}(0 :_M \langle\mathfrak{m}\rangle) = \mathfrak{a}(0 :_M \mathfrak{a}) = 0$.

Therefore let $0 < r < d$ and assume that $x_1, \ldots, x_r$ is a part of a system of parameters of M contained in $\mathfrak{a}$. By the induction hypothesis, $\mathfrak{a}\big((y_1, \ldots, y_s)\, M : \langle\mathfrak{m}\rangle\big) \subseteq (y_1, \ldots, y_s)\, M$ for every part $y_1, \ldots, y_s$ of a system of parameters of M contained in $\mathfrak{a}$ with $0 \leq s < r$.

Choose an element $y \in \mathfrak{a}$ such that $x_1, \ldots, x_r, y$ is again a part of a system of parameters of M. Set $\mathfrak{q} := (x_1, \ldots, x_r)\, A$, $\mathfrak{q}' := (x_1, \ldots, x_r, y)\, A$. Consider the commutative diagram of Lemma 0.1.5:

$$\begin{array}{ccc} H^r(\mathfrak{a}; M) & \longrightarrow & H^r(\mathfrak{q}'; M) \\ \downarrow{\scriptstyle \lambda^r_M} & & \downarrow{\scriptstyle \lambda'^r_M} \\ H^r_{\mathfrak{a}}(M) & \xrightarrow{\mu} & H^r_{\mathfrak{q}'}(M) \end{array}$$

By Lemma 22, μ is an isomorphism and since λ_M^r is surjective, λ'^r is surjective. Now consider the right-hand part of the first commutative diagram of Corollary 0.1.7:

$$\begin{array}{ccccc} H^r(\mathfrak{q}'; M) & \to & H^0(y; H^r(\mathfrak{q}; M)) & \to & 0 \\ \downarrow \lambda'^r & & \downarrow \bar{\lambda}^r & & \\ H^r_{\mathfrak{q}'}(M) & \to & H^0_{yA}(H^r_{\mathfrak{q}}(M)) & \to & 0. \end{array}$$

It shows us that $\bar{\lambda}^r$ is surjective. By the remarks made in Chapter 0, § 1, 3., we get

$$H^0_{yA}(H^r_{\mathfrak{q}}(M)) \cong H^0_{yA}(\varinjlim_n M/(x_1^n, \ldots, x_r^n) M) \cong \varinjlim_n H^0_{yA}(M/(x_1^n, \ldots, x_r^n) M)$$
$$= \varinjlim_n [(x_1^n, \ldots, x_r^n) M : \langle yA \rangle / (x_1^n, \ldots, x_r^n) M].$$

Take $m \in \mathfrak{q}M : \langle \mathfrak{m} \rangle$, $\bar{m} := m \bmod \mathfrak{q}M$. Then the image of $\bar{m}$ in $H^0_{yA}(H^r_{\mathfrak{q}}(M))$ can be represented by an element $\bar{\lambda}^r(\nu)$ with $\nu \in H^0(y; H^r(\mathfrak{q}; M)) \cong H^0(y; M/\mathfrak{q}M) = \mathfrak{q}M : y/\mathfrak{q}M$. Let $\nu = \bar{m}'$ with $m' \in \mathfrak{q}M : y$. Then there is an integer $n > 0$ with

$$x_1^n \cdot \ldots \cdot x_r^n(m - m') \in (x_1^{n+1}, \ldots, x_r^{n+1}) M,$$

hence $m - m' \in \sum_{i=1}^{r} ((x_1, \ldots, \hat{x}_i, \ldots, x_r) M : \mathfrak{a}) \subseteq \mathfrak{q}M : \mathfrak{a} \subseteq \mathfrak{q}M : y$ by Lemma 23. Therefore $m \in \mathfrak{q}M : y$.

Now take an $(M/\mathfrak{q}M)$-basis $y_1, \ldots, y_t$ of $\mathfrak{a}$. Then $m \in \bigcap_{i=1}^{t} (\mathfrak{q}M : y_i) = \mathfrak{q}M : \mathfrak{a}$, i.e. $\mathfrak{a}(\mathfrak{q}M : \langle \mathfrak{m} \rangle) \subseteq \mathfrak{q}M$, q.e.d.

Now we are able to prove our Theorem 20.

Proof of Theorem 20: (i) $\Rightarrow$ (iii) is trivial and (iv) $\Rightarrow$ (i) is a consequence of Theorem and Definition 17 since for any system of parameters $x_1, \ldots, x_d$ of M contained in $\mathfrak{a}$, $x_1^{n_1}, \ldots, x_d^{n_d}$ is an $(x_1, \ldots, x_d) A$-weak M-sequence for all $n_1, \ldots, n_d \geq 1$ if it is an $\mathfrak{a}$-weak M-sequence. To prove the implication (iii) $\Rightarrow$ (ii) we use by applying Theorem and Definition 17 word for word the proof of the implication (iii) $\Rightarrow$ (ii) of Theorem I.2.15 replacing only $\mathfrak{m}$ by $\mathfrak{a}$.

Finally we prove (ii) $\Rightarrow$ (iv). The surjectivity of λ_M^i implies that $l(H^i_{\mathfrak{m}}(M)) \leq l(H^i(\mathfrak{a}; M)) < \infty$ for all $i < d$. Let $x_1, \ldots, x_d$ be a system of parameters of M contained in $\mathfrak{a}$ and let $1 \leq i \leq d$. Then

$$\mathfrak{a}((x_1, \ldots, x_{i-1}) M : x_i) \subseteq \mathfrak{a}((x_1, \ldots, x_{i-1}) M : \langle \mathfrak{m} \rangle) \subseteq (x_1, \ldots, x_{i-1}) M$$

by Lemma 24, i.e. $x_1, \ldots, x_d$ is an $\mathfrak{a}$-weak M-sequence and the proof is finished, q.e.d.

Bibliography

Abhyankar, S. S.

[1] Algebraic space curves. Les Presses de l'Université de Montreal, Montreal **1971**.

Achilles, R., P. Schenzel and W. Vogel

[1] Bemerkungen über normale Flachheit und normale Torsionsfreiheit und Anwendungen. Period. Math. Hungar. **12** (1981), 49–76.

Achilles, R., und W. Vogel

[1] Über vollständige Durchschnitte in lokalen Ringen. Math. Nachr. **89** (1979), 285–298.

Amasaki, M.

[1] Preparatory structure theorem for ideals defining space curves. Publ. Res. Inst. Math. Sci. **19** (1983), 493–518.

[2] On the structure of arithmetically Buchsbaum curves in $\mathbf{P}_k^3$. Publ. Res. Inst. Math. Sci. **20** (1984), 793–837.

[3] Examples of nonsingular irreducible curves which give reducible singular points of $\mathrm{red}(H_{d,g})$. Publ. Res. Inst. Math. Sci. **21** (1985), 761–786.

Aoyama, Y.

[1] On the depth and the projective dimension of the canonical module. Japan. J. Math. **6** (1980), 61–66.

[2] Some basic results on canonical modules. J. Math. Kyoto Univ. **23** (1983), 85–94.

Aoyama, Y., and S. Goto

[1] On the endomorphism ring of the canonical module. J. Math. Kyoto Univ. **25** (1985), 21–30.

[2] Special cases of a conjecture of Sharp. R.I.M.S. Kokyuroku **543** (Kyoto Univ.), 13–52 (in Japanese). (English version is in preparation, 1985.)

Apery, R.

[1] Sur certain caractères numériques d'un idéal sous composant impropre. C. R. Acad. Sci. Paris, Sér. A–B, **220** (1945), 234–236.

[2] Sur les Courbes de première espèce de l'espace à trois dimensions. C. R. Acad. Sci. Paris, Sér. A–B, **220** (1945), 271–272.

Artin, M., and M. Nagata

[1] Residual intersections in Cohen-Macaulay rings. J. Math. Kyoto Univ. **12** (1972), 307–323.

Atiyah, M. F., and I. G. Macdonald

[1] Introduction to Commutative Algebra. Addison-Wesley, Reading, Mass., 1969.

Auslander, M.

[1] Remarks on a theorem of Bourbaki. Nagoya Math. J. **27** (1966), 361–369.

Auslander, M., and D. A. Buchsbaum

[1] Codimension and multiplicity. Ann. of Math. **68** (1958), 625–657.

Baclawski, K.

[1] Whitney numbers of geometric lattices. Adv. in Math. **16** (1975), 125–138.

[2] Galois connections and the Leray spectral sequences. Adv. in Math. **25** (1977), 191–215.

[3] Cohen-Macaulay ordered sets. J. Algebra **63** (1980), 226–258.
[4] Canonical modules of partially ordered sets. J. Algebra **83** (1983), 1–5.
[5] Combinatorial decompositions of rings and almost Cohen-Macaulay complexes. J. Algebra **83** (1983), 6–19.

Barth, W.
[1] Kummer surfaces associated with the Horrocks-Mumford bundle. In: Algebraic geometry, Angers 1979, 29–48. Sijthoff and Noordhoff, Alphen aan den Rijn, The Netherlands, 1980.

Bass, H.
[1] On the ubiquity of Gorenstein rings. Math. Z. **82** (1963), 8–28.

Bayer, M. M.
[1] Facial enumeration in polytopes, spheres and other complexes. Ph. D. Thesis, Cornell University, Ithaca, NY, 1983.

Bayer, M. M., and L. J. Billera
[1] Generalized Dehn-Sommerville relations for polytopes, spheres and Eulerian partially ordered sets. Invent. Math. **79** (1985), 143–157.

Bertin, M.-J.
[1] Anneaux d'invariants d'anneaux de polynômes en caractéristique p. C. R. Acad. Sci. Paris, Sér. A, **264** (1967), 653–656.

Bezout, É.
[1] Sur le degré des équations résultantes de l'évanouissement des inconnus, Mémoires présentés par divers savants à l'Académie des sciences de l'Institut de France, 1764.
[2] Sur le degré résultant des méthodes d'élimination entre plusieurs équationes. Mémoires présentés par divers savants à l'Académie des sciences de l'Institut de France, 1764.
[3] Sur le degré des équations résultantes de l'évanouissement des inconnus. Histoire de l'Académie Royale des sciences, année 1764, 288–338, Paris 1767.
[4] Théorie générale des équations algébriques. Paris 1779.

Boda, E., and W. Vogel
[1] On systems of parameters, local intersection multiplicity and Bezout's Theorem. Proc. Amer. Math. Soc. **78** (1980), 1–7.

Bolondi, G.
[1] Liaison and maximal rank. Preprint, University of Trento, Italy. Povo (Trento) 1985.
[2] On the classification of curves linked to two skew lines. Talk at the International Conference "Algebraic Geometry", Berlin, November 13–19, 1985.

Bolondi, G., and J. Migliore
[1] Classification of maximal rank curves in the liaison class L_n. Preprint. May 1986.

Bresinsky, H.
[1] Minimal free resolutions of monomial curves in $\mathbf{P}_k^3$. Linear Algebra and Appl. **59** (1984), 121–129.
[2] Monomial Buchsbaum ideals in $\mathbf{P}^r$. Manuscripta Math. **47** (1984), 105–132.
[3] On the Cohen-Macaulay property for monomial curves in $\mathbf{P}_k^3$. Monatsh. Math. **98** (1984), 21–28.

Bresinsky, H., and B. Renschuch
[1] Basisbestimmung Veronesescher Projektionsideale mit allgemeiner Nullstelle ($t_0^m, t_0^{m-r}t_1^r, t_0^{m-s}t_1^s, t_1^m$). Math. Nachr. **96** (1980), 257–269.

Bresinsky, H., P. Schenzel and W. Vogel
[1] On liaison, arithmetical Buchsbaum curves and monomial curves in $\mathbf{P}^3$. J. Algebra **86** (1984), 283–301.

Brodmann, M.
[1] Kohomologische Eigenschaften von Aufblasungen an lokal vollständigen Durchschnitten. Thesis, University of Münster, Münster (Westf.) 1980.

[2] Local cohomology of certain Rees- and form-rings I. J. Algebra **81** (1983), 29–57.
[3] Local cohomology of certain Rees- and form-rings II. J. Algebra **86** (1984), 457–493.
[4] Einige Ergebnisse aus der lokalen Kohomologietheorie und ihre Anwendung. Osnabrücker Schriften zur Mathematik, Heft **5**, Osnabrück 1983.
[5] Some remarks on blow-up and conormal cones. In: Commutative algebra (Trento, 1981), 23–37, Lecture Notes in Pure and Appl. Math., 84, Dekker, New York 1983.

Bruggesser, H., and P. Mani
[1] Shellable decompositions of cells and spheres. Math. Scand. **29** (1971), 197–205.

Buchsbaum, D. A.
[1] Complexes in local ring theory. In: Some aspects of ring theory, 223–228. C.I.M.E. Roma 1965.
[2] Homological and commutative algebra. In: Categories and commutative algebra. 13–93, C.I.M.E. Roma 1973.

Buchsbaum, D. A., and D. Eisenbud
[1] Some structure theorems for finite free resolutions. Adv. in Math. **12** (1974), 84–139.
[2] Generic free resolutions and a family of generically perfect ideals. Adv. in Math. **18** (1975), 245–301.

Cartan, H., and S. Eilenberg
[1] Homological algebra. Princeton Univ. Press., Princeton, N.J., 1956.

Catanese, F.
[1] Babbage's conjecture, contact of surfaces, symmetric determinantal varieties and applications. Invent. Math. **63** (1981), 433–465.

Cavaliere, M. P., and G. Niesi
[1] Sulle equazioni di una curva monomiale proiettiva. Ann. Univ. Ferrara Sez. VII, Sc. Mat. **30** (1984), 89–96.

Cayley, A.
[1] Note sur les hyperdéterminants. J. Reine Angew. Math. **34** (1847), 148–152.

Chow, W.-L.
[1] On unmixedness theorem. Amer. J. Math. **86** (1964), 799–822.

Cuong, N. T., and N. V. Trung
[1] Über schwache Sequenzen. Period. Math. Hungar. **12** (1981), 77–80.

Daepp, U., and A. Evans
[1] A note on Buchsbaum rings and localizations of graded domains. Canad. J. Math. **32** (1980), 1244–1249.

Daepp, U., and W. E. Kuan
[1] A note on the localization of a graded ring. Bull. Inst. Math. Acad. Sinica **11** (1983), 337–342.

Dehn, M.
[1] Die Eulersche Formel im Zusammenhang mit dem Inhalt der nicht-Euklidischen Geometrie. Math. Ann. **61** (1905), 561–586.

Dieudonné, J.
[1] The historical development of algebraic geometry. Amer. Math. Monthly **79** (1972), 827–866.

Eisenbud, D., and S. Goto
[1] Linear free resolutions and minimal multiplicity. J. Algebra 88 (1984), 89–133.

Eisenreich, G.
[1] Zur Definition der Perfektheit von Polynomidealen. Arch. Math. (Basel) **21** (1970), 571–573.

Ellingsrud, G.
[1] Sur le schéma de Hilbert des variétés de codimension 2 dans $\mathbf{P}^e$ à cone de Cohen-Macaulay. Ann. Sci. École Norm. Sup. (4) 8 (1975), 423–431.

Evans. E. G. Jr., and P. A. Griffith
[1] Local cohomology modules for normal domains. J. London Math. Soc. **19** (1979), 277–284.
[2] Syzygies. London Mathematical Society Lecture Note Series. Vol. 106. Cambridge—London—New York—New Rochelle—Melbourne—Sydney: Cambridge University Press 1985.

Faltings, G.
[1] Über Macaulayfizierung. Math. Ann. **238** (1978), 175–192.

Ferrand, D., and M. Raynaud
[1] Fibres formelles d'un anneau local Noethérian. Ann. Sci. École Norm Sup. (4) **3** (1970), 295 to 311.

Flenner, H.
[1] Die Sätze von Bertini für lokale Ringe. Math. Ann. **229** (1977), 97–111.

Fiorentini, M.
[1] On relative regular sequences. J. Algebra **18** (1971), 384–389.
[2] Esempi di curve di Buchsbaum che non sono di Macaulay. Univ. Bologna, Seminario di variabili complesse, 73–84, Bologna, Instituto di Geometria, 1981, CNR, Rome 1982.
[3] Intersections résiduelles dans les anneaux de Cohen-Macaulay et un critère de régularité relative. In: Teubner-Texte zu Mathematik, Band 40, 50–62. Teubner-Verlag, Leipzig 1981.

Fossum, R.
[1] Duality over Gorenstein rings. Math. Scand. **26** (1970), 165–176.

Fossum, R., H.-B. Foxby, P. Griffith and I. Reiten
[1] Minimal injective resolutions with applications to dualizing modules and Gorenstein modules. Publ. Math., I.H.E.S. **45** (1976), 193–215.

Fossum, R., and P. A. Griffith
[1] Complete local factorial rings which are not Cohen-Macaulay in characteristic p. Ann. Sci. École Norm. Sup. (4) **8** (1975), 189–200.

Foxby, H.-B.
[1] Bounded complexes of flat modules. J. Pure Appl. Algebra **15** (1979), 149–172.

Freitag, E.
[1] Stabile Modulformen. Preprint, Universität Heidelberg, 1976.

Freitag, E., and R. Kiehl
[1] Algebraische Eigenschaften der lokalen Ringe in den Spitzen der Hilbertschen Modulgruppen. Invent. Math. **24** (1974), 121–148.

Fulton, W.
[1] Intersection theory, Springer-Verlag, New York—Heidelberg—Berlin—Tokyo 1984.

Gaeta, F.
[1] Quelquesprogrès récents dans la classification des variétés algébriques d'un espace projectif. Deuxième Colloque de Géométrie Algébrique Liège. C.B.R.M., 1952.

Geramita, A. V., and C. A. Weibel
[1] On the Cohen-Macaulay and Buchsbaum property for unions of planes in affine space. J. Algebra **92** (1985), 413–445.

Geramita, A. V., P. Maroscia and W. Vogel
[1] On curves linked to lines in $\mathbf{P}^3$. In: The curves seminar at Queen's. Vol. II, B1–B26. Queen's papers in pure and applied mathematics, No. 61, Kingston, Ontario, Canada, 1982.
[2] A note on arithmetically Buchsbaum curves in $\mathbf{P}^3$. Matematiche (Catania) (to appear).

Gopalakrishnan, N. S.
[1] Commutative algebra. Oxonian Press Pvt. Ltd., New Delhi 1984.

Goto, S.
[1] On Buchsbaum local rings. R.I.M.S. Kôkyuroku **374** (1980), 58–74. Report of the R.I.M.S. Symposium on commutative rings. Oct. 31–Nov. 2, 1979, at Research Institute of Math. Sciences of Kyoto Univ. (in Japanese).
[2] On the Cohen-Macaulayfiction of certain Buchsbaum rings. Nagoya Math. J. **80** (1980), 107–116.
[3] On Buchsbaum rings. J. Algebra **67** (1980), 272–279.
[4] On Buchsbaum rings obtained by glueing. Nagoya Math. J. **83** (1981), 123–135.
[5] Blowing-up of Buchsbaum rings. In: Commutative Algebra: Durham 1981. London Math. Soc. Lecture Note Series 72, 140–162.
[6] Rings with linear resolution. In: Conference on commutative algebra, 4–11. Proc. of the 9th Intern. Symposium, Division of Math. of the Taniguchi Foundation. Katata 1981.
[7] Buchsbaum rings of maximal embedding dimension. J. Algebra **76** (1982), 383–399.
[8] Buchsbaum rings with multiplicity 2. J. Algebra **74** (1982), 494–508.
[9] Approximately Cohen-Macaulay rings. J. Algebra **76** (1982), 214–225.
[10] A note on quasi-Buchsbaum rings. In: Study of Buchsbaum rings and generalized Cohen-Macaulay rings, 27–32. Proc. of a Symposium held a the Research Institute for Math. Sciences, Kyoto Univ., Kyoto 1982.
[11] A note on standard systems of parameters for generalized C-M modules. In: Study of Buchsbaum rings and generalized Cohen-Macaulay rings, 181–182. Proc. of a Symposium held at the Research Institute for Math. Sciences, Kyoto Univ., Kyoto 1982.
[12] On the associated graded rings of parameter ideals in Buchsbaum rings. J. Algebra **85** (1983), 490–534.
[13] Noetherian local rings with Buchsbaum associated graded rings. J. Algebra **86** (1984), 336–384.
[14] A note on quasi-Buchsbaum rings. Proc. Amer. Math. Soc. **90** (1984), 511–516.

Goto, S., and T. Ogawa
[1] A note on rings with finite local cohomology. Tokyo J. Math. **6** (1983), 403–411.

Goto, S., and Y. Shimoda
[1] On Rees algebras over Buchsbaum rings. J. Math. Kyoto Univ. **20** (1980), 691–708.
[2] On the Rees algebras of Cohen-Macaulay local rings. In: Commutative algebra (analytic methods), 201–231. Lecture notes in Pure and Applied Math., Vol. 68. Marcel Dekker Inc., New York 1982.
[3] On the Gorensteinness of Rees and form rings of almost complete intersections. Nagoya Math. J. **92** (1983), 69–88.

Goto, S., and N. Suzuki
[1] Index of reducibility of parameter ideals in a local ring. J. Algebra **87** (1984), 53–88.

Goto, S., N. Suzuki and K. Watanabe
[1] On affine semigroup rings. Japan J. Math. **2** (1976), 1–12.

Goto, S., and K. Watanabe
[1] On graded rings, I. J. Math. Soc. Japan **30** (1978), 172–213.
[2] On graeed rings, II ($\mathbf{Z}^n$-graded rings). Tokyo, J. Math. **1** (1978), 237–261.

Gräbe, H.-G.
[1] Über den Stanley-Reisner-Ring eines simplizialen Komplexes. Thesis, University of Halle, Halle 1982.
[2] The Gorenstein property depends on characteristic. Beitr. Algebra Geom. **17** (1984), 169–174.
[3] The canonical module of a Stanley-Reisner ring. J. Algebra **86** (1984), 272–281.
[4] A dualizing complex for Stanley-Reisner rings. Math. Proc. Cambridge Philos. Soc. **96** (1984), 203–212.

Griffiths, P., and J. Harris
[1] Principles of algebraic geometry. John Wiley, New York 1978.

Gröbner, W.
[1] Moderne algebraische Geometrie. Die idealtheoretischen Grundlagen. Springer-Verlag, Wien—Innsbruck 1949.
[2] Über Veronesesche Varietäten und deren Projektionen. Arch. Math. (Basel) **16** (1965), 257 to 264.
[3] Algebraische Geometrie. I, II. Mannheim: BI. 1968 und 1970.

Grothendieck, A.
[1] Cohomologie locale des faisceaux cohérents et théorèms de Lefschetz locaux et globaux. SGA 2, Paris 1962.
[2] Eléments de géométrie algébrique. Publ. Math., No. 11, No. 17, No. 24, I.H.E.S. Paris 1961—65.
[3] Local cohomology (notes by R. Hartshorne). Lecture Notes in Math. **41**, Springer-Verlag, Berlin—Heidelberg—New York 1967.

Grünbaum, B.
[1] Convex polytopes. London—New York—Sydney 1967.

Gruson, L., and C. Peskine
[1] Genre des courbes de l'espace projectif. In: Algebraic geometry, 31—59. Lecture Notes in Math. 687, Springer-Verlag, Berlin—Heidelberg—New York 1978.

Hartshorne, R.
[1] Complete intersections and connectedness. Amer. J. Math. **84** (1962), 497—508.
[2] Residues and Duality. Lecture Notes in Math. 20, Springer-Verlag, Berlin—Heidelberg—New York 1966.
[3] Ample subvarieties of algebraic varieties. Lecture Notes in Math. 156, Springer-Verlag, Berlin—Heidelberg—New York 1970.
[4] Algebraic geometry. Springer-Verlag, New York—Heidelberg—Berlin 1977.

Herrmann, M., and S. Ikeda
[1] Remarks on lifting of Cohen-Macaulay property. Nagoya Math. J. **92** (1983), 121—132.

Herrmann, M., and R. Schmidt
[1] Remarks on generalized Cohen-Macaulay rings and singularities. J. Math. Kyoto Univ. **19** (1979), 33- 40.

Herrmann, M., R. Schmidt and W. Vogel
[1] Theorie der normalen Flachheit. Teubner-Texte zur Mathematik, Teubner-Verlag, Leipzig 1977.

Herzog, J., E. Kunz et al.
[1] Der kanonische Modul eines Cohen-Macaulay-Rings. Lecture Notes in Math. 238, Springer-Verlag, Berlin—Heidelberg—New York 1971.

Herzog, J., M. E. Rossi and G. Valla
[1] On the depth of the symmetric algebra. Preprint, University of Essen and Genova, 1985.

Herzog, J., A. Simis and W. V. Vasconcelos
[1] On the arithmetic and homology of algebras of linear type. Trans. Amer. Math. Soc. **283** (1984), 661—683.

Hilton, P. J., and S. Wylie
[1] Homology theory. Cambridge University Press, London 1960.

Hironaka, H.
[1] Resolution of singularities of an algebraic variety over a field of characteristic zero. Ann. of Math. **79** (1964), 109—326.

Hoa, L. T.
[1] Classification of the tripel projections of Veronese varieties. Math. Nachr. (to appear).
[2] On Segre product of affine semigroup rings. Preprint, Institute of Mathematics, Hanoi, Vietnam, 1985.

Hochster, H.
[1] Rings of invariants of tori, Cohen-Macaulay rings generated by monomials, and polytopes. Ann. of Math. **96** (1972), 318–337.
[2] Deep local (preliminary version). Aarhus University preprint series, 1973.
[3] Topics in the homological theory of modules over commutative rings. C.B.M.S. Regional Conference Ser. in Math. No. 24, A.M.S. Providence, 1975.
[4] Cohen-Macaulay rings, combinatorics, and simplicial complexes. Proc. Oklahoma Ring Theory Conference, 171–223. Lecture Notes in pure and appl. math., Vol. 26, Marcel Dekker, New York–Basel 1977.
[5] Invariant theory of commutative rings. Preprint, University of Michigan, Ann Arbor, MI, 1985.
[6] Canonical elements in local cohomology modules and the direct summand conjecture. J. Algebra **84** (1983), 503–553.

Hochster, M., and J. L. Roberts
[1] Rings of invariants of reductive groups acting on regular rings are Cohen-Macaulay. Adv. in Math. **13** (1974), 115–175.
[2] The purity of the Frobenius and local cohomology. Adv. in Math. **21** (1976), 117–172.

Horrocks, G., and D. Mumford
[1] A rank 2 vector bundle on $\mathbf{P}^4$ with 15,000 symmetries. Topology **12** (1973), 63–81.

Huneke, C.
[1] The theory of d-sequences and powers of ideals. Adv. in Math. **46** (1982), 249–279.
[2] On the symmetric and Rees algebra of an ideal generated by a d-sequence. J. Algebra **62** (1980), 268–275.

Ikeda, S.
[1] The Cohen-Macaulayness of the Rees algebras of local rings. Nagoya Math. J. **89** (1983), 47–63.

Ishida, M.-N.
[1] Torus embeddings and dualizing complexes. Tôhoku Math. J. **32** (1980), 111–146.
[2] Tsuchihashi's cusp singularities are Buchsbaum singularities. Tôhoku Math. J. (2) **36** (1984), 191–201.

Iversen, B.
[1] Noetherian graded modules I. Aarhus University preprint series, 1971–72.

Kaplansky, I.
[1] Commutative rings. Boston 1970.

Keller, O.-H.
[1] Vorlesungen über algebraische Geometrie. Akad. Verlagsgesellschaft, Leipzig 1974.

Kiehl, R.
[1] Beispiele von Buchsbaum-Ringen und -Moduln. Preprint of the University of Mannheim (unpublished).
[2] Letter to W. Vogel dated 7 October, 1976.

Kirby, D., and Hefzi A. Mehran
[1] A note on the coefficients of the Hilbert-Samuel polynomial for a Cohen-Macaulay module. J. London Math. Soc. (2) **25** (1982), 449–457.

Klee, V.
[1] The number of vertices of a convex polytope. Canad. J. Math. **16** (1964), 701–720.

Kleiman, S. L.
[1] Problem 15. Rigorous foundation of Schubert's enumerative calculus. In: Proc. of symposia in pure math., 445—482. Amer. Math. Soc. Series PS PM 28, Providence, R.I., 1976.
[2] Motives. In: Algebraic geometry, 53—82. Wolters-Noordhoff Publ., Groningen, 1972.
[3] Letter to W. Vogel dated 27 August, 1979.

Kuan, W.-E.
[1] On the hyperplane sections through two given points of an algebraic variety. Canad. J. Math. **22** (1970), 128—133.
[2] On the hyperplane section through a rational point of an algebraic variety. Pacific J. Math. **36** (1971), 393—405.
[3] Specialization of a generic hyperplane section through a rational point of an algebraic variety. Ann. Mat. Pura Appl. **94** (1972), 75—82.

Kunz, E.
[1] Einführung in die kommutative Algebra und algebraische Geometrie. F. Vieweg und Sohn, Braunschweig—Wiesbaden 1980.

Lorenzini, Anna
[1] Graded rings. In: The curves seminar at Queen's, Vol. II, D1—D41. Queen's papers in pure and applied mathematics, No. 61. Queens' University, Kingston, Ontario, Canada, 1982.
[2] Buchsbaum rings. In: The curves seminar at Queen's, Vol. III, A1—A35. Queen's papers in pure and applied mathematics, No. 67. Queen's University, Kingston, Ontario, Canada, 1984.

Macaulay, F. S.
[1] The algebraic theory of modular systems. Cambridge 1916.

Macdonald, I. G., and R. Y. Sharp
[1] An elementary proof of the non-vanishing of certain local cohomology modules. Quart. J. Math. Oxford (2) **23** (1972), 197—204.

MacLane, S.
[1] Homology. Springer-Verlag, Berlin—Heidelberg—New York 1963.

Maroscia, P., J. Stückrad and W. Vogel
[1] Upper bounds for the degrees of the equations defining locally Cohen-Macaulay schemes. Preprint, University of Rome, Leipzig and Halle 1986.

Maroscia, P., and W. Vogel
[1] On the defining equations of points in general position in $\mathbf{P}^n$. Math. Ann. **269** (1984), 183—189.

Matlis, E.
[1] Injective modules over Noetherian rings. Pacific J. Math. 8 (1958), 511—528.

Matsumura, H.
[1] Commutative algebra. Second Edition. Benjamin, London 1980.

McMullen, P.
[1] The maximum numbers of faces of a convex polytope. Mathematika **17** (1970), 179—184.

McMullen, P., and G. C. Shephard
[1] Convex polytopes and the upper bound conjecture. London Math. Soc. lecture note ser. 3. Cambridge Univ. Press, 1971.

Migliore, J.
[1] Topics in the theory of liaison of space curves. Thesis, Brown University, Providence, R.I., 1983.
[2] Geometric invariants for liaison of space curves. J. Algebra **99** (1986), 548—572.
[3] On linking double lines. Trans. Amer. Math. Soc. **294** (1986), 177—185.
[4] Buchsbaum curves in $\mathbf{P}^3$. Preprint, Drew University, Madison, New Jersey, USA, 1986.
[5] Liaison of a union of skew lines in $\mathbf{P}^4$. Preprint, Drew University, Madison, New Jersey, USA, 1986.

Mori, S.
[1] On affine cones associated with polarized varieties. Japan. J. Math. **1** (1975), 301—309.
Mumford, D.
[1] Geometric invariant theory. Ergebnisse der Math. Springer-Verlag, Berlin—Heidelberg—New York 1965.
[2] Lectures on curves on an algebraic surface. Ann. of Math. Studies 59, Princeton Univ. Press. Princeton, N.J., 1966.
[3] Abelian varieties. Tata studies in Math. 5, Oxford, 2nd ed., 1974.
[4] Curves and their Jacobians. The Univ. of Michigan Press, Ann. Arbor, 2nd ed., 1976.
[5] Algebraic geometry I. Grundlehren. Springer-Verlag, Berlin—Heidelberg—New York 1976.
Munkres, J.
[1] Topological results in combinatorics. Michigan Math. J. **31** (1984), 113—128.
Nagata, M.
[1] Local rings. Interscience Tracts in Pure and Appl. Math. 13, J. Wiley, New York 1962.
Noether, M.
[1] Zur Grundlegung der Theorie der algebraischen Raumkurven. Abh. Königl. Preuss. Akad. Wiss. Berlin, Verlag der Königlichen Akademie der Wiss., Berlin 1883.
Northcott, D. G.
[1] Lessons on rings, modules and multiplicities. Cambridge at the Univ. Press, 1968.
Ogoma, T.
[1] Existence of dualizing complexes. J. Math. Kyoto Univ. **24** (1984), 27—48.
Ooishi, A.
[1] Castelnuovo's regularity of graded rings and modules. Hiroshima Math. J. **12** (1982), 627—644.
[2] Castelnuovo's regularity of graded rings and generic Cohen-Macaulay algebras. In: Study of Buchsbaum rings and generalized Cohen-Macaulay rings, 120—129. Proc. of a Symposium held at the Research Institute for Math. Sciences, Kyoto Univ., Kyoto 1982.
[3] On seminormal rings (general survey). In: Lecture Notes in RIMS, Kyoto Univ. **374** (1980), 1—7.
[4] Noetherian property of symbolic Rees algebras. Hiroshima Math. J. **15** (1985), 581—584.
[5] Reductions of graded rings and pseudo-flat graded modules. Preprint, Hiroshima University, Hiroshima, Japan, 1986.
Peskine, C., and L. Szpiro
[1] Dimension projective finie et cohomologie locale. Publ. Math., I.H.E.S. **42** (1973), 49—119.
[2] Syzygies et multiplicités. C. R. Acad. Sci. Paris, Sér. A, **278** (1974), 1421—1424.
[3] Liaison des variétés algebriques I. Invent. Math. **26** (1974), 271—302.
Rao, A. P.
[1] Liaison among curves in $\mathbf{P}^3$. Invent. Math. **50** (1979), 205—217.
[2] On self-linked curves. Duke Math. J. **49** (1982), 251—273.
Reisner, G. A.
[1] Cohen-Macaulay quotiens of polynomial rings. Adv. in Math. **21** (1976), 30—49.
Renschuch, B.
[1] Verallgemeinerungen des Bezoutschen Satzes. Sitz.-ber. Sächs. Akad. Wiss. Leipzig, Math.-nat. Kl., Bd. 107, Heft 4, Leipzig 1966.
[2] Elementare und praktische Idealtheorie. VEB Deutscher Verlag der Wissenschaften, Berlin 1976.
Renschuch, B., and W. Vogel
[1] Zum Nachweis arithmetischer Cohen-Macaulay-Varietäten. Monatsh. Math. **85** (1978), 201—210.
[2] Über den Bezoutschen Satz seit den Untersuchungen von B. L. van der Waerden. Beitr. Algebra Geom. **13** (1982), 95—109.

Renschuch, B., J. Stückrad and W. Vogel
[1] Weitere Bemerkungen zu einem Problem der Schnitttheorie und über ein Maß von A. Seidenberg für die Imperfektheit. J. Algebra **37** (1975), 447–471.

Roberts, P.
[1] Two applications of dualizing complexes over local rings. Ann. Sci. Ecole Norm. Sup. (4), **9** (1976), 103–106.

Roberts, L. G., and B. Singh
[1] Seminormality and cohomology of projective varieties. Queen's Mathematical Preprint, No. 21. Kingston, Ontario, Canada, 1984.

Rohn, K.
[1] Die Raumkurven auf den Flächen 3. Ordnung. Ber. Königlichen Sächsischen Gesellschaft Wiss. Leipzig Math.-Phys. Kl. **46** (1894), 84–119.
[2] Die Raumkurven auf den Flächen IVter Ordnung. Ber. Königlichen Sächsischen Gesellschaft Wiss. Leipzig Math.-Phys. Kl. **49** (1897), 631–663.

Rohn, K., and L. Berzolari
[1] Algebraische Raumkurven und abwickelbare Flächen. Encyklopädie der Math. Wiss., Band 3, Teil 2, zweite Hälfte/Teilband A, 1229–1426, B. G. Teubner, Leipzig 1921–28.

Roloff, H., and J. Stückrad
[1] Bemerkungen über Zusammenhangseigenschaften und mengentheoretische Darstellung projektiver algebraischer Mannigfaltigkeiten. Beitr. Algebra Geom. **8** (1979), 125–131.

Rota, G.-C.
[1] On the foundations of combinatorial theory I. Theory of Möbius functions. Z. Wahrsch. Verw. Gebiete **2** (1964), 340–368.

Sally, J. D.
[1] Numbers of generators of ideals in local rings. Marcel Dekker, Inc., New York and Basel 1978.

Salmon, G.
[1] Cambridge Dublin J. **5** (1849).

Samuel, P.
[1] Méthodes d'algèbre abstraite en géométrie algébrique. Ergebnisse der Math., Springer-Verlag, Berlin–Göttingen–Heidelberg 1955.

Sauer, T.
[1] Smoothing projectively Cohen-Macaulay space curves. Math. Ann. **272** (1985), 83–90.

Schenzel, P.
[1] Theorie und Anwendungen dualisierender Komplexe. Thesis, University of Halle, Halle, 1979 = Lecture Notes in Math., No. 907, Springer-Verlag, Berlin–Heidelberg–New York 1982.
[2] On Buchsbaum rings and their canonical modules. In: Seminar D. Eisenbud/B. Singh/W. Vogel, Vol. 1, 65–77. Teubner-Texte zur Mathematik, Band 29, Teubner-Verlag, Leipzig 1980.
[3] Standard systems of parameters and their blowing-up rings. J. Reine Angew. Math. **344** (1983), 201–220.

Schenzel, P., and Š. Šolčan
[1] On ideals generated by square-free monomials in regular sequences. Preprint, University of Bratislava and Halle 1986.

Schenzel, P., Ngo Viet Trung und Nguyen Tu Cuong
[1] Verallgemeinerte Cohen-Macaulay-Moduln. Math. Nachr. **85** (1978), 57–73

Schubert, H.
[1] Kategorien I. Akademie-Verlag, Berlin 1970.

Schur, F.
[1] Über die durch collineare Grundgebilde erzeugten Curven und Flächen. Math. Ann. **18** (1881), 1–32.

Schwartau, P.
[1] Liaison addition and monomial ideals. Thesis, Brandeis University, Waltham, MA, 1982.

Seidenberg, A.
[1] The hyperplane sections of normal varieties. Trans. Amer. Math. Soc. **69** (1950), 357–386.
[2] The hyperplane sections of arithmetically normal varieties. Amer. J. Math. **94** (1972), 609–630.

Serre, J.-P.
[1] Faisceaux algébriques cohérents. Ann. of Math. **61** (1955), 197–278.
[2] Algèbre Locale-Multiplicités (rédigé par P. Gabriel). Lecture Notes in Math. **11**, Springer-Verlag, Berlin–Heidelberg–New York 1965.

Shafarevich, I. R.
[1] Basic algebraic geometry. Springer Study Edition. Springer-Verlag, Berlin–Heidelberg–New York 1977.

Sharp, R. Y.
[1] Local cohomology theory in commutative algebra. Quart. J. Math. Oxford (2) **21** (1970), 425–434.
[2] Some results on the vanishing of local cohomology modules. Proc. London Math. Soc. (3) **30** (1975), 177–195.
[3] Dualizing complexes for commutative Noetherian rings. Math. Proc. Cambridge Philos. Soc. **78** (1975), 369–386.
[4] A commutative Noetherian ring which possesses a dualizing complex is acceptable. Math. Proc. Cambridge Philos. Soc. **82** (1977), 197–213.

Sharp, R. Y., and M. A. Hamich
[1] Lengths of certain generalized fractions. J. Pure Appl. Algebra **38** (1985), 323–336.

Sharp, R. Y., and H. Zakeri
[1] Local cohomology and modules of generalized fractions. Mathematika **29** (1982), 296–306.
[2] Generalized fractions, Buchsbaum modules and generalized Cohen-Macaulay modules. Math. Proc. Cambridge Philos. Soc. **98** (1985), 429–436.

Shimoda, Y.
[1] A note on Rees algebras of two-dimensional local domains. J. Math. Kyoto Univ. **19** (1979), 327–333.
[2] On Rees algebras of ideals generated by a subsystem of parameters. J. Math. Kyoto Univ. **21** (1981), 231–238.
[3] On the syzygy part of Koszul homology on certain ideals. J. Math. Kyoto Univ. **24** (1984), 83–90.

Šolčan, Š.
[1] Über die Abhängigkeit der Buchsbaum Eigenschaft von der Charakteristik des Grundkörpers. Math. Slovaca **31** (1981), 437–444.

Sommerville, D. M. Y.
[1] The relations connecting the anglesums and volume of a polytope in space of n dimensions. Proc. Roy. Soc. London, Ser. A, **115** (1927), 103–119.

Spanier, E. H.
[1] Algebraic topology. McGraw-Hill, New York 1966.

Stanley, R. P.
[1] The upper bound conjecture and Cohen-Macaulay rings. Stud. Appl. Math. **54** (1975), 135–142.

[2] Cohen-Macaulay complexes. In: Higher Combinatorics (M. Aigner, ed.), 51–62. Reidel, Dordrecht and Boston 1977.
[3] Hilbert functions and graded algebras. Adv. in Math. **28** (1978), 57–83.
[4] Finite lattices and Jordan-Hölder sets. Algebra Universalis **4** (1974), 361–371.
[5] Combinatorial reciprocity theorems. Adv. in Math. **14** (1974), 194–253.
[6] Balanced Cohen-Macaulay complexes. Trans. Amer. Math. Soc. **249** (1979), 135–157.
[7] The number of faces of a simplicial convex polytope. Adv. in Math. **35** (1980), 236–238.
[8] Combinatorics and commutative algebra. Birkhäuser Boston, Inc., 1983.

Steiner, J.
[1] Über die Flächen dritten Grades. J. Reine Angew. Math. **53** (1857), 133–141.

Steurich, M.
[1] On rings with linear resolutions. J. Algebra **75** (1982), 178–197.
[2] A characterization of unconditioned weak sequences. Proc. Amer. Math. Soc. **87** (1983), 189–199.

Stückrad, J.
[1] Grothendieck-Gruppen abelscher Kategorien und Multiplizitäten. Math. Nachr. **62** (1974), 5–26.
[2] Zur Theorie der Buchsbaum-Moduln. Thesis, University of Leipzig, Leipzig 1979.
[3] Über die kohomologische Charakterisierung von Buchsbaum-Moduln. Math. Nachr. **95** (1980), 265–272.
[4] On the Buchsbaum property of Rees and form modules. Beitr. Algebra Geom. **19** (1985), 83–103.

Stückrad, J., and W. Vogel
[1] Ein Korrekturglied in der Multiplizitätstheorie von D. G. Northcott und Anwendungen. Monatsh. Math. **76** (1972), 264–271.
[2] Eine Verallgemeinerung der Cohen-Macaulay-Ringe und Anwendungen auf ein Problem der Multiplizitätstheorie. J. Math. Kyoto Univ. **13** (1973), 513–528.
[3] Über das Amsterdamer Programm von W. Gröbner und Buchbaum Varietäten. Monatsh. Math. **78** (1974), 433–445.
[4] Toward a theory of Buchsbaum singularities. Amer. J. Math. **100** (1978), 727–746.
[5] On Segre products and applications. J. Algebra **54** (1978), 374–389.
[6] On the number of equations defining an algebraic set of zeros in n-space. In: Seminar D. Eisenbud/B. Singh/W. Vogel, Vol. 2, 88–107. Teubner-Texte zur Mathematik, Band 48, Teubner-Verlag, Leipzig 1982.
[7] An Euler-Poincaré characteristic for improper intersections. Math. Ann. **274** (1986), 257–271.
[8] Castelnuovo bounds for locally Cohen-Macaulay schemes. Preprint, University of Leipzig and Halle 1986.
[9] Castelnuovo bounds for certain subvarieties in $\mathbf{P}^n$. Preprint, University of Leipzig and Halle 1986

Suzuki, N.
[1] On applications of generalized local cohomology. Report of the first symposium on commutative algebra (1978), 88–96 (in Japanese).
[2] On the generalized local cohomology and its duality. J. Math. Kyoto Univ. **18** (1978), 71–85.
[3] On the Koszul complex generated by a system of parameters for a Buchsbaum module. Bull. Dept. Gen. Ed. Shizuoka College Pharmacy 8 (1979), 27–35.
[4] The Koszul complex of Buchsbaum modules. R.I.M.S. Kôkyruroku **446** (1981), 15–25. Report of the R.I.M.S. Symposium on commutative algebra and algebraic geometry at Research Institute of Math. Sciences of Kyoto Univ., Kyoto 1981.
[5] On a basic theorem for quasi-Buchsbaum modules. Bull. Dept. Gen. Ed. Shizuoka College Pharmacy **11** (1982), 33–40.

[6] Canonical duality for Buchsbaum modules. — An application of Goto's lemma on Buchsbaum modules. In: Study of Buchsbaum rings and generalized Cohen-Macaulay rings, 136—139. Proc. of a Symposium held at the Research Institute for Math. Sciences, Kyoto Univ., Kyoto 1982.
[7] Canonical duality for Buchsbaum modules. Bull. Dept. Gen. Educ. Shizuoka Coll. Pharmacy **13** (1984), 47—60.

Takeuchi, Y.
[1] Filter-regular sequences, quasi-Cohen-Macaulay rings and Buchsbaum rings. Math. Sem. Notes Kobe Univ. **9** (1981), 531—543.

Tamone, G.
[1] Sugli incollamenti di ideals primi. Boll. Un. Mat. Ital. A(5), **14** (1977), 810—825.

Traverso, C.
[1] Seminormality and Picard group. Ann. Scuola Norm. Sup. Pisa Sci. Fis. Mat. **24** (1970), 585—595.

Trung, N. V.
[1] Über die Übertragung der Ringeigenschaften zwischen R und $R[u]/(F)$. Math. Nachr. **92** (1979), 215—229.
[2] Some criteria for Buchsbaum modules. Monatsh. Math. **90** (1980), 331—337.
[3] Spezialisierungen allgemeiner Hyperflächenschnitte und Anwendungen. In: Seminar D. Eisenbud/B. Singh/W. Vogel, Vol. 1, 4—43. Teubner-Texte zur Mathematik, Band 29, Teubner-Verlag, Leipzig 1980.
[4] A characterization of two-dimensional unmixed local ring. Math. Proc. Cambridge Phil. Soc. **89** (1981), 237—239.
[5] On the associated graded ring of a Buchsbaum ring. Math. Nachr. **107** (1982), 209—220.
[6] Standard systems of parameters of generalized Cohen-Macaulay modules. Report at Karuisawa Symposium on commutative algebra 1982, 1—17.
[7] Zur Theorie der Buchsbaum-Ringe. Thesis, University of Halle, Halle 1983.
[8] Classification of the double projections of Veronese varieties. J. Math. Kyoto Univ. **22** (1983), 567—581.
[9] From associated graded modules to blowing-ups of generalized Cohen-Macaulay modules. J. Math. Kyoto Univ. **24** (1984), 611—622.
[10] Toward a theory of generalized Cohen-Macaulay modules. Nagoya Math. J. (to appear).
[11] Absolutely superficial sequence. Math. Proc. Cambridge Philos. Soc. **93** (1983), 35—47.
[12] Projections of one-dimensional Veronese varieties. Math. Nachr. **118** (1984), 47—67.
[13] Local cohomology modules of affine semigroup rings. Preprint, Hanoi 1984.
[14] A class of imperfect prime ideals having the equality of ordinary and symbolic powers. J. Math. Kyoto Univ. **21** (1981), 239—250.
[15] Bounds for the minimum numbers of generators of generalized Cohen-Macaulay ideals. J. Algebra **90** (1984), 1—9.
[16] Reduction exponent and degree bound for the defining equations of graded rings. Preprint, Institute of Mathematics, Hanoi, Vietnam, 1986.

Trung, N. G., and L. T. Hoa
[1] Affine semigroups and Cohen-Macaulay rings generated by monomials. Trans. Amer. Math. Soc. (to appear).

Trung, N. G., and S. Ikeda
[1] When is the Rees algebra Cohen-Macaulay. Preprint, Institute of Mathematics, Hanoi and Nagoya 1984.

Trung, N. V., and G. Valla
[1] Degree bounds for the defining equations of arithmetically Cohen-Macaulay and Buchsbaum varieties. Preprint, Institute of Mathematics, Hanoi, Vietnam and University of Genova, Italy, 1986.

Tsuchihashi, H.
[1] Higher dimensional analogues of periodic continued fractions and cusp singularities. Tôhoku Math. J. (2) **35** (1983), 607–639.

Vogel, W.
[1] Grenzen für die Gültigkeit des Bezoutschen Satzes. Monatsb. Deutsch. Akad. Wiss. Berlin 8 (1966), 1–7.
[2] Eine Bemerkung zur idealtheoretischen Multiplizitätstheorie. Math. Ann. **166** (1966), 64–75.
[3] Idealtheoretische Schnittpunktsätze in homogenen Ringen über Ringen mit Vielfachkettensatz. Math. Nachr. **34** (1967), 277–295.
[4] Über eine Vermutung von D. A. Buchsbaum. J. Algebra **25** (1973), 106–112.
[5] Probleme bei der Weiterentwicklung des Vielfachheitsbegriffes aus dem Fundamentalsatz der klassischen Algebra. In: Mitt. Math. Gesellsch. der DDR, Heft 3/4, 1975, 134–147.
[6] On Bezout's Theorem. In: Seminar D. Eisenbud/B. Singh/W. Vogel, Vol. 1, 113–144. Teubner-Texte zur Mathematik, Band 29, Teubner-Verlag, Leipzig 1980.
[7] A non-zero-divisor characterization of Buchsbaum modules. Michigan Math. J. **28** (1981), 147–152.
[8] Lectures on results on Bezout's theorem. (Notes by D. P. Patil). Lectures notes, Tata Institute of Fundamental Research, Bombay, No. 74, Springer-Verlag, Berlin–Heidelberg–New York –Tokyo 1984.

Vogel, W., and L. Marki
[1] On the theory of local rings, I. On a problem of D. A. Buchsbaum. (in Hungarian). Magyar Tud. Akad. Mat. Fis. Oszt. Kösl. **22** (1973), 55–78.
[2] On the theory of local rings, II. On a problem of D. A. Buchsbaum, (in Hungarian). Magyar Tud. Akad. Mat. Fis. Oszt. Kösl. **23** (1974), 69–100.

van der Waerden, B. L.
[1] On Hilbert's function, series of composition of ideals and a generalization of the theorem of Bezout. Proc. Roy. Acad. Amsterdam **31** (1928), 749–770.
[2] Eine Verallgemeinerung des Bézoutschen Theorems. Math. Ann. **99** (1928), 497–541, and **100** (1928), 752.
[3] Einführung in die algebraische Geometrie. 2. ed., Springer-Verlag, Berlin–New York 1973.

Watanabe, J.
[1] A note on Gorenstein rings of embedding codimension three. Nagoya Math. J. **50** (1973), 227–232.

Weibel, C. A.
[1] Reisner's Theorem. In: The curves seminar at Queen's, Vol. III, K1–K25. Queen's Papers in Pure and Appl. Math. **67**. Queen's Univ., Kingston, Ont., Canada, 1984.

Weil, A.
[1] Foundations of algebraic geometry. 2nd ed., Amer. Math. Soc., Providence, R.I., 1962.

Wright, D. J.
[1] General multiplicity theory. Proc. London Math. Soc. (3) **15** (1965), 269–288.
[2] A characterization of multiplicity. Monatsh. Math. **79** (1975), 165–167.

Yamagishi, K.
[1] Quasi-Buchsbaum rings obtained by idealizations. In: Study of Buchsbaum rings and generalized Cohen-Macaulay rings, 183–189. Proc. of a Symposium held at the Research Institute for Math. Sciences, Kyoto Univ., Kyoto 1982.

Zariski, O., and P. Samuel
[1] Commutative algebra, Vol. I and II. D. van Nostrand Company, Princeton, N.J., 1958 and 1960.

Notations

Index